International Conference on Science in Engineering and Technology (ICoSiET 2020)

IOP Conference Series: Materials Science and Engineering
Volume 1212

Palu, Indonesia
21 - 22 October 2020

Editors:

Anita Ahmad Kasim
Diana Contrias Mojica
Roman Voliansky
Leonel Hernandez
Rafal Drezewski
Andri Pranolo
Joni Adiansyah
Yun Arifatul Fatimah
Aji Prasetya Wibawa
Haviluddin

ISBN: 978-1-7138-4671-0
ISSN: 1757-8981

Printed from e-media with permission by:

Curran Associates, Inc.
57 Morehouse Lane
Red Hook, NY 12571

Some format issues inherent in the e-media version may also appear in this print version.

Institute of Physics
Dirac House, Temple Back
Bristol BS1 6BE UK

Phone: 44 1 17 929 7481
Fax: 44 1 17 920 0979

techtracking@iop.org

Additional copies of this publication are available from:

Curran Associates, Inc.
57 Morehouse Lane
Red Hook, NY 12571 USA
Phone: 845-758-0400
Fax: 845-758-2633
Email: curran@proceedings.com
Web: www.proceedings.com

TABLE OF CONTENTS

Preface

Welcome to The 2020 International Conference on Science in Engineering and Technology (ICoSiET), online held 21-22 October, 2020. The 2020 International Conference on Science in Engineering and Technology (ICoSiET) is an international conference which covers Science and applied engineering and technology. ICoSiET 2020 will be held on October 21-22, 2020 in Palu, Indonesia. This conference hosted by Universias Tadulako and is jointly organized with Association for Scientific Computing Electronics and Engineering (ASCEE), Universitas Islam Alauddin Makassar, Universitas Negeri Malang, Universitas Mulawarman, Universitas Muslim Indonesia, and Universitas Lambung Mangkurat. This conference is IOP conference so that papers accepted and presented will be forwarding for consideration to be published in the IOP Material Science and Engineering. Thanks are due to Keynote Speakers, Prof. Dr. Zhen-jiang Shen (Faculty of Geosciences and civil Engineering, Institute of Science and Engineering Kanazawa University, Japan), Prof. Dr. Richard Sliuzas (Department of Urban and Regional Planning and Geo-Information Management Twente University, Netherland), Prof. Yandi Andri Yatmo., Ph.D (Department of Architecture Faculty of Engineering Universitas Indonesia, Indonesia). Prof. Amar Akbar Ali (Department of Architecture Universitas Tadulako, Indonesia), and Dr. Diana Contrias Mojica, New Castle University, UK)

This year, the ICOSIET conference received 110 papers submissions from 106 countries such as, Bangladesh, China, Colombia, Germany, Hong Kong, India, Indonesia, Iraq, Japan, Jorda, Peru, Philippines, Russia, Saudi Arabia, Thailand, United Kingdom, United States and Viet Nam. But only 68 papers were accepted for presentations for online sessions (the acceptance ratio is 65%). We are very grateful for the extensive efforts of many individuals who worked diligently to ensure a successful and high quality conference. We would like to thank to IPO, ASCEE, TPC Committee and Faculty of Engineering Tadulako University for supporting this event.

Once again we would like to say Welcome to ICOSIET 2020. Congratulations for your papers have been accepted. We invite all participants to actively participate in the conference activities to learn from one another. Thank you for choosing ICOSIET as your conference reference. We hope to have your pleasant supports and participations in the next year 2021 7th ICOSIET.

Thank you

General Chair
Dr. Eng. Rifai Mardin., ST., MT., M.Sc
Department of Information Technology
Faculty of Engineering Tadulako University

IOP Conf. Series: Materials Science and Engineering 1212 (2022) 011001 doi:10.1088/1757-899X/1212/1/011001

ICoSIET 2020 Committee

Patron

Prof. Dr. Ir. Mahfudz, MP. (Rector of Universitas Tadulako, Indonesia)

Advisor

1. Dr. Lukman. M.Hum (Vice Rector 1 of Universitas Tadulako, Indonesia)
2. Dr. Muhammad Nur Ali, M.Si (Vice Rector 2 of Universitas Tadulako, Indonesia)
3. Dr. Ir. Sagaf, MP. (Vice Rector 3 of Universitas Tadulako, Indonesia)
4. Prof. Dr. Amar, ST., MT. (Vice Rector 4 of Universitas Tadulako, Indonesia)

Steering Commitee

1. Dr. Eng. Ir. Andi Rusdin, ST., MT., MSc. (Dean of Faculty of Engineering, Universitas Tadulako, Indonesia)
2. Ir. Andi Arham., ST., M.Sc., Ph.D (Vice Dean 1 of Faculty of Engineering, Universitas Tadulako, Indonesia)
3. Dr. Ir. *Tutang Muhtar* Kamaludin ST.,M.Si (Vice Dean 2 of Faculty of Engineering, Universitas Tadulako, Indonesia)
4. Dr. Ir. Rusli ., ST., MT (Vice Dean 3 of Faculty of Engineering, Universitas Tadulako, Indonesia)
5. Dr. Fuad Zubaidi, ST., MT. (Head of Department of Architecture, Universitas Tadulako, Indonesia)

Organizing Committee

General Chair :Ir. Rifai Mardin, ST.,M.Si.,M.Sc.,Ph.D Eng
Co-Chair : Dr. Burhanuddin, ST.,M.Sc
Dr. Ir. Muhammad Bakri, ST.,MT
Ir. H. Syarifuddin, MT
Secretary : Rezki Awalia, ST.,MT
Sri Mulyati.,S.Si.,M.Sc
Hariyadi Salenda, S.PdT.,M.Sc

Technical Program Chair
Dr. Eng Puteri Fitriaty, ST.,MT
Dr. Ir. Anita Ahmad Kasim, S.Kom.,M.Cs
Gator Timbang, ST.,MT
Khairin Rachmat, ST.,MT
Ardiansyah Winarta,ST.,M.Si

Layout TCP
Moch Rachmat Syahrullah, ST.,M.Sc
Sari Widyaningsih, ST.,MT
Deltri Dikwardi Eisenring, ST.,M.SP

Publication and Website
Sutrati Melisa Malik, ST.,MT
Supriadi Takwim, ST.,M.Eng

Financial dan Sponsorship
Koordinator : Luthfiah, ST.,M.Sc
Anggota : Khaerunnissa, ST.,M.T
Muhammad Ismail, S.Pd.,M.Sc
Budi Andresi, S.Pd.,M.Eng

Communication and Relationship
Koordinator : Ir. Irdinal Arief, ST.,MT

IOP Conf. Series: Materials Science and Engineering 1212 (2022) 011001 doi:10.1088/1757-899X/1212/1/011001

Anggota : Fratika Julia, ST.,MT
Rizkhi, ST.,MT

ICoSIET 2020 Editors

1. Anita Ahmad Kasim (Tadulako University, Indonesia)
2. Diana Contrias Mojica (New Castle University, UK)
3. Roman Voliansky (Dniprovsky State Technical University, Ukraine)
4. Leonel Hernandez (ITSA University, Colombia)
5. Rafal Drezewski (AGH University of Science and Technology, Poland)
6. Andri Pranolo (Universitas Ahmad Dahlan, Indonesia)
7. Joni Adiansyah (Universitas Muhamadiyah Mataram, Indonesia)
8. Yun Arifatul Fatimah (Universitas Muhammadiyah Mageleng, Indonesia)
9. Aji Prasetya Wibawa (Universitas Negeri Malang, Indoensia)
10. Haviluddin (Universitas Mulawarman, Indonesia)

IOP Conf. Series: Materials Science and Engineering 1212 (2022) 011002 doi:10.1088/1757-899X/1212/1/011002

Peer review declaration

All papers published in this volume of IOP Conference Series: Materials Science and Engineering have been peer reviewed through processes administered by the Editors. Reviews were conducted by expert referees to the professional and scientific standards expected of a proceedings journal published by IOP Publishing.

- **Type of peer review: Single-blind**
  Each submitted paper reviewed by two minimum of reviewers after meet the minimum criteria. The review based on the following aspects: 1) Technical Criteria (Scientific merit, Clarity of expression, and Sufficient discussion of the context of the work, and suitable referencing); 2) Quality Criteria (Originality, Motivation, Repetition, Length); and 3) presentation criteria (Title, Abstract, Diagram, figures, tables and captions, Text and mathematics, and Conclusion). We also used iThenticate for plagiarism detection.

- **Conference submission management system:** Easychair

- **Number of submissions received:** 125

- **Number of submissions sent for review:** 117

- **Number of submissions accepted:** 90

- **Acceptance Rate (Number of Submissions Accepted / Number of Submissions Received X 100):** 90/125 = 72%

- **Average number of reviews per paper:** 2 Reviewers

- **Total number of reviewers involved:** 36

- **Any additional info on review process:** No

- **Contact person for queries:**
  Name : Dr. Anita Ahmad Kasim
  Affiliation: Universitas Tadulako, Indonesia
  Email : info@icosiet.org

An opportunity for using constructed wetland technology in hospital wastewater treatment: a preliminary study

A G Akhmad[a], S Darman[b], Aiyen[b], W Pingkan S. Hamsens[b]

[a] Faculty of Engineering, Tadulako University, Palu, Central Sulawesi, Indonesia
[b] Doctoral Study Program of Agriculture, Postgraduate of Tadulako University, Palu, Central Sulawesi, Indonesia.

[a] gani9367@yahoo.co.id

Abstract. The problem faced by many hospitals in Indonesia is the low efficiency of Waste Water Treatment Plant (WWTP) hospitals. If it does not get the attention, it will have an impact on health and environmental pollution. Various technologies have been used but experienced many obstacles. Constructed wetlands based on aquatic plants that have been used in developed countries are very prospects of being developed in regions such as Indonesia with a tropical climate. A preliminary study is needed with the first step to investigate the quality of hospital wastewater in Palu City, Indonesia, as a sample. This study aims to obtain a description of hospital wastewater characteristics and to evaluate the performance of the hospital WWTP in Palu. Data collection was done by taking data indirectly through laboratory test results during 2015-2019. Tabulating data using Excel software to illustrate statistics, then presented in the form of bar charts, interpreting according to the quality standards. Investigation results showed that the characteristics of hospital inlet wastewater in Palu are parameter values varying with four high concentration parameters: total coliform, TSS, Ammonia Nitrogen, and COD. The efficiency level of WWTP hospitals in Palu is relatively low in removing pollutants.

Key words: WWTP hospitals, wastewater treatment, constructed wetlands,

1. Introduction

Hospitals' presence with the complexity of their activities is expected not to add to the negative burden of environmental pollution. Infrastructure requirements such as hospital waste management were regulated in Act Number 44 of 2009. However, at this time, the number of hospitals that conduct medical waste management according to standards has only reached 33.63% ("Data Dan Informasi Profil Kesehatan Indonesia 2018," 2019).

Regarding hospital wastewater management, Regulation of the Minister of Health of the Republic of Indonesia Number 7 of 2019 requires hospitals to have a Waste Water Treatment Plant (WWTP) with a permit, and the results of liquid waste treatment meet quality standards. Nevertheless, in reality, many large hospitals do not have wastewater treatment facilities (Santoso, 2015). Also, several studies have reported the low efficiency of WWTP hospitals, especially in Indonesia (B. and Anwar, 2018; Harlisty, Akili, and Kandou, 2016; Kusuma, Yanuwiadi, and Laksmono, 2013; Rahmawati and Azizah, 2005).

Wastewater treatment technologies that are commonly used in hospitals in Indonesia include trickling filters or biofilter, rotating biological contactor (RBC), contact aeration/oxidation, and others (Djohan & Halim, 2014), the operation of this wastewater treatment technology generally requires many artificial resources in the form of chemicals and energy sources of fuel oil through electricity or generators (Mangkoedihadjo and Samudro, 2010). Also, the operational costs are high, and the complexity of the treatment of wastewater treatment plants (Zulkifli, 2014).

The research, development, and application of wastewater treatment technology based on water plant species were currently receiving attention in developed and developing countries such as America, Australia, Europe, Thailand, and Malaysia. For a long time, it has been known that wetland ecosystems have the natural ability to eliminate organic and inorganic pollution. This ability is mainly due to the presence of water plant species that act as waste processors. Constructed Wetlands technology is one example of the application of wastewater treatment systems based on water plant species that are easy to operate and eco-friendly (Khiatuddin, 2003).

According to Noor (2007), Indonesia has a tropical climate, where various types of water plant species thrive throughout the year. This potential will be useful in resolving the low current efficiency of WWTP hospitals. Because the Constructed Wetlands technology has never been explicitly applied to hospitals in Indonesia, it is essential to conduct a study of hospital wastewater treatment systems in Indonesia that adopt Constructed Wetland technology.

The efforts mentioned above will provide many benefits such as: supporting government programs in realizing environmentally friendly hospitals (Green Hospital); providing an efficient, inexpensive, easy to operate and environmentally friendly WWTP hospital; and, Constructed Wetland with Horizontal Sub Surface Flow type can be used as a hospital garden that has aesthetic value.

With the issue of the development of Constructed Wetland for hospital wastewater treatment, a preliminary study has been carried out with the first step investigating the quality of hospital wastewater during 2015-2019 by selecting two hospitals in the City of Palu in Indonesia as a sample.

This preliminary study's main objective is to obtain a description of the characteristics of hospital wastewater in Palu City, Indonesia. The results of this preliminary research will be input in the design of hospital wastewater treatment systems that adopt the Horizontal Sub Surface Flow Constructed Wetland (HSSF-CW) technology.

2. Methods

This preliminary study was conducted in December 2019 by selecting two hospitals in the city of Palu in Indonesia as the research location. Data collection of hospital wastewater characteristics was done by taking data indirectly through laboratory test results during the 2015-2019 period, which are the results of the laboratory tests of the UPT Central Sulawesi Provincial Health Laboratory and the UPTB Environmental Laboratory of Donggala Regency. Tabulating data using EXCEL software to illustrate statistics (average, minimum and maximum), then presented in the form of bar charts, interpreting according to the quality standards of the Minister of Environment RI No. 5 of 2014.

3. Results and Discussion

Wastewater Treatment Plants (WWTP) of the two hospitals, where the study was conducted, used the Aerobic Anaerobic Biofilter system.

The first hospital (RS-1) is a type B education general hospital that belongs to the city government, which has a capacity of 505 beds with an average number of inpatient visits of 27,827 patients per year. The second hospital (RS-2) is a type B education general hospital that belongs to the provincial government, which has a capacity of 357 beds with an average number of inpatient visits of 15,800 patients per year.

3.1 Characteristic of hospital wastewater

Overview of RS-1 and RS-2 inlet wastewater quality is the parameter concentration value varies. However, there are four parameters of concern because the concentration is high, namely: total coliform, total suspended solids (TSS), Ammonia Nitrogen, and Chemical Oxygen Demand (COD).

3.1.1 Total coliform.

The average range of total coliform concentrations of RS-1 and RS-2 inlet wastewater during 2015 - 2019 were 45,000 - 213,000 and 204,000 - 676,000 MPN/100 ml (Figure 1). The total Coliform standard for hospital wastewater is 5,000 MPN/100 ml.

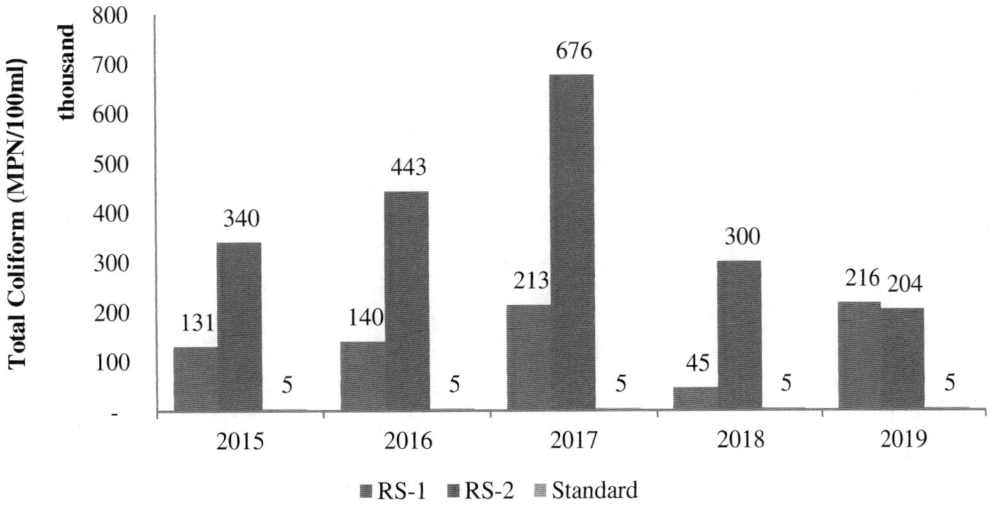

Figure 1. Average Total Coliform of RS-1 and RS-2 Inlet Wastewater Period 2015 - 2019

The results of the study at several hospitals also reported the condition of high total coliform parameters. The total coliform of wastewater from Nganjuk District Hospital in 2003 was reported to reach 240,000 MPN/100 ml (Rahmawati & Azizah, 2005). The study results at Tulehu District Hospital recorded an average total coliform of 4,186,028 MPN/100 ml (Kerubun, 2014). The wastewater from the Babol University Medical Sciences Hospital, Babol, Iran, understudy was recorded at 5.4×10^8 MPN/100 ml (Amouei et al., 2015). The total coliform content of inlet wastewater in Bitung City Hospital is 160,000 MPN/100 ml (Harlisty et al., 2016).

The coliform total in RS-1 and RS-2 inlet wastewater was confirmed to be sourced from Blackwater. The presence of fecal coliform bacteria in the aquatic environment indicates that water has been contaminated with human feces containing pathogenic bacteria. The increase of coliform total in RS-2 during 2015 - 2017 could be caused by WWTP's condition not functioning optimally because it was only operated in 2015.

Most of the digestive tract pathogens become the cause of various outbreaks of enteric diseases such as Salmonella typhi (typhoid fever), Shigella spp. (shigellosis), Salmonella paratyphi (salmonellosis), Vibrio cholerae (cholera), Campylobacter jejuni (dysentery), and pathogenic Escherichia coli (diarrhea).

3.1.2 TSS.

During 2015 - 2019, TSS of RS-1 inlet wastewater was higher than the standard with an average range of 59 - 147 mg/l as shown in Figure 2. TSS in RS-2 is not available. The TSS standard for hospitals is 30 mg/l.

IOP Conf. Series: Materials Science and Engineering 1212 (2022) 012001 doi:10.1088/1757-899X/1212/1/012001

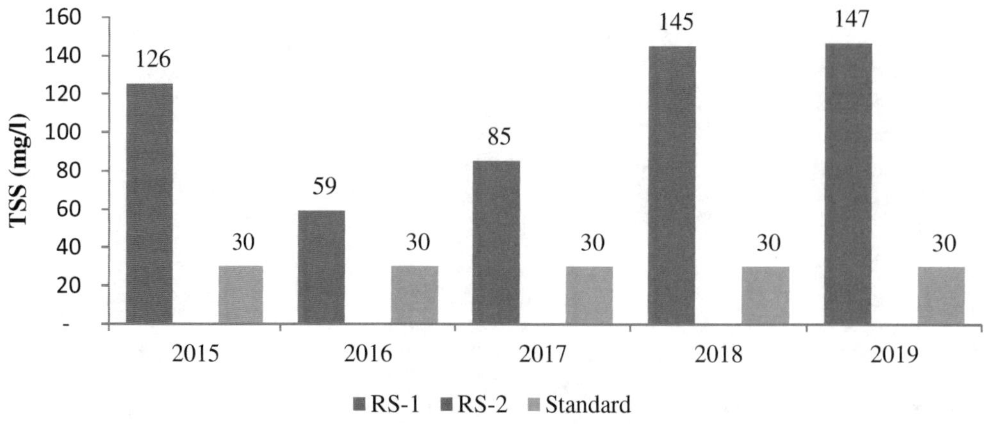

Figure 2. Average TSS RS-1 Inlet Wastewater for Period 2015 – 2019

TSS consists of precipitated components, suspended material, and suspended colloidal components. Suspended solids contain organic and inorganic substances. Suspended solids affect water turbidity and brightness. Moersidik (1993) examined hospital wastewater quality in Indonesia. TSS concentrations were in the range of 36-269 mg / l. Wangsaatmaja (1997) reported that the TSS concentration of hospital wastewater in Bangkok was 90 mg / l.

High TSS concentrations in RS-1 wastewater may be related to high COD (organic matter) and total coliform. According to Slamet (1996), organic matter can be a portion of food for bacteria to support their breeding. This bacterium is also a suspended organic substance so that the addition will add to the turbidity of the water. Turbid water is difficult to disinfect. Because these suspended substances protect microbes, this is undoubtedly dangerous for health, if microbes are pathogens.

3.1.3 Ammonia nitrogen.
Ammonia nitrogen concentrations of RS-1 inlet wastewater recorded high occurred in 2015, 2016, 2017, and 2018 with an average value of 3.93, 4.18, 15.46, and 2.68 mg / l. Whereas the RS-2 recorded high occurred in 2015, 2016, 2018, and 2019 with an average value of 2.00, 2.29, 2.02, and 1.05 mg / l, respectively (see Figure 3). The standard ammonia nitrogen concentration for hospital wastewater is 1 mg / l.

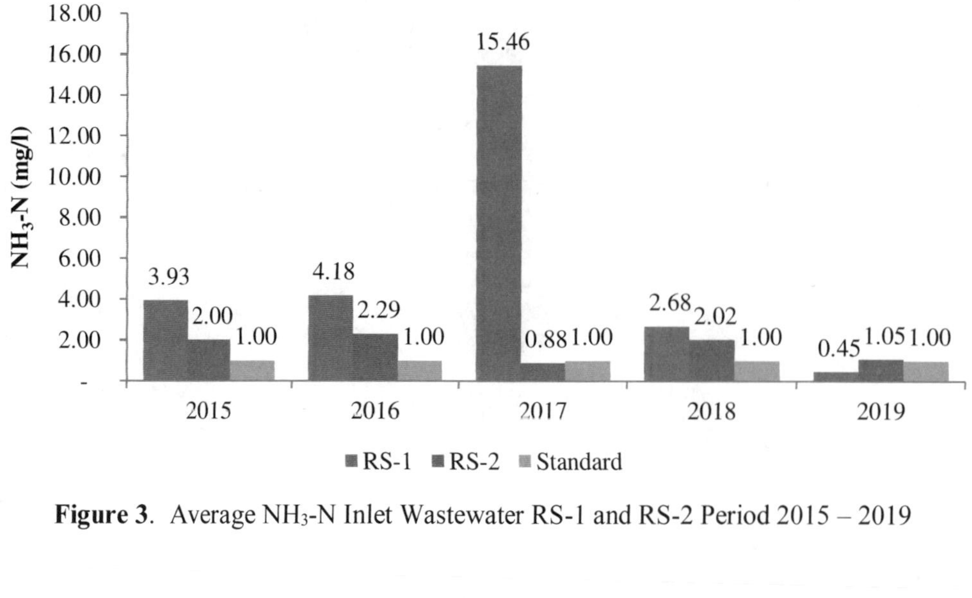

Figure 3. Average NH_3-N Inlet Wastewater RS-1 and RS-2 Period 2015 – 2019

In another study, Djaja and Maniksulistya (2006) reported high ammonia content in the inlet wastewater in Hospital X Jakarta, 35.5 mg / l. Kolibu and Tewal (2011) also reported high ammonia levels in the inlet wastewater of the GMIM Hospital Bethesda Tomohon, 28.25 mg / l.

Aquatic organisms are susceptible to ammonia-nitrogen, and also eutrophication occurs because of the high load of nitrogen in wastewater. The lethal concentration of ammonia-N for fish is 0.44 mg / l (Yadu et al., 2018).

High concentrations of NH_4^+-N in RS-1 and RS-2 wastewater may be caused by detergents, drugs, and urine, as reported from several previous studies. Seiler *et al.*, (1999) reported that the use of detergents in laundry services and drugs contributed significantly to nitrate waste. For example, these drugs (such as chlorpropamide, phensuximide dan carbamazepine) are known to be excreted in the urine in an unmetabolized form. They, therefore, survive in the environment after channeling it into any body of water. Amanfo, Nyarko and Jaime (2014) report that nitrate content in hospital waste is sourced from detergents used in cleaning floors. Liquid floor cleaners generally use the necessary ingredients of a detergent type that contains SLES (sodium lauryl ether sulfate) or better known by the trade names texapone and NI (Nitrate). Beler-Baykal *et al.* (2004) revealed that the main components of nitrogen in human urine are ammonia and urea, which in turn are converted into ammonia itself.

3.1.4 COD.

The high average COD concentration of RS-1 inlet wastewater recorded only occurred in 2015 with 154 mg / l. In RS-2, it occurred in 2015 and 2016 with levels of 355 and 183 mg / l (Figure 4). The COD standard for hospital wastewater is 80 mg / l.

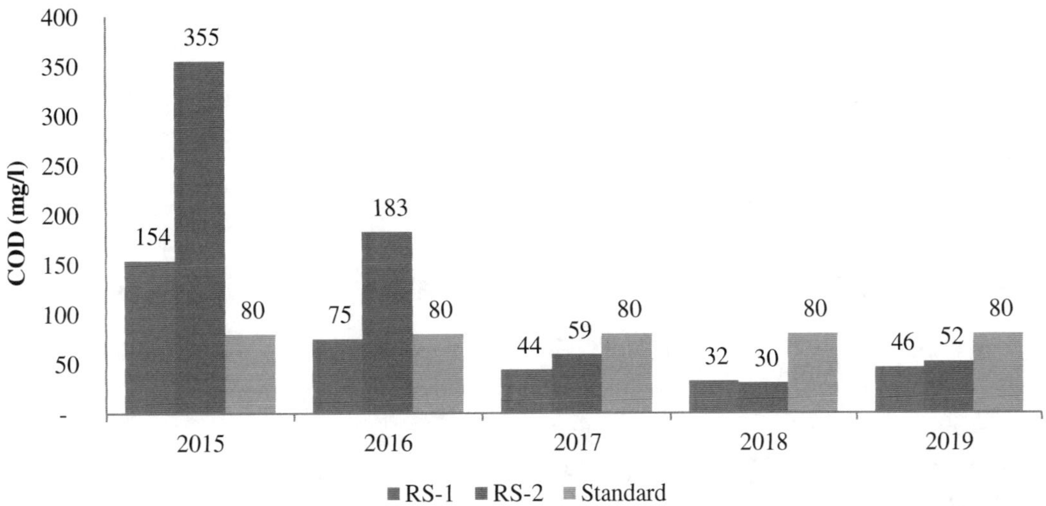

Figure 4. Average COD of RS-1 and RS-2 Inlet Wastewater Period 2015 – 2019

Some other research results also report high COD concentrations in several hospitals. The COD concentration in Valiasr Hospital was 435 mg / l, Bahrami Hospital was 1362 mg / l (Mesdaghinia et al., 2009). The COD concentration of hospital raw wastewater in Thailand is 232 mg / l (Wangsaatmaja, 1997). COD concentrations of raw wastewater from 70 hospitals in Iran have been reported 527 mg / l (M. Majlesi Nasr & A.R., 2008).

The content of organic pollutants in RS-1 and RS-2 wastewater can be sourced from Domestic liquid waste, such as a bathroom, kitchen, and laundry waste; Clinical liquid waste, such as wound washing water, blood washing, and others; Laboratory liquid waste originates from the washing of laboratory equipment and material from examinations such as blood, urine, and others.

Organic waste will experience degradation and decomposition by aerobic bacteria (using oxygen in water) so that the oxygen that gets in the water will be significantly reduced over time. In the condition of reduced oxygen, only certain species of organisms that can live.

3.1.5 The other parameters.

Except for the parameters COD, TSS, ammonia nitrogen, and total coliform, the conditions of the other parameters are still in the normal range (Table 1, Table 2).

Table 1. Average of other Parameters for RS-1 Inlet Wastewater Period 2015 – 2019

Parameters	Highest concentration		Average of Parameters				
	Value	Unit	2015	2016	2017	2018	2019
Iron, dissolved (Fe)	5	mg/l			0,18	0,13	
Manganese, dissolved (Mn)	2	mg/l				0,58	
Copper, (Cu)	2	mg/l			0,02	<0,01	
Zinc, (Zn)	5	mg/l			<0,01	<0,01	
Cadmium, (Cd)	0,05	mg/l			<0,01	<0,01	
Lead, (Pb)	0,1	mg/l			<0,01	<0,01	
Nickel, (Ni)	0,2	mg/l				,<0,01	
Cyanide, (CN)	0,05	mg/l			<0,002	<0,002	
Sulfide, (S =)	0,05	mg/l			0,04	0,18	
Flourida, (F-)	2	mg/l				0,10	
Free chlorine, (Cl2)	1	mg/l				Nol	
Nitrate (NO3-N)	20	mg/l				0,66	
Nitrite (NO2-N)	1	mg/l			0,55	0,38	
MBAS	5	mg/l				0,01	0,01
Phenol	0,5	mg/l				0,13	

Table 2. Average of other Parameters for RS-2 Inlet Wastewater Period 2015 – 2019

Parameters	Highest concentration		Average of Parameters				
	Value	Unit	2015	2016	2017	2018	2019
Iron, dissolved (Fe)	5	mg/l	0,05	0,11			
Manganese, dissolved (Mn)	2	mg/l	<0,01	0,03			
Copper, (Cu)	2	mg/l	0,01	0,02			
Zinc, (Zn)	5	mg/l	<0,01	<0,01			
Chrome valence six, (Cr6 +)	0,1	mg/l	<0,01	<0,01			
Cadmium, (Cd)	0,05	mg/l	<0,01	<0,01			
Lead, (Pb)	0,1	mg/l	<0,01	<0,01			
Nickel, (Ni)	0,2	mg/l	0,11	<0,01			
Cobalt, (Co)	0,4	mg/l	0,01	0,01			
Cyanide, (CN)	0,05	mg/l	<0,002	<0,002			
Sulfide, (S =)	0,05	mg/l	<0,02	1,11			
Free chlorine, (Cl2)	1	mg/l	Nol	Nol			
Nitrate (NO3-N)	20	mg/l	0,19	0,29			
Nitrite (NO2-N)	1	mg/l	<0,003	<0,003			
MBAS	5	mg/l	0,04	0,05	0,02	0,02	0,02
Phenol	0,5	mg/l	<0,025	<0,025			

According to Djohan and Halim (2014), laboratory wastewater generally contains a lot of various chemical compounds as reagents when examining blood samples and other materials, contains antiseptic agents and antibiotics that are toxic to microorganisms, and contain heavy metals. Low levels of chemical compounds and heavy metals in wastewater RS-1 and RS-2 can occur due to dilution factors. The volume of laboratory wastewater is much smaller than the volume of domestic wastewater.

3.2 WWTP hospitals existing

The performance of WWTP RS-1 and RS-2 was generally low in the removal of Total Coliform, TSS, and Ammonia Nitrogen, except COD. Total coliform removal efficiency is generally quite high, in the range of 54% - 97%, but its effect on the removal of total coliform is still above the quality standard. It may be caused by the total coliform burden that is too large than the hospital WWTP capacity.

The efficiency of WWTP RS-1 on TSS removal is low, with a range of 21% - 53%, resulting in TSS outlet levels still above the quality standard. Also, the efficiency of WWTP RS-1 and RS-2 on NH3-N removal is generally low, with a range of 7% - 62%, so that NH3-N outlet levels are still above the quality standard.

Table 3. Efficiency of WWTP RS-1 and RS-2 on Removal of Total Coliform, TSS, Ammonia Nitrogen, and COD Period 2015 - 2019

	Parameters	**2015**		**2016**		**2017**		**2018**		**2019**	
		RS-1	**RS-2**	**RS-1**	**RS-2**	**RS-1**	**RS-2**	**RS-1**	**RS-2**	**RS-1**	**RS-2**
Coliform	Inlet (MPN/100ml)	130,808	340,000	140,183	443,083	213,333	676,083	45,045	300,218	216,375	204,400
	Outlet (MPN/100ml)	27,900	10,333	41,358	205,545	11,508	51,000	7,391	13,891	6,875	5,600
	Efficiency	79%	97%	70%	54%	95%	92%	84%	95%	97%	97%
	Std (MPN/100ml)	5,000	5,000	5,000	5,000	5,000	5,000	5,000	5,000	5,000	5,000
	Indication	over	over	over	over	Over	over	over	over	over	over
TSS	Inlet (mg/l)	126	-	59	-	85	-	145	-	147	-
	Outlet (mg/l)	100	-	31	-	40	-	103	-	99	-
	Efficiency	21%	-	48%	-	53%	-	29%	-	32%	-
	Std (mg/l)	30	-	30	-	30	-	30	-	30	-
	Indication	over	-	over	-	Over	-	over	-	over	-
NH3-N	Inlet (mg/l)	3.93	2.00	4.18	2.29	15.46	0.88	2.68	2.02	0.45	1.05
	Outlet (mg/l)	0.78	1.23	1.57	1.41	14.42	0.61	2.10	1.28	0.13	1.16
	Efficiency	80%	39%	62%	38%	7%	30%	22%	36%	72%	-10%
	Std (mg/l)	1	1	1	1	1	1	1	1	1	1
	Indication	low	over	over	over	Over	low	over	over	low	over
COD	Inlet (mg/l)	154	355	75	183	44	59	32	30	46	52
	Outlet (mg/l)	52	126	42	108	37	41	26	26	57	43
	Efficiency	67%	64%	44%	41%	14%	31%	21%	15%	-24%	17%
	Std (mg/l)	80	80	80	80	80	80	80	80	80	80
	Indication	low	over	low	over	Low	low	low	low	low	low

3.3 Overview of constructed wetlands with HSSF

The Vymazal study (2019) indicate that Horizontal Surb Surface Flow Constructed Wetlands (HSSF-CWs) can provide a sustainably excellent treatment performance for 20+ years despite being designed according to a simple formula used in the early 1990s. The present study evaluated treatment performance of 114 Czech HSSF-CWs with special attention to 17 systems that have been in operation for at least 20 years. The study clearly revealed that if the HSSF-CWs are properly loaded, their treatment performance is very steady with outflow concentrations < 15 mg l^{-1} of BOD_5 and TSS and < 50 mg l^{-1} COD. The removal efficiency of 17 systems that have been in operation for at least 20 years amounted to 91.7%, 82.9% and 88.3% for BOD5, COD and TSS, respectively during the last five years of operation. Removal of both organics and TSS does not depend on the season.

Haddis, Bruggen, and Smets (2020) concluded that Horizontal Sub Surface Flow Constructed Wetlands could function well under variable flow conditions and be the technology of choice in low-income countries, particularly in tropical climates. Two selected plant species, i.e., Cyperus papyrus

(papyrus) and Scirpus Validus (bulrush), were tested using the wastewater from a dormitory at Jimma University Campus in Ethiopia. BOD_5 removal was 81% in papyrus, 76% in bulrush, and 48% in Control. TSS removal was 76% and 75% in papyrus and bulrush compared to 54% in the Control. COD removal showed 65% and 62% in the planted beds and 32% in the Control.

García-Ávila *et al.* (2019) studied of the purification of domestic wastewater using HSSF-CWs with Cyperus Papyrus on a small scale received municipal wastewater with primary treatment. The results obtained in the experimental tests indicated that the HSSF-CWs with Cyperus Papyrus presented a high capacity of pollutants removals such as BOD (80.69%), COD (69.87%), ammoniacal nitrogen (69.69%), total phosphorus (50%), total coliforms (98.08%) and fecal coliforms (95.61%).

Typha with the broken brick was significantly improved ($P<0.05$) the treatment performance of the Horizontal Sub Surface Flow Constructed Wetland systems for the removal of TSS, BOD5, COD, TKN, NH4-N, NO3-N, and phosphate from hospital wastewater, respectively, 93.2%, 90.4%, 83.7%, 64%, 64.3%, 52.1% and 56.1% in the dry season and 89.7%, 85.8%, 82.9%, 66%, 62.7%, 56.1% and 59.5% in the rainy season (Dires, Birhanu, and Ambelu, 2019).

Dires *et al.* (2018) reveal that in general horizontal subsurface flow constructed wetlands could help to solve the problem of the cost-effective disposal of hospital wastewater, being a suitable treatment for reducing indicator, pathogenic, and antibiotic-resistent bacteria which is comparable to activated sludge treatment system and be promoted in a strategy to reduce water pollution in low-income countries like Ethiopia.

Tabel 4 Performance of Constructed Wetlands with Horizontal Sub Surface Flow

Parameters	Performance (%)	Location	Reference
BOD_5	91.70		
COD	82.90	Czech	Vymazal, (2019)
TSS	88.30		
BOD_5	81.00		
COD	65.00	Ethiopia	Haddis, Bruggen, and Smets, (2020)
TSS	76.00		
BOD_5	80.69		
COD	69.87		
Amonia-N	69.69		García-Ávila *et al.* (2019)
Total Coliform	98.08		
Fecal Coliform	95.61		
BOD_5	90.40		
COD	83.70		
TSS	93.20		
Amonia-N	64.30	Ethiopia	Dires, Birhanu, and Ambelu, (2019)
TKN	64.00		
NO3-N	52.10		
Phosphate	56.10		

4. Conclusions and Recommendations

The investigation results have shown that the characteristics of wastewater hospitals inlet in Palu are variable values with four high concentration parameters: total coliform, TSS, Ammonia Nitrogen, and COD. The hospital WWTP's efficiency level in Palu is relatively low in removing pollutants such as total coliform, TSS, Ammonia Nitrogen, and COD. Some research results describe the performance of CW-HSSF in wastewater treatment in tropical areas. Therefore HSSF-CW has the opportunity to be used for hospital wastewater treatment in Indonesia.

References

[1] Amanfo, N. N. Y., Nyarko, B. E., & Jaime, G. J. (2014). *Physico-Chemical Characteristics of Wastewater at Saint Patrick's Hospital, Offinso, Ghana*. Kwame Nkrumah University of Science and Technology College of Science.

[2] Amouei, A., Asgharnia, H., Fallah, H., Faraji, H., Barari, R., & Naghipour, D. (2015). Characteristics of Effluent Wastewater in Hospitals of Babol University of Medical Sciences, Babol, Iran. *Health Scope*, *4*(2). https://doi.org/10.17795/jhealthscope-23222

[3] B., R., & Anwar, M. (2018). Studi Karakteristik dan Kualitas BOD dan COD Limbah Cair Rumah Sakit Umum Daerah Lanto Dg. Pasewang Kabupaten Jeneponto. *Jurnal Nasional Ilmu Kesehatan*, *1*. http://journal.unhas.ac.id/index.php/jnik/article/view/4285

[4] Beler-Baykal, B., Bayram, S., Akkaymak, E., & Cinar, S. (2004). Removal of ammonium from human urine through ion exchange with clinoptilolite and its recovery for further reuse. *Water Science and Technology*, *50*(6), 149–156. https://doi.org/10.2166/wst.2004.0371

[5] Data dan Informasi Profil Kesehatan Indonesia 2018. (2019). In *Profil Kesehatan Indonesia 2018*.

[6] Dires, S., Birhanu, T., & Ambelu, A. (2019). Use of broken brick to enhance the removal of nutrients in subsurface flow constructed wetlands receiving hospital wastewater. *Water Science and Technology*, *79*(1), 156–164. https://doi.org/10.2166/wst.2019.037

[7] Dires, S., Birhanu, T., Ambelu, A., & Sahilu, G. (2018). Antibiotic resistant bacteria removal of subsurface flow constructed wetlands from hospital wastewater. *Journal of Environmental Chemical Engineering*, *6*(4), 4265–4272. https://doi.org/10.1016/j.jece.2018.06.034

[8] Djaja, I. M., & Maniksulistya, D. (2006). Gambaran Pengelolaan Limbah Cair di Rumah Sakit X Jakarta Februari 2006. *Makara*, *10*(2), 60–63.

[9] Djohan, A. J., & Halim, D. (2014). *Pengelolaan Limbah Rumah Sakit*. Salemba Medika.

[10] García-Ávila, F., Patiño-Chávez, J., Zhinín-Chimbo, F., Donoso-Moscoso, S., Flores del Pino, L., & Avilés-Añazco, A. (2019). Performance of Phragmites Australis and Cyperus Papyrus in the treatment of municipal wastewater by vertical flow subsurface constructed wetlands. *International Soil and Water Conservation Research*, *7*(3), 286–296. https://doi.org/10.1016/j.iswcr.2019.04.001

[11] Haddis, A., Bruggen, V. der B., & Smets, I. (2020). Constructed wetlands as nature based solutions in removing organic pollutants from wastewater under irregular flow conditions in a tropical climate. *Ecohydrology & Hydrobiology*, *20*(1), 38–47. https://doi.org/10.1016/j.ecohyd.2019.03.001

[12] Harlisty, B. F., Akili, R. H., & Kandou, G. D. (2016). *Analisis Kandungan Amoniak dan Bakteri Coliform Total pada Limbah Cair Rumah Sakit Umum Daerah (RSUD) Kota Bitung Pada Tahun 2016*. http://medkesfkm.unsrat.ac.id/wp-content/uploads/2016/10/JURNAL-Bima-Febrian-Harlisty.pdf

[13] Kerubun, A. A. (2014). Kualitas Limbah Cair di Rumah Sakit Umum Daerah Tulehu. *JURNAL MKMI*, *10*(3), 180–185. https://doi.org/http://dx.doi.org/10.30597/mkmi.v10i3.500

[14] Khiatuddin, M. (2003). *Melestarikan Sumber Daya Air dengan Teknologi Rawa Buatan*. Gadjah Mada University Press.

[15] Kolibu, F. K., & Tewal, F. B. (2011). Kualitas Limbah Cair Rumah Sakit GMIM Bethesda Tomohon. *Fakultas Kesehatan Masyarakat Universitas Sam Ratulangi Manado*. https://www.scribd.com/document/425553430/21makan-docx

[16] Kusuma, Z., Yanuwiadi, B., & Laksmono, R. W. (2013). Study of Hospital Wastewater Characteristic in Malang City. *Research Inventy: International Journal Of Engineering And Science Issn Www.Researchinventy.Com*, *2*(2), 13–16. https://s3.amazonaws.com/academia.edu.documents/31244050/C022013016.pdf?response-content-disposition=inline%3B
filename%3DResearch_Inventy_International_Journal_o.pdf&X-Amz-Algorithm=AWS4-HMAC-SHA256&X-Amz-Credential=AKIAIWOWYYGZ2Y53UL3A%2F20191217%2Fus-e

[17] M. Majlesi Nasr, & A.R., Y. (2008). Study on Wastewater Treatment Systems In Hospitals of Iran. *Iran. J. Environ. Health. Sci. Eng.*, *5*(3), 211–215. file:///C:/Users/TOSHIBA/Downloads/STUDY_ON_WASTEWATER_TREATMENT_SYSTEMS_IN_HOSPITALS.pdf

[18] Mangkoedihadjo, S., & Samudro, G. (2010). *Fitoteknologi Terapan*. Graha Ilmu.

[19] Mesdaghinia, A., Naddafi, K., Nabizadeh, R., Saeedi, R., & Zamanzadeh, M. (2009). Wastewater Characteristics and Appropriate Method for Wastewater Management in the Hospitals. *Iranian J Publ Health*, *38*(1), 34–40.

[20] Moersidik, S. S. (1993). Technology alternatives for hospital waste management. *Seminar Limbah Rumah Sakit. Jakarta. 14 September 1993. Universitas Indonesia. Indonesia*.

[21] Noor, M. (2007). *Rawa Lebak; Ekologi, Pemanfaatan, dan Pengembangannya*. PT. Raja Grafindo Persada.

[22] Rahmawati, A. A., & Azizah, R. (2005). Perbedaan Kadar Bod, Cod, Tss, dan Mpn Coliform pada Air Limbah, Sebelum dan Sesudah Pengolahan di RSUD Nganjuk. *Jurnal Kesehatan Lingkungan Unair*, *2*(1), 97–110. http://www.journal.unair.ac.id/filerPDF/KESLING-2-1-10.pdf

[23] Santoso, I. (Ed.). (2015). *Inspeksi Sanitasi Tempat-Tempat Umum*. GosyenPublishing.

[24] Seiler, R. L., Zaugg, S. D., Thomas, J. M., & Howcroft, D. L. (1999). Caffeine and Pharmaceuticals as Indicators of Waste Water Contamination in Wells. *Ground Water*, *37*(3), 405–410.

[25] Slamet, J. S. (1996). *Kesehatan Lingkungan*. Gadjah Mada University Press.

[26] Vymazal, J. (2019). Is removal of organics and suspended solids in horizontal sub-surface flow constructed wetlands sustainable for twenty and more years? *Chemical Engineering Journal*, *378*, 122117. https://doi.org/10.1016/j.cej.2019.122117

[27] Wangsaatmaja, S. (1997). *Environmental Action Plan For A Hospital*. Asian Institute of Technology School of Environment, Resources and Development Bangkok, Thailand.

[28] Yadu, A., Sahariah, B. P., & Anandkumar, J. (2018). Influence of COD/ammonia ratio on simultaneous removal of NH 4 + -N and COD in surface water using moving bed batch reactor. *Journal of Water Process Engineering*, *22*, 66–72. https://doi.org/10.1016/j.jwpe.2018.01.007

[29] Zulkifli, A. (2014). *Pengelolaan Limbah Berkelanjutan*. Graha Ilmu.

IOP Conf. Series: Materials Science and Engineering 1212 (2022) 012002 doi:10.1088/1757-899X/1212/1/012002

Tsunami evacution building model (TEBM) in Teluk Palu (Palu Bay) area

M Bakri[1], R Awalia[1], S Mulyati[1], F Zubaidi[1], M Rusydi[2], A A Kasim[3], R Prawita[4]

[1]Department of Architecture, Faculty of Engineering, Tadulako University, Palu, Indonesia
[2]Department of Physics, Faculty of Mathematics and Natural Sciences, Tadulako University, Palu, Indonesia
[3]Department of Informatics, Faculty of Engineering, Tadulako University, Palu, Indonesia
[4]Study Program of Architecture, Faculty of Engineering, Tadulako University, Palu, Indonesia

Email : embakri@gmail.com

Abstract. The research was conducted in the coastal area of Palu Bay, Lere Village. Data were collected using observation, interviews, and literature study techniques. The data were analyzed using qualitative descriptive methods and focused on the concept of the design of the tsunami evacuation building on the coast of Palu Bay. Data analysis was carried out to obtain an overview of the problems that occurred at the planning location which was analyzed with the concept of architectural design by considering the existing conditions of the site and the surrounding environment, as well as utilizing the potential of the site so that it has the potential for building functions. The conclusion obtained is based on the results of data analysis, it is necessary to have a building that can function as an evacuation container for the community when a disaster occurs in the coastal area of Palu Bay, thereby reducing the number of casualties. The benefit of this research is that it is a solution for local people to save themselves when the tsunami disaster.
Keywords: Tsunami, Evacuation, Building, Palu Bay

1. Introduction

Tsunamis are a threat of disaster for many coastal areas in Indonesia. Tsunamis are generally triggered by an earthquake in the sea which causes a vertical shift in the seabed. Tsunami threat analysis is carried out to determine the character of tsunamis that may have occurred or will occur by considering the source mechanism, location, wave propagation, tsunami wave propagation, and the height of the tsunami inundation. A tsunami is a disaster with the character of a fast-onset disaster or a type of disaster with a fast process. Tsunamis can occur from a near field location with a propagation time of fewer than 30 minutes from the source to the monitoring coastline and far-field locations with the propagation time to the monitoring shore area longer than 30 minutes or the tsunami source has a distance further than 1000 km. Tsunami threat characters tend to be site-specific. This leads to the need to analyze these threats and avoid the generalization process. The character of tsunami events in Indonesia is generally local, where the distance to the source of the tsunami is relatively close so that it only has a short time to anticipate or evacuate[1]. Tsunami disaster mitigation efforts in coastal areas require a building model that can be used as a place for community evacuation when a tsunami occurs.

One of the evacuations carried out in tsunami disaster mitigation is vertical evacuation where the community will look for upland areas.

Development of evacuation structures in coastal areas that are at risk of a tsunami must pay attention to whether the buildings remain safe during the earthquake and tsunami and whether people can reach the evacuation building in a short time and find an evacuation space[2]. One of the evacuation structures developed in Japan in 2011 is the Tsunami Vertical Evacuation (TVE). The TVE structure saved many lives in the 2011 Great East Japan earthquake. The Palu Bay area as one of the areas that are at risk of having a tsunami impact in Palu City requires various tsunami disaster mitigation efforts. This study will analyze the Tsunami Evacuation Building (TEB) model which can be used as an evacuation building in the Palu Bay area. This research will focus on an evacuation building model that can be built in the Palu Bay area with a structural design that is safe for earthquakes and tsunamis as well as an easy to reach and safe location for evacuation in the Palu Bay area.

2. Methods

Evacuation, whether initiated by individuals or managed by public officials has an important role to play in saving lives from natural disasters. There are many studies to determine evacuation models such as demographic, contextual, and warning messages that influence and motivate individuals to evacuate[3]–[7]. For example, demographic attributes related to gender, ethnicity, household size, caregiver status, education level, and age. These factors have been documented to predict evacuation behavior, such as other factors such as risk perception, length of stay in an area, previous disaster experience, and trust in warning sources[8].

The tsunami vertical evacuation strategy is designed to provide safe protection in tsunami-stricken areas by offering a sufficient height above the maximum water level. In the event of a tsunami near the field, vertical evacuation can be an alternative protective measure against horizontal evacuation [9]. Safe heights can be provided in artificially raised open ground or towers specially designed for evacuation. Research related to the construction of a vertical evacuation building for Tsunami Vertical Evacuation (TVE) Building has been carried out in various countries with the potential for a tsunami, one of which is Japan. The Japanese government has led an initiative in the preparation of vertical evacuation guidelines for the construction and management of tsunami vertical evacuation buildings (TVEB), in early 1982 in Kesennuma City, Japan [10]. Tsunamis present interesting challenges for evacuation planners with a range of evacuation characteristics. Local earthquakes can create tsunami waves that arrive within tens of minutes of an earthquake occurring, are preceded by significant ground shaking that alter the evacuation landscape, and represent a significant threat to life safety. As a result, emergency evacuations usually emphasize individually initiated pedestrian-based evacuations. This research has shown that individuals at risk have used vehicles to evacuate during the local tsunami disaster. To achieve the research objectives, data processing and analysis was carried out in four stages.

The first stage is editing, namely checking / checking the completeness and correctness of the data that has been obtained either from the results of observations, interviews, questionnaires, and documentation techniques. The second stage is data reduction, including the allowance for considered data less related to research, so the remaining data can be compiled in a simpler form. The third stage is the grouping of primary and secondary data and then comparatively analyzed to determine the function of the data further, whether as a concept support or as a conceptualization. The final step is data analysis using macro and micro analysis methods. The flow of data processing techniques in this study can be seen in Figure 1.

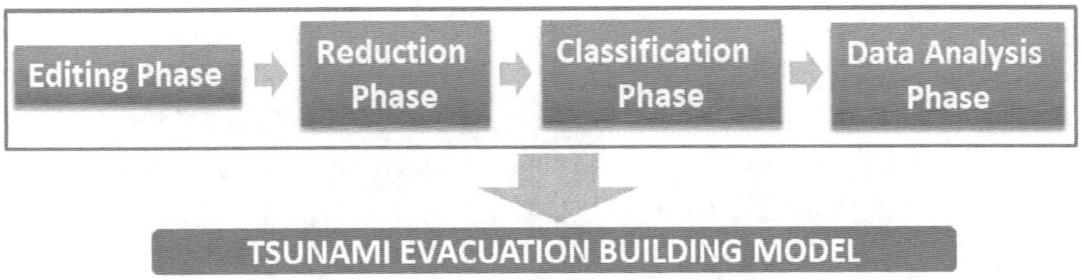

Figure 1. Proposed method for building the Tsunami Evacuation Building Model (TEBM)

Data analysis that has been collected during the study was carried out by using macro and microanalysis methods. The macro analysis includes site analysis of sun and wind, circulation, noise, view, and outdoor layout. Meanwhile, macro data analysis includes architectural design approaches, namely the function of space, the activity of actors, space requirements, the amount of space, the pattern of space organization, structure, and utility as well as the image of the building, namely the shape of the building. The synthesis stage interprets and concludes the results of the analysis that lead to the right concept and design as an architectural solution with an effective structural system. The analytical steps to produce a Tsunami evacuation building model can be seen in Figure 2.

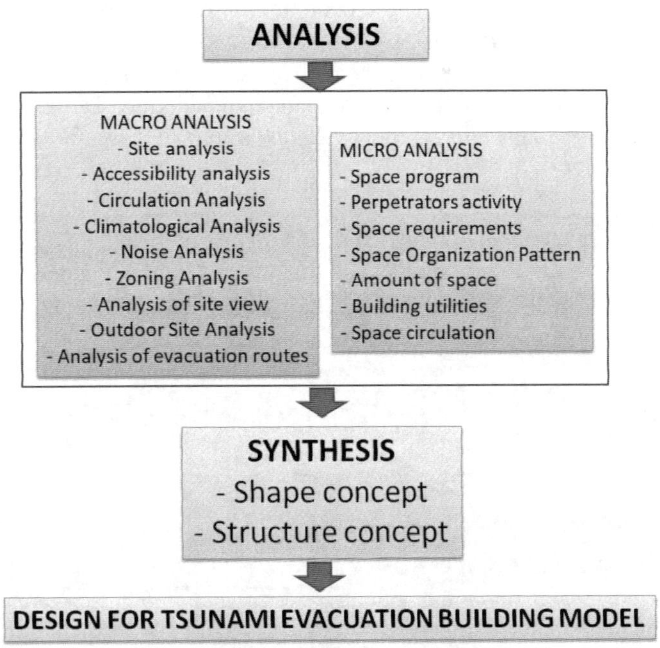

Figure 2. Stages of the Tsunami Evacuation Building Model (TEBM) Analysis

3. Result and Discussion

Selection of site locations for the planning of Tsunami Evacuation Buildings. On the coast of Palu Bay, an alternative location was planned based on the potential and problems that have occurred around the site. The first alternative, the site is located in the East Palu District, West Besusu Village, Raden Saleh street, Hotel Palu Golden, the location description can be seen in Figure 3.

Figure 3. Image alternative location 1 (source: google earth map, 2019)

The potential at the location has the potential to support the function of the building as an evacuation building, a short distance from the position around the coast, a densely populated environment, and provincial road access. The second alternative, the chosen location is in the area of the State Islamic Institute (IAIN) Palu, precisely located between Diponegoro Street and Squid Street. This location is the right location to build a tsunami evacuation building on the Pesisi Teluk Palu, because looking at the description of the location and considering the incident that occurred on September 28, 2018, which resulted in many casualties around the location. The second alternative location can be seen in Figure 4.

Figure 4. Image alternative location 2 (source: google earth map, 2019)

The concept of a site zoning system in a tsunami evacuation building aims to organize and create space groups according to the activities of the building so that there are no errors in the arrangement between the zones and the creation of harmonious functions of existing facilities on the site. In this zoning site, the potential and site conditions in the area greatly influence the zoning results so as to produce a good layout that supports the overall function.

Based on the existing criteria, the zoning system on the site is divided into:

a. Zone of impact
   The affected zone is the area or place that is affected by the tsunami in the vicinity of the site and how to handle it.
b. Zone of Tsunami Evacuation
   The evacuation zone is the main object in the site that involves all the people who come during a tsunami emergency.
c. Zone countermeasures plan
   This zone includes a disaster mitigation zone using green zone safety systems and breakwaters. The green zone is a plant that can prevent coastal abrasion and tsunamis and can withstand the crashing of seawater along the coast which is positioned inside the building while the breakwater is a pile of certain formations made of Benton rebates positioned outside the site on the seafront. Zone mapping can be seen in Figure 5.

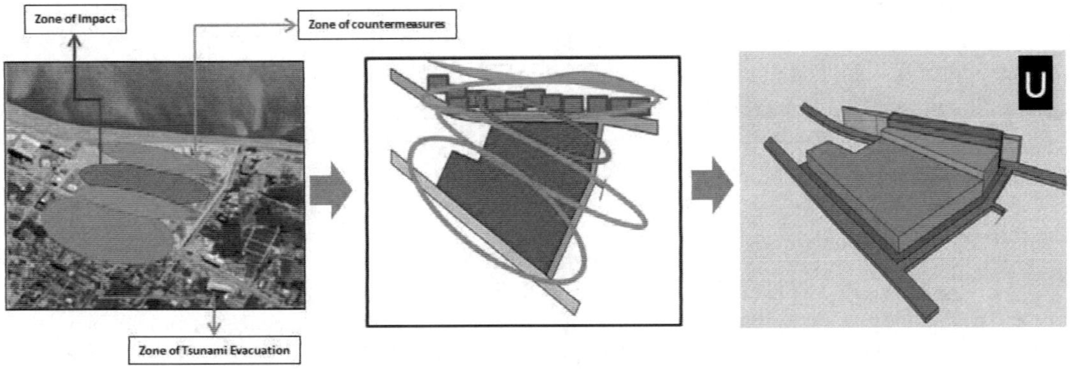

Figure 5. Zone Mapping of TEBM

To obtain a building design that is following the function of the tsunami evacuation building, adjustments to the shape concept are made by transforming the formations of triangles, circles, squares and iconic natural forms that are used as shapes that are under their main function and are generally attractive to the eye. The transformation model can be seen in Figure 6.

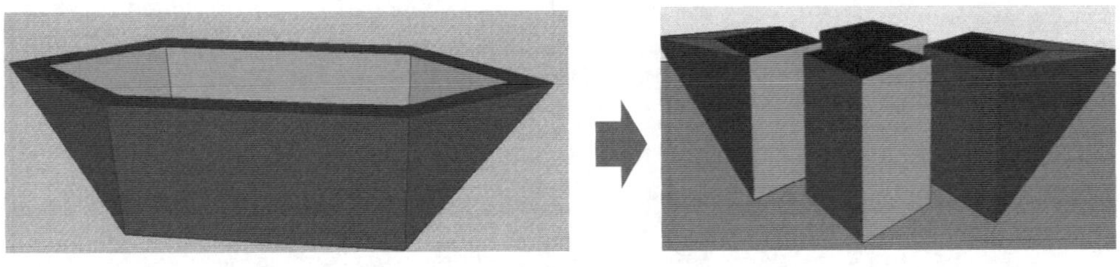

Figure 6. Model Transformation of TEBM

The square shape has added shapes on the right and left sides which can be seen to resemble the shape of a ship, because this building functions as an evacuation building, the function of this building adapts to the shape of the building so that the merging of the two transformations of the square and the shape of the boat produces a simple and iconic design. The shape of the tsunami evacuation building can be seen in Figure 7.

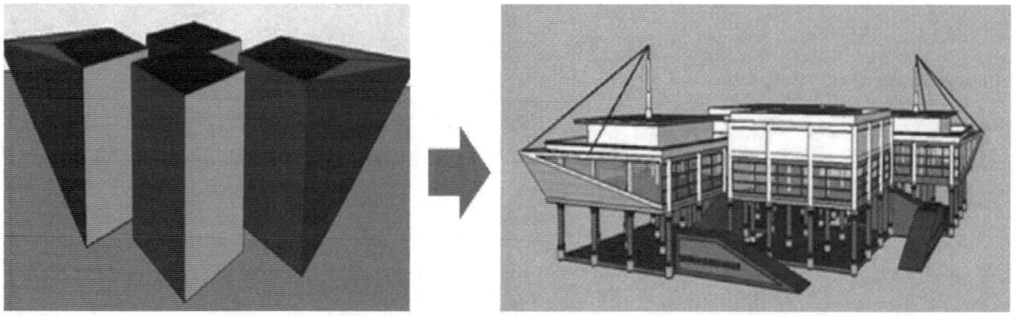

Figure 7. Tsunami Evacuation Building Model (TEBM)

4. Conclusion

The Tsunami Evacuation Building on the Teluk Palu (Palu Bay) is realized with a micro and macro analysis approach that plans based on the analysis and synthesis of problems for human needs and utilizes all information obtained from sources, media, and so on to produce designs. Apart from its main function as an evacuation site, this building can also be a place for educational education about disasters. The concept of shape by transforming triangles, circles, squares, and iconic natural shapes.

References

[1] M. R. Amri *et al.*, *Risiko Bencana Indonesia*. 2006.

[2] J. W. McCaughey, I. Mundir, P. Daly, S. Mahdi, and A. Patt, "Trust and distrust of tsunami vertical evacuation buildings: Extending protection motivation theory to examine choices under social influence," *Int. J. Disaster Risk Reduct.*, vol. 24, no. June, pp. 462–473, 2017, doi: 10.1016/j.ijdrr.2017.06.016.

[3] J. LIM, M. LIM, and M. PIANTANAKULCHAI, "Factors Affecting Flood Evacuation Decision and Its Implication to Transportation Planning," *J. East. Asia Soc. Transp. Stud.*, vol. 10, pp. 163–177, 2013, doi: 10.11175/easts.10.163.

[4] D. Sun, J. E. Kang, R. Batta, and Y. Song, "Optimization of evacuationwarnings prior to a hurricane disaster," *Sustain.*, vol. 9, no. 11, pp. 1–29, 2017, doi: 10.3390/su9112152.

[5] S. Lamb, D. Walton, K. Mora, and J. Thomas, "Effect of authoritative information and message characteristics on evacuation and shadow evacuation in a simulated flood event," *Nat. Hazards Rev.*, vol. 13, no. 4, pp. 272–282, 2012, doi: 10.1061/(ASCE)NH.1527-6996.0000070.

[6] W. Soviana and H. A. Rani, "Study of alternative building for tsunami evacuation in Kuta Alam sub-district Banda Aceh," *IOP Conf. Ser. Mater. Sci. Eng.*, vol. 674, no. 1, 2019, doi: 10.1088/1757-899X/674/1/012021.

[7] K. Penuel and M. Statler, "Evacuation Planning," *Encycl. Disaster Reli.*, 2014, doi: 10.4135/9781412994064.n76.

[8] N. Wood, J. Jones, J. Peters, and K. Richards, "Pedestrian evacuation modeling to reduce vehicle use for distant tsunami evacuations in Hawai'i," *Int. J. Disaster Risk Reduct.*, vol. 28, no. November 2017, pp. 271–283, 2018, doi: 10.1016/j.ijdrr.2018.03.009.

[9] A. Mostafizi, H. Wang, D. Cox, and S. Dong, "An agent-based vertical evacuation model for a near-field tsunami: Choice behavior, logical shelter locations, and life safety," *Int. J. Disaster Risk Reduct.*, vol. 34, no. December 2018, pp. 467–479, 2019, doi: 10.1016/j.ijdrr.2018.12.018.

[10] S. Fraser, G. S. Leonard, H. Murakami, and I. Matsuo, "Tsunami Vertical Evacuation Buildings – Lessons for International Preparedness Following the 2011 Great East Japan Tsunami," *J. Disaster Res.*, vol. 7, no. October 2011, p. 12, 2012.

Acknowledgement

This work supported and funding by DIPA BLU Fakultas Teknik Universitas Tadulako in scheme Penelitian Unggulan in 2020 (Grant No. 647.i/UN28.2/PL/2020)

Collective space on residential as a resilience measure to disaster mitigation earthquakes and liquefaction

A Mulyati, M Najib, A M Y Astha

Department of Architecture Engineering, Faculty of Engineering, University of Tadulako
Soekarno Hatta Street No. Campus of Bumi Tadulako Tondo Palu, Central Sulawesi, Indonesia

ahdamulyati@gmail.com

Abstract. Disasters have just hit Palu, Sigi, and Donggala, resulting in the fall of so many victims, not only humans but also city and village physicals (residential buildings, facilities, and infrastructure). Several villages hit by the disaster were hamlet 3 Vou, Langaleso village, Dolo sub-district, Jono Oge village, Biromaru sub-district, and Sigi district. The village's condition was damaged, but most of the buildings and plants were lost to the mud. Settlement neighborhoods that should be full of activity are nowhere to be seen, and people become apathetic. They just accept this situation and are always waiting for help. Huntara, a form of assistance provided by funders, temporarily occupy the settlers' land because the condition of the land or land needs to be cleared for the rebuilding process.The provision of inclusive spaces is one of the efforts to mitigate disasters in residential areas, namely public spaces that can be used by all settlers. These spaces must be safe and provide comfort, easy to reach, and become part of the residential environment. Village arrangement is one way that can be done to minimize the impact of a disaster. Arrangement of buildings, facilities, and environmental infrastructure are necessary not to create chaos in development. Also, aspects of environmentally friendly residential (village) spatial patterns, structural systems, and building construction that meet the requirements (SNI), the application of local and humane wisdom are no less important to consider in the process of procuring inclusive spaces in settlements, structuring villages, especially after a disaster.

Key Words : Space, Collective, Settlement, Resilience, Mitigation, Disaster.

1. Introduction

Hamlet 3 Vou Langaleso village and hamlet 3 Jono Oge Village are part of Dolo sub-district, Sigi regency which experienced the impact of the disaster/calamity on September 28, 2018. Each hamlet has a population of approximately 650 people or 130 households with farming (gardening) livelihoods. and raise livestock, so that it is known as a supplier of vegetables to the city of Palu and other areas around it. Land conditions are very fertile, in some parts of the hamlet there are water sources and irrigation channels flowing well. This is supported by the existence of the Gumbasa river water source which is part of the Sigi district. People live well, it can be seen from several permanent buildings surrounded by a large yard and garden with various types of plants. Coconut, banana, corn, lowland rice fields and various types of vegetables are found along the village roads. They have lived their lives in an atmosphere of harmony and simplicity for decades.

The disaster had devastated Langaleso village and Jono Oge village, the land that was originally fertile was cracked due to the earthquake and buried in mud due to liquefaction and brought all their belongings. The community is grieving, activities are totally paralyzed, all that is seen are the mountains of several roofs of buildings, piles of building materials that are still left behind, and their belongings that have also been destroyed without shape. Some others lost their families without any word of their whereabouts, so that the survivors had to struggle to continue their lives.

After the earthquake they did not have any shelter. Most of the settlers used evacuation tents, sometimes without food and drink, wearing volunteer clothes. Land for setting up tents only uses empty land around the settlement without paying attention to its suitability, both land conditions and health considerations, so it is often flooded when it rains heavily. Another problem arises because there are no available places for bathing, washing and toileting, resulting in various diseases due to the absence of medicines and medical personnel.

2. Research Method

This study uses a "qualitative-phenomenological" approach so that it will find patterns of residential space utilization [1];[2].

2.1. Research location : Sigi Biromaru sub-district

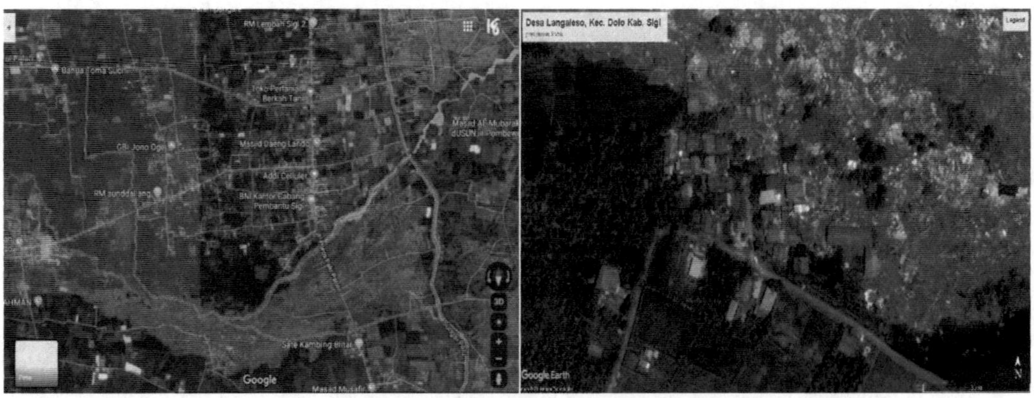

Figure 1. Map of Jono Oge Village (left) and Langaleso Village, Sigi Regency
(Source: Google Earth, March 2020)

Figure 2. Condition of Langaleso village (left) and Jono Oge village after the earthquake and liquefaction (Source: field data, October 2018)

2.2 Research tools and materials
Research tools, namely researchers as the main instrument equipped with assistive tools, in the form of writing tools and drawing tools, portrait tools, tape recorders, measuring instruments, GPS, and basic settlement maps. The research material consisted of: settlements and the environment, namely residential units, community activities, and community behavior in settlement spaces.

2.3. Research design
The first step in conducting the research begins with preparing maps of the research location. Then make observations of the research area as a whole to determine the study case as the first "sample". The focus of the research is specific things, namely housing, settlement layout, settler activities, behavior towards the residential space. This specific environmental condition was taken as the first research sample. Through interviews and in-depth observation, information themes are obtained. Themes are emphasized on settlement units, in relation to social, cultural and historical aspects. Basically, the theme is not limited, but is still related to the focus of the study. Then identify the spatial pattern of residential land use, in order to obtain the most dominant spatial use pattern. The spatial pattern describes the 'settlement system' practiced by the people who live in the area (Sigi Biromaru and Dolo districts). Data analysis used the 'inductive' method through descriptions of cases as samples [3]. The cyclic iteration process was developed as the basis for a temporary research design, determining other cases (by purposive sampling) until they reached 'saturation' or to the extent agreed upon. Analysis by means of categorization, description and explanation is carried out at every observation and interview. These results are 'research findings', are provisional in nature, which are continuously tested through observation and interviews. 2.3 Method of interpretation and theoretical dialogue Meaning is carried out by interpreting and explaining the findings between substantive concepts in order to obtain the findings of 'local concepts', namely the values of local knowledge in the use of settlement spaces. The form of theoretical dialogue is to position local concepts on other concepts / theories. The next stage is the discussion of research findings, with the "explanatory" technique, namely explaining the findings and confirming the related theories, as well as other conditions in different areas. The results of the discussion are the conclusions of research which are 'local theories' or 'substantive theories', which are specific to the research area and are temporary.

3. Discussion

3.1 Settlements as shelter and places to carry out post-earthquake and liquifaction activities
The residential environment in both 3 Jono Oge and 3 Langaleso hamlets is a community settlement environment whose daily life carries out activities both as a vegetable garden farmer and raising chickens, goats and cows. Therefore, their settlement is not only a residential environment [4] but also surrounded by gardens and cattle sheds as a source of their livelihoods [5]. The two hamlets are known by the community as suppliers of vegetables and chickens for the city of Palu and its surroundings. The fertile land, climate and clean water sources really support their activities, so that their income is not only sufficient for their daily life, but also their economic condition can be said to be good (the economic level of the community is moderate) [6]. In several environmental units, several supporting facilities or facilities are found, be it for education, health, religious activities (a mosque in Langaleso village and a church in Jono Oge village), and other social activities [6].

The settlement conditions in the two hamlets changed after the earthquake and liquefaction. Most of the houses and settlement facilities were carried away by the mud, so that the settlers no longer had a place to take shelter, a place to carry out social activities or a place to earn a living, especially in the 3 Vou hamlet of Langaleso village. Unlike in hamlet 3, Jono Oge village, residential buildings and other social facilities were cracked to pieces, so that many settlers left their houses to move to other places and a small portion still occupied their homes. Shelters (temporary shelters) are built on vacant lands or residual lands in residential areas as well as supporting facilities, such as places of worship, schools, washing baths, and others. Their land is not fully cultivated due to limited sources of clean water that can flow through their gardens, as well as livestock sheds which require money to procure seeds and repair stables.

3.2 Settler behavior towards their settlement spaces

Living is living and living together, therefore the settlement is not only a place to live but also as a place to interact. Socio-cultural aspects greatly influence the formation of a place to live (space arrangement and placement) to suit their needs. Factors that are considered in establishing a place to live (occupancy) include religion, social relations, livelihoods and the natural environment [6]. Usually, people build their living quarters around places of worship, markets or shopping centers. As the smallest part of a village (ngata), settlements tend to be near the main neighborhood roads and a spreading pattern (not arranged in specific centers) and close to their livelihood places, namely gardens and livestock pens, making it easier for them to carry out activities.

Things like this did not change drastically after the earthquake, some settlers continued to build houses on their land, a small proportion received housing assistance from the government which was located far from the original settlement (landless settlers in the two hamlets). Likewise, land for gardens and livestock, settlers continue to work on their land to provide for their families, with some assistance provided by the private sector, including water sources (wells and pumps), so that they take turns using it to irrigate garden lands. Even though the results are still minimal, they remain and will not leave their hamlet. There are still very few supporting facilities for activities in the form of schools, places of worship, roads, places of interaction but they still do it in the remaining spaces or other empty land. Some settlers built a small garden in front of the house and made a shop to sell in the shelter space (huntara), while the small children played in the spaces between their shelters [2] [6].

Figure 3. Settlers continue to build huntara and huntap on their respective lands as a process of adaptation to post-earthquake settlements and liquefaction (Source: Analysis based on field data, March and June 2020)

The residential environment has local community values as local values and supports the cultural aspects of the people who live in the hamlet. Culture will be interrelated with cosmic entities or social strata, so that each will occupy these spaces. This can also be found in the two hamlets, land use, aid distribution, land ownership, coloring the life of the settlers and the formation of residential spaces [2] [3]

3.3 Adaptation of settlers to the settlement environment

Culture must be able to be adaptive, that is, make adjustments to the social environment of its settlers. Some actions and ideas will develop in a social system, because culture as a result of actions and ideas will develop in a social system. Therefore actions and ideas always adjust to the development of the social environment [7]. This means that the culture owned by a certain community is used as an adaptation strategy in a certain environment so that the pollinators are still able to carry on their lives. This is also experienced by the settlers in the two hamlets, namely 3 Jono Oge hamlet and 3 Langaleso hamlet, the settlers are still trying to be able to adapt to their residential environment after the earthquake. The settlement that was previously comfortable, safe as a place to live, work and interact, is now just a stretch full of ruins of houses and buildings that can no longer be used by settlers.

In some lands, residential houses have started to be built but they do not use the space in the original area but use the remaining land in their yard. An atmosphere like this illustrates that they still feel part of an environment that previously provided a calm atmosphere and will still provide the same atmosphere even though the conditions are far different. The garden plots and stables remain in their original position without any desire to change them or move them. The adaptation process is also influenced by the settlers' perceptions and interpretations of objects and then refers to a categorization system in the form of a response to an environment both for residential buildings, access, building technology, building materials and other living activities [7]. Love for the house and its environment is a way to survive in the settlement which is influenced by the physical condition of the house, the environment, socio-culture, kinship, proximity to work, economy and psychologically to feel the living space [7] [8].

Figure 4. Farm land and livestock land as a source of livelihood they still do on their respective land, namely in hamlet 3 Langaleso village and hamlet 3 Jono Oge village (Source: Analysis based on field data, March and June 2020)

The structure is essentially a process of change, be it in the socio-cultural, economic and physical aspects. Even though the environmental conditions have changed, local or traditional values are still preserved, they still maintain kinship, religion, rural life and are not capital oriented [8], because the consequences of changes will manifest in norms, values, cultural adaptation, because of the existence of outside influence on their lives and will affect other elements in their lives. Settlers still survive to live and live, especially because this is where they were born and raised, so they feel the need to restore their village function as it should be, and it is arranged according to land conditions and its designation [9] and the safety factor against disasters (earthquakes and liquefaction) starting from structuring The area as a whole, occupancy and its environment, proper and disaster-friendly buildings according to standards (SNI) and based on local wisdom [10] [11] [12].

3.4 Collective space as a room for resilience and disaster mitigation

The settlement is a form of living space, an architectural work formed by the settlers who occupy it, consisting of a collection of houses, equipped with facilities and infrastructure to support the settlers' life [4]. Settlements have a spatial layout [5] as a living space including the behavior of the settlers and their social life [6]. Settlers will interact with each other in the residential spaces both in the micro and macro housing in the residential space. In addition to living, settlers also need work or business land, where the settlers try or work to support their families and meet other needs. Business spaces should also complement a settlement, such as in the two hamlets. The settlement consists of a collection of dwellings equipped with garden land and cattle sheds as a source of livelihood.

The procurement of collective space in settlements is a resilience space that should exist in every settlement that is prone to disasters, especially in Palu City and its surroundings and Central Sulawesi Province in general.

A settlement does not only consist of houses, equipped with facilities and infrastructure according to the needs of the settlers. Based on Indonesian national standards, a settlement should be equipped with public spaces, where humans can interact with fellow settlers for various activities [15] [17] [18]. These spaces consist of various facilities and are easily accessible (have adequate access) by the settlers from their residence. Facilities in the form of learning places, recreation areas, places of worship, shopping places, and various other activities, so that this public space is the center of a settlement. These spaces are very important for settlers because they are multi-functional and can be used by all settlers in a settlement as a common space [13] [14]. Procurement of public spaces known as inclusion spaces is tailored to the needs of each. settlements, as well as in the villages of Jono Oge and Langaleso villages which were affected by the earthquake and liquefaction disaster [15] [16]. After the disaster, the settlers in the two villages still used the remaining spaces from their houses, empty land from the yard of their homes, or empty land from their settlements. interaction has not been fully fulfilled because environmental conditions have not fully recovered.

Figure 5. Interaction spaces in the form of residential terraces and empty land in settlements (Source: Analysis based on field data, March and June 2020)

4. Conclusion

A. Settlers make use of the remaining space and empty spaces in the residential area as a place to live in the form of a shelter (temporary shelter) or a place to carry out daily activities.
B. Supporting activities (studying, praying, bathing and washing, etc.) use aid buildings both by the local government and private sector which are also built on the remaining land which is not used by the settlers as a place to live.
C. The land for gardens and livestock still occupies the original land, which the owners are working on to fulfill their daily needs.
D. The remaining space becomes a public space for activities that require a large enough area and meet the requirements for the implementation of these activities.
E. Therefore a collective space is needed in a settlement as a multi-functional space for disaster resilience, namely as a public open space equipped with adequate, humane, easily accessible, environmentally friendly facilities and infrastructure that pay attention to the local residents' activism to support their activities.
F. The arrangement and design of settlements should be habitable and sustainable and take into account the placement of Collective Space in a safe zone, according to its designation in the settlement.

References

[1] Groat, L dan D. Wang, 2002. Architectural Research Methods, John Wiley and Sons, New York.in Mulyati, A. 2015.
[2] Mulyati, A., 2015. Sama'di Lao as the Basic Forming Architecture of the Bajo Tribe Settlements in Sulawesi. Dissertation of PPS Universitas Gadjah Mada Yogyakarta.
[3] Muchamad, BN, 2013. Residential Architecture of the Dayak Bukit Community, Dissertation of PPS-UGM, Yogyakarta.
[4] Rapoport, A, 1969. House Form and Culture, Prentice Hall Inc, New York.
[5] Purbadi, YD, 2010. Spatial Architecture of Traditional Kaenbaun Settlements on Timor Island, Dissertation of PPS Gadjah Mada University Yogyakarta.
[6] Doxiadis, C.A, 1967, 1968. Ekistic : An Introduction to The Science of Human Settlement, in Mulyati, A., 2014. The Influence of Cultural Life on Typo-Morphology of Aquatic Settlements in Tojo Una-Una Regency. Research. Univ. Tadulako, Palu.
[7] Naing, N., 2011. Moving Settlements in a Floating House System as a Process of Human Adaptation to the Environment. ITS Dissertation, Surabaya.
[8] Hamidah, Noor, etc., 2015. Analysis of Sustainable Riverside Settlements. Inersia Vol. XII No. 1 Mei 2016.
[9] Mulyati, A., M. Najib, N.R. Burhany, 2018. Local Wisdom in Architecture of Vernacular Water Settlement of Bajo People in Central Sulawesi. Paper. UKDW, Yogyakarta.
[10] Winarto, Giri, 2017. Natural Disaster Emergency Response. Gosyen Publishing, Yogyakarta.
[11] Rahmawati, Turningtyas Ayu etc., 2018. Spatial - Based Disaster Risk Reduction. UB Press, Malang.
[12] Idham, Noor Chalis, 2014. Principles of Earthquake Resistant Architectural Design. Andi Offset, Yogyakarta.
[13] Ksatria, G.R., Kusumaningdyah N.H., Pitana, T.S., 2016. Settlements Grow on Principle Metabolism Above the Lapindo Sidoarjo Mud Disaster Land Arsitektura, Vol.14 No.2 Oktober 2016.
[14] Nurfatimah, C dan Q. Pramukanto, 2011. Settlement Landscape Planning for Disaster Mitigation Earthquake in Pangalengan District, Bandung Regency. Tesis. IPB Bogor.
[15] Pratiwi W.D. etc., 2012. Environmental Management and Post-Disaster Settlement Transformation in Sukabumi, West Java. School of Architecture and Planning and Policy Development ITB, Bandung.
[16] Giyarsih, S.R. dan S.A. Dalimunthe, 2013. Factors Affecting the Quality of Settlements After the Earthquake in Pleret District, Bantul Regency. Tata Loka Vo. 15 No. 1 February 2013. p. 28-38.
[17] Cici Nurfatimah, 2011. Settlement Landscape Planning for Earthquake Disaster Mitigation. Skripsi. IPB Landscape Architecture Department. Bogor.
[18] Watung, CHT., etc., 2018. The Level of Resilience and Resilience of the City of Manado to Disasters. Spatial Journal Vol. 5 No. 1 year 2018. ISSN 2442 3262 p 46-60.

Acknowledgments

This research was carried out with DIPA funding from the Faculty of Engineering for the 2020 budget year, therefore our research team would like to thank the faculty, LPPM Tadulako University and the entire community and the Village Heads of Jono Oge and Langaleso Village for their participation in helping to complete this research.

Life cycle energy analysis of a green building in Vietnam

D L Le, T Q Nguyen*, and H C Pham

National University of Civil Engineering, 55 Giai Phong Road, Hai Ba Trung District, Hanoi, Vietnam

*E-mail: quannt@nuce.edu.vn

Abstract. The paper presents the life cycle energy analysis (LCEA) of an office green building in Hanoi, Vietnam to prove the advantages of green buildings regarding energy efficiency and environmental effects. The case study building is a concrete structured one, which consists of 3 basements, 17 floors, and 1 attic with a gross area of 14,112 m^2. In the study, the building's embodied energy is determined based on the contained energy coefficient of the ith material and its quantity needed. Whereas, the operating energy is computed according to the annual energy consumption of the building, which is stimulated by the EnergyPlus simulation software. Relying on the relative share of the demolition energy with the life cycle energy that has been proposed by previous publications, this category will be estimated. Results showed that the initial embodied energy contributed the largest share to the life cycle energy (61.37%), followed by operational energy (27.61%). It also indicated that the percentage share of the operational energy of a green building is much lower than that of other buildings. The primary reason for this is associated with the usage of environmentally friendly materials and energy-saving equipment in the design option of the green building. Therefore, it can be convincing evidence that may help to change the mindset of decision-makers in Vietnam about green buildings.

Keywords: green building; life cycle energy analysis (LCEA); embodied energy; operating energy; demolition energy.

1. Introduction

Vietnam is one of the nations that has been suffered severely from severe climate conditions, especially climate change and global warming. It is predicted that most areas of the Red River and Mekong River Delta are under the sea level by 2050, and this will affect more than 31 million Vietnamese people [1]. Buildings sector consumes a tremendous amount of energy, which could be one of the most primary causes of these problems [2, 3]. Some recent studies have shown that nearly 40% of the world's produced energy was consumed by this sector [4, 5]. In Vietnam, the buildings sector used up to 36% of the total electricity expenses [6]. Moreover, because of the rapid growth of population and economic progress in Vietnam, the energy requirement in the next ten years is predicted to triple [7]. By the year 2025, it is forecasted that approximately 49% of the amount of energy needed for commercial purposes in Vietnam will need to import [7]. Therefore, reducing energy consumption in buildings is a viable solution to ensure energy security and limit the negative impacts on the environment in Vietnam.

Life cycle energy analysis (LCEA) is one of the Life Cycle Analysis (LCA) methods, which can use to analyze the energy consumption of buildings. Particularly, literature shows that LCEA can be applied to examine the relationship between embodied energy and operational energy of buildings [4, 5, 8]. This method also can be able to evaluate the impacts of a variety of components such as floors, walls, foundation, etc., on the total needed energy in the lifetimes of several houses in India [4, 5] and Turkey [8]. LCEA can use to analyze the energy demand of residential buildings throughout their entire life span [9] or to calculate the amount of energy per square meter on average, and evaluate the influence of buildings' lifetimes on life cycle energy [10], then solutions for potential energy-saving can be proposed [11]. Several methods for calculating embodied energy of construction materials have been proposed in New Zealand [12] and the United Kingdom [13]. In order to assist LCEA, it is necessary to develop a database on energy consumption and embodied energy for common types of materials [14, 15].

In Vietnam, LCEA has not been widely practiced because of several different factors. There are (i) this approach has not been introduced appropriately to both scholars and practitioners nationwide, (ii) the lack of standard data as input for the calculation process such as embodied energy coefficient and service life of materials, and (iii) formulas used in other countries to estimate life-cycle energy (LCE) have not been revised to reflect the local practice. Therefore, this study with the aim to present a LCEA step by step in the Vietnam context may help to address these above issues. Moreover, by using a green building as the case study, this study will estimate its LCE and compare with standard buildings to prove the dominant of green building from the lifecycle perspective. It could be convincing evidence to change the mindset of decision-makers in Vietnam who are more familiar with the "traditional buildings" and focus more on their design functions and costs. The findings of this research also can be a good reference for policymakers and practitioners in Vietnam when developing policies or making decisions in the buildings' design stage.

2. Life Cycle Energy Analysis Method

A building consumes a tremendous amount of energy over its entire life, which comprises of embodied energy, operating energy, and demolition energy (figure 1). Embodied energy consists of initial embodied energy and recurrent embodied energy related to the usage of materials and machinery in both construction and maintenance activities [4, 16, 17]. During the operation stage, energy is needed for operating heating, ventilation, and air conditioning (HVAC), domestic hot water (DHW), lighting systems, and household devices, etc. This type of energy is named as operational energy [4, 16, 17]. At the building's end-of-life stage, demolition energy is needed for demolishing the building and in transporting construction waste to disposal areas and recycling factories [4, 16, 17]. When estimating energy uses, some elements of energy consumption, especially the embodied energy may be omitted due to the lack of needed data. The following sections discuss these types of energy in detail.

2.1. Embodied energy

As mentioned above, embodied energy of buildings comprises initial embodied energy and recurrent embodied energy [16].

Initial embodied energy (E_{IE}) includes (i) energy used to produce and transport materials to construction sites, and (ii) energy needed to operate machines during the construction phase [18]. It can be estimated using formula (1) below [16]:

$$E_{IE}=E_{MT}+E_{OS} \tag{1}$$

where: E_{OS} is the energy used for on-site processes; E_{MT} is the embodied energy of materials, which can be calculated as the following formula:

$$E_{MT}=\sum m_i \times EC_i \tag{2}$$

where: m_i is the amount of the ith material; EC_i is the embodied energy coefficients of the ith material per unit on average.

Recurring embodied energy (E_{RE}) is associated with the usage of materials and machines for renovating the building and regular annual maintenance. This type of energy can be determined by the following formula [16]:

$$E_{RE} = \sum_{i=1}^{n} m_i \times EC_i \times (\frac{L_{building}}{L_{material\ (i)}} - 1) \quad (3)$$

where: m_i and EC_i are similar to the formula (2); $L_{building}$ is the lifetime of the building; $L_{material\ (i)}$ is the service duration of the ith material.

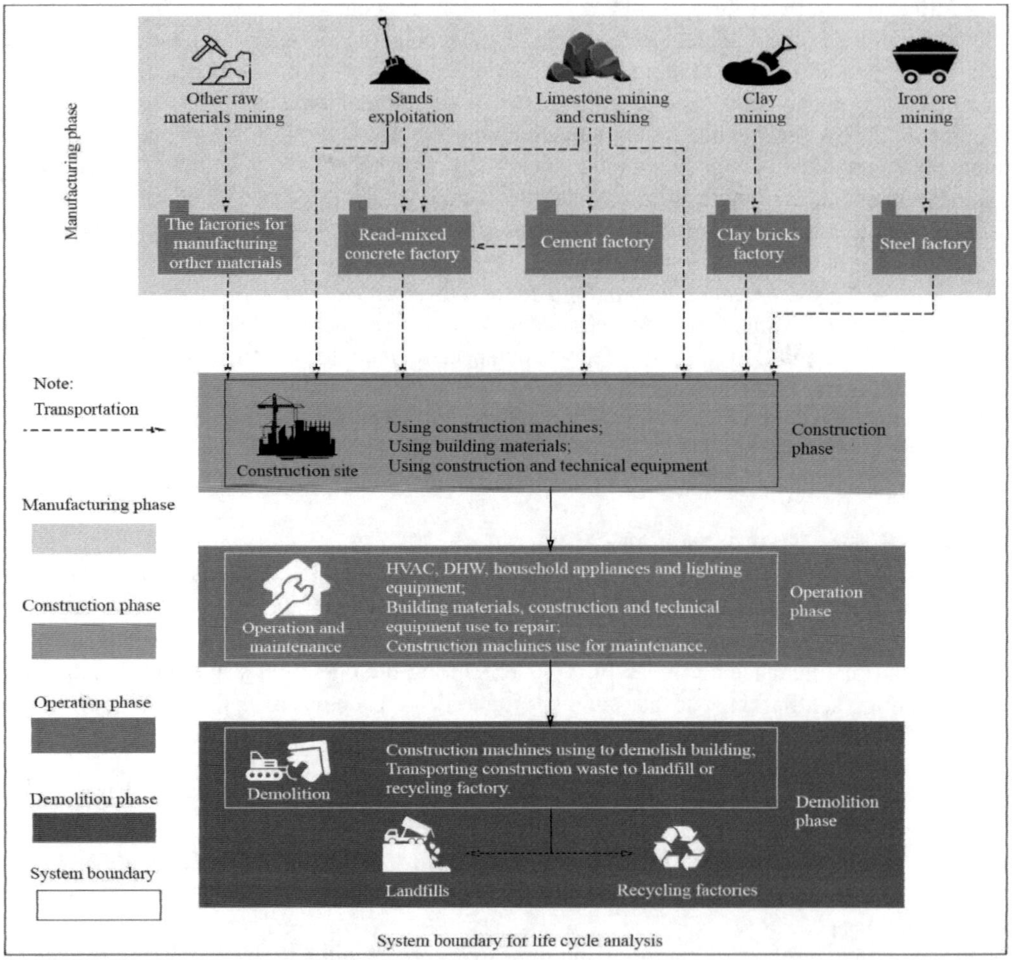

Figure 1. The LCEA's system boundary (adapted from [16])

2.2. Operating energy

Operating energy (E_O) is attributed to the operation of HVAC, DHW systems, and other energy-consumed appliances during the operation stage, and it can be computed by using equation (4) as below [16]:

$$E_O = E_{AO} \times L_{building} \quad (4)$$

where: E_{AO} is the annual operating energy of buildings; $L_{building}$ is similar to formula (3).

The annual operating energy can be estimated according to the types, quantity and wattage of energy-consumed equipment, and their average using duration. Besides, energy simulation software can also use to estimate this source of energy by simulation based on design models. In this research, the second approach will be used to simulate the annual operating energy of the case study building.

2.3. Demolition energy

This category of energy is associated with the use of machines and trucks to demolish buildings and transport waste to landfills or recycling areas, and it can be expressed as formula below [16]:

$$E_D = E_{MD} + E_{CR} \quad (5)$$

where: E_D is demolition energy; E_{MD} is the energy required to demolish the building; E_{CR} refers to the energy used to carry construction waste.

2.4. Life cycle energy (LCE)

LCE is the total of all categories of energy that a building consumed over its entire lifetime, and it can be calculated as the formula below [16]:

$$E_{LC}= E_{IE} + E_{RE} + E_O + E_D \quad (6)$$

3. The Case Study

3.1. The description of the case study building

The case study in this paper is a green high-rise building in Vietnam, which is the headquarter of the Vietnam Securities Depository Center. This building consists of 3 basements, 17 floors, and 1 attic. The building's gross floor area is 14,112m2. Table 1 shows the key figures and features of the building

Table 1. Key figures and features of the case study building

Contents	Specifications
Basements	3 floors
Upper floors	17 floors and 1 attic above ground
Service duration	75 years
Total floor area	14,112m^2
Load - bearing structure	Reinforced concrete
Covering envelope	Combination of brick walls and glass curtain walls
Foundation structure	Reinforced concrete bored piles and concrete foundation slab, and walls
Internal walls	Unburnt brick
External walls	Unburnt brick and glass curtain wall combination
Floors	Cast-in-place concrete
Floor finish	Concrete, ceramic tiles, gypsum board
Doors	Fire-resistant steel doors, Ironwood doors, MDF doors, Aluminum glass doors
Roof	Flat roof, concrete

Aiming to save energy, the building uses the following architectural solutions: (i) louvers placing along the south facade as sunlight-preventing panels outside the building; (ii) reduce the window/wall area ratio to the West, North, and East sides of the building; and (iii) use Low-E glass which has two layers and wall made of unburnt materials (light concrete blocks). An energy-saving central air-conditioning system (VRF) and an effective artificial lighting system have been installed for the same purpose. Moreover, the building also uses renewable energy with 8kwp solar panels and solar hot water systems with 37kw heating capacity. Regarding the water systems, the building is equipped with saving-water toilets and faucets as well as applying the reuse of treated sewage to meet 100% of irrigation needs.

With the target to conduct a LCEA, a variety of input data is required, such as:

a. the quantity of materials each type being used to construct the building;
b. the energy embodies of materials in each type per unit;

c. the types of equipment and duration using them in the construction phase;
d. the total amount of energy using in the operating phase; and
e. data to determine demolition energy.

The data that are not available in Vietnam but are necessary for calculating LCEA will be replaced with data that has been published worldwide for the convenience of the research. Moreover, in Vietnam, there are no available sources of information for energy consumption for the workers' living activities on the construction sites during the construction phase, and for energy used to operate construction machines in the maintenance activities. Consequently, such amount of energy will not be included in the calculation. This will not affect much on the outcomes due to the relative insignificance of the energy consumed for those activities in comparison to other types of energy uses.

3.2. Determining the initial embodied energy

3.2.1. Embodied energy of building materials (E_{MT}).

In this study, 20 key materials are taken into consideration for calculating this type of energy, since they are used in key components of the building such as frames, slabs, floors, staircases, foundation, walls, and even finishing works. The quantity of these materials is estimated based on the building's design. By using formula (2), the calculated results of this sort of energy are displayed in table 2. The total embodied energy of building materials in this case study is 102,142 GJ.

Table 2: Embodied energy of main building materials and their relative distribution

Materials	Quantity	Unit	EC_i (MJ/unit)	E_{MT} (GJ)	Relative contribution (%)	Source
Ready-mixed concrete	21,717,356	kg	1.11	24,106	23.60	[15]
Reinforcement steel	1,268,635	kg	24.6	31,208	30.55	[15]
Structural steel	255,350	kg	36.8	9,397	9.20	[15]
Sand	970,777	kg	0.1	97	0.10	[11]
Aggregate	454,511	kg	0.083	38	0.04	[4]
Cement	363,859	kg	4.6	1,674	1.64	[15]
Cement mortar	422,736	kg	1.4	592	0.58	[15]
Concrete bricks	1,704,795	kg	0.6	1,023	1.00	[19]
Galvanized iron	44,552	kg	38	1,693	1.66	[9]
Plasterboard	56,845	kg	4.4	250	0.24	[9]
Glass	106,205	kg	23.5	2,496	6.38	[15]
Paint	9,810	kg	30.6	300	0.29	[14]
Ceramic tiles	66,842	kg	10	668	0.77	[8]
Marble stone	36,328	kg	2	73	0.07	[8]
Wood (work form)	140,979	kg	8.5	1,198	1.17	[15]
PVC pipe	319,809	kg	67.5	21,587	21.13	[14]
Steel doors	2,213	kg	48.4	107	0.10	[14]
MDF (doors)	626	kg	11	7	0.01	[14]
Timber (doors)	3,688	kg	16	59	0.06	[14]
Doors and windows (Aluminium Framed)	376	m^2	3798.6	1,428	1.40	[14]
Total				102,142	100	

3.2.2. Embodied energy in the construction phase (Ec).

Three major types of energy to be used in this phase include petrol, diesel, and electricity. To determine the energy consumption in this stage, figures on the total number of working shifts of the ith machine, the amount of fuel consumed per shift of the ith machine, and the heat generation capacity of each fuel have been collected and manipulated. The number of working shifts for each machine will be determined based on the workload and the machine norms, which are issued by the Government. The amount of fuel that each machine consumes per shift on average was taken from the databook attached to the Decision 1134/QD-BXD of the Ministry of Construction of Vietnam dated October 8, 2015. These figures were put into the calculating formula and the amount of this type of energy was calculated to be 6,731GJ.

3.2.3. Initial embodied energy.,.;

This category is the total of embodied energy of materials and the energy used to construct the building. Based on the equation (1), this category of energy was calculated as 108,873 GJ, which equates to 7.71 GJ/m^2. The analysis illustrates that reinforcement steel, ready-mixed concrete, PVC pipe, and structural steel are the most significant contributors to the initial embodied energy of the building (figure 2). Reinforcement steel displays the largest portion (30.55%), followed by ready-mixed concrete (23.6%), PVC pipe (21.13%), and structural steel (9.2%) (Table 2). The load-bearing structure as well as the water supply and drainage system are responsible for up to 84.49% of the initial embodied energy of the building.

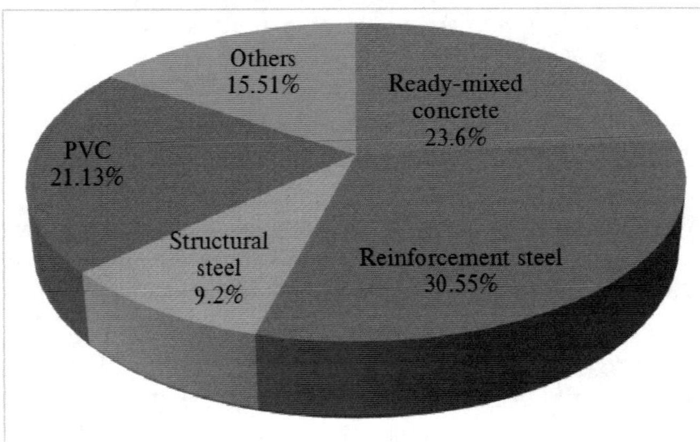

Figure 2. The contribution of primary materials to the embodied energy (%)

3.3. Recurring embodied energy

This category of energy is related to the energy contained in materials and equipment used to maintain and refurbish the building over the entire 75-year service life. Because the case study building has just moved from the construction to operation stages, no statistics exist about the materials' lifetime or the information about the percentage at each type of material will be replaced throughout the building's life span. Hence, this paper has referenced the required figures from a variety of publications to estimate this kind of energy of the building.

According to equation (3) and relevant data, this sort of energy was calculated, and the results were displayed in table 3.

Table 3. The recurring embodied energy (E_{RE}) of the building

Material	Quantity	Unit	EC_i (MJ/unit)	Building's life span (year)	Materials' life span (year) [5, 21]	E_{RE} (GJ)
Plasterboard	56,845	kg	4.4	75	30	375
Paint	9,810	kg	10.2	75	10	650
Ceramic tiles	66,842	kg	10	75	25	1,576
Motar	422,736	kg	1.4	75	50	296

Doors and windows (Aluminium Framed)	376	m^2	3798.6	75	50	714
Timber – hardwood (doors)	3,688	kg	16	75	25	118
Glass	277,315	kg	23.5	75	50	3,258
PVC	319,809	kg	67.5	75	50	10,794
Total						17,782

Although the HVAC, DWH systems, and lighting equipment as well as other appliances need replacements several times over the lifetime of the building, the recurring embodied energy related to these systems and appliances is excluded in this study. The primary reason for this is the unavailability of their embodied energy coefficients. Therefore, this study only considers several key finishing materials.

3.4. Operating energy

According to equation (4), the annual operation energy of the building is needed to compute the operating energy. This study uses EnergyPlus software to simulate the energy consumption of the building based on its 3D models and other information, such as (i) the climatic conditions, (ii) parameters of envelope materials, (iii) data on energy-consumed systems and appliances, etc. The simulated result is shown in figure 3. With energy equivalent to one kWh of electricity is 3.6 MJ/kWh, the total operating energy of the building was estimated as 48,989 GJ for the entire 75-year life span.

3.5. Demolition energy

At the end stage of the building's lifetime, energy is needed in order to operate machines in the demolishing process of the building and to transport waste to dumps or recycling factories. Because of the lack of data related to the energy consumption of demolition activity in Vietnam, this type of energy cannot be calculated exactly. Based on a study by Crawford RH [21], the amount of demolishing energy required is unnoticeable in comparison with the total LCE (only under 1%). Aiming to provide a comprehensive analysis of the LCE of a green building, this paper has expected that the percentage of this energy is equal to 1% of the total of embodied energy and operation energy (175,644 GJ). Therefore, the demolition energy is equal to 1,756.44 GJ.

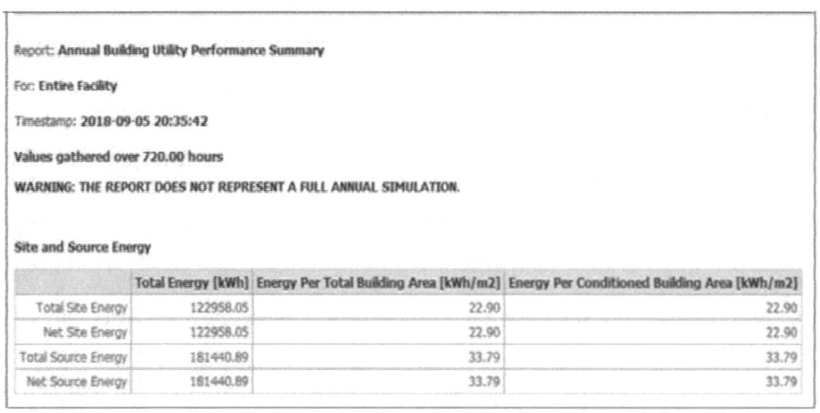

Report: Annual Building Utility Performance Summary

For: Entire Facility

Timestamp: 2018-09-05 20:35:42

Values gathered over 720.00 hours

WARNING: THE REPORT DOES NOT REPRESENT A FULL ANNUAL SIMULATION.

Site and Source Energy

	Total Energy [kWh]	Energy Per Total Building Area [kWh/m2]	Energy Per Conditioned Building Area [kWh/m2]
Total Site Energy	122958.05	22.90	22.90
Net Site Energy	122958.05	22.90	22.90
Total Source Energy	181440.89	33.79	33.79
Net Source Energy	181440.89	33.79	33.79

Figure 3. The building's energy consumption annually simulating by EnergyPlus software

3.6. Life cycle energy

The total building's LCE was calculated as 177,400 GJ (equal to 12.57 GJ/m^2). The results show that the largest contributor to the energy demand of the building is the embodied energy (61.37%), following by the operational energy (27.61%). On the contrary, the share of recurrent embodied energy is only

10.02%, and equal to about 16.33% of the initial embodied energy of the building. In comparison with other research, this result is only about one-fourth of the figure proposed by Crawford RH [21] (60%). The main reason for this difference is that this study does not calculate the recurrent energy of HVAC, DHW, and lighting systems. The energy demand in the demolition phase is only about 1% (figure 4).

The results display that the proportion of the operational energy of the green building (27.61%) is much lower than the figures in [5] (89.06%), and [22] (62% to 77%). The primary reason for this difference is that the case study is a green building. By using environmentally friendly materials like unburnt bricks, energy-saving systems, combining with other design solutions, the building consumes less energy in the operation phase than other buildings. This finding can be proven the dominant of the green buildings compared to standard buildings. Thus, it may be a significant basis for policymakers, decision-makers, and practitioners reference when evaluating design scenarios in the design phase or developing policies regarding the energy efficiency of the buildings sector in Vietnam.

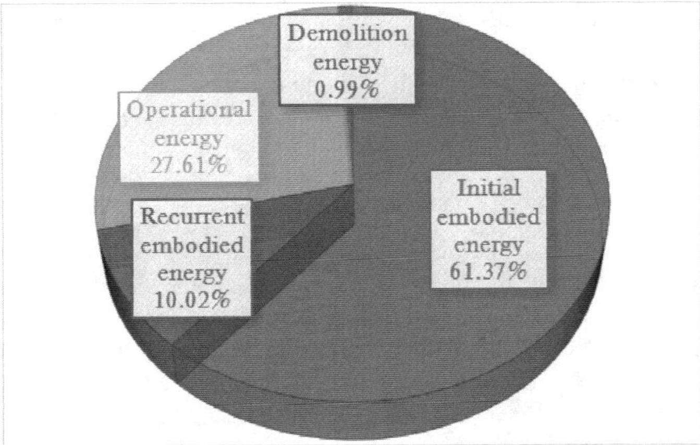

Figure 4. Proportion share of energy by phases over the building's lifetime

4. Conclusion

This paper has given a holistic LCEA of a green building in Vietnam. The paper demonstrates that the proportion share of energy by phases of a green building is a big difference compared to other residential houses which have been shown in other papers. Several articles have shown that operation energy is much higher than other sorts of energy over houses' life span, while this category of a green building only takes more than a half of the initial embodied energy. Particularly, the calculated results indicate that the initial embodied energy takes the most significant share to LCE (61.37%), whereas the operating energy is much lower than that, with only 27.61%. Recurrent embodied energy is only 10.02% of the total energy demand of the building due to this paper does not take HVAC, DHW, lighting systems, and other appliances into account since lacking needed data. In terms of initial embodied energy, reinforcement steel, ready-mixed concrete, and PVC pipe are the most significant contributors, contributing up to 84.49% of the total amount of this category of energy. The LCE consumption of the green building is estimated to be 177,400 GJ or 12.57 GJ/m2.

This paper provides decision-makers a much more dependable and holistic understanding one the total amount of energy that a green building demands over its lifetime. Although this paper only focuses on a specific green building, it can give a clear indication of the significant contributors to the total energy demand of a green building over its service life. By comparing the figures of the case study green building with other residential buildings that have been taken as case studies in previous publications, the paper has discussed the reason why the operational energy could be saved. These discussions can change the public's awareness in Vietnam about the LCE demand of a building and provides essential information to decision-makers in order to develop strategies and make decisions in the design stage to boost the development of sustainable construction in Vietnam.

Although the LCEA plays a very important role, it could be better if the LCEA is simultaneously considered with other lifecycle assessments such as lifecycle CO_2 and lifecycle cost analyses. Future research should consider the combination of those assessments for a more comprehensive and effective assessment of the sustainability of green buildings.

References

[1] Climate Central 2019 *Flooded Future: Global vulnerability to sea level rise worse than previously understood,* last accessed 10-6-2020, https://www.climatecentral.org/news/report-flooded-future-global-vulnerability-to-sea-level-rise-worse-than-previously-understood.

[2] Ma Z, Cooper P, Daly D and Ledo L 2012 Existing building retrofits: Methodology and state-of-the-art. *Energy and Buildings*. **55** 889-902.

[3] Pérez-Lombard L, Ortiz J and Pout C 2008 A review on buildings energy consumption information *Energy and Buildings*. **40(3)** 394-398.

[4] Pinky Devi L and Palaniappan S 2018 Life cycle energy analysis of a low-cost house in India *International Journal of Construction Education and Research*. **15(4)** 256-275.

[5] Talakonukula R, Ravi P and Karunesh KS 2013 Life cycle energy analysis of a multifamily residential house: a case study in Indian context *Open Journal of Energy Efficiency*. **2** 34-41.

[6] Vietnamplus.vn 2015 *Construction sector goes green,* last accessed 15-6-2020 https://cn.vietnamplus.vn/construction-sector-goes-green/71445.amp.

[7] Duc Luong N 2015 A critical review on Energy Efficiency and Conservation policies and programs in Vietnam *Renewable and Sustainable Energy Reviews*. **52** 623-634.

[8] Atmaca A and Atmaca N 2015 Life cycle energy (LCEA) and carbon dioxide emissions (LCCO2A) assessment of two residential buildings in Gaziantep, Turkey *Energy and Buildings*. **102** 417-431.

[9] Guan L, Walmsely M and Chen G 2015 Life Cycle Energy Analysis of Eight Residential Houses in Brisbane, Australia *Procedia Engineering*. **121** 653-661.

[10] Fay R, Treloar G and Iyer-Raniga U 2000 Life-cycle energy analysis of buildings: a case study *Building Research & Information*. **28(1)** 31-41.

[11] Kofoworola OF, Gheewala SH 2009 Life cycle energy assessment of a typical office building in Thailand *Energy and Buildings*. **41(10)** 1076-1083.

[12] Alcorn J and Baird G 1996 *Embodied energy analysis of New Zealand building materials—methods and results. Proceedings of the embodied energy: the current state of play* (Geelong, Australia, Deakin University) 61-71.

[13] Harris DJ 1999 A quantitative approach to the assessment of the environmental impact of building materials *Building and Environment*. **34(6)** 751-758.

[14] Hammond G, Jones C, Lowrie F and Tse P 2008 *Inventory of carbon & energy: ICE* vol. 5 Department of Mechanical Engineering (University of Bath, UK).

[15] Hammond GP and Jones CI 2008 *Embodied energy and carbon in construction materials. Proceedings of the Institution of Civil Engineers - Energy*. **161(2)** 87-98.

[16] Ramesh T, Prakash R, Shukla KK 2010 Life cycle energy analysis of buildings: An overview *Energy and Buildings*. **42(10)** 1592-1600.

[17] Chen T, Burnett J, Chau C 2001 Analysis of embodied energy use in the residential building of Hong Kong *Energy*. **26(4)** 323-340.

[18] Davies PJ, Emmitt S and Firth SK 2014 Challenges for capturing and assessing initial embodied energy: a contractor's perspective *Construction Management and Economics*. **32(3)** 290-308.

[19] Mehta S, Chandur A and Palaniappan S 2017 Life Cycle Energy Assessment of a Multi-storey Residential Building *Journal of The Institution of Engineers (India): Series A*. **98(1-2)** 155-162.

[20] Reddy BV and Jagadish K 2003 Embodied energy of common and alternative building materials and technologies *Energy and buildings*. **35(2)** 129-137.
[21] Crawford RH 2013 Post-occupancy life cycle energy assessment of a residential building in Australia *Architectural Science Review*. **57(2)** 114-124.
[22] Devi LP and Palaniappan S 2014 A case study on life cycle energy use of residential building in Southern India *Energy and Buildings*. **80** 247-259.

IOP Conf. Series: Materials Science and Engineering 1212 (2022) 012005 doi:10.1088/1757-899X/1212/1/012005

Using virtual reality (VR) simulation to help developing a smart meeting space with AI Product to support moderator

X Guo, Z Shen, X Teng, and Z Zhao

School of Natural Science and Technology, Environmental Design Division, Kanazawa University, Kanazawa City 920-1192, Japan

Email : shenzhe@se.kanazawa-u.ac.jp

Abstract. Smart building has a pivotal role in promoting the industrialization of buildings. Meeting room is a common type of public space. Developing a smart meeting space can play an important role in addressing the issue of research on smart building. In this paper, we develop a smart meeting space with AI product to support moderator. We create 3D model of the meeting room and import it into VR headset to having an experiment. 20 participants are engaged to answer questionnaire and have interview. The results of this study show that most of the participants think the meeting space we develop is smart than usual one. We also found that participants expect functions of smart meeting room in AI assistance, energy saving, smart display solution, remote meeting. And participants expect devices of smart meeting room in AI assistance product, sensors, and devices for remote meeting.

1. Introduction

Smart building has a pivotal role in promoting the industrialization of buildings. It can be divided into two parts, smart residential building and smart public building. There are many studies about smart residential building, but very few studies on smart public building. Meeting room is a common type of public space, developing a smart meeting space can play an important role in addressing the issue of research on smart building. Meeting room is a public place to communicate information and generate ideas. People also can reach agreements in meeting room. However, in current meeting process, someone needs to take minutes or arrange meeting room devices. Usually, moderator can not have an accurate overview of the meeting so that they can not making a rational decision.

Several attempts have been made to help achieving smart meeting in an ambient environment, remote meeting, meeting analysis and so on. Freitas, Carlos Filipe, et al developed a context aware middleware for supporting idea generation meetings in smart decision rooms (Freitas et al., 2011). Sfikas et al using IoT approach to implement and manage a smart room (Sfikas, Akasiadis, and Spyrou, 2016). Venkataraman, Vinay, et al empowered remote user to interact with the capabilities of an agent-based smart conference room. It supports remote participants while also enabling them to control most aspects of the smart conference room (Venkataraman, 2016). Serdaroglu and Baydere proposed an edge computing solution for real time meeting room attendance control system. Attendance decision obtained from the tracking system can be used by a meeting room planner to decide if the planned meetings have started, ongoing or completed (Serdaroglu and Baydere, 2019). In order to provide a clearer picture of the papers reviewed and the functions and demand in smart meeting space discussed in each one of them, Figure 1 has been created.

IOP Conf. Series: Materials Science and Engineering 1212 (2022) 012005 doi:10.1088/1757-899X/1212/1/012005

	Attendee Anaylsis	Touch Table	Ambient Intelligent Environment	AI assistance	Meeting Record/Summary	Detection of Meeting Room Occupancy	Energy Saving	Meeting Anaylsis	Device/Environment Management	Access Control	Remote Meeting Room Booking	Auto-hospitality system	Smart sanitation system	meeting room assignment	Smart Display Solution	Hearing impaired friendly	Remote Meeting	Attendee Recognition
[1]	✓																	
[2]					✓													
[3]		✓																
[4]			✓															✓
[5]				✓	✓													
[6]						✓												
[7]						✓	✓											
[8]								✓										
[9]									✓									
[10]	✓																	
[11]							✓		✓	✓	✓	✓	✓					
[12]						✓												
[13]	✓								✓									
[14]			✓														✓	
[15]							✓							✓				
[16]			✓				✓		✓									
[17]	✓				✓													
[18]			✓		✓			✓										
[19]	✓																	
[20]																		✓
[21]					✓			✓										
[22]															✓			
[23]	✓																	
[24]			✓		✓													
[25]	✓																	
[26]	✓				✓													
[27]	✓				✓													✓
[28]			✓						✓									
[29]															✓			
[30]															✓			
[31]			✓															
[32]								✓								✓		
[33]			✓		✓				✓								✓	
[34]					✓													
[35]			✓		✓			✓										
[36]	✓				✓													
[37]	✓				✓													
[38]	✓																	
[39]										✓								
[40]	✓																	
[41]	✓				✓			✓										✓
[42]	✓		✓		✓													
[43]									✓									

Figure 1. Smart meeting space literature approaches: 2010-2019

Most of the previous studies focus on the product aspects, few of them see this topic form the view of architecture. The specific objective of this study is to develop a smart meeting space to support moderator that can keep track of meeting situation and make rational decisions and organize the meeting more efficiently. Smart meeting space in this research is defined as a meeting space that have automatic device and environment management in an ambient intelligent environment to support moderator. We created a 3D model in SketchUp and using Unity to make VR application to evaluate this smart meeting space.

2. Method

In order to develop a smart meeting space to support moderator, the AI product information and building material information should be collected, and the model of the smart meeting space should be created and evaluated. This section describes AI product and building material information collecting, model creating, and conducting a survey to evaluate the smart meeting space.

2.1. Product and building material information collecting

Some text. From the perspective of product information collecting, we mainly considered four aspects, namely, remote meeting system, access control system, meeting management system, meeting environment control system. The function and product picture can be found in Appendix C.

To achieve remote meeting, we chose the following products: Surface Hub 2S, Steelcase Roam, APC Charge Mobile Battery, Logitech Tap, Rally camera, Rally speaker, Rally mic, Rally display hub, and RALLY remote meeting system. To achieve access control, we chose the following products: 8-inch temperature measuring face recognition machine, and KOB 280kg Electronic magnetic lock. To achieve meeting management, we chose Microsoft Black Tower AI assistant (prototype). To achieve meeting environment control, we chose the following products: Electric control dimming glass, Electric roller blinds, Temperature and humidity sensor, Light sensor, Human body infrared sensor, Smart curtain control panel, Smart infrared sensor, Smart thermostat, and RGBW color changing light bulb.

From the perspective of building material information collecting, we mainly considered 3 aspects, namely, metal stud of wall and ceiling, insulation materials, board of wall and ceiling. The Type and material picture can be found in Appendix D.

After obtained the product and building material information, we can be combined those products and meet the function and demand of smart meeting (Figure 2). Based on these functions, we create three scenes to support moderator in smart meeting space. The first scene is moderator can keep track of meeting situation and make rational decisions. The combination of human body infrared sensor and AI camera can help capturing, recognizing, and visualizing human semantic interactions in a smart meeting (Bhattacharya, Eshed, and Radke, 2017). For instance, with the interaction recognition result, it is possible to make index about "the persons who proposed a lot of ideas", "the persons who were critical", "the topics that were common interest", "the topics that were not agreed by all the members", etc (Yu et al., 2017). Therefore, moderator can keep track of meeting situation by watching the analysis result in Logitech Tap and making rational decisions. The second scene is a real-time meeting record and a summary can be displayed on the screen. Hearing-impaired persons can see the conversations of others through the screen to understand what other people are talking about. Also, those who are late can keep up with the meeting based on the meeting overview show on the screen. The third scene is device and equipment manage automatically and it is beneficial to reduce energy consumption. For example, in summer, when the sensor detects the meeting room is vacant, the electronic blind will automatically roll down to save energy form HVAC. When the sensor and access control terminal detects the meeting room is occupied, the electronic blind will automatically roll up and the thermostat will be turned on. Moreover, when speakers display slides on the screen or they are going to have a remote meeting, the electronic blind will roll down to prevent attendee from discomfort glare.

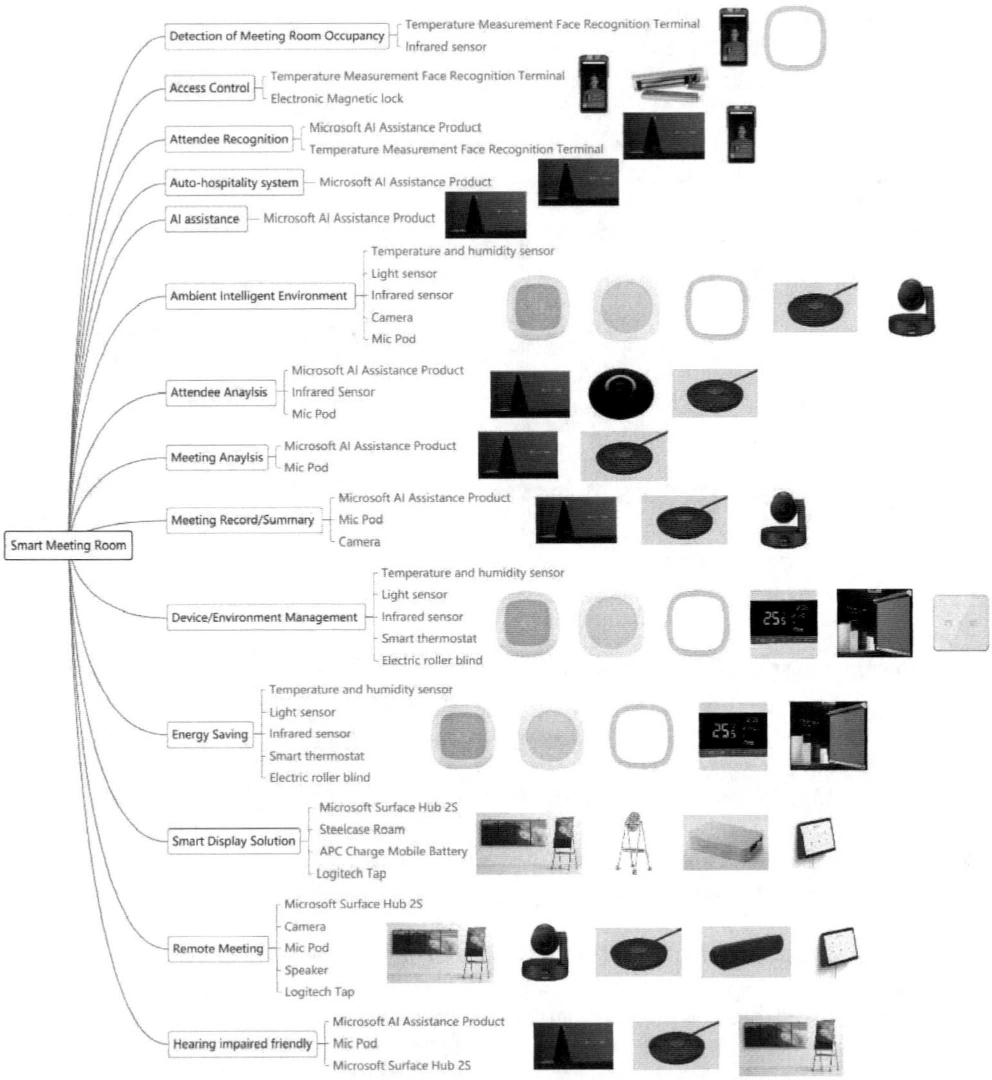

Figure 2. Using AI product to meet smart meeting space demand and functions

2.2. Model creating

Pre-construction evaluation of smart meeting space is necessary, so we started to create 3D model in SketchUp. For the purpose of making this smart meeting space suitable for both traditional meetings and remote meetings, we designed two layouts to address this problem. Figure 3 (a) shows the layout that suitable to have a remote meeting. The U-shaped table can let the camera capture every participant face. Figure 3 (b) shows the layout that suitable to have a traditional meeting. All the devices of two layouts are the same, but the position to put the table is different, which can be changed flexibly to fit the meeting style.

We put the main camera and speakers under four pieces of Surface Hub 2S and put the Logitech Tap in front of the moderator. Then we let two participants use one mic so that the voice of each one can be heard clearly and loud enough. If people need to discuss something privately, they can simply press the button on the mic to mute it. In order to analyse every participant's video and voice, the Microsoft Black Tower AI assistant should be put in the middle of the table. Most sensors are integrated and installed on the ceiling. Temperature measuring face recognition machine is put in the hallway to have access control. Figure 4 (a) shows the structure of a metal stud wall and ceiling and plasterboard with a sensor embedded. Figure 4 (b) shows the detail of the sensor container. We designed this sensor container to integrate different kinds of sensors and put them into plasterboard with a prefabricated hole that filled with glass wool around. This structure can reduce the heat exchange and reduce energy consumption.

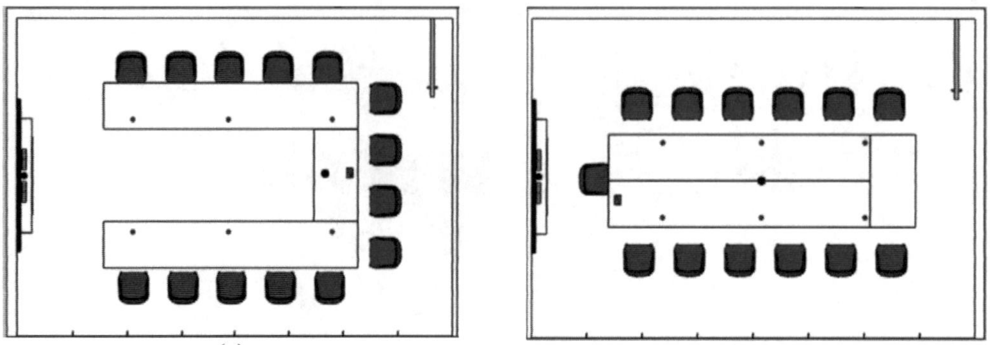

(a) (b)

Figure 3. (a) Remote meeting layout; (b) Traditional meeting layout

Because the pre-construction evaluation will be conduct in VR, so we need to import the SketchUp model into Unity to build a VR application. In the beginning, we set up the Unity environment with Virtual Reality Supported setting and switch to the Android platform to make it suitable for VR application development. Then we import the SketchUp model into the Unity project and drag it to the scene. To develop mobility and simple interaction with the model, we import VR Beginner: The Escape Room as a tool kit. To move in the smart meeting space, teleporters were added and configured to move to a new location. To make some object interactable, adding Rigidbody, Mesh Collider, and XR Grab Interactable component to make the interactable object can be reached and picked up by press and hold the grip button on the controller. After all the function is added, we can build and run the VR application in Oculus Quest, which is a VR headset (Figure 5).

Finally, we design a questionnaire and make a survey to evaluate the smart meeting space. The full-text of the questionnaire is shown in Appendix E. Out of a large group of Kanazawa University students, 20 students ages, 20-35, from different division volunteered to participate in the experiment. We approached each participant individually. We asked them using a VR headset to experience the smart meeting space and fill out the questionnaire. We collected the questionnaire to be examined and analysed them. Then we interviewed 10 out of 20 students about which part of this smart meeting space can be improved and how to improve it.

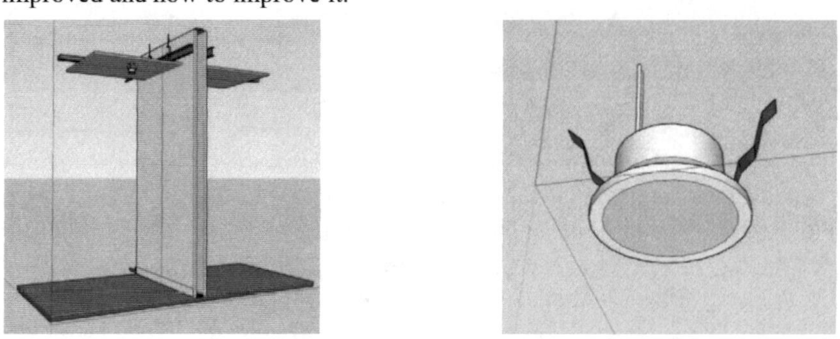

(a) (b)

Figure 4. (a) Structure of metal stud wall and ceiling and plasterboard with sensor embedded; (b) Sensor container

Figure 5. Smart meeting space in VR

3. Result and Discussion

The first set of questions aimed to compare the difference between the current meeting room and this smart meeting room in satisfaction aspect. The analysis of the result is shown in Figure 6. The average rating is 6.11 for the current meeting room and 8.47 for this smart meeting room. This result shows that the smart meeting room we developed did better than the usual ones.

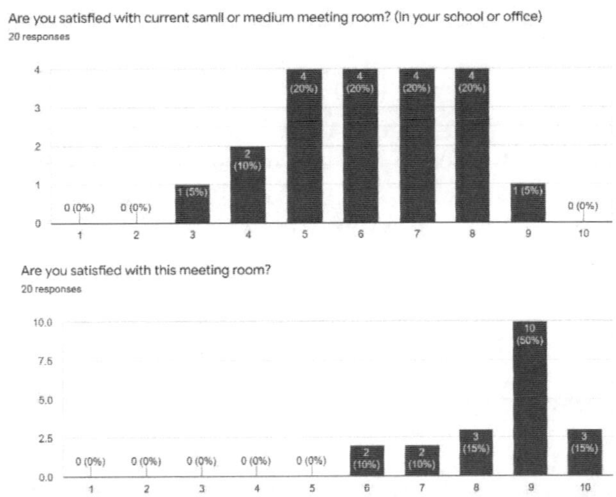

Figure 6. Satisfaction of current meeting room and our smart meeting room

The second set of questions aimed to evaluate this smart meeting space in different aspects. Figure 7 (a) and Figure 7 (b) show that participants are more satisfied with the functions and devices of this smart meeting space, and suggest the layout of this smart meeting space needs to be improved. Figure 7 (c) shows that most of the participants think this meeting space can be called "smart meeting room". As shown in Figure 7 (d), participants are more interest in scene 1, which is "moderator can keep track of meeting situation and make rational decisions".

The third set of questions aimed to investigate the expected functions and demand for smart meeting space. As shown in Figure 8 (a), 10 out of 20 participants think the current meeting room has remote meeting function, with the help of computer, projector, projector screen. 7 out of 20 participants think the current meeting room has access control function by using IC card. As shown in in Figure 8 (b), 12 out of 20 participants think this smart meeting space have remote meeting function, smart display solution, energy saving, and AI assistance. 9 out of 20 participants think this smart meeting space have attendee function, remote meeting room booking function. In in Figure 9, we can see that most participants think AI product and sensor are crucial for a smart meeting space.

After the survey, we modified the 3D model in the layout and environment based on the feedback of the questionnaire and interview. As shown in Figure 10, we made the aisle side bigger and modified the configure of the light on the ceiling to improve the environment.

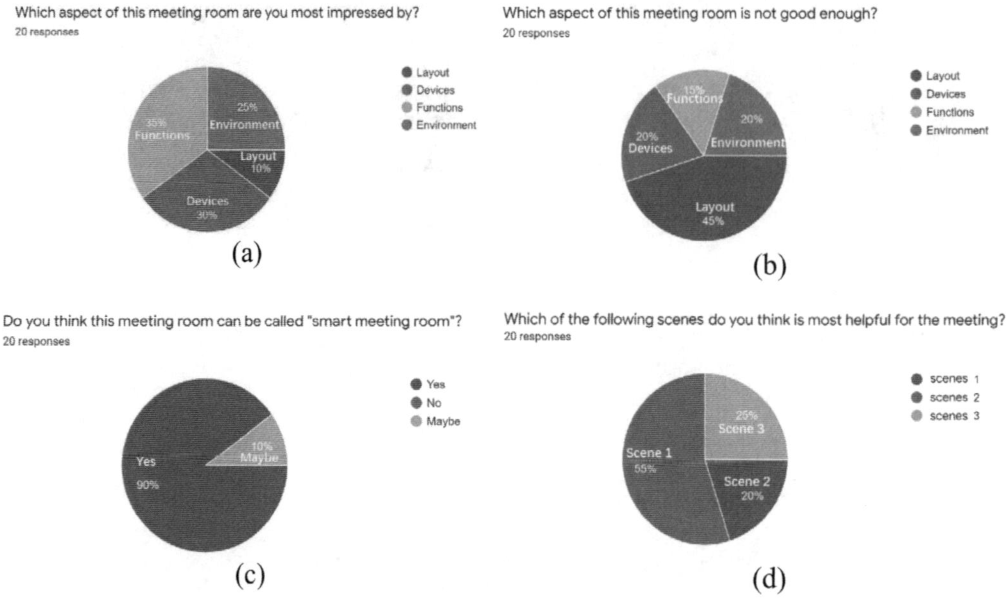

Figure 7. (a) Result of "Which aspect of this meeting room participants most impressed by?"; (b) Result of "Which aspect of this meeting room is not good enough?"; (c) Result of "Do you think this meeting room can be called smart meeting room?"; (d) Result of "Which of the following scenes participants think is most helpful?"

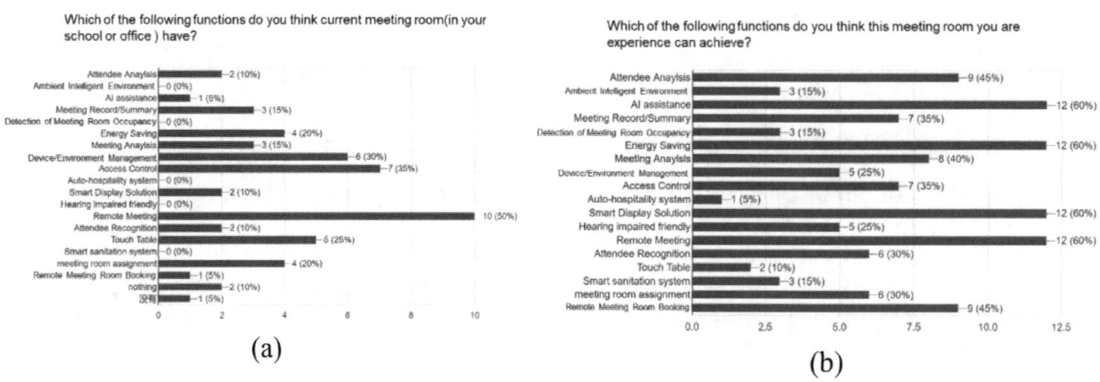

Figure 8. (a) Structure of metal stud wall and ceiling and plasterboard with sensor embedded; (b) Sensor container

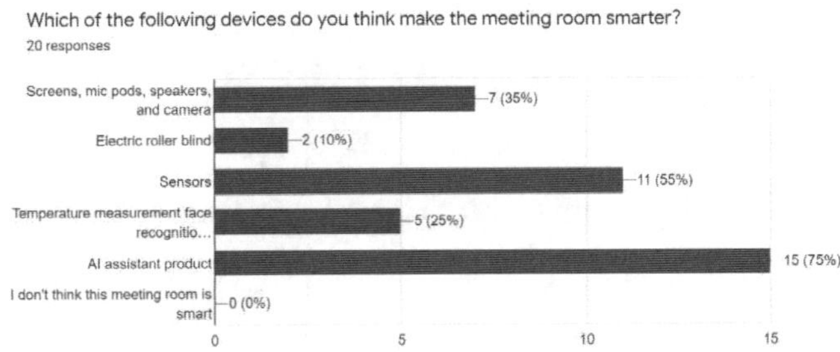

Figure 9. Expected products in smart meeting space

Figure 10. Modify the model of smart meeting space

4. Conclusion

This research using AI product to develop a smart meeting space to support moderator organizing the meeting. We create 3D model and import it into VR headset to having an experiment. After taking a survey about this smart meeting space in VR, we found that participants expect functions of smart meeting room in AI assistance, energy saving, smart display solution, remote meeting. And participants expect devices of smart meeting room in AI assistance product, sensors, and devices for remote meeting.

References

[1] Freitas CF, Meireles A, Figueiredo L, Barroso J, Ramos C. Context aware middleware for supporting idea generation meetings in smart decision rooms. *In 2011 Fourth International Conference on Ubi-Media Computing* 2011 Jul 3 (pp. 53-58). IEEE.

[2] Sfikas G, Akasiadis C, Spyrou E. Creating a Smart Room using an IoT approach. InProceedings of the Workshop on AI and IoT (AI-IoT), *9th Hellenic Conference on Artificial Intelligence*, Thessaloniki, Greece 2016 May (pp. 18-20).

[3] Venkataraman V, Lenchner J, Trewin S, Ashoori M, Guo S, Dholakia M, Turaga P. Ceding control: Empowering remote participants in meetings involving smart conference rooms. *In Workshops at the Thirtieth AAAI Conference on Artificial Intelligence* 2016 Mar 29.

[4] Serdaroglu KC, Baydere S. Edge Computing based Smart Meeting Room Controller. *In 2019 International Symposium on Networks, Computers and Communications (ISNCC)* 2019 Jun 18 (pp. 1-4). IEEE.

[5] Bhattacharya I, Eshed N, Radke RJ. Privacy-preserving understanding of human body orientation for smart meetings. *In Proceedings of the IEEE Conference on Computer Vision and Pattern Recognition Workshops* 2017 (pp. 90-98).

[6] Yu Z, Yu Z, Aoyama H, Ozeki M, Nakamura Y. Capture, recognition, and visualization of human semantic interactions in meetings. *In 2010 IEEE International Conference on Pervasive Computing and Communications (PerCom)* 2010 Mar 29 (pp. 107-115). IEEE.

Bioclimatic and sustainable features on vernacular architecture in Ternate

M Rahim

Department of Architecture, Khairun University, Ternate-Indonesia.

E-mail: mustamin_rahim@yahoo.co.id

Abstract. This study aims to identify the response of the vernacular building to climate and sustainability by literature study and field observations. The results show that the vernacular building in Ternate Island develops based on the characteristics of the region. The typology of building differs in each tribe following the natural condition and local culture, it has a line of equality in the building philosophy as an embodiment of the human body: feet (foundation), body (wall and room), and head (roof). Vernacular architecture in Ternate contains bioclimatic and sustainable features and can be developed to the new building concept; the basic concept is the combination of spiritual/cultural harmony, human harmony, environmental harmony, and harmony with green technology/innovation for comfort and protection of the disasters by bioclimatic architecture design strategy. This study provides insight and inspiration for sustainability in the future especially in the context of archipelago areas.

Keywords: Vernacular Architecture, Bioclimatic, Sustainable, Archipelago

1. Introduction

The Maluku Islands are located in the eastern part of Indonesia and commonly also called as Spice Islands. The Maluku Islands comprise some of the most geologically complex and active regions in the world [1]. The Moluccas Islands formed a single province. It was split into two provinces on October 12, 1999: Maluku and North Maluku (North Moluccas). Ternate is a small island in the North Maluku (Moluccas Island) and dominated by the area of volcanic mount Gamalama with a height reaches 1715 m. The population is fewer than 200,000 with an area of 111.39 km^2 [2]. Ternate City is an archipelago region characterized by a tropical climate. Indonesian vernacular architecture grows and expands according to the culture of traditional society. Its characteristic is influenced by the cultural manner where the architecture masterpiece resides [3]. Characteristics of traditional houses differ in each area; they are strongly influenced by the natural environment and local culture. Vernacular architecture is rich in form, traditional technology and meaning especially in the North Moluccas area including in Ternate Island, cultural influences are very strong on traditional architecture in Ternate and the surrounding area because there are different tribes yielding architecture masterpieces in the various forms of the traditional house [4]. Several studies before suggest that vernacular architecture is a model of sustainable architecture [5-7]. Vernacular architecture is the content of sustainable architecture [8-9]. The basis of sustainable construction is adopted from vernacular features[10]. The vernacular architecture successfully overcomes the challenges of adaptation to extreme local conditions and serves as a model for a sustainable design approach [11]. The bioclimatic strategy is a sustainable system of vernacular architecture. The traditional architecture uses a bioclimatic building strategy and can be considered as the basic concept for developing solar architecture in the future [12]. Traditional houses in the tropics refer to the concept of bioclimatic design with passive solar strategies for natural lighting and ventilation, and also usage of natural materials for protection to the hot and humid weather conditions [13]. Moreover, features of vernacular architecture have been adopted for new building concepts according to the current condition. The usage of traditional methods has been manifested in contemporary architecture in several ways such as using elements, characteristics, styles, and expressions [14]. Therefore it is very important to explore the bioclimatic and sustainable features in the past for the strategy of sustainability in the future. On the other hand, According to Gangwar [15] that it is not important to imitate the whole traditional architecture in one project but only part of the elements to inspire conformity. Several study results suggest that the materials of vernacular architecture have low environmental effects[16], even some politicians and people view vernacular architecture as a representation of an underdeveloped past and contrary to modern ideas and expectations [17]. Traditional techniques are considered inadequate and substandard, while modern materials are seen as civilized technology and a symbol of prosperity so that detailed research from various context is needed to explore useful features for modern buildings and further reason to elevate the image of vernacular architecture from the label "low quality"[18]. Sundarraja [19] argues that the study of vernacular architecture provides useful insights into contemporary building design by taking design inspiration from vernacular buildings following local conditions. Different climatic conditions require different responses to meet different needs so that climatic variations cause variations in architectural response in vernacular buildings [20-21]. Therefore, this study will identify the characteristics of vernacular architecture and its response to climate and sustainability in the context of the archipelago to explore useful sustainability features in the future.

2. Research Methods

This study is a qualitative research that emphasizes the identification of bioclimatic and sustainable features especially natural material, natural ventilation, orientation, adaptation to climate, and local wisdom ornament on the traditional building in the Ternate. The study area is located in eastern Indonesia in Ternate Island, North Moluccas (see figure 1). The object of study is focused on several traditional buildings, namely Falakanci house, Siko house, Sigilamo mosque, Keraton Ternate (Palace of Ternate Sultanate). The research was performed by the literature study, field observation, and depth-interview. (1) Literature study: for collecting data about the development and history of the traditional building. (2) Field observation: a direct survey in the traditional building about the material, construction, and bioclimatic features. (3) Depth-Interview: to clarify data and complement data obtained from study literature and field observations, especially about the building construction process and cultural activities related to the traditional building. The data were analyzed using architectural analysis by comparison of the building condition in past and current situations to see their change and development, and analyze the consistency of bioclimatic and sustainability features in past and present conditions.

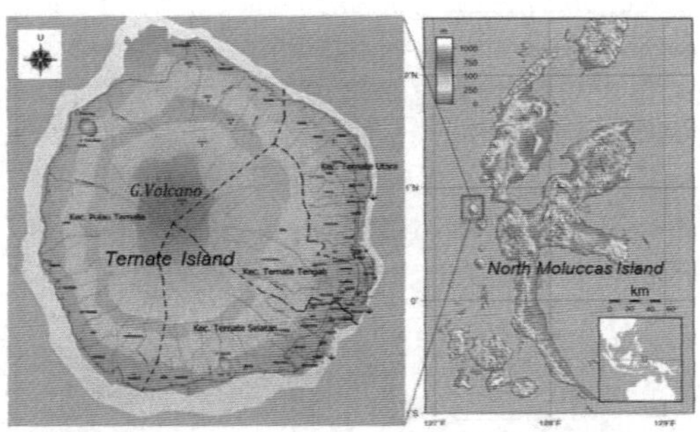

Figure 1. Study area in Ternate Island.

3. Result and Discussion

3.1. *Characteristics of Traditional Building*

Vernacular architecture refers to structures that are built using locally available materials with functional styles designed to meet the needs of lay people in their time and place. Most vernacular architecture responds to regional climate [22]. This type of structure is found to be modified and evolved over time through feedback mechanisms that already exist in the system, to reflect the environmental, cultural, and historical context in which they are there [23]. Vernacular architecture developed over the centuries has many original and interesting design practices and technologies [24]. Vernacular architecture is a climate-responsive architecture, which is an important factor in sustainability [25]. Figure 2 shows the traditional house of *falakanci* uses a button system without the usage of nails or materials made of iron. It's only using straight hooked lip joints and pens which are also made from wood or bamboo and *gamutu* ropes (palm fiber) all of which function as reinforcing joints (Figure 3a). The room size has no specific standard, it may differ depending on the wishes of the occupants, but in general, a rectangular floor plan with the main building separate from the kitchen and there are four pillars on the porch, using natural materials namely bamboo, wood, roof of thatched leaves tied by *gamutu* rope (Fig. 3b). The roof model is a pyramid-shaped roof with a high *tiang raja* (king pillar) (Figure 3c). The house consists of a terrace, living room, bedroom, and family/dining room, while the kitchen is separate from the main building.

Figure 2. Traditional house of *falakanci* (Source: author).

Figure 3. Building construction: (a) Button System (b) Bond system, (c) *Tiang raja* (king pillar), (d) *gaba-gaba* (sago stem), (Source: author).

Figure 4 shows a traditional house in Siko Village (Siko House). Siko Village is located at North Ternate Subdistrict, which is inhabited by many tribes from Sulawesi. The dimension and orientation of buildings had important and sacred meaning in the past, especially during the reign of the Sultanate. The meaning is related to norms and ancestral values. For this reason, a long time ago, home locations had the following requirements: house orientation must be facing mt. gamalama because the mountain is considered to have a special and sacred value. the size of the house plan must consist of an even and odd size, as a symbol of balance in life. Building material uses natural material, namely stone foundation, wall material from *gaba-gaba* (sago stem) (Fig. 3d), wall and roof frames from the bamboo, roof material from sago palm leaves.

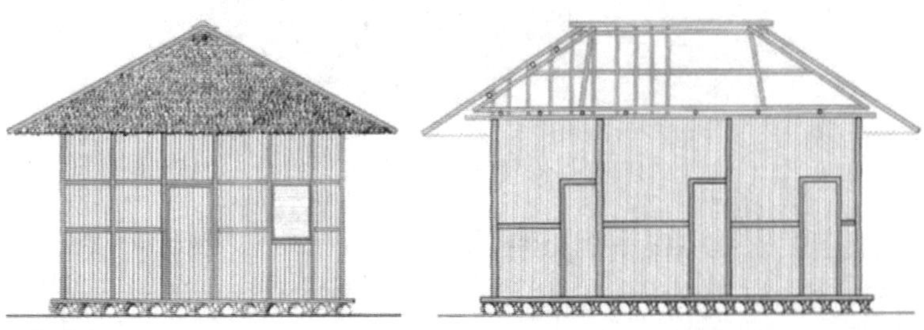

Figure 4. Traditional house of Siko (Source: author)

Figure 5 shows the Keraton Ternate building (Palace of Ternate Sultanate). This palace was built in 1257 AD using natural materials from the surrounding area. The roof material uses sago palm leaves and is then replaced with zinc. This building applies natural lighting and ventilation with an open-close window model. Figure 6 shows the Sigilamo mosque. It was built around 1606 and then rebuilt in 1818 due to fire, before the fire this mosque has also been reconstructed several times. Initially, building material used stone for wall and sago palm leaves for the roof before and then replaced with zinc in 1983. Since its inception, this building has been applying passive solar concepts by maximizing natural lighting and ventilation systems.

Figure 5. Keraton Ternate (Source: author)

Figure 6. Sigilamo mosque (Source: author).

Sigilamo mosque uses a multi-story roof model without a ceiling for maximizing cross ventilation and roof ventilation systems (see figure 7c). According to Mustamin [26] that by application of cross ventilation systems a nd roof ventilation at Sigilamo, the indoor thermal conditions can reach a comfortable level. It confirms that the vernacular buildings in Ternate have been applying the principle of sustainability. Also, these buildings survive from extreme climatic conditions (mountain and ocean climate) and earthquake disasters where the Ternate region is often shaken by earthquakes. This provides a valuable lesson that those local technique has good strength and is useful in the future. This result is different from the one conveyed by Oliver [17] that vernacular architecture is considered a representation of an underdeveloped past and contradicts modern ideas and expectations.

a b c

Figure 7. Ventilation types (a) Falakanci, (b) Keraton (c) Sigilamo (Source: author).

3.2. Response to Climate and Sustainable

The concept of sustainable development was described by the 1987 Bruntland Commission Report as "development that meets the needs of the present without compromising the ability of future generations to meet their own needs [27]. Economic, environmental, social, and cultural sustainability are the pillars of sustainable development [28-29]. The bioclimatic design of buildings is one strategy for sustainable development, as it contributes to reduce energy consumption, further, reduce air pollution and greenhouse gas emissions from conventional energy generation [30]. Bioclimatic architecture refers to the design of buildings and spaces (interior - exterior - outdoor) according to local climatic conditions for visual and thermal comfort by utilizing natural environmental sources such as sun, air, wind, vegetation, water, land, and sky. The passive solar design strategy is to integrate building elements with environmental resources naturally without mechanical devices and energy supply, which consist of three categories: passive solar heating systems, passive cooling systems, natural lighting systems [31]. Among bioclimatic features in Ternate are traditional buildings always adapted to environmental conditions such as the building location is in the potential areas of living resources, using natural materials, daylighting, natural ventilation, local construction systems, adapted to local wisdom, protection to the natural disasters with the simple construction methods. Building ornaments are usually from local characters such as flowers, birds, animals, and others.

Figure 8 illustrates the basic concepts of the traditional architecture of Ternate, which is the analysis result of the study. In the construction process is always based on the ritual ceremony to the almighty (spiritual) as a symbol of permission to build houses and the blessing to the occupants, and also usually religious values are transformed into the philosophy of building elements, such as the philosophy of king's pillar illustrated like a good relationship between occupants with the almighty so that the material of king's pillar must be from a strong and goodness wood material. Humans are respectable creatures; they must live in harmony, respect each other, and mutual help in society. The settlement is a place to build a life together and society harmony based on the culture and local conditions, and nature as a source of life that must be preserved for long-term survival.

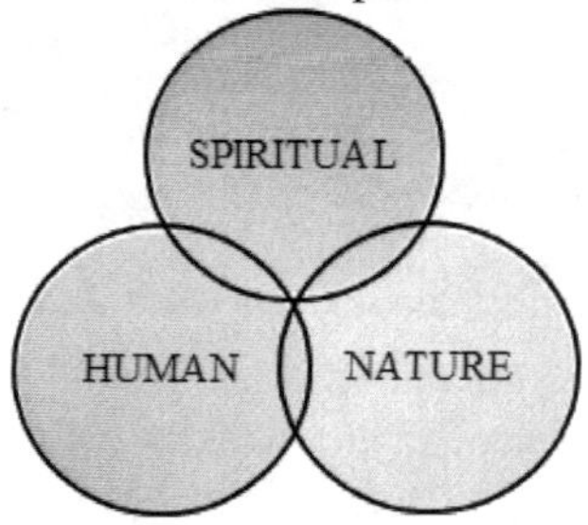

Figure 8. The basic concept of vernacular architecture in Ternate Island.
(Source: Author elaboration)

Table 1 shows traditional architectural values in Ternate in the past. It indicates that traditional houses have many architectural values related to spiritual, human, and environmental aspects. The spiritual aspect or vertical relationship to God is a form of worship and a request for blessings. Spiritual values are very strong in the construction process; such as the ritual when starting to build a house is a form of permission and blessings, ritual constructs the king pillar or roof frame, and the ritual of entering the house. Also, spiritual value is manifested in building elements, such as the philosophy of the king pillar is a symbol of the vertical relationship to the almighty. The human aspect or human harmonious relationship that is manifested in a culture of mutual respect and care for each other. The human aspect is also realized to the philosophy of building elements as a symbol of respect for occupants such as building dimensions use human anatomy, the shaped roof in the terrace is lower than the main roof contains meaning that all guests should bow as a symbol of respect to the occupants. Natural aspects or a harmonious relationship with the environment which is manifested in the form of houses must respond to climatic and environmental conditions, utilize natural resources without damaging the environment. House location criteria are based on natural resource potential and safe from disasters because Ternate and surrounding is an area prone to natural disaster and volcano (gamalama volcano). For the building construction, the society tries to meet the safety and comfort aspects even though in a simple construction according to the conditions at the time, where the house was built using natural materials, which are available in the surrounding environment. The application of button and bond construction techniques is one of their efforts for protection from the earthquake hazards, they believe that those techniques are more resistant to earthquake shocks and are easy to reconstruct and move. The existence of people's lives is very dependent on natural resources so that they take good care of nature. They hold strongly the philosophy of life that when humans care for nature, nature gives life to humans, when humans destroy nature, nature will bring disaster to humans. It indicates that the Ternate community is very responsive to sustainability in the past. These are very valuable sustainability values that need to be transferred to current and future generations.

Table 1. Features of a traditional house in the Past.

	Values in Past
Spiritual	Religion ceremony for permission and blessing Philosophy building element from religious values Tradition and ritual
Human	Culture and local wisdom Building philosophy is the human body: Feet (foundation), body (wall-room), and head (roof) Occupant safety (wild animal and disaster) Dimension: human anatomy for harmony occupant with house Entrance roof lower: guests stoop to respect the occupants Built and opening house ceremonies, break bread ceremony Custom and community ceremonies
Nature	Building location in resource potential area (mountains, valleys, and coastal areas) Site adapted to the topography condition Natural material without iron and nail Response to the climate and natural condition Natural ventilation and lighting Rainwater for life and farm

Sources: Author Elaboration

Table 2 shows the architectural features that can be adopted from traditional houses in the past to be developed into a new concept for sustainability. It portrays that traditional architecture contains features of sustainable architecture. Traditional architectural characteristics must be used as guidelines in the design of settlement and building contemporary, which combined with a green building design approach to meet the needs and expectations of today's society so that it contributes to the creation of a more sustainable built environment. Those features show that building technology is needed to realize an earthquake-resistant building in the modern day; where Ternate Island is an earthquake-prone region. Therefore, the recommendation of the basic concept for new buildings or contemporary buildings is the combination of spiritual/cultural harmony, human harmony, environmental harmony, and harmony with green technology/innovation for comfort and protection of the disasters by bioclimatic architecture design strategy.

Respect to the social, economic, environmental, and cultural conditions as pillars of sustainable development can be easily understood in traditional buildings in the Ternate Island and surrounding area. It indicates that the archipelago vernacular architecture in the North Moluccas Island offers important lessons about the relationship of harmony with nature from various perspectives of local wisdom and become real examples of sustainable architecture with the basic concept "spiritual-human-nature". The footprint of sustainability in vernacular architecture was noticed by the local population consciously. The people built their dwellings have focused on the principles of spiritual and sustainable design to improve their quality of life and safety with a simple construction system according to the limited knowledge and economic conditions at that time. Even though the conditions were limited in the past, these conditions highlight the parameters of sustainability which are very valuable compared to today, where the buildings emphasize more prestige or luxury and ignore the parameters of sustainability. Therefore, it is very important to transfer knowledge and sensitivity of sustainability from the past to the present and the future to create a life that is harmonious with nature, environment, and humans.

Table 2. Adaptation of architectural values for further.

	Values for Future
Spiritual	Religious activity Improve worships
Human	Local wisdom and culture sustainability. Mutual help, cooperation and social harmony. Integrated construction: foundation, wall, roof in a solid structure unity. Safety & comfortable by local character building. Disaster mitigation: community-based.
Nature	Building responsive: climate & natural disaster (earthquake, volcano, and others). Environmental friendly / sustainable Green material, low energy, and low carbon Natural resources and sustainable economy

Sources: Author Elaboration

Conclusion

The characteristics of Ternate's vernacular architecture are a response to local climatic conditions (mountain and ocean climates effect) using natural materials and a passive solar system with a "button house" construction system for earthquake hazard protection. Building typology differs in each region following the natural condition and local culture of each tribe, but it has a line of equality in the building philosophy as an embodiment of the human body: feet (foundation), body (wall and room), and head (roof). This philosophy symbolizes designating high appreciation to human beings and as a symbol of strength and stability for the comfort and protection of occupants. The material of traditional houses uses local material from the surrounding area, namely; sago palm leaves for roof material, wall material from wood, bamboo and sago palm stem, and stack stone for the foundation. Vernacular architecture in Ternate contains bioclimatic and sustainable features and can be developed to the new building concept; the basic concept is the combination of spiritual/cultural harmony, human harmony, environmental harmony, and harmony with green technology/innovation for comfort and protection of the disasters by bioclimatic architecture design strategy.

References:

[1] Monk, K.A.; Fretes, Y.; Reksodiharjo-Lilley, G. 1996. *The Ecology of Nusa Tenggara and Maluku*. Hong Kong: Periplus Editions Press.

[2] Indonesia: Administrative division (provinces, regencies and cities) - population statistics, charts and map. *www.citypopulation.de*. (Retrieved 2 October 2020).

[3] Soeroto, M. 2003. *From Traditional Architecture to the National Architecture of Indonesia*. Ghalia Indonesia Press, Jakarta.

[4] Rahim M. 2009. *Traditional Architecture of Moloku Kie Raha*. Bina Ilmu Press, Surabaya.

[5] Salgın, B., Bayram, Ö. F., Akgün, A., Agyekum, K. 2017. Sustainable features of vernacular architecture: Housing of Eastern Black Sea Region as a case study. *J. Arts* 6 (3) 1-11.

[6] Gamón, M. R. 2020. Parameters of sociocultural sustainability in vernacular architecture. *The International Archives of Photogrammetry, Remote Sensing and Spatial Information Sciences*, 44, 227-231.

[7] Convertino, F., Di Turi, S., Stefanizzi, P. 2017. The color in the vernacular bioclimatic architecture in Mediterranean region. *Energy Procedia*, 126, 211-218.

[8] Salman Al-Zubaidi, M.S. 2017. The sustainability potential of traditional architecture in the Arab world with reference to domestic buildings in the UAE. *Doctoral Thesis, University of Huddersfield Press*.

[9] Liu, Q., Liao, Z., Wu, Y., Degefu, D.M., Zhang, Y. 2019. Cultural sustainability and vitality of Chinese vernacular architecture: A pedigree for the spatial art of traditional villages in Jiangnan region, *J. Sustainability*, 11 (6898), 1-27.

[10] Fernandes, J., Dabaieh, M., Mateus, R., Bragança, L. 2014. The influence of the Mediterranean climate on vernacular architecture: a comparative analysis between the vernacular responsive

architecture of southern Portugal and north of Egypt. *In Proceedings: World Sustainable Buildings SB14.*

[11] Philokyprou, M., Michael, A., Malaktou, E., Savvides, A. 2017. Environmentally esponsive design in Eastern Mediterranean. The case of vernacular architecture in the coastal, lowland and mountainous regions of Cyprus. *Building and Environment*, 111, 91-109.

[12] Petrasincu, N., and Fara, L. 2006. Bioclimatic elements for traditional Romanian houses. *In PLEA Proc*, 85-88.

[13] Victoria, J., Mahayuddin, S. A., Zaharuddin, W. A. Z. W., Harun, S. N., Ismail, B. 2017. Bioclimatic design approach in Dayak traditional longhouse. *Procedia Engineering*, 180, 562-570.

[14] Dhepe, S.S., Valsson, S. 2017. Traditional Approach towards contemporary design: A case study, *J. Eng. Res. and App*, 7 (3), 12-22.

[15] Gangwar, G. 2016. Sustainability lesson learnt from the Traditional and Vernacular Architecture. Journal of Civil Engineering and Environmental Technology, 3 (2), 106-111.

[16] Fernandes, J. E. P., Mateus, R., Bragança, L. 2014. The potential of vernacular materials to the sustainable building design. In *Vernacular Heritage and Earthen Architecture: Contributions for Sustainable Development – Correia*, Carlos & Rocha Eds, 625-629.

[17] Oliver, P. 2006. *Built to Meet Needs: Cultural Issues in Vernacular Architecture*. UK: Architectural Press,pp. 21-27.

[18] Wahid, A. 2012. Adaptive vernacular options for sustainable architecture. *J Int Soc Study Vernac Settl*, 2, 74-85.

[19] Sundarraja, M., Radhakrishnan, S. Priya, R. 2009. Understanding Vernacular Architecture as a tool for Sustainability,*10th National Conference on Technological Trends*, Trivanduram.

[20] Soleymanpour, R., Parsaee, N., Banaei, M. 2015. Climate Comfort Comparison of Vernacular and Contemporary Houses of Iran. In *Proceedings of the Asian Conference on Environment-Behaviour Studies, AcE-Bs2015*, 49–61.

[21] Sahebzadeh, S., Heidari, A., Kamelnia, H., Baghbani, A. 2017. Sustainability features of Iran's vernacular architecture: a comparative study between the architecture of hot–arid and hot–arid–windy regions. *Sustainability*, 9(5), 749.

[22] Albatici, R., 2009. Elements and strategies for sustainable intervention in the residential building sector: a case study. *Indoor and Built Environment*, 18 (5), 447–453.

[23] Lau, C.C.S., Lam, J.C., Yang, L. 2007. Climate classification and passive solar design implications in China. *Energy Conversion and Management*, 48 (7), 2006–2015.

[24] Singh, M.K., Mahapatra, S., Atreya, S.K. 2009. Bioclimatism and vernacular architecture of North-East India. *Building and Environment*, 44 (5), 878–888.

[25] Motealleh, P., Zolfaghari, M., Parsaee, M. 2018. Investigating climate responsive solutions in vernacular architecture of Bushehr city. *HBRC Journal*, 14 (2), 215-223.

[26] Rahim, M., Marasabessy, F. 2019. Evaluation of natural ventilation characteristics on the Sultanate of Ternate Mosque. In *IOP Conference Series: Materials Science and Engineering*, 506 (1), p.012035. IOP Publishing.

[27] UN 1987. *Report of the World Commission on Environment and Development*: Our Common Future.

[28] Attia, A.S. 2020. Traditional multi-story house (tower house) in Sana'a City, Yemen: An example of sustainable architecture, *Alexandria Eng. J.*, 59 (1), 381-387.

[29] Mahmoud Bayoumi, O.A. Nubian vernacular architecture and contemporary Aswan buildings' enhancement, *Alexandria Eng. J.* 2018, 57 (2), 875-883.

[30] Hansen, V.g., Esteves, A., Pattini, A. 2002. Passive solar systems for heating, daylighting and ventilation for rooms without an equator facing facade. *Renewable Energy*, 26, 91–111.

[31] *http://www.cres.gr/kape/energeia_politis/energeia_politis_bioclimatic_eng.htm.* (Retrieved 2 October 2020).

Contemplating the traditional rural architecture of Nalanda: A case study

R Kumari

Department of Architecture, National Institute of Technology Hamirpur, India

email id: rashmi.archi.nitp@gmail.com

Abstract. The fundamental purpose of a building is to provide a comfortable and safe living environment, which protects from the extremes of surrounding environment, as well as give an opportunity of physical and psychosomatic development. The traditional buildings evolved gradually to meet the eventually changing human needs and lifestyle over a period of time, integrates the environmental and socioeconomic characters of the society in a well manner. These buildings have attained deep reconciliation with site surroundings, and thus, have minimal environmental impact. In spite of rapidly growing concept of advanced technological approach towards sustainability, the concept of indigenous technology and traditional architecture plays an active role. The main objective of this paper is to understand the principles and strategies for sustainability from traditional rural architecture and indigenous technologies. The methodology adopted is documentation of a traditional rural house of Nalanda, India, and analyse the environmental and socioeconomic aspects. The data has been collected from surveys, site measurements, literature, and other secondary sources. The thermal performance of a traditional house is analysed through computer aided simulations in the present investigation. This paper concludes with an appreciation of principles of traditional rural architecture and advocates their integration in the present scenario.

1. Introduction

India is a vast sub-continent comprising more of villages than cities. The culture and traditions of Indian villages are primitive and fundamental, and have been far secluded from the modern civilisation and culture [1]. The hut, unit of rural architecture, has almost preserved its ingenuity since inception, thousands of years ago, and the same is believably true even for the whole pattern of rural architecture in India. This sustained tradition of rural architecture of India, may claim to be the living and organic architecture.

The India rural architecture is characteristically and archetypally primitive and has maintained its distinct rudimentary character, with common use of mud, bamboo, locally burnt bricks and tiles, and thatch in house construction, and with the sense of openness, unity, and space. The physical and psychological needs of the users are at the core of evolution of rural architectural expression, which is intimately related to the nature of amity and tranquillity of the village [2]. The villages have relished isolation from various external philosophies due to remoteness, and have been thus able to preserve a tradition flourished from the roots since the time immemorial [3]. The findings of the present research can further be utilised to discover strategies for improving energy efficiency and thermal comfort in consonance with the preservation of the traditional features, having their own thermal benefits, in the rural housing.

2. Sustainable development and rural architecture

Apart from the country's political, social and economic conditions, the pursuit of knowledge and art is the basis of shaping the country. The creative architecture must, therefore satisfy both the tangible and the intangible requirements of life [4]. The integrated design system, thus, must incorporate the eight subsystems of the building design features, which are, functional workability; aesthetical value; economic feasibility; comfortable liveability; environmental sustainability; social acceptability; structural safety; and technical viability, to achieve overall sustainable development (Figure 1).

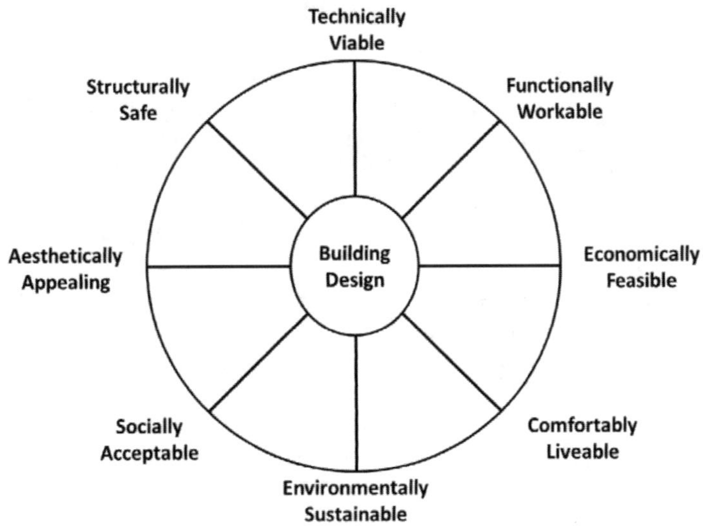

Figure 1. Dynamics of Building Design.

The rural architecture of India has been in consonance with its surrounding environment and evolved as a response to the social requirements as well as the unique interaction between the human cognizance and the natural phenomena [5-8]. The effective building forms have developed overtime that assures the attainment of high level of comfort both inside and outside space of built environment without employing alternate methods [2, 9, 10]. The studies suggest that the aspects such as socio-economic structure, physiographical condition, environmental condition, infrastructure availability, as well as institutional structure contributed to the development of traditional building construction methods [11]. The traditional vernacular buildings are providing edifying illustrations of methods of achieving sustainable building construction. Further, the dependence on modern design methods and materials has resulted in increased cost of construction in recent years, whereas, the traditional building construction methods enable cost effectiveness in terms of locally available materials, indigenous techniques, and simple innovative expertise of the local builders [12]. In addition, the built structure needs maintenance and renovation, which would be easier without consulting an expert that is necessary with the use of unacquainted materials and techniques. The traditional rural architecture can be an inspiration for advancements in environmentally and socio-economically sustainable planning and design [13].

3. Methodology

The methodology adopted in the present paper is to examine the energy performance of an existing traditional rural building, which adopts the traditional construction method, prevailing in the rural Nalanda. The selected case study was intended to determine the extent to which the traditional architecture incorporates the parameters of the building design. The major emphasis has been on examining the environmental sustainability of the tradition building by evaluation the energy performance of the building to achieve thermal comfort. The data has been collected from primary and secondary sources. The climatic data have been collected from the website of ASHRAE (ASHRAE

90.1, 2004; USAID ECO-III Project, 2009) [14]. The data about the study area has been collected from the government bodies. The field measurement, photography, photogrammetry, and surveys have been conducted to collect onsite information. The thermal performance of a traditional house has been analysed through modelling and simulations in the present investigation. The data have been analysed rigorously to find the inferences and conclusion.

4. About the study area

Nalanda, a district in Bihar state, India, is famous all over the world for the ancient International Monastic University established in 5th century BC, which was a learning centre for the subjects of Vedas, Logic, Grammar, Medicine, Meta-Physics, Rhetoric, etc. Nalanda district is located within the Mid-Ganga basin, in the southern margin of the Gangetic plains of Patna. It lies between latitude 24°57' and 25°27' North and longitude 85°18' and 85°56' East and represents mainly flat alluvium terrain. The climate of this region is composite, hot during the summers, and cold during the winters.

A little less than one-third of the houses are in good condition, whereas one-tenth of the houses are dilapidated condition (Figure. 2). One third of the houses have roof made of grass/ thatch, mud, bamboo, wood, mud, or handmade tiles (Figure. 3). Most of the (three-fourth) of the houses of Nalanda have mud floors (Figure. 4). A little less than two-third of the houses have walls made of burnt bricks and less than one-third of the houses have walls made of mud or unburnt bricks (Figure. 5).

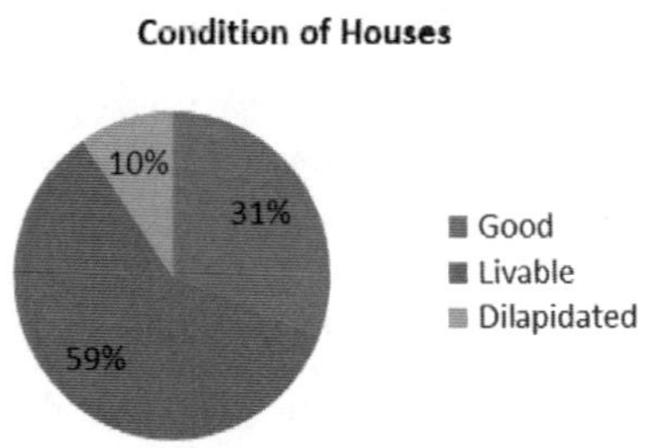

Figure 2. The percentage of houses in different conditions in Nalanda.

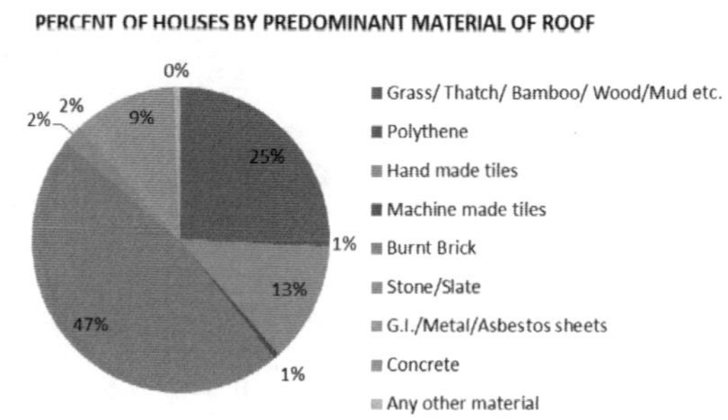

Figure 3. The percentage of houses by predominant material of roof in Nalanda.

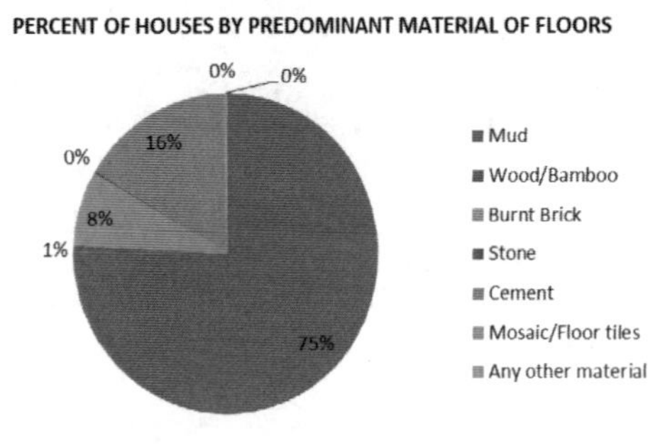

Figure 4. The percentage of houses by predominant material of floors in Nalanda.

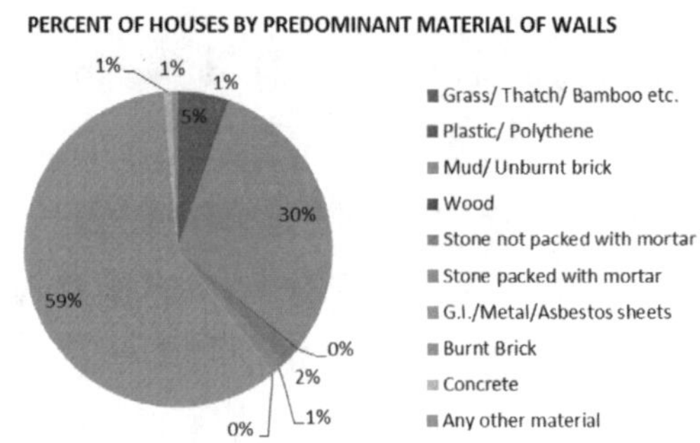

Figure 5. The percentage of houses by predominant material of walls in Nalanda.

5. Case study

A small residential single storied typical traditional house, built before 1950 AD, has been selected for the present investigation. This house, an example of vernacular architecture, is situated in the village Mohammadpur, a small village situated on the bank of river Goithwa in Nalanda district, India, having 40 Hectares of land area with total population of 369 people. The west side of the house has a lot of vegetation including the paddy fields on the agricultural land of the village. About 80.5 per cent of total land area is agricultural land. There are 71 households in the village. Local materials available in the region, mostly come from the agricultural by-products, are wood, hay, husk, straw, reed, clay, mud, and cow dung. This is a typical single storied rural traditional house built of load bearing burnt brick walls and tiled roof, on the base of earth. The wooden structure of beams and lintels bears the thatched roof, which is covered by burnt clay tiles. The wall and roof are the major cost consuming component of this house. About two-third of the cost of construction of the house has been invested for the construction of walls, and about one-fourth of the cost is invested in the construction of roofs.

5.1 Plan form

The building has rectangular plan form (Figure 2). The building is facing north. The four rooms are in a row, opens in a common veranda, which is open from the north side. The heat gain or loss depends on the surface and volume ratio, smaller the surface and volume ratio, lesser the heat gain or heat loss will be. For the present case, the surface and volume ratio is higher as compared to the modern buildings.

5.2 Layout

This is oriented on east-west direction, and the long elevations are facing north and south, which reduces exposure to the sun (Figure 6). The façade, interface between indoors and outdoors, is facing the north direction. The high boundary walls and a room planned on the opposite side of the house create a courtyard type place. This open space works as a thermal regulator.

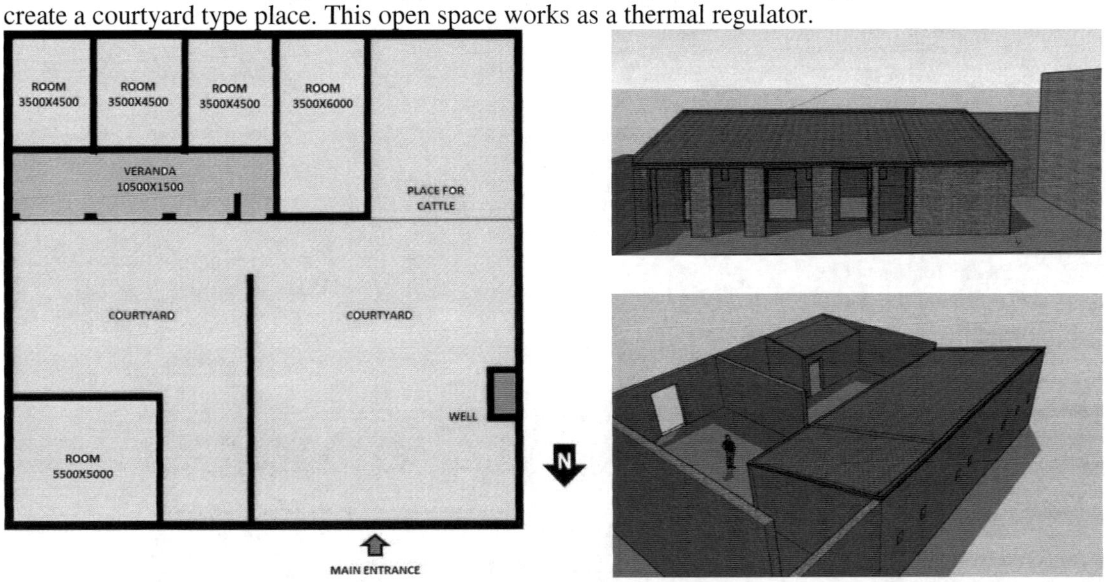

Figure 6. Plan layout and views of the building including the open spaces within the compound.

5.3 Walls

The solid load bearing 250 mm thick burnt brick walls are constructed using mud mixed with husk and straw as mortar, with columns made of brickwork as well (Figure 6). The walls are heavy with high thermal capacity (U-value is 3.25 W/m^{2}K). The walls have rough texture of exposed brick. The rough texture causes self-shading of walls. This is one of the reasons for low temperature in the traditional building. Brick making is an activity that had been done by the household, which makes the house vernacular and cost effective as well. In the countryside it is often part of the local economy; the locally available soil is mixed with water and sawdust, moulded, and left to dry for several days. Then the green bricks (fired bricks are red) are stacked with thin layers of sawdust and coal between them, resulting in a stack that is a kiln at the same time.

5.4 Floors

The earth floors are sealed by arranging burnt brick, and then plastered with layers of mud and cow dung. This plasterwork is essential in keeping snakes, insects, and rodents out. Cow dung is sticky and has a binding effect, but is easily damaged by termites. Mud is termite resistant, but brittle. By combining these characteristics, a sturdy termite resistant plasterwork is reached through the application of alternating layers of cow dung and mud. The cow dung is dissolved in water and then used to wash the earth floor, by which cracks get repaired, dirt is bound, and the floor becomes more solid over time. The result is a smooth dirt free floor, easy to sweep and pleasant to the feet. It also contributes to quality of the interior climate. This technique requires frequent maintenance and consequently the house demands time from its occupants.

5.5 Roof

The traditional building has red clay tiled roofing over thatched roof, locally available material, with 2.5 m high ceiling (Figure 7). The portion of the building studied has ridge line running east west and the roof slopes to the north. This is a light but insulated lean-to roof, with low thermal capacity. The whole Roof is not exposed to sun due to trees on the south side. The building stores less heat because of lower thermal mass as compared to the concrete roof building (U-value is 0.12 W/m^{2}K). Even though, the inside temperature in the building is high in the noon, it does not extend up to midnight. The tiled roof helps the building in cooling down much faster resulting in higher comfort during nights.

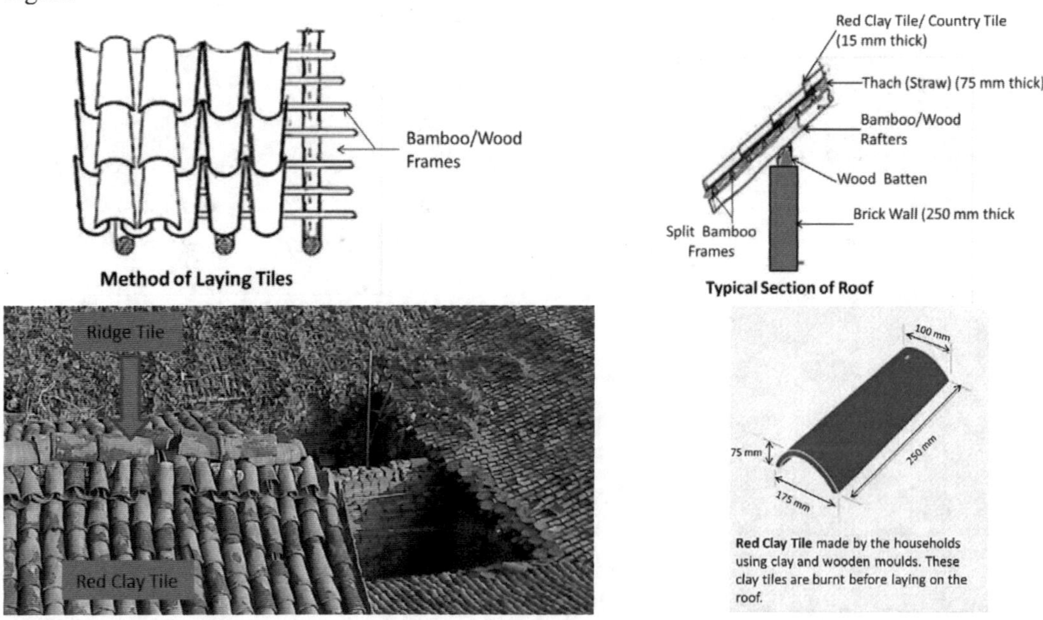

Figure 7. Details of pitched roofs made of red clay tiles.

5.6 Veranda

The 1.5 m wide covered veranda expresses the hybrid notion of inside and outside very clearly. It is a sheltered outside and still part of the home, protected from harsh weather.

5.7 Openings

The rooms are single banked with small windows, of 0.25m x 0.50m size, in the south wall to ensure air movement. The near absence of windows in the rural architecture is an indicator of this outdoor oriented lifestyle. A small window is enough to allow daylight where needed. The windows in a wall are considered to be a weakening of the protective function of that wall. The percentage of door openings on the north facing side is most. The fenestration is completely shaded in traditional building with deep overhanging eaves.

IOP Conf. Series: Materials Science and Engineering 1212 (2022) 012007 doi:10.1088/1757-899X/1212/1/012007

5.8 Outdoor Space

The area adjoining the building becomes an extension of the indoor spaces, and is treated with equal care. The building is so arranged as to surround a central open space, courtyard or patio, partly, and remaining part is surrounded by the high boundary wall. The courtyard is divided in two sections, private open space and public open space, by a brick wall. In the climate of Nalanda, many of the activities like, sleeping, cooking, eating, playing, and working, are performed in the outdoor spaces. The people, who come to meet and chat from outside, gathers in the public open space created in the compound. The shelter of the building is sought only when the need for privacy or unfavourable weather conditions demand it, and for the safe keeping of belongings. Therefore, there is limited need for indoor space. The balance of indoor and outdoor activities is an important aspect of the way of life of rural people of Nalanda. The house is mainly a place to keep things safe, whereas home activities stretch out around the house. The land around it is part of the realm of the home. The quite high boundary walls surrounding the courtyard or outdoor space not only protects the space from scorching hot wind, dust, stray dogs and goats or other animals, trespassers or the limitation of evaporation and retention of an air pool cooled by evaporation [15]; but also serve the psychological need of privacy. There is a shade for cattle also at the enclosed outdoor space.

5.9 Landscape

The surroundings of the site have dense vegetation with lots of trees and agricultural land plantations. The entire area has similar nature of vegetation. In the compound, an evergreen guava tree is in the middle of the courtyard. There are shrubs and creepers, and the earth is covered by small grass contrasting the barren pathways in between. The creepers are supported by bamboo frames, giving an overhead cover, and pleasant dappled shade. There is an open well used for fetching water, also serves for the evaporative cooling in the enclosed outdoor space. The vertical elements like, walls and the building itself, provide shade only in the morning and late afternoon hours. The plants and creepers give the most pleasant shade, reducing the contrast between bright light and solid shadow with its soft half-light.

6. Result and Discussion

Thermal performance modelling and simulation for the investigated case, as well as comparative analysis of the results, has been done in the present study. The analysis of the measured energy consumption values for different months, from January to December, of the year 2017 with the climatic data of the region for each month, from January to December, of the year 2017, has been done (Figure 8 & Figure 9). It shows that the months of February, April, October, and November are physically most comfortable, as the temperature rarely exceeds the comfort level.

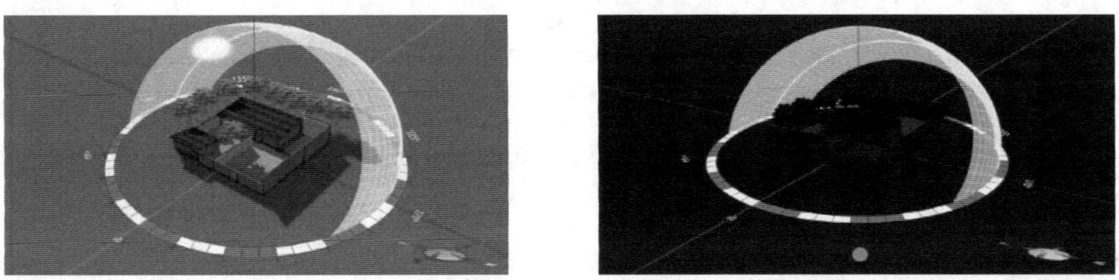

Figure 8. The sun path diagram and the pattern of light and shadow during a day and night time.

The month of January and February are a little colder; the temperate rarely exceeds the upper comfort level during day time, while the temperature reaches below the comfort level, up to 10^0 C in between 5 pm to 8 am. The daily energy consumption during this month is comparatively low (about 17 Kwh), in the present investigation, and most of the energy consumed in heating. The month of March is almost comfortable during the whole day and night as well. The heating loads decreases to zero, and the energy consumption for cooling, only during 1:00 pm to 2:00 pm, is required. The month of April and May becomes warmer due to intense radiation of sun, and the temperature rises up to 30^0 C to 36^0 C, during the day time, 12:00 pm to 5:00 pm. The daily energy consumption increases up to 34 Kwh in the month of May. Most of the energy is consumed in cooling the air temperature during day time. The month of June can be characterised by high temperature, up to 44^0 C, and moderate humidity; whereas the months of July, August, and September have high temperature and dense humidity during the day time. Thus, the discomfort level reaches its peak during the month of June, July, August, and September, and the daily energy consumption for cooling and circulating the air, is comparatively more (about 34 Kwh). The month of October and November are quite comfortable with moderate temperature during day and night, varying between 17^0 C and 30^0 C. The energy consumed per day is less during the month of October and November, about 28 Kwh and 19 Kwh respectively. The temperature starts lowering and the humidity increases during the month of December. The daily energy consumption, mostly for heating, is low (about 19 Kwh) (Figure 10). It is evident from the analysis that most of the energy required is during the day time; however most of the occupants do not use the building during day time. This study can be used as a reference in efforts to improve energy efficiency and thermal comfort in the traditional rural housing.

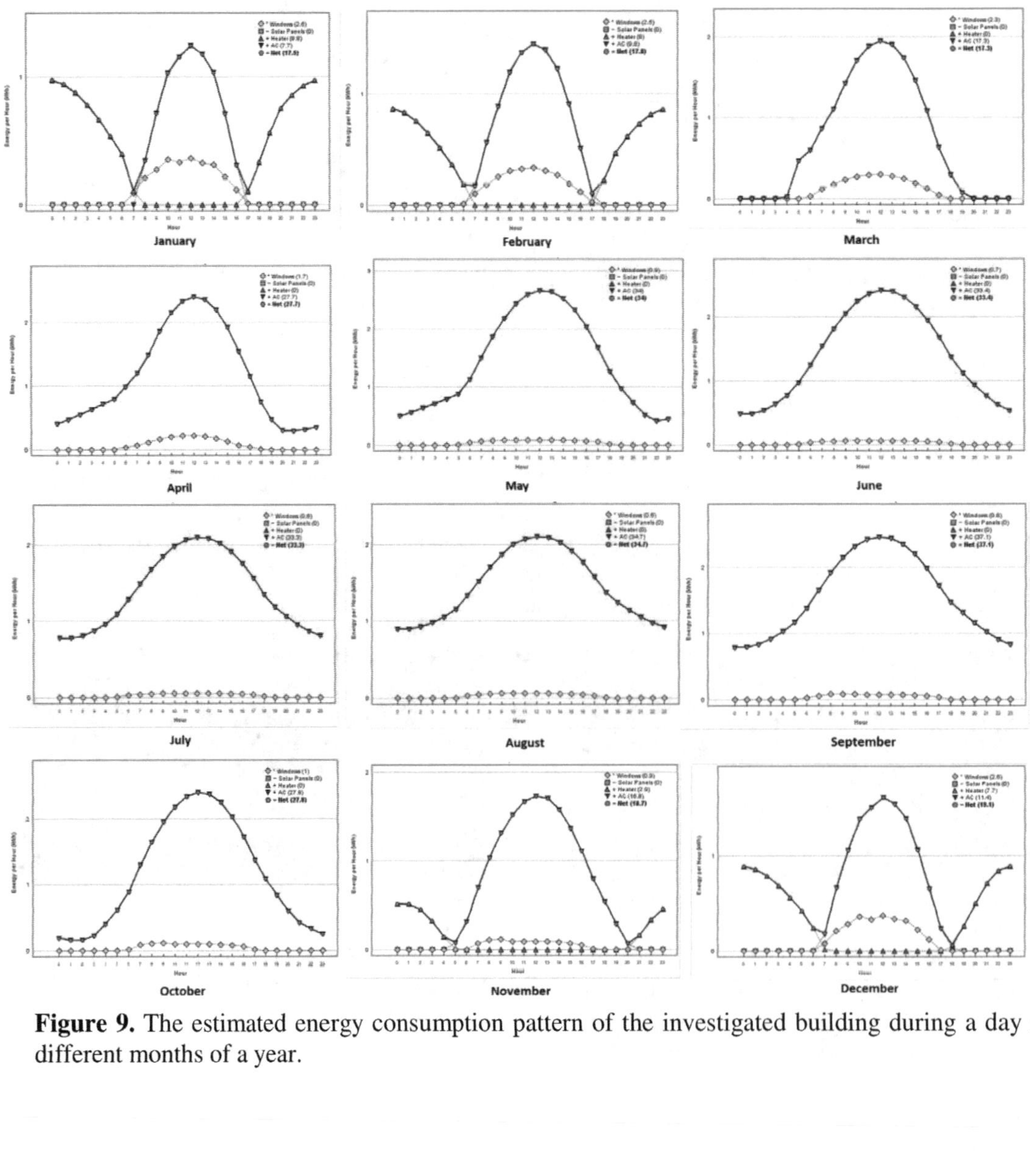

Figure 9. The estimated energy consumption pattern of the investigated building during a day in different months of a year.

7. Conclusion

This paper discusses the features of traditional rural building, and examines the level of thermal comfort and wellbeing which is being achieved through these architectural practices. The result shows, in the climatic condition of Nalanda, decreasing heat gain is an important aspect in striving to achieve thermal comfort and energy efficiency of the buildings. The results of the study provide arguments, to navigate the insignificant and trivial trends in house designs, towards the sustainable path. The need is to integrate the principles of traditional rural architecture with the significant contemporary knowledge and technology.

8. References

[1] Bansal N K, Hauser G and Minke G 1994 *Passive Building Design, a Handbook of Natural Climate Control* (Amsterdam: Elsevier Science)
[2] Previtali J M, & Zhai Z 2016 A taxonomy of vernacular architecture An addendum to "Ancient vernacular architecture: Characteristics categorization and energy performance evaluation" *Energy and Buildings.* **110** 71–78
[3] Singh I and Michaelowa A 2004 Indian urban building Sector: CDM potential through energy efficiency in electricity consumption (Hamburg Institute of International Economics) p 289
[4] Dasgupta P 1995 An Inquiry in to Well-Being and Destitution (New York: Oxford University Press)
[5] Bansal N K and Bhattacharya A 2009 Parametric equations for energy and load estimations for buildings in India *Appl Therm Eng Dec.* **29** (17e18:3710e5)
[6] Motealleh P, Zolfaghari M, & Parsaee M 2016 Investigating climate responsive solutions in vernacular architecture of Bushehr city *HBRC Journal.*
[7] Eilouti B 2018 Concept evolution in architectural design: an octonary framework *Frontiers of Architectural Research* **7(2)** 180-196
[8] Kumari R and Devadas V 2017 Modelling the dynamics of economic development driven by agricultural growth in Patna Region, India. *Economic Structures* **6,15** https://doi.org/10.1186/s40008-017-0075-x
[9] Philokyprou M, Michael A, Malaktou E, & Savvides A 2017 Environmentally responsive design in Eastern Mediterranean. The case of vernacular architecture in the coastal, lowland and mountainous regions of Cyprus *Building and Environment* **111** 91-109
[10] Alves S 2017 The Sustainable Heritage of Vernacular Architecture: the Historic Center of Oporto *Procedia Environmental Sciences* **38** 187-95
[11] Rashida M, & Ara D R 2015 Modernity in tradition: Reflections on building design and technology in the Asian vernacular *Frontiers of Architectural Research* **4** 46–55
[12] Opoko A P 2001 Low energy features of traditional buildings in the Hot-Dry Climatic zones of Nigeria *Nigerian Institute of Architects Journal* **2** 8-12
[13] Kirbas B, & Hizli N 2015 Learning from Vernacular Architecture: Ecological Solutions in Traditional Erzurum Houses *Procedia - Social and Behavioral Sciences* **216** 788-99
[14] ASHRAE 90.1 2004 Energy standard for buildings except low-rise residential buildings. (Inch-Pound: ASHRAE)
[15] Koenigsberger O H, Ingersoll T G, Alan M and Szokolay S V 2009 *Manual of tropical housing & building* (Delhi: Orient Longman Private Limited) p 320

Simulation of thermal comfort on public space and buildings around river in Banjarmasin-Indonesia

A Rahman[1,2]

[1]Program Studi Arsitektur Fakultas Teknik Universitas Lambung Mangkurat
[2]Co-Founder Zero Waste Kalimantan Selatan

arzhi_teks@ulm.ac.id

Abstract. The climate and environmental aspects are one of the things that affect architectural products. The city as a gathering place that interacts for a particular purpose has influenced the shape and visual of the city. The density of buildings in a city has affected the urban microclimate. Urban get hotter than rural areas. Urban planners need to pay attention to several aspects related to the solution to the design of the humid tropics. The concept of greening the city evenly and thoroughly, so that the thermal is not too high that can affect comfort. In this research used Rhinoceros 5, Grasshopper, Ladybug, and ladybug for simulation and validation data of wet bulb temperature on a psychometric chart and CBE Thermal Comfort Tool from ASHRAE-55 standard. The purpose of this study used simulation is to facilitate and predict the thermal conditions of buildings and the environment. This application is also used by researchers and architect designers. Based on the simulation, the indoor maximum effective temperature and standard effective temperature are always uncomfortable zones. The building condition with wooden construction is higher of thermal comfort compared with concrete construction buildings and the wood construction is faster reaches maximum value compared to concrete construction.

1. Introduction

The climate and environmental aspects are one of the things that affect architectural products [1]. The climate, or average weather, is primarily a function of the sun. Climate word is used by scientists to divide the earth region based on the different seasons experienced. The city as a gathering place that interacts for a particular purpose has influenced the shape and visual of the city. The increasing population living in a city, causing the complexity of the problem, is no exception to the architecture and urban design. The density of buildings in a city has affected the urban microclimate. Urban get hotter than rural areas [2][3]. Urban planners need to pay attention to several aspects related to the solution to the design of the humid tropics. The relationship between design and rules is the first half of what architects and planners need to know, that is, how design meets the requirements of the rules [4][5]. For example, urban design should pay attention to the comfort of pedestrians. Pedestrians do not need to use an umbrella, shelter when it rains and the sun shines. The pedestrian ways should be provided with corridors that are protected at the top. Buildings must be connected, so the activity does not stop when it rains or the sun shines [6][7][8][9].

The hard materials used as surface coatings will absorb much heat, but then the heat will reflect into the air, resulting in the warming of the surrounding air [10]. Thus, the concept of greening the city evenly and thoroughly, so that the thermal is not too high that can affect comfort. The building mass arrangement by optimizing the airflow around the building needs to be a concern. One way to improve thermal comfort in humid tropical climates is to optimize the airflow around (outdoor) and indoor buildings [11][12]. Wind movement will be better in public space if the airflow is not blocked. In the concept of arrangement of building mass, public space between buildings is important so that the airflow around the building. This research used Rhinoceros 5, Grasshopper, Ladybug, and ladybug for simulation and validation data of wet bulb temperature on the psychometric chart and CBE Thermal Comfort Tool from ASHRAE standard 55 [13][14]. The purpose of this study used simulation is to facilitate and predict the thermal conditions of buildings and the environment. The thermal input of the location is based on the recorded Energy + data, and the shape of the building and site is drawn according to real conditions. Also besides, to support calculations at CBE Thermal Comfort Tool, in this study also used NOAA's Solar Calculator to calculate solar altitude or elevation for simulation thermal comfort.

2. Research Methods

This research is quantitative research based on simulation data collected from Energy Plus to view thermal comfort in the field (Figure 1). This research used the standard thermal comfort is effective temperature and standard effective temperature by ASHRAE-55. This research was conducted in two sub-districts or three urban villages in Banjarmasin city. The sub-district is a Central Banjarmasin sub-district and the East Banjarmasin sub-district. The village located in the Central Banjarmasin sub-district, namely: Gadang village and Sasirangan village and the village in the East Banjarmasin sub-district is Bilu River village. The simulation was conducted from August 20 to September 20, 2020. The software used for simulation is shortwave solar calculator, solar position calculator, Rhinorceros 5 software, grasshopper software, ladybug software, and honeybee software. The ladybug and honeybee software are simulation-based on ASHRAE international standards. This application is also used by researchers and architect designers [15]. In addition, this application can use the energy plus data bank so that it can observe the thermal comfort conditions of a place in more detail. Step by step of the simulation method is as follows:

- Survey results in the form of maps: area and location, then processed in grasshopper software based on observations in the field.
- The sketch of buildings form based on survey results and photographs in the field, then processed in grasshopper software.
- After the drawings are complete. The maps and building images are collaborated to provide information about conditions in the location and study place.
- Import Energy+ data in grasshopper software.
- Run the honeybee application in grasshopper software, and input coordinate points based on the research place.
- Setting building and environment conditions in honeybee application, like material, orientation, and position.
- Render, and result simulation show in grasshopper software based on model design.

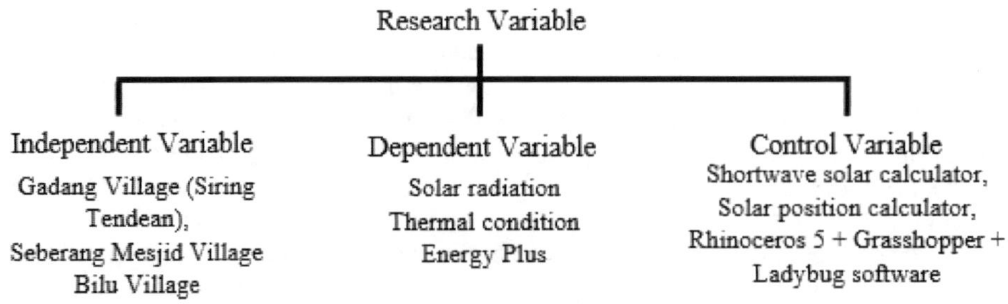

Figure 1. The research variable scheme

3. Result and Discussion

3.1. Sunlight Hours Based on Simulation

The duration of solar radiation during the day affects thermal comfort in the study area. Simulation with Rhinoceros 5 + Grasshopper application can help to determine the duration of solar radiation so that it can be known the causes of differences in thermal and thermal comfort conditions. Building elements such as roof and wall affect indoor thermal comfort, due to longer solar radiation times. Based on the simulation results (Figure 2), the duration of solar radiation on the roof is between 10 and 11.5 hours while the wall is only 4 hours. If observed, the area covered by a roof in Seberang Mesjid village and Bilu village is around 50% while in Siring Tendean only 7.5% while the rest is green open space, open space or river. The roof surface with a long duration of solar radiation causes high thermal and thermal comfort in the building. The dominant roof that affects indoor thermal comfort, outdoor conditions affect such conditions around buildings, green open spaces, or without vegetation and rivers affects. If seen from the simulation results, the duration of solar radiation in open spaces without protection (vegetation) and river area between 11.5 to 12 hours or a full day. The duration of this solar radiation affects indoor and outdoor around the river. Meanwhile, the open space around the building has a lower sun exposure duration of between 8 and 9 hours due to the influence of building shadows. Likewise, the green open space gets much lower solar radiation in the outdoor, especially under the tree.

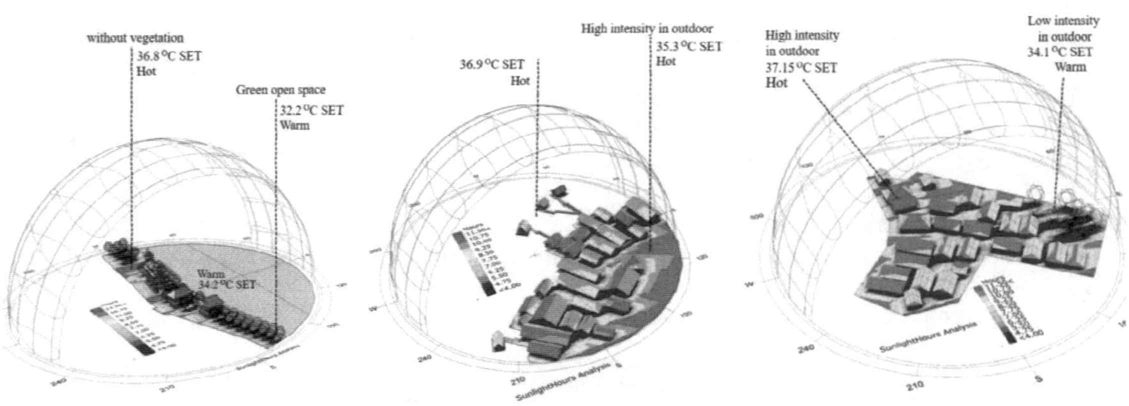

Public Space Sasirangan Village Bilu Village

Figure 2. Simulation of sunlight hours

Based on the simulation results it can be seen (Table 1), thermal comfort in Public Space is better in green open spaces because the duration of solar radiation is lower than open space without protection or vegetation. Thermal comfort in Sasirangan village is more influenced by the duration of sun irradiation on the roof and river surfaces that affect environmental thermal and of course thermal comfort, especially in floating homes. Similar to Sasirangan village, thermal comfort in Bilu village is also affected by sun exposure on the roof surface and open space in the river area. However, the green open space around the settlements has a lower duration of solar radiation and affects the surrounding buildings.

Table 1. Sunlight hours and the percentage covered

Place	Roof	Wall	Around building	Around the tree	Under the tree	Without protection or vegetation	Without protection (river)
Public place	10-11.5 (7.5%)	4	8-9 (7.5%)	7 (25%)	4 (25%)	11.5-12 (10%)	11.5-12 (25%)
Seberang Mesjidvillage	10-11.5 (50%)	4	8-9 (10%)	-	-	11.5-12 (15%)	11.5-12 (25%)
Bilu village	10-11.5 (50%)	4	8-9 (5%)	7 (3%)	4 (7%)	11.5-12 (10%)	11.5-12 (25%)

3.2. Characters of Thermal Comfort around River

The indoor effective temperature and standard effective temperature was higher during the day in floating houses at Sasirangan village and houses aside pedestrian ways at Bilu village. The Outdoor minimum effective temperature just past the threshold of thermal comfort SNI 03-6572 and almost all places have a minimum effective temperature outdoor around 20°C ET or cool comfort condition in the morning, the same time based on ASHRAE-55, the standard effective temperature around 22°C SET or slightly cool, but based this standard is not comforted as shown Figure 3. These conditions indicate that the outdoor or environmental conditions in the morning are cool. The green open space is the best condition based effective temperature and standard effective temperature because always in the comfort zone and neutral comfort until warm comfort in ASHRAE-55. This proves the open space with vegetation or green open space, the better for micro-climatic conditions. The effective temperature and standard effective temperature around the river are lower than the mainland region, this condition can be seen from the different thermal comfort between floating houses, houses aside pedestrian ways, and settlement in Sasirangan village and Bilu village. The indoor and outdoor thermal comfort are highest on minimum conditions. Meanwhile, the maximum thermal comfort is quite varied.

Sun radiation effect to thermal comfort, mean of maximum sun radiation is from 1065 W/m2-1207 W/m2 occurs at 1:00 PM. If it is correlated between the mean of maximum and minimum standard effective temperature, it is known that the indoor effective temperature and standard effective temperature not so much different but the outdoor conditions are so much different because the solar radiation includes calculation in standard effective temperature. Besides, the maximum value on the time of indoor mean effective temperature and standard effective temperature occurs from 12:30 AM to 1:30 AM, except for indoor buildings in green open spaces located: slower to reach a maximum point and mean of maximum conditions are lower than houses in Sasirangan village and Bilu village. Meanwhile, the mean of outdoor maximum effective temperature and effective temperature occurs from 12:30 PM to 2:30 PM.

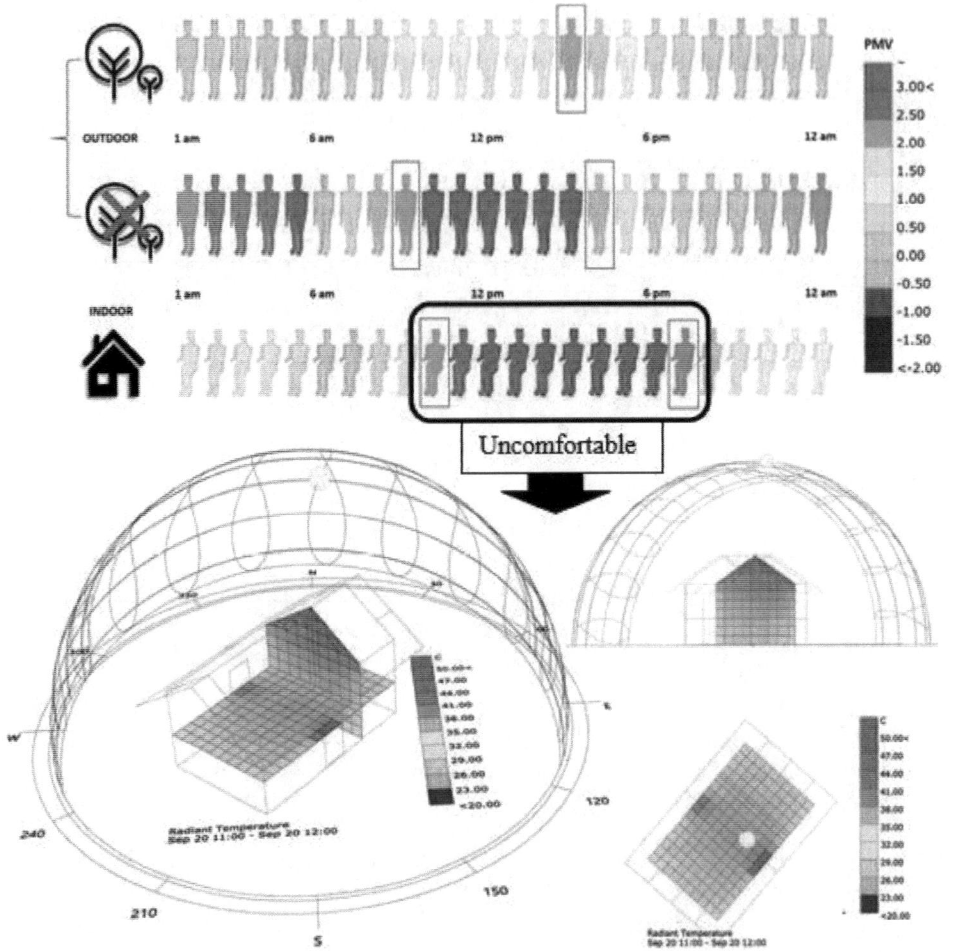

Figure 3. Comparison of thermal comfort between public space and building around the river

4. Conclusion

Based on the previous explanation, thermal comfort conditions are more convenient in buildings in public space. The high solar radiation affects the dry bulb temperature and relative humidity. The thermal comfort conditions are more comfortable in public space with vegetation than buildings in Siring Tendean. Meanwhile, the public space without vegetation during the day tends to hot and early morning is cool based on ASHRAE-55. The presence of trees in public space can be control in thermal comfort because the green open space is not low at night or minimum conditions, these conditions can be concluded the vegetation affects the thermal environment and thermal comfort conditions in public space. The open space conditions that have lots of vegetation is better thermal conditions than open space without vegetation. Open space with trees has been able to reduce the thermal environment. As a public space, these places should provide comfort for visitors. Thermal comfort during the day needs to be considered and the direct sunlight barrier is required in open spaces, buildings, or houses. Based on the simulation, the indoor maximum effective temperature and standard effective temperature are always an uncomfortable zone. The building condition with wooden construction is higher of thermal comfort compared with concrete construction buildings and the wood construction is faster reaches maximum value compared to concrete construction. The green open space in public space has been the best thermal comfort compared to the others. Meanwhile, open space without vegetation at minimum conditions an uncomfortable zone, and effective temperature and standard effective temperature are faster reach to maximum condition, compared to the open space with vegetation.

References

[1] Rapoport A 1969 *House Form And Culture* Prentice-Hall Englewood Cliffs New Jersey.
[2] Roth M 2013 *Handbook of Environmental Fluid Dynamics: Urban Heat Islands*. CRS Press/Taylor and Francis Group Vol 2 143-160
[3] Aaron R Y Sarah E D Ryan A. M 2020 *Evidence for the evolution of thermal tolerance, but not desiccation tolerance, in response to hotter, drier city conditions in a cosmopolitan, terrestrial isopod* Evolutionary Applications
[4] Al-Bishawi M Ghadban S 2011 *International Journal of Architectural Research: A Methodological Approach for Reading Urban Open Space* ArchnetIJAR Vol 5 No 1 73-85.
[5] Silvia G David S Greg M Karine D Christhina C Simon T 2020 *Public health and well-being in public open spaces through climate responsive urban planning and design* The Journal of Public Space Vol **5** No 2 1-6
[6] Karyono T H, 2001 *Dimensi Teknik Arsitektur: Wujud Kota Tropis di Indonesia: suatu pendekatan iklim, lingkungan dan energi* Vol **29** No 2 141 – 146
[7] Akbar R Shoichi K 2017 *Proc. Int. Conf. on Chemical, Biological and Environment Engineering: Analysis of thermal comfort SNI-6390 in the Lanting (floating house)* **100** 54-58
[8] Satoshi S Ivan F Darko R 2020 *Public-private interaction in low-rise, high-density Tokyo* The Journal of Public Space Vol **5** No 2 63-88
[9] Kaoru M Rui I Shihona A Akiko T Yusuke H Riki N 2020 *Research on the relationship between human behaviour and climatic characteristics in a public open space* The Journal of Public Space Vol **5** No 2 111-128
[10] Akbar Rahman 2010 *Proc. Int. Conf. on Sustainable Environmental Architecture (SENVAR): The Different Air Temperature in The Shadowed and Unshadowed Condition* ITS Surabaya
[11] NazhatulzalkisJ Nurul I M Mohd F K Suriani N A 2015 *Thermal Comfort of Residential Building in Malaysia at Different Micro-climates* Procedia - Social and Behavioral Sciences Vol **170** 613-623
[12] Latif S Hamzah B Ihsan. 2016 *Pengaliran Udara untuk Kenyamanan Termal Ruang Kelas dengan Metode Simulasi Computational Fluid Dynamics* Sinektika Vol **14** No 2 209-216
[13] ASHRAE Standard 55 P 2003 *Thermal Environmental Conditions for Human Occupancy*
[14] Fabbri K 2015 *Indoor Thermal Comfort Perception* Switzerland Springer International Publishing
[15] Zacepins A Egils S 2012 *Proc. Int. Conf. on International Carpathian Control Conference: Architecture of automatized control system for honey bee indoor wintering process monitoring and control* Slovakia **13**

IOP Conf. Series: Materials Science and Engineering 1212 (2022) 012009 doi:10.1088/1757-899X/1212/1/012009

Revealing circular material flow in Terrazzo making process

V A Ujung, A R Wahid, P Atmodiwirjo

Department of Architecture, Faculty of Engineering, Universitas Indonesia

arifrahman@ui.ac.id

Abstract. This paper investigates the value in material expression that reveals the material resource flow through terrazzo making process. The potential of the patterning process as a vital practice in terrazzo making is posed through attention to the use of salvaged elements and fragmented pieces of the material. They are an essential part of circular economy practice knowledge. The patterning process, such as reinforcement and recombination of salvaged materials is decreasing carbon emissions produced by fabricating new components of the terrazzo. Besides, the patterning process enables the materialization of the design intention and locals' interests and particularities; in which it adds value to the material. This study was conducted through a workshop in Lombok, Indonesia, as part of an exhibition of architectural materials that were produced locally from earth-based ingredients. It is found that the value of sustainability lies in its ability to promote such circular strategies that can enable improved material resource efficiency as well as generate material value.

1. Introduction

In recent years the architectural design and construction industry has begun to focus on the environmental impact of building materials deliberately. A wide range of significant concerns includes the material value and longevity, material composition, and maintenance has risen towards circularity design thinking and practices as it enhances sustainable development [1]. The proliferating industrially produced material involves the ever-accelerating flow of materials [2]. The materials' flow that is understood in a broad and critical sense related to circularity thinking and practices has gained increasing reflection with regards to the potential of material value generation and resource-life extension practices [1]. It is the understanding of circular material flow that will improve material resource efficiency as well as generate material value.

Understanding the circularity mechanism encompasses an in-depth understanding of the environmental, social and economic challenges and interconnections. Within a sustainability framework, circularity thinking, and practices addresses strategies that are related to the restorative and regenerative acts. Therefore, the circularity mechanism promotes the urgency to understand the material resource flow to generate resource efficiency [1,3] and also to respond to resource depletion. The idea of using salvaged elements and fragmented pieces of material is being highlighted in this study to "bring all resource flows back into a closed loop" [4]. It emphasizes the benefits of a circular economy related to the material resource efficiency that reduces the environmental impact in comparison to a traditional linear economy that requires new material production [1].

This paper addresses the critical role of local and traditional value and labor of terrazzo in the establishment of the circular-economy practice. It particularly highlights the importance of understanding the journey from material expression to the material value that represents material resources flow and shapes what we now understand as the circular practices. Focusing on the material resources flow as continuous and optimal processes within the energy and resource system, this study reveals the approaches to circular resource flow to determine what circular operations are essential within the terrazzo making process.

2. Context and Methods

This paper explores the patterning process of the terrazzo through a traditional hand-casting, labor-intensive method. In particular, the patterning process of terrazzo is identified within the three states: particles or materials, parts or components, and products or finished goods. The patterning process involves material extraction, part or component shaping, and product assembly.

The data in this study are collected from a workshop in Suradadi Village, Lombok, West Nusa Tenggara, Indonesia. This workshop is a part of an exhibition of architectural materials that were produced locally from earth-based ingredients [5,6]. The artisan is a Lombok-native who previously worked in Bali. We selected this workshop to be studied during our visit because there were only a few terrazzo workshops in Lombok. The workshop is also in the working process on several terrazzo sample types for a homestay construction project in Kuta, Lombok. Therefore, we had the opportunity to see the lengthy process of developing terrazzo. Photographs, videos, and interview excerpts are used to record the patterning process of the terrazzo. We look closely at what are the steps required to produce terrazzo; tools that are used in each stage; what kind of salvaged material is used in the mix of the terrazzo ingredients; and also the final sample from the different main ingredients. This information is then interpreted to unfold the flow from material expression to reach its value.

3. Revealing the Value in Expression

Incorporation of naturally occurring aggregates [7], recycled glass or plastic, as well as stone or marble that has been salvaged from other buildings makes terrazzo considered as one of the 'green concrete' [8] that responds to the resource and energy saving. The National Terrazzo and Mosaic Association (NTMA) stated that aggregates used in terrazzo making processes are the most common sources of recycled content [7]. The aggregates such as post-consumer recycling of colored glass, mirror, metal, brass, aluminum, seashells, broken bottles, bricks, porcelain and concrete are crushed and ground and capable of forming various unique finishes [7,9]. The offcuts leftovers are used for expression of material handling and formation [10] in the terrazzo making process. The potential of the way material is handled, formatted and exposed is significance to the understanding of the material value. Value or significance arises not from things, but from the ways they (materials) are handled, worked, or treated [11]. The notion that material value is an integral part of material expression and labor is being promoted in this study.

The material expression of the terrazzo exposed in the form of surface, color and texture, relies on a mixture of embedded elements, particles and parts. Exposed aggregate in a cementitious or polymeric binder expresses the specificity of materials through the ways it is handled, worked or treated as we know as material labor [10]. "Material value is not in the material itself, but in the labor necessary for its formation" [10]. Material labor generates material 'expression of assemblage' [10] of the terrazzo making process and further creates material value and significance. This 'expression of assemblage' [10] is achieved through the visual articulation of each particle, parts or components and products of the terrazzo through the patterning process. The patterning process is a medium to highlight the journey from material expression of assemblage to the material value.

The circular economy's benefits of having the ability to use salvaged and recycled materials means increased flexibility and optimal quality in supply due to access to multiple sources of raw materials. The potential of the terrazzo as building materials being refurbished, rather than replacing existing flooring, stairs, and cladding contributes to reducing waste and environmental impact. "Terrazzo tiles can be reused if they are bedded in a way that makes them easy to remove. They can, for example, be laid in the sand and given a weak lime cement mortar joint [4]." This process enables material expression disassembly to create the new material expression of assemblage. This condition represents substitution flows "in which the new input flow is being replaced with an outgoing flow from another process that has been poorly exploited or not at all. For example, replacing flow of "new" raw materials consumed within a process with the flow of waste or by-products; this type of synergy can reduce supply costs or the cost of treating an outgoing flow [3]." Figure 1 shows the material resources flow in terrazzo making process from which the continual reuse loops [12] is being circulated within multiple states from particles, parts to products. In this way, substituted and dissembled material expression leads to the condition in which multiple values are being implemented and repurposed.

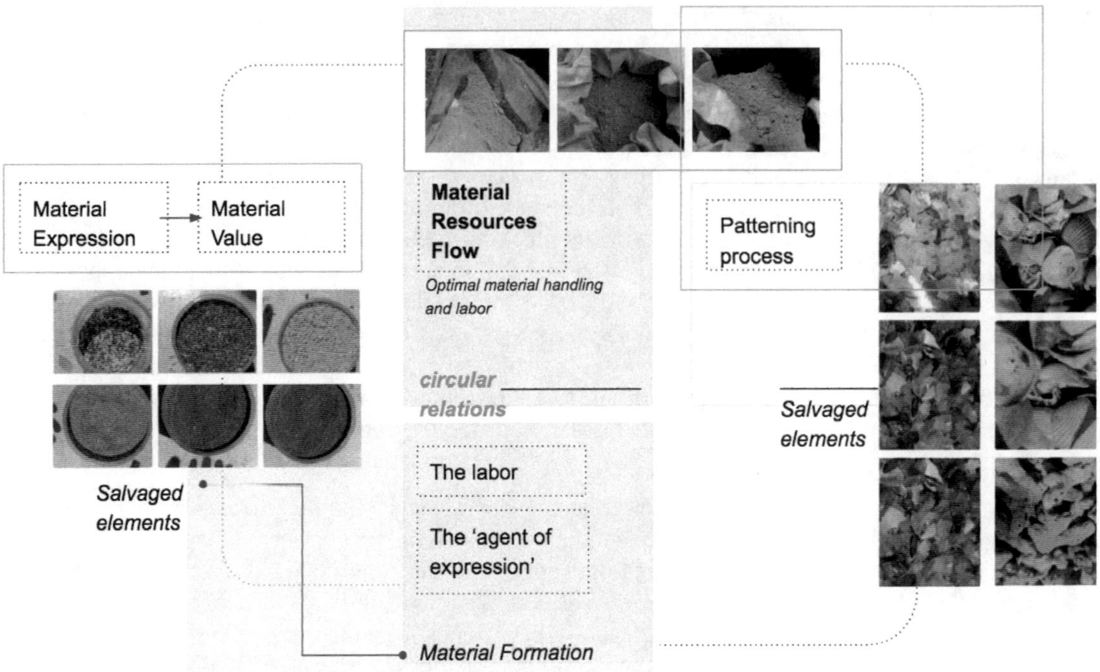

Figure 1. The material resources flow in terrazzo making

The study highlights the patterning process as a medium that is "limiting environmental impacts and improving the economic competitiveness and attractiveness of the territories" [3] and enhancing sustainability development. "Sustainability is not a utopian concept as within something that is always changing there is no ideal state. What it does demand is the ability to measure resources flows and how resources are used by people to ensure that these are achieved without damaging the ecosystems that supply them [13]." Understanding the journey from material expression to material value indeed acquires instrumental knowledge about sustainable functioning and performance. Moreover, it also requires building the competence to link material resources flow meaningfully together with sustainability factors in an environmentally, socially and economically responsible way.

4. Revealing Circular Material Flow and Operations

Optimizing the flows of resources used must consider the consistency, density, water absorption, and abrasion resistance [14] in the patterning process of the terrazzo production. Compared to the study on the incorporation of aggregates in green concrete production [8] it has shown that it is advantageous to increase the workability of concrete and produce durability properties. Terrazzo is one of the durable materials which "causes less dust problems than a plain concrete floor, and offers a hard, hygienic surface [4]." Particles or materials and parts or components of terrazzo production in particular in the patterning process require careful attention to ensure consistency of mix and coloration. From the workshop, it was apparent that the basic process of terrazzo production is mixing all the ingredients (broken marbles, white cement, calcium) with additional water. Comparison of mixed marble, calcium and white sand fragments is 2:2:1. Water quality and quantity determines the color produced. The mixture composition depends on its use. Color is added based on the project's specific order and requirements. Then, sigma dyes are mixed to produce colored terrazzo.

The process of embedding different particles or ornaments is usually done when the new terrazzo dough is poured. When it is almost dry, the terrazzo will be compacted again. After it becomes solid, the terrazzo must be allowed to stand for 1-2 weeks until completely dry. The terrazzo mixture is poured over a mixture of dry and leveled castings using a cement grinder. Based on the material formation, size, type, and color of particles used in the mixture as well as the type of finish used, terrazzo offers a wide range of material expression options.

Figure 2 shows the terrazzo patterning process that denotes the marks of human hands and enables the materialization of the design intention, locals' interests and particularities. In this way, the process of patterning, embedding and revealing in the terrazzo making process promotes the marks of human hands and the evidence of uncoerced craft [11] that are expressed in the material surface of the terrazzo

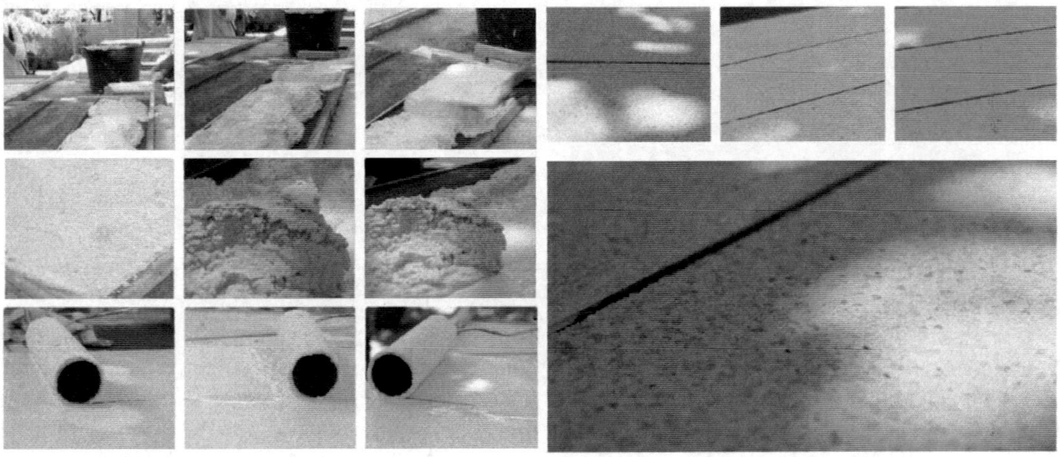

Figure 2. Terrazzo patterning process denotes the marks of human hands

Revealing the potential of the patterning process enables the materialization of particularities which add the material value. "One aspect of this essence lies in local geography, climate, and customs, involving the use and transformation of local, 'natural' materials... Clearly, the engagement with particularities is one way the domination of technology can be interrupted [10]." In the process of embedding waste particles into the composition of terrazzo, the craftsmen's intimate handling of the material requires the engagement with particularities, locality and specificity [10,15]. "Intimate handling of the material, the sense of the presence of the maker in the artefact, had considerable consequence on the evocative power of the work and perhaps, by association, its emotional accessibility [15. At the micro-level, the design, the choice and formation of materials, the manner of its treatment, the tools, the color, the pattern, the texture, and all their compositions reflects the character of the local craftsmen and the depth of surface in the cultural sense [16] that can be seen in Figure 3.

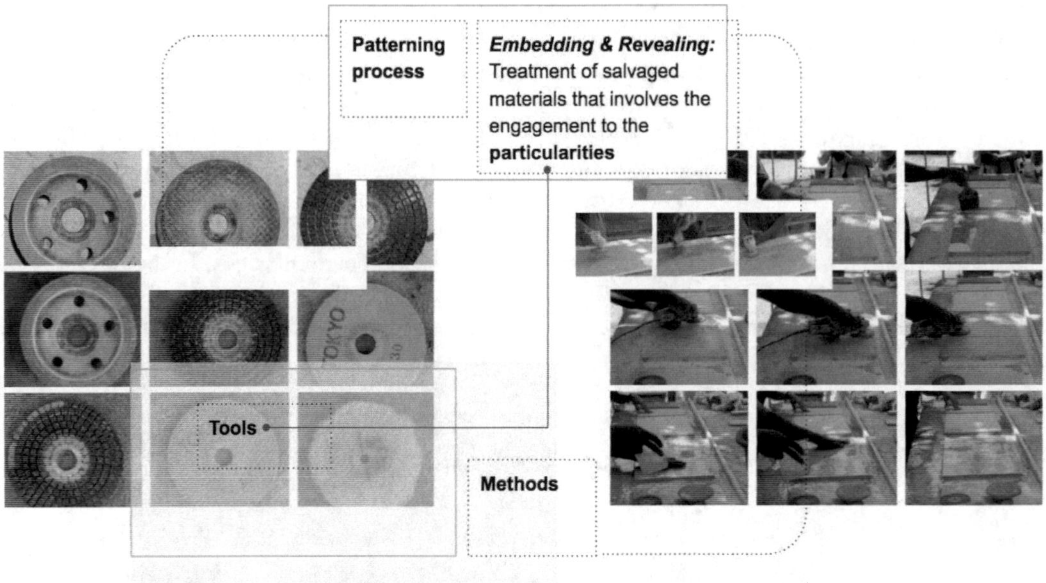

Figure 3. Tools and methods in terrazzo making process

At a macro level, sustainable performance permeated the built environment of the terrazzo making process to the accommodation of socio-economic relations. Investigating the circular resource flow of terrazzo making process to the accommodation of socio-economic relations also involves the engagement with particularities as Leatherbarrow suggested [10]. The collection system of salvaged materials is one of the keys to circular practice in the terrazzo making process. Due to the availability of the local craftsmen and terrazzo workshop in Lombok is limited, traditional labor has relied on large quantities of cheap and easily accessible materials and energy. In the collection system of salvaged materials, repurposing the waste as a resource is related to the circular supplies as well as resource recovery [17]. The identification of all the active and relevant 'agents of expression'[18] around the terrazzo making process defines the actor-network through which the value in an expression is enhanced.

5. Conclusion

The understanding of circular material resource flow expands to take account of the value in the material expression and formation. Understood in a broad and critical sense, terrazzo making process concerns material expression that impact on material value. Through the micro and macro level understanding of circular material resources flow within the three states of terrazzo making process, the important characteristics of the circular value chain are established, to retain the material value in use and ensure an enhanced sustainable performance. The value of sustainability lies in its ability to promote such circular strategies that can enable improved material resource efficiency as well as generate material value. It is found that the patterning process in the terrazzo making process promotes the circular material flow which adds value to the material, to the socioeconomic relations and could optimize the quality of circular supplies and continual resource flow.

References

[1] Charter M 2018 Designing for the Circular Economy (Abingdon, Oxon ; New York, NY : Routledge, 2018.: Routledge)
[2] Hagan S 2007 The New and the Renewed Material matters: architecture and material practice ed K Lloyd Thomas (London ; New York: Routledge) pp 250–7
[3] Gallaud D and Laperche B 2016 Circular economy, industrial ecology and short supply chain

[4] Berge B 2009 The ecology of building materials (Oxford: Architectural Press)
[5] Atmodiwirjo P and Yatmo Y A 2020 "Tanahku Indonesia": On Materialscape as the Materiality of a Nation Archit. Cult. 1–22
[6] Johanes M and Wahid A R 2018 Tanahku Indonesia: Celebrating the Indigenous Interior Interiority 1 79–86
[7] The National Terrazzo & Mosaic Association, Inc. 2018 Green Terrazzo: The Environmental Impact of the Use of Cement and Epoxy Terrazzo
[8] Garas G L, Allam M E and Bakhoum E S 2014 Studies undertaken to incorporate marble and granite wastes in green concrete production ARPN J. Eng. Appl. Sci. 9
[9] North American Terrazzo, Inc. 1999 Terrazzo Specifications & Design Guide
[10] Leatherbarrow D and Mostafavi M 2002 Surface architecture (Cambridge, Mass: MIT Press)
[11] Leatherbarrow D 2009 Architecture oriented otherwise (New York: Princeton Architectural Press)
[12] Durmišević E 2018 Reversible building design Designing for the Circular Economy ed M Charter (Abingdon, Oxon ; New York, NY : Routledge, 2018.: Routledge) pp 344–59
[13] Garcia E and Vale B 2017 Unravelling sustainability and resilience in the built environment (London ; New York: Routledge, Taylor & Francis Group)
[14] Karam G and Tabbara M 2009 Properties of pre-cast terrazzo tiles and recommended specifications Cerâmica 55 84–7
[15] Shotton E 2007 Material Imprecision Material matters: architecture and material practice ed K Lloyd Thomas (London ; New York: Routledge) pp 92–102
[16] Eichinger G and Tröger E 2011 Touch Me! das Geheimnis der Oberfläche ; Gregor Eichinger spricht über die Benutzeroberfläche in der Architektur (Baden, Schweiz: Müller)
[17] McDonough W and Braungart M 2002 Cradle to cradle: remaking the way we make things (New York: North Point Press)
[18] Lloyd Thomas K 2007 Material matters: architecture and material practice (London ; New York: Routledge)

Acknowledgement
This paper is funded by the PUTI Proceedings 2020 research grant from the Directorate of Research and Community Engagement Universitas Indonesia (DRPM UI).

Extraction of Zn (II) ions in an polypropylene glycol 425 – sodium chloride - water aqueous two-phase system

M I Fedorova*, I V Zinov'eva

Kurnakov Institute of General and Inorganic Chemistry of the Russian Academy of Sciences, 31 Leninsky Prospect, Moscow, 119991, Russia

*Corresponding author: mif@igic.ras.ru

Abstract. Liquid-liquid extraction is one of the most used separation methods in chemical technology for recovery and separation of metal ions, other inorganic and organic substances. It is known that for extraction of Zn(II) the most frequently used extractants are D2EHPA, Aliquat 336, etc., diluted in an organic solvent. The use of these reagents does not meet the principles of "green" chemistry. Thus, in the present work, the extraction system based on polypropylene glycol 425 and sodium chloride for the extraction of Zn(II) ions from aqueous solutions is proposed. Equilibrium values of the distribution coefficient in the proposed aqueous two-phase system have been determined. Dependence of metal distribution coefficient on time of phase contact is obtained, time to reach equilibrium was 10 minutes. The isotherm of Zn(II) extraction obtained in the proposed system is a straight line, which indicates the independence of the distribution coefficient from the initial concentration of metal in the solution. The received experimental data can be used at the creation of "green" schemes of processing of Ni-MH batteries.

Keywords: aqueous two-phase system, extraction, polypropylene glycol, sodium chloride, Ni-MH battery

1. Introduction

Nickel-metal hydride (Ni-MH) batteries are widely used as energy sources [1]. However, improper disposal of electronic waste can lead to environmental pollution as well as harm to human health. The products of primary processing of Ni-MH batteries contain a large number of strategically important metals (Ni, Co, Mn, Zn, La, Ce, etc.) [2,3]. The hydrometallurgical method is the most promising way to extract metals from used chemical power sources, due to the advantages of high metal recovery efficiency and high purity of the product. The leaching process allows obtaining a solution containing a large number of metals for subsequent separation and recovery.

Extraction is a popular method used in chemical and technological processes for:

1) recovery of biologically active compounds (proteins, cells, DNA/RNA, vitamins, etc.) from aqueous media [4,5];
2) purification of hydrocarbons from sulfur, nitrogen, and aromatic substances [6,7];
3) recovery of organic acids from fermentation solutions [8-10];

4) extraction and separation of rare-earth [11], precious [12], non-ferrous [13], radioactive elements [14,15] and others.

Recent works present the results of metal extraction using neutral and ion-exchange extractants – TBP, Cyanex 272, D2EHPA, PC 88A, ionic liquids, binary extractants, and others [16, 17]. Gupta et al. studied Cd extraction in the presence of accompanying metals (Ni, Co, Fe, Mn, Al, Hg, etc.) using commercial extractant Cyanex 923 [18]. Belova et al. studied the extraction of rare-earth metals from chloride and nitrate solutions using binary extractants based on Aliquat 336, D2EHPA, Cyanex 272, etc. [19,20]. The effective extraction of lanthanides in such systems is shown.

Aghdam et al. studied the extraction of Zn(II) from chloride media by solvent extraction using D2EHPA. Quantitative extraction of zinc was achieved at double excess of D2EHPA and medium acidity (pH) from 3 to 6 [21]. Spathariotis et al. studied the extraction of Zn(II) from deep eutectic solvent using Aliquat 336. Using Aliquat 336, which was not diluted with solvent, 80 % zinc extraction was achieved [22].

However, the use of toxic and explosive organic solvents negatively affects the environment and humans [20]. Nowadays such direction of extraction chemistry as aqueous two-phase systems based on ecologically safe water-soluble polymers meeting the principles of "green" chemistry is intensively developing [23]. Aqueous two-phase systems also find application for the extraction and purification of biologically active substances [24], organic acids [25-27], metal salts [28].

Extraction and recovery of zinc from hydrometallurgical processing solutions is an urgent task for today, as the emission of such waste is regulated by environmental legislation. Previously, zinc extraction from aqueous solutions using ATPS based on polypropylene glycol was not carried out. In this work, we studied for the first time the interphase distribution of Zn(II) in the eco-safety extraction system of PPG 425 - NaCl - H_2O. The time of metal ion equilibrium achievement between polymeric and salt phases has been determined. The extraction isotherm Zn(II) in the proposed system was obtained. The results of this work show that the system based on polypropylene glycol is no less effective than the classical extraction systems.

2. Experimental

2.1. Reagents and instruments

The tools and materials were used in this work: 50 ml measuring beakers, measuring pipettes (from 100μ to 5ml), 20 ml separation funnels, 15 ml graduated plastic tubes, analytical Balance (AND HR-100AZ), centrifuge (ELMI CM-6MT), desalinated filter paper, Enviro-Genie thermostatically controlled shaker (Scientific Industries, Inc.), spectrophotometer Cary-60 device (Agilent Tech.), quartz cuvettes (*l=10*мм).

The ATPS was made by mixing precisious weigh of sodium chloride with the qualification of “chemical grade”, precisious volume of polypropylene glycol 425 and precisious volume of distilled water.

The initial solutions of metal chlorides or nitrates were prepared by dissolving precise weights of $ZnCl_2$ with the qualification of “chemical grade” in distilled water suspended on an analytical balance (AND HR-100AZ).

2.2. Research methods.

The extraction of metal ions was carried out using an aqueous two-phase system based on polypropylene glycol 425 (30 wt.%) – NaCl (8 wt.%) - H_2O with an initial metal concentration of 0.01mol L^{-1}. The ratio of polymer and salt phases was 1:3. To prevent the formation of insoluble zinc hydroxide in the system was introduced hydrochloric acid in the amount of 0.00012 mol. The extraction of metal ion (Zn (II)) was carried out at a temperature of 25°C in graduated plastic tubes in a thermostatic Enviro-Genie shaker (Scientific Industries, Inc.) at a rotation speed of 30 rpm to establish thermodynamic equilibrium (30 min).

The evaluation of extraction efficiency was carried out using characteristics such as distribution coefficient (D) calculated using formulas (1):

$$D = \frac{C_{t.ph}}{C_{b.ph}}, (1)$$

where $C_{t.ph}$ – metal concentration in top (polymer) phase, $C_{b.ph}$ – metal concentration in bottom (salt) phase.

The concentration of Zn ion in the initial solutions and the top and bottom phases after extraction was determined spectrophotometrically using 4-(2-pyridylazo)resorcinol, which forms complexes with metals that absorb in the visible spectrum at the wavelength of 492 mn. The optical density values were determined using a Cary-60 device (Agilent Tech.) In quartz cuvettes l = 10 mm.

The presented experimental data are the result of a series of experiments and processed by methods of mathematical statistics.

3. Results and discussion

We have carried out a study on the interphase distribution of Zn(II) ions in the polypropylene glycol 425 (30 wt %) - NaCl (8 wt %) - H_2O system. We studied the effect of phase contact time on zinc ion extraction. Figure 1 shows the kinetic dependence of the Zn(II) ions distribution coefficient. As can be seen from Fig. 1, phase contact time practically does not affect the interphase distribution of $ZnCl_2$. In this case, the equilibrium is achieved already at mixing time equal to 10 minutes.

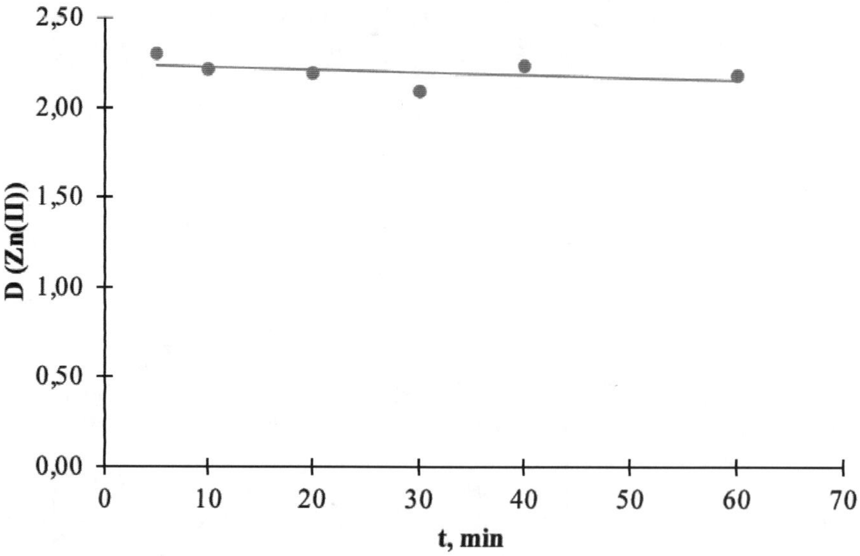

Figure 1. Dependence of Zn(II) ion distribution coefficient on phase contact time in PPG 425 (30 wt. %) - NaCl (8 wt. %) - H_2O system, $C(Zn(II))_{init.}$= 0.01 mol L^{-1}.

Further, we studied the influence of initial metal concentration on the distribution coefficient. For this purpose, we carried out an experiment on the variation of the initial concentration of $ZnCl_2$ from 0.001 to 0.01 mol L^{-1} in PPG 425 (30 wt. %) - NaCl (8 wt. %) - H_2O system. Figure 2 shows the extraction isotherm. As can be seen from Figure 2, the extraction isotherm has a straightforward character, which suggests that the initial concentration of metal (in the range under study) does not affect the distribution coefficient.

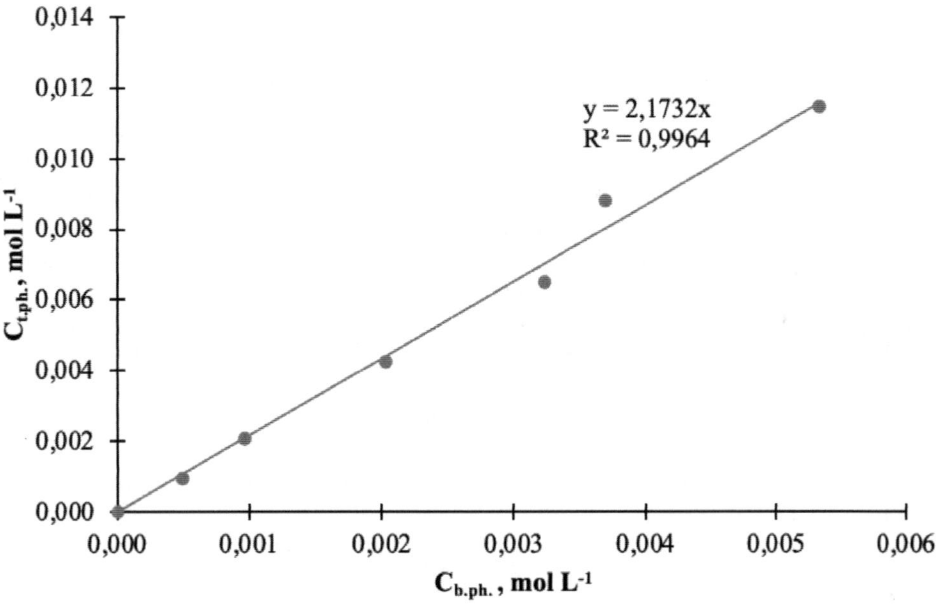

Figure 2. Isotherm of extraction of Zn(II) ions in PPG 425 (30 wt. %) - NaCl (8 wt. %) - H_2O system.

It should be noted that similar results on metal extraction in this system were obtained earlier for Fe(III) Zakhodyaeva and others [28]. In the work of Zakhodyaeva et al. [29], the distribution of zinc and other non-ferrous metals ions in the system with polyethylene glycol 1500 was studied. In the PEG 1500 (16.3 wt.%) - $NaNO_3$ (36 wt.%) - H_2O system phase contact time and initial metal concentration also do not influence the zinc distribution coefficient. In the works of Fedorova et al. the extraction of Zn(II) ions is described in PPG 425 (30 wt. %) - NaCl (8 wt. %) - H_2O system with the introduction of complex-forming agents of organic nature [30-31]. In these works, the mechanism by which zinc is extracted from hydrochloric acid solutions was proposed. Extraction occurs due to the formation of chloride anion complexes, composition $[ZnCl_4]^{2-}$.

4. Conclusion

In the course of the research, we obtained quantitative characteristics of Zn(II) ion extraction in PPG 425 (30 wt. %) - NaCl (8 wt. %) - H_2O system. It was found that the contact time of phases, as well as the initial concentration of Zn(II) ions, do not affect the degree of metal extraction. It was shown that this system is promising for extracting this metal from acidic aqueous solutions. The data obtained can be used to solve the problems of zinc extraction from hydrochloric acid solutions.

References

[1] Fan E S, Li L, Wang Z P, Lin J, Huang Y X, Yao Y, Chen R J and Wu F 2020 Sustainable Recycling Technology for Li-Ion Batteries and Beyond: Challenges and Future Prospects *Chemical Reviews* **120** 14 7020-63
[2] Maroufi S, Nekouei R K, Hossain R, Assefi M and Sahajwalla V 2018 Recovery of Rare Earth (i.e., La, Ce, Nd, and Pr) Oxides from End-of-Life Ni-MH Battery via Thermal Isolation *ACS Sustainable Chem. Eng.* **6** 9 11811-8
[3] Assefi M, Maroufi S, Yamauchi Y and Sahajwalla V 2020 Pyrometallurgical recycling of Li-ion, Ni-Cd and Ni-MH batteries: A mini-review *Current Opinion in Green and Sustainable Chemistry* **24** 26-31
[4] Banožić M, Babić J and Jokić S 2020 Recent advances in extraction of bioactive compounds

from tobacco industrial waste-a review *Ind. Crops Prod.* **144** 112009

[5] Zinov'eva I V, Zakhodyaeva Y A and Voshkin A A 2019 Interphase Distribution of Caffeine and Coumarin in Extraction Systems with Polyethylene Glycol and Sodium Sulfate *Theor. Found. Chem. Eng.* **53** 6 996–1000

[6] Zakhodyaeva Y A, Solov'ev V O, Zinov'eva I V, Rudakov D G, Timoshenko A V and Voshkin A A 2019 Interphase Distribution of Thiophene, Toluene, and o-Xylene in the Hexane–Polymer–Water Extraction System *Theor. Found. Chem. Eng.* **53** 4 550-5

[7] Solov'ev V O, Zakhodyaeva Y A and Voshkin A A 2020 О влиянии добавок полимера, нитрата натрия и 1-метил-2-пирролидона на экстракцию тиофена в системе н-гексан – вода *Theor. Found. Chem. Eng.* **54** 5 623-8

[8] Khol'kin A I, Zakhodyaeva Y A, Voshkin A A and Belova V V 2013 Interphase distribution of weak acids in systems with binary extractants *Theor. Found. Chem. Eng.* **47** 4 453-60

[9] Belova V V, Zakhodyaeva Y A and Voshkin A A 2017 Extraction of carboxylic acids with neutral extractants *Theor. Found. Chem. Eng.* **51** 5 786-94

[10] Zakhodyayeva Y A, Voshkin A A, Belova V V, Kostanyan A E and Khol'kin A I 2010 Extraction of lactic acid from technological (concentrated) solutions *Theor. Found. Chem. Eng.* **44** 5 782-5

[11] Egorova N S, Belova V V, Voshkin A A, Zhilov V I, Khol'kin A I 2005 Extraction of lanthanide chlorides by binary extractants based on phosphinic acid derivatives *Russ. J. Inorg. Chem.* **50** 11 1781-4

[12] Nguyen V T, Lcc J, Kim M, Kim S, Chagnes Λ and Cote G 2017 Sustainable extraction and separation of precious metals from hydrochloric media using novel ionic liquid-in-water microemulsion *Hydrometallurgy* **171** 344-54

[13] Belova V V, Voshkin A A and Khol'kin A I 2011 Liquid-liquid distribution of metals in systems with trioctylmethylammonium dinonyl naphthalenesulfonate *Theor. Found. Chem. Eng.* **45** 5 764-8.

[14] Kostanyan A E, Egorova N S, Voshkin A A and Zonov S S 2008 Extraction of uranyl, ytterbium, and lanthanum nitrates in a three-compartment multiphase extractor *Theor. Found. Chem. Eng.* **42** 5 718-23

[15] Belova V V, Egorova N S, Voshkin A A and Khol'kin A I 2015 Extraction of rare earth metals, uranium, and thorium from nitrate solutions by binary extractants *Theor. Found. Chem. Eng.* **49** 4 545-9

[16] Yi X T, Huo G S and Tang W 2020 Removal of Fe(III) from Ni-Co-Fe chloride solutions using solvent extraction with TBP *Hydrometallurgy* **192** 105265

[17] Wang Y, He Y, Yin S, Long H and Li S 2020 Research on extraction of zinc from spent pickling solution using Aliquat 336 *Hydrometallurgy* **193** 105322

[18] Gupta B, Deep A and Malik P 2001 Extraction and recovery of cadmium using Cyanex 923 *Hydrometallurgy* **61** 1 65-71

[19] Belova V V, Voshkin A A, Kholkin A I and Payrtman A K 2009 Solvent extraction of some lanthanides from chloride and nitrate solutions by binary extractants *Hydrometallurgy* **97** 3-4 198-203

[20] Belova V V, Voshkin A A, Egorova N S and Kholkin A I 2012 Solvent extraction of rare earth metals from nitrate solutions with di(2,4,4-trimethylpentyl)phosphinate of methyltrioctylammonium *J. Mol. Liq.* **172** 144-6

[21] Aghdam, A. A. B., Yoozbashizadeh, H. and Moghaddam, J. Simple separation method of Zn(II) and Cd(II) from brine solution of zinc plant residue and synthetic chloride media using solvent extraction 2020 *Chin. J. Chem. Eng.* **28** 4 1055-61

[22] Spathariotis, S., Peeters, N., Ryder, K. S., Abbott, A. P., Binnemans, K. and Riano, S. 2020 Separation of iron(iii), zinc(ii) and lead(ii) from a choline chloride-ethylene glycol deep eutectic solvent by solvent extraction *RSC Advances* **10** 55 33161-70

[23] Zakhodyaeva Y A, Rudakov D G, Solov'ev V O, Voshkin A A and Timoshenko A V

2019 Liquid–Liquid Equilibrium of Aqueous Two-Phase System Composed of Poly(Ethylene Oxide) 1500 and Sodium Nitrate *J. Chem. Eng. Data* **64** 3 1250-5

[24] Diederich P, Amrhein S, Hämmerling F and Hubbuch J 2013 Evaluation of PEG/phosphate aqueous two-phase systems for the purification of the chicken egg white protein avidin by using high-throughput techniques *Chem. Eng. Sci.* **104** 945-56

[25] Zinov'eva I V, Zakhodyaeva Y A and Voshkin A A 2019 Extraction of monocarboxylic acids from diluted solutions with polyethylene glycol *Theor. Found. Chem. Eng.* **53** 5 871–4

[26] Zinov'eva I V, Zakhodyaeva Y A and Voshkin A A 2020 Data on the extraction of benzoic, salicylic and sulfosalicylic acids from dilute solutions using PEG-based aqueous two-phase systems Data in Brief **28** 105033

[27] Voshkin A A, Zakhodyaeva Y A, Zinov'eva I V and Shkinev V M 2018 Interphase distribution of aromatic acids in the polyethylene glycol-sodium sulfate-water system *Theor. Found. Chem. Eng.* **52** 5 890-3

[28] Zakhodyaeva Y A, Zinov'eva I V and Voshkin A A 2019 Extraction of Iron(III) Chloride Complexes Using the Polypropylene Glycol 425-NaCl-H2O *Theor. Found. Chem. Eng.* **53** 5 735-40

[29] Zakhodyaeva Y A, Zinov'eva I V, Tokar E S and Voshkin A A 2019 Complex Extraction of Metals in an Aqueous Two-Phase System Based on Poly(Ethylene Oxide) 1500 and Sodium Nitrate *Molecules* **24** 22 4078

[30] Fedorova M I, Zakhodyaeva Y A and Voshkin A A 2020 Interphase Distribution of Fe(III) and Zn(II) in Chloride Systems with Aliquat 336 in Polypropylene Glycol 425 *Theor. Found. Chem. Eng.* **54** 3 425–30

[31] Fedorova M I, Zinov'eva I V, Zakhodyaeva Y A and Voshkin A A 2020 Extraction of Fe(III), Zn(II), and Mn(II) Using a System with a Green Solvent for Trioctylmethylammonium Thiocyanate *Theor. Found. Chem. Eng.* **54** 2 313–8

Acknowledgments
The reported study was funded by RFBR and Moscow city Government according to the research project № 19-33-70011.

IOP Conf. Series: Materials Science and Engineering 1212 (2022) 012011 doi:10.1088/1757-899X/1212/1/012011

Liquid-liquid extraction of Pt(IV) from hydrochloric acid solutions using PPG 425 – NaCl – H_2O system

I V Zinov'eva

Kurnakov Institute of General and Inorganic Chemistry of the Russian Academy of Sciences, 31 Leninsky Prospect, Moscow, 119991, Russia

iz@igic.ras.ru

Abstract. Today, metal extraction from e-waste is beneficial from both an environmental and economic point of view. Natural resources of metals, especially platinum group metals, are limited. At the same time, the amount of waste containing many valuable elements continues to grow. In this work, we studied the extraction of Pt (IV) from hydrochloric acid solutions using polypropylene glycol 425 (PPG 425). In the course of the experimental work, the dependence of the platinum extraction degree on the influence of hydrochloric acid concentration and medium pH in the polypropylene glycol 425 - sodium chloride - water system have been established. The maximum recovery (distribution coefficient > 3,5) has been achieved in the presence of 2 M HCl in an aqueous two-phase system (ATPS), which allows the use of the proposed system for the extraction of platinum from leaching solutions.

Keywords: extraction, aqueous two-phase system, platinum, green chemistry, e-waste

1. Introduction

The growing attention to the problem of environmental conditions associated with the continuous increase in the production of electronic equipment, such as computers, cell phones, household appliances, prompted researchers to develop new technologies for the processing of huge amounts of waste electronic and electrical equipment [1]. Essential components of e-waste, particularly in the field of information technology and communication equipment, are non-ferrous metals (Fe, Zn, Al, Mn, etc.) [2], as well as precious metals (Au, Ag, Pt, Pd) [3], and rare elements (In, Sb, Te, Ta, rare earth elements (REE), etc.) [4]. It is interesting to note that electronic waste may be secondary "ores", as they may contain higher concentrations of valuable metals than primary natural resources [5]. The hydrometallurgical method is the most promising one for metal extraction from electronic waste. Traditional liquid extraction has proven its effectiveness in extracting organic acids from aqueous media [6-8], metals from nitrate [9,10], and chloride [11,12] leaching solutions. Belova et al. proposed extraction systems based on binary extractants for the effective extraction of rare-earth elements from aqueous solutions [13-15]. Xu Z-X et al. used imidazolium-based task-specific polymeric ionic liquid (CP-AMIN) for separation of Pt(IV), Pt(II), and Pd(II) and associated metal ions from hydrochloric acid media [16].

In recent years, systems that do not contain any flammable, toxic, or volatile compounds have been developed and used in the extraction process [17,18]. Several works are devoted to the application of such systems in the extraction of biological compounds [19], organic acids [20,21], known work on the production of nanomaterials in aqueous two-phase systems [22,23]. ATPS based on polymers as a new type of environmentally safe extraction systems have been proposed in the extraction and separation technology of rare-earth [24], heavy [25], precious [26], nonferrous [27-29], and other metals [30,31] by many scientific groups.

For example, Bulgariu L and Bulgariu D offered ATPS based on PEG and sulfate salts in the presence of iodide for selective extraction of Hg(II), Cd(II), and Zn(II) ions from the aqueous solution. The efficiency of metal ion extraction under study depends on many factors, including the nature of inorganic salt, the amount of introduced iodine, pH of the initial salt solution, etc [32]. Zakhodyaeva Yu A et al. introduced potassium rhodanide into the PEG 1500 - $(NH_4)_2SO_4$ - H_2O system to separate a mixture of metals contained in the leaching solution of used batteries [33]Water-soluble polymers (polyethylene glycols, polypropylene glycols, copolymers) and inorganic salts are usually used to form aqueous two-phase systems. Polyethylene (propylene) glycols are non-ionic, biodegradable, non-toxic polymers with low volatility and are widely used in the cosmetic, food, pharmaceutical, and chemical industries. Unlike PEG, polypropylene glycol can form ATPS with sodium chloride at room temperature. This system allows for the extraction of metals from acidic chloride solutions. The elements of the platinum group, especially Pt, are important metals that are used in the manufacture of catalysts, technological devices, reactors, jewelry, electronics. In the present work, we studied for the first time the extraction of platinum(IV) from model hydrochloric acid solutions with the use of PPG 425 without introducing additional extracting agents. The results of studies of the effect of hydrochloric acid concentration on Pt^{4+} distribution coefficients are presented.

2. Experimental details

2.1. Reagents and equipment

The following reagents were used in the work: Pt dissolved in 5% HCl (Sigma-Aldrich, St. Louis, MO, USA), hydrochloric acid (chemically pure Chimmed, Moscow, Russia), poly(ethylene glycol) with a molecular weight of 1500 produced by Fluka (Shanghai, China), sodium chloride (chemically pure, Chimmed, Moscow, Russia). Distilled water was used for the preparation of solutions.

The following equipment was used in the work: temperature-controlled shaker (Enviro-Genie SI-1202, Scientific Industries, Inc., Bohemia, NY, USA), analytical balance with an accuracy of 0.0001 g (HR-100AZ, AND Company, Seoul, Korea), centrifuge (CM-6MT, SIA ELMI, Riga, Latvia), Cary-60 spectrophotometer (Agilent Tech., Santa Clara, CA, USA), pH-meter (Starter 5000, OHAUS, Parsipanny, NJ, USA) with a combined STMICRO5 RU glass electrode calibrated against buffers with the pH-values 1.68, 4.01, 7.00, and 10.01 (at 298.15 K).

2.2. Experiment method

All experiments were performed at a temperature of 298.15 K and an atmospheric pressure of ~100kPa. For platinum extraction, PPG 425 (30 wt %) - NaCl (8 wt %) - H_2O system was chosen. Precisely weighed amounts of polymer and salt, as well as necessary amounts of hydrochloric acid and extractable metal, were placed in graduated tubes, dissolved in distilled water, and mixed until a two-phase system was formed. The tubes were then placed in a temperature-controlled shaker and stirred at 45 rpm at a constant temperature for 30 minutes to achieve thermodynamic equilibrium. The samples were centrifuged at 2500 rpm for 10 minutes for complete phase separation. The volumes of the upper and lower phases were fixed, then the pH and metal concentration were determined in the phases. Figure 1 shows the scheme of the extraction experiment.

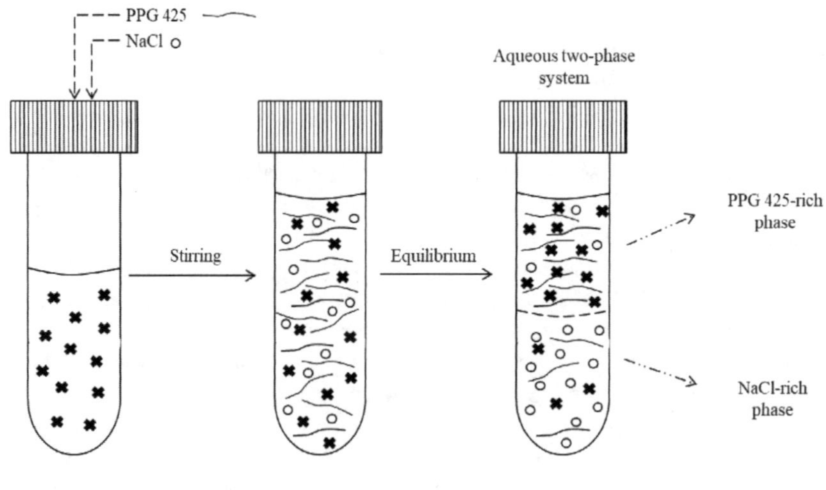

Figure 1. Schematic of the platinum(IV) extraction process in PPG 425 – NaCl – H_2O ATPS.

The concentration of metal in the initial solution and the equilibrium polymer and salt phases after extraction was determined using a spectrophotometric method based on the formation of a platinum complex with tin chloride, which has an absorption band at a wavelength of 403 nm. The error of determination was <5%.

The distribution coefficient (D) of platinum was calculated by the formula (1):

$$D = \frac{[Pt]_{t.ph.}}{[Pt]_{b.ph.}}, (1)$$

where $[Pt]_{t.ph.}$ - concentration of metal in the upper phase, $[Pt]_{b.ph.}$ - is the concentration of metal in the lower phase. All experiments were repeated 3 times each.

3. Results and Discussion

3.1. Interphase distribution of Pt(IV) in PPG 425 – NaCl – H_2O system.

When studying the extraction of platinum in an aqueous two-phase system, the influence of the salt phase acidity on the interphase distribution of the metal under study is of interest. Thus, the influence of equilibrium values of a salt phase pH on the efficiency of platinum(IV) extraction in the (PPG 425 (30 wt %) - NaCl (8 wt %) - H_2O) system in a range of pH 0-2 (figure 2) is studied.

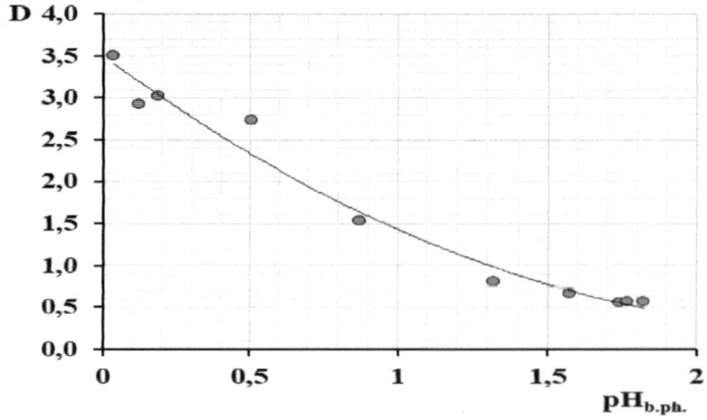

Figure 2. Dependence of Pt(IV) distribution coefficient on equilibrium pH values of the bottom phase in the presence of HCl: $[Pt]_{init.}$=0.0001 mol/L.

From the received extraction data, it is visible, that at a decrease in the value of a salt phase acidity increase in distribution coefficients Pt^{4+} occurs. A decrease in the pH value of the salt phase leads to an increase in hydrophobicity of the polymer phase [34], which in turn contributes to the extraction of hydrophobic anion $PtCl_6^{2-}$. The maximum distribution coefficient (> 3.5) was obtained at pH ~ 0. In hydrometallurgical processes, platinum group metals are contained in hydrochloric acid leaching solutions, where the HCl content may vary depending on the conditions of the process, the initial concentration of metals, and acid. Figure 3 shows the results of the influence of hydrochloric acid initial concentration on the platinum (IV) extraction in the range from 0 to 2 mol/L in the PPG 425 - NaCl - H_2O system.

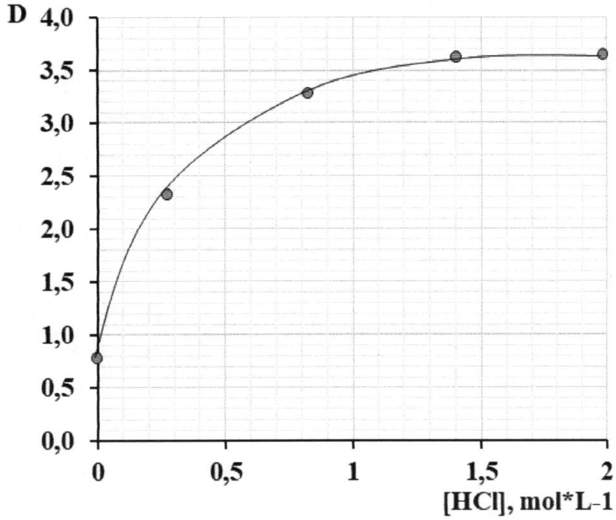

Figure 3. Influence of hydrochloric acid concentration on platinum(IV) extraction in PPG 425 - NaCl - H_2O system: $[Pt]_{init.}$=0.0001 mol/L.

Further increase in acid concentration leads to the homogenization of the extraction system. The figure shows that the maximum distribution coefficient is achieved at HCl concentration ~2 mol/l. A similar effect was observed in Mhaske A A et al. in the work on extraction of this metal, where Cyanex 921 was used as an extractant [35].

Conclusions

Thus, we studied the interphase distribution of Pt(IV) in the aqueous two-phase PPG 425 - NaCl - H_2O system. We have studied the influence of the acidity of the salt phase and the initial concentration of hydrochloric acid on the metal extraction. It is established that a reduction of the medium pH the metal distribution coefficient increases; at acid concentration ~1.5 mol/l the maximum distribution of platinum in the polymer phase is achieved. It has been shown that an aqueous two-phase system based on polypropylene glycol 425 can be used for the effective extraction of Pt ions from chloride solutions.

References

[1] Islam A, Ahmed T, Awual M R, Rahman A, Sultana M, Aziz A A, Monir M U, Teo S H and Hasan M 2020 Advances in sustainable approaches to recover metals from e-waste-A review

J. Clean. Prod. **244** 118815.

[2] Belova V V, Voshkin A A and Khol'kin A. I. 2011 Liquid-liquid distribution of metals in systems with trioctylmethylammonium dinonyl naphthalenesulfonate *Theor. Found. Chem. Eng.* **45** 5 764–8.

[3] Masilela M and Ndlovu S 2019 Extraction of ag and au from chloride electronic waste leach solutions using ionic liquids *J. Environ. Chem. Eng.* **7** 1 102810.

[4] Kostanyan A E, Egorova N S, Voshkin A A and Zonov S S 2008 Extraction of uranyl, ytterbium, and lanthanum nitrates in a three-compartment multiphase extractor. *Theor. Found. Chem. Eng.* **42** 5 718–23.

[5] Schaeffer N, Passos H, Billard I, Papaiconomou N and Coutinho J A P 2018 Recovery of metals from waste electrical and electronic equipment (WEEE) using unconventional solvents based on ionic liquids *Crit. Rev. Environ. Sci. Technol.* **48** 13-15 859-922.

[6] Belova V V, Kulichenkov S A, Voshkin A A, Khol'kin A I, Kuvaeva Z I and Soldatov V S 2007 Extraction of mineral acids with methyltrioctylammonium dinonylnaphthalenesulfonate *Russ. J. Inorg. Chem.* **52** 3 460–4.

[7] Zakhodyaeva Y A, Voshkin A A, Belova V V and Khol'kin A I 2012 Extraction of monocarboxylic acids by trioctylmethylammonium di(2-ethylhexyl)phosphate *Theor. Found. Chem. Eng.* **46** 4 413–8.

[8] Zakhodyaeva Y A and Voshkin A A 2013 Extraction and separation of carboxylic acids by liquid-liquid extraction and chromatography *Theor. Found. Chem. Eng.* **47** 4 461–6.

[9] Belova V V, Voshkin A A, Kholkin A I and Payrtman A K 2009 Solvent extraction of some lanthanides from chloride and nitrate solutions by binary extractants *Hydrometallurgy* **97** 3-4 198–203.

[10] Ahmed I M, Nayl A A and Daoud J A 2011 Extraction of palladium from nitrate solution by CYANEX 471X *Int. J. Miner. Process.* **101** 1-4 89–93.

[11] Egorova N S, Belova V V, Voshkin A A, Zhilov V I and Khol'kin A I 2005 Extraction of lanthanide chlorides by binary extractants based on phosphinic acid derivatives *Russ. J. Inorg. Chem* **50** 11 1781-4.

[12] Preston J S and du Preez A C 2002 Solvent extraction of platinum-group metals from hydrochloric acid solutions by dialkyl sulphoxides *Solvent Extr. Ion Exch.* **20** 3 359–74.

[13] Belova V V, Voshkin A A, Egorova N S and Kholkin A I 2012 Solvent extraction of rare earth metals from nitrate solutions with di(2,4,4-trimethylpentyl)phosphinate of methyltrioctylammonium *J. Mol. Liq.* **172** 144–6.

[14] Belova V V, Voshkin A A, Egorova N S and Khol'kin A I 2010 Extraction of rare earth metals from nitrate solutions with a binary extractant based on Cyanex 272 *Russ. J. Inorg. Chem.* **55** 4 629–33.

[15] Belova V V, Egorova N S, Voshkin A A and Khol'kin A I 2015 Extraction of rare earth metals, uranium, and thorium from nitrate solutions by binary extractants *Theor. Found. Chem. Eng.* **49** 4 545–9.

[16] Xu Z X, Zhao Y L, Wang P Y, Yan X Q, Cai M M and Yang Y 2019 Extraction of Pt(IV), Pt(II) and Pd(II) from acidic chloride media using imidazolium-based task-specific polymeric ionic liquid *Ind. Eng. Chem. Res.* **58** 5 1779–86.

[17] Zakhodyaeva Y A, Rudakov D G, Solov'ev V O, Voshkin A A and Timoshenko A V 2019 Liquid–Liquid Equilibrium of Aqueous Two-Phase System Composed of Poly(Ethylene Oxide) 1500 and Sodium Nitrate *J. Chem. Eng. Data* **64** 3 1250-5.

[18] Zakhodyaeva Y A, Rudakov D G, Solov'ev V O, Voshkin A A and Timoshenko A V 2019 Liquid–Liquid Equilibrium in an Extraction System Based on Polyvinylpyrrolidone-3500 and Sodium Nitrate *Theor. Found. Chem. Eng.* **53** 2 159–65.

[19] Desai R K, Streefland M, Wijffels R H and Eppink M H M 2014 Extraction and stability of selected proteins in ionic liquid based aqueous two phase systems *Green Chem.* **16** 5 2670–9

[20] Zinov'eva I V, Zakhodyaeva Y A and Voshkin A A 2019 Extraction of Monocarboxylic Acids

from Diluted Solutions with Polyethylene Glycol *Theor. Found. Chem. Eng.* **53** 5 871–4.

[21] Voshkin A A, Zakhodyaeva Y A, Zinov'eva I V and Shkinev V M 2018 Interphase Distribution of Aromatic Acids in the Polyethylene Glycol–Sodium Sulfate–Water System *Theor. Found. Chem. Eng.* **52** 5 890–3.

[22] Kinhal K V, Sinha S, Ravisankar A, Bhatt N P and Pushpavanam S 2020 Simultaneous synthesis and separation of nanoparticles using aqueous two phase systems *ACS Sustainable Chem. Eng.* **8** 7 3013-25.

[23] Shkinev V M, Zakhodyaeva Y A, Dzhenloda R K, Mokhodoeva O B and Voshkin A A 2017 Synthesis of magnetic iron oxide nanoparticles at the interface of the polyethylene glycol–ammonium sulfate–water extraction system *Mendeleev Commun.* **27** 5 485–6.

[24] de Oliveira W C M, Rodrigues G D, Mageste A B and de Lemos L R 2017 Green selective recovery of lanthanum from Ni-MH battery leachate using aqueous two-phase systems *Chem. Eng. J.* **322** 346–52.

[25] Hamta A and Dehghani M R 2017 Application of polyethylene glycol based aqueous two-phase systems for extraction of heavy metals *J. Mol. Liq.* **231** 20–4.

[26] Bulgariu L and Bulgariu D 2011 Extraction of gold(III) from chloride media in aqueous polyethylene glycol-based two-phase system *Sep. Purif. Technol.* **80** 3 620–5.

[27] Zakhodyaeva Y A, Zinov'eva I V, Tokar E S and Voshkin A A 2019 Complex Extraction of Metals in an Aqueous Two-Phase System Based on Poly(Ethylene Oxide) 1500 and Sodium Nitrate *Molecules* **24** 22 4078.

[28] Fedorova M I, Zakhodyaeva Y A and Voshkin A A 2020 Interphase Distribution of Fe(III) and Zn(II) in Chloride Systems with Aliquat 336 in Polypropylene Glycol 425 *Theor. Found. Chem. Eng.* **54** 3 425–30.

[29] Fedorova M I, Zinov'eva I V, Zakhodyaeva Y A and Voshkin A A 2020 Extraction of Fe(III), Zn(II), and Mn(II) Using a System with a Green Solvent for Trioctylmethylammonium Thiocyanate *Theor. Found. Chem. Eng.* **54** 2 313–8.

[30] Pan Y, Sun X, Qi M, Qin R, Che X and Zhang Y 2020 A clean and efficient method for separation of vanadium and molybdenum by aqueous two-phase systems *J. Mol. Liq.* 113540.

[31] Fedorova M I, Levina A V, Zakhodyaeva Y A, Voshkin A A 2020 Interphase Distribution of V(IV) in the Polyethylene Glycol 1500–Sodium Nitrate–Water System *Theor. Found. Chem. Eng.* **54** 4 604–9.

[32] Bulgariu L and Bulgariu D 2013 Selective extraction of Hg(II), Cd(II) and Zn(II) ions from aqueous media by a green chemistry procedure using aqueous two-phase systems *Sep. Purif. Technol.* **118** 209–16.

[33] Zakhodyaeva Y A, Izyumova K V, Solov'eva M S and Voshkin A A 2017 Extraction separation of the components of leach liquors of batteries *Theor. Found. Chem. Eng.* **51** 5 883–7.

[34] Zakhodyaeva Y A, Zinov'eva I V and Voshkin A A 2019 Extraction of iron(III) chloride complexes using the polypropylene glycol 425–NaCl–H2O system *Theor. Found. Chem. Eng.* **53** 5 735-40.

[35] Mhaske A A and Dhadke P M 2001 Extraction separation studies of Rh, Pt and Pd using Cyanex 921 in toluene—a possible application to recovery from spent catalysts *Hydrometallurgy* **61** 2 143–50.

Acknowledgments
The reported study was funded by RFBR according to the research project № 18-29-24170.

Extraction of Eu(III), Gd(III), and Tb(III) in aqueous two-phase systems based on polyethylene glycol 1500-NaNO3-H2O with the addition of extractants (D2EHPA, TBP, TOMAN)

A Ya Fedorov[1*], A V Levina[1], M I Fedorova[1]

1 Kurnakov Institute of General and Inorganic Chemistry of the Russian Academy of Sciences, Leninsky prospect, 31, 119991, Moscow, Russia

*Corresponding author: alfedwork@gmail.com

Abstract. Luminophores that used, for example, in fluorescent lamps, contain a large number of rare earth elements. Therefore, the processing of waste equipment containing luminophores is a rational approach to the obtaining of rare-earth metals, firstly, from the economics point of view, since they have a high cost, and secondly, from the ecological point of view, since environmental pollution will not occur. The cheapest way to extract rare earth elements from waste products is to dissolve them in strong acids and the following reprocessing by liquid extraction methods. In this case, neutral or ion exchange extractants (tributyl phosphate, di(2-ethylhexyl)phosphoric acid and quaternary ammonium salts) are used, which show high extraction ability and, in some cases, selectivity. Their applying is associated with the use of non-polar organic solvents, which contradicts the principles of «green» chemistry. A good and promising alternative to organic solvents can be aqueous two-phase systems, which have already proved themselves as low-toxic, but highly effective systems for the separation of a number of metals. Thus, in this work, we performed an experimental study of the interphase distribution of Eu(III), Gd(III), and Tb(III) in two-phase systems based on water-soluble polymers with or without the introduction of traditional organic extractants as an additive. The possibility of using such ATPS as a «green» solvent for traditional extragents for Eu(III), Gd(III), and Tb(III) extraction has been shown.

Key words: luminophores, liquid-liquid extraction, «green» chemistry, aqueous two-phase systems, waste, rare earth elements.

1. Introduction

Rare earth elements (REEs) are widely used, they can be found in smartphones and PCs, in energy-saving lamps and displays, in permanent magnets and electric motors, in radars and lasers [1]. Since the demand for REE is growing, their production is also growing, as a result, prices for this already expensive raw material are growing. Consequently, the demand for alternative sources of these elements is increasing. This is e-waste, the increase in which is directly proportional to the growth in demand for electronic equipment. However, at the moment, only about 1% of the used REE is returned from secondary raw materials, which is explained by the difficulties associated with waste collection [2].

Waste fluorescent lamps, which contain luminophores, are considered the most economical light sources and are promising for processing in order to extract REE. Luminophores that include rare-earth elements in their composition can be divided into five main types: red luminophores - $Y_2O_3:Eu^{3+}$, green luminophores - $LaPO_4$: Ce^{3+}, Tb^{3+}, blue luminophores $BaMgAl_{10}O_{17}$: Eu^{2+}, $GdMgB_5O_{10}$: Ce^{3+}, $TbMgAl_{11}O_{19}$ [3]. Of interest is the fact that the listed luminophores often contain more REE than ores.

The traditional method based on the leaching of phosphor scrap using mineral acids (sulfuric, nitric, hydrochloric and their mixtures) with subsequent extraction and separation of REEs [4]. Researchers Tunsu and others [5] leached Y and Eu with a mixture of hydrochloric and nitric acids (within 24 hours), De Carolis and others [6] leached them with HCl (6 $mol{\cdot}L^{-1}$) for a day, Takahashi and others [7] succeeded in doing this using 1.5 $mol{\cdot}L^{-1}$ H_2SO_4 at 70°C per one hour.

Organophosphate acids are effective extractants of REE, among which D2EHPA is of the greatest practical application [8]. The extraction proceeds according to the cation-exchange mechanism and is characterized by high distribution coefficients. The work [9] studied the extraction of Y, Dy, and Nd.

Salts of quaternary ammonium bases (QAB) are extracted with REE by an anion exchange mechanism [10]. They have the highest selectivity among primary, secondary, tertiary amines, and make it possible to extract REE not only from nitrate, but also from thiocyanate solutions [11]. QAB salts can effectively extract REE from solutions of mineral acids (in the form of synergistic mixtures with other extractants) and from weakly acidic or neutral solutions (QAB salts with inorganic anions) [12].

The listed methods imply the use of strong acids, which entails strong economic losses due to the need for neutralization of reagents and disposal of waste water [13]. In addition, organic solvents are often used, which are not only toxic and environmentally unsafe, but also flammable and explosive [14]. Also, often, traditional extraction systems have low selectivity, which reduces the efficiency of the entire process [15].

These problems can be avoided by following the principles of "green" chemistry and making a partial or full transition to "green" solvents and reagents [16]. In addition, many modern approaches allow using and combining existing ineffective methods, increasing their effectiveness. A striking example is the hybrid method of liquid-liquid chromatography - this is one of the most promising methods for separating organic and inorganic substances, combining the ideology of liquid extraction and high efficiency of chromatography [17]. At present, using this method, the processes of separation of rare-earth metals and a number of other metals [18] have been implemented, both with the use of column equipment and industrial extractors [19]. Another promising approach to replacing organic solvents with "green" reagents is the use of deep eutectic solvents (DES) [20]. Being a eutectic mixture of solids (Lewis and Bronsted acids and bases) linked by hydrogen bonds, DES are most often liquid at room temperature and show high efficiency for the extraction of rare earth metals and their compounds from various media [21]. Another approach is the use of aqueous two-phase systems, including those based on water-soluble polymers [22]. These heterogeneous systems are already used for the separation and extraction of organic [23, 24] and inorganic compounds, including metals [25, 26]. Replacing the organic solvent with non-toxic polymers will have a positive effect in the environmental aspect, as well as change the selectivity of the processes. In addition, in studies [27, 28] it was found that the use of salts of quaternary ammonium bases has a positive effect on the extraction efficiency, and in some cases contributes to the selective extraction of certain metals [29]. On the contrary, the use of aqueous two-phase systems is not a complete solution to the problem of extraction metals, since the system can have low selectivity and simultaneously extract a number of metals [30]. The initial solution medium is also important, for example, sulfate [31] or chloride. In addition, attention is paid to liquid-liquid equilibria of aqueous two-phase systems, which can subsequently be used for extraction processes [32].

In this regard, the aim of this study was the replacement of organic solvents with aqueous two-phase systems, in the processes of extraction processing of acidic solutions of leaching of phosphors. Promising luminophores containing Eu, Gd, and Tb were chosen as model solutions.

2. Experimental

The tools were used in this work: 50 ml measuring beakers, measuring pipettes (from 10μ to 5ml), 20 ml separation funnels, 15 ml graduated plastic tubes, drop tube, analytical Balance (AND HR-100AZ), Magnetic stirrer (IKA C-MAG HS4) with stirrer bar, centrifuge (ELMI CM-6MT), desalinated filter paper, Enviro-Genie thermostatically controlled shaker (Scientific Industries, Inc.).

Reagents were used in this research: sodium nitrate ($NaNO_3$, distilled water, poly(ethylene glycol) 1500 (PEG-1500), Trioctylmethylammonium chloride (Aliquat 336), bis(2-Ethylhexyl)hydrogen phosphate (D2EHPA), Tributyl phosphate (TBP), Ethylenediamine tetraacetate sodium (EDTA-Na_2), Europium nitrate hexahydrate ($Eu(NO_3)_3 \cdot 6H_2O$), Gadolinium nitrate hexahydrate ($Gd(NO_3)_3 \cdot 6H_2O$), Terbium nitrate hexahydrate ($Tb(NO_3)_3 \cdot 6H_2O$), xylenol orange, nitric acid (HNO_3).

The initial solutions of metal nitrates were obtained by dissolving the sample weights of $Eu(NO_3)_3 \cdot 6H_2O$, $Gd(NO_3)_3 \cdot 6H_2O$ and $Tb(NO_3)_3 \cdot 6H_2O$ of the qualification "chemical grade" suspended on an analytical balance (AND HR-100AZ) in distilled water. Trioctylmethylammonium nitrate (TOMAN) was prepared as follows: 1.6 mol·L-1 sodium nitrate solution was mixed with Aliquat 336 in a 1:1 ratio [33] Stirring was performed until the chloride ions were completely replaced by nitrate ions. The mixing time was 20 minutes.

The metal ions extraction was carried out using nitrate two-phase aqueous systems, based on polyethylene glycol 1500 and with an initial metal concentration of $0.01 mol \cdot L^{-1}$. Phase composition were chosen based on phase diagrams, PEG-1500 - $NaNO_3$ [34]. The acid concentration in the salt phase was equal $0.01 mol \cdot L^{-1}$ (for the hydrolyze avoiding).

The metal ions Eu(III), Gd(III) and Tb(III) extraction was carried out at a temperature of 25°C in graduated plastic tubes in an Enviro-Genie thermostatically controlled shaker (Scientific Industries, Inc.) at a rotation speed of 30 rpm for it was chosen as optimal time for the effective extraction of both metals.

The complexometric titration using EDTA and xylenol orange at pH=6 was used to determine the metal ions concentration in the initial solutions and in the aqueous phases after extraction.

The presented experimental data are the result of a series of experiments and processed by methods of mathematical statistics.

3. Results and discussion

We carried out a series of experiments aimed at studying the interphase distribution of metal ions Eu(III), Gd(III), Tb(III) in various systems PEG-1500 (16.3 wt.%) - $NaNO_3$ (36 wt.%) - H_2O, PEG -1500 (16.3 wt.%) - $NaNO_3$ (36 wt.%) - H_2O with the addition of TOMAN, PEG-1500 (16.3 wt.%) - $NaNO_3$ (36 wt.%) - H_2O with the addition of TBP, PEG-1500 (16.3 wt%) - $NaNO_3$ (36 wt.%) - H_2O with the addition of TBP and D2EHPA in the ratios of 1:4, 2:3, and 4:1. The addition of TOMAN, TBP, and D2EHPA was motivated by the fact that the introduction of traditional extractants into the system intensifies the extraction process in the internal combustion engine. As a result, quantitative characteristics of extraction were obtained, such as distribution coefficients (Fig.1), degrees of extraction (Fig.2) and the number of theoretical stages (Fig.3) of extraction for metal ions Eu(III), Gd(III) and Tb(III). The number of theoretical stages of extraction was calculated by the formula (1):

$$n = \frac{\lg\left[q\left(1-\frac{\lambda}{D}\right)\right]}{\lg\left(\frac{D}{\lambda}\right)}, \tag{1}$$

where are n the number of theoretical extraction steps; q is the ratio X_o / X_n; X_o is the initial concentration of the metal introduced into the system, $mol \cdot L^{-1}$; X_n is the final specified metal concentration, $mol \cdot L^{-1}$; λ is the ratio V_n / V_w; V_n - volume of the lower vase, ml; V_b - volume of the upper phase, ml; D - distribution coefficient. When using this formula, the following assumptions were made:

- distribution coefficient D = const;
- the volumes of the phases Vn and Vw are constant, therefore, their ratio (λ = const);
- the original extractant does not contain the substance to be distributed.

With the initial metal concentration Xo = 0.01 mol·L-1, the final metal concentration Xn = 1·10-6 mol·L-1, therefore, q = 10000, thus, the extraction of metal ions will occur with a degree of 99.99%.

As can be seen in Figures 1-3, with the addition of TOMAN (Fig. 1-3, b), in comparison with the first experiment (Fig. 1-3, a), the distribution coefficients increased: for Eu(III) by 1.78, Gd(III) by 1.15 and Tb(III) by 1.65. The number of theoretical stages of extraction has noticeably decreased (by about

2 times). Consequently, such an extraction system with the addition of TOMAN can be used in the extraction of Eu(III), Gd(III), and Tb(III), but is impractical due to low distribution coefficients.

With the addition of TBP (Fig. 1-3, c) to the upper phase, in comparison with the first experiment (Fig. 1-3, a), the distribution coefficients increased: for Eu(III) by 1.12, Gd(III) by 2.74 and Tb(III) at 1.96. The number of theoretical stages of extraction, as well as in the system with the addition of TOMAN, decreased by about 2 times. This extraction system can be used in the extraction of Eu(III), Gd(III) and Tb(III), however, it is impractical due to low distribution coefficients.

It is known from the literature [35] that the combined use of TBP and D2EHPA gives rise to a synergistic effect, and the extraction of Nd(III) from acidic media, the distribution coefficients increased several times. It was decided to conduct a series of experiments to find the most effective ratio of TBP and D2EHPA for extraction (Fig. 1-3, d, f, e). The synergistic effect was calculated by the formula (2):

$$S_k = \frac{D}{D_1+D_2}, \tag{2}$$

where are D the distribution coefficient for extraction with a mixture of extractants; D_1, D_2 - distribution coefficients for extraction with each extractant separately.

From the data obtained (Figs. 1-3, d, e, f), a synergistic effect is observed only in the case of an extraction system with the addition of TBP and D2EHPA in a ratio of 1: 4 and only in the case of Eu(III): $Sk(Eu)$ = 1.99, that is, this system can improve the extraction by almost two times. The distribution coefficients are high in all cases, which makes it possible to solve the problem of element extraction. Accordingly, such extraction systems with the addition of TBP and D2EHPA are excellent for use in the extraction of Eu(III), Gd(III) and Tb(III).

The system PEG-1500 (16.3 wt.%) - $NaNO_3$ (36 wt.%) - H_2O + TBP:D2EHPA (1:4) is the most effective for separating Eu(III) from its mixture with Gd (III) and Tb (III). We have calculated the separation coefficients based on experimental data by the formula (3):

$$\beta_{Eu/Me} = \frac{D_{Eu}}{D_{Me}}, \tag{3}$$

where, $\beta_{Eu/Me}$ – is the separation coefficient of Eu(III) in relation to another metal; D_{Eu} – is the distribution coefficient of Eu(III); D_{Me} – is the distribution coefficient of Eu(III).

Thus, $\beta_{Eu/Gd}$=2.97 and $\beta_{Eu/Tb}$=3.36 are high values for the separation coefficients for metal [36], which suggests that this system is excellent for solving the problem of selective extraction of Eu(III) from this mixture of metals.

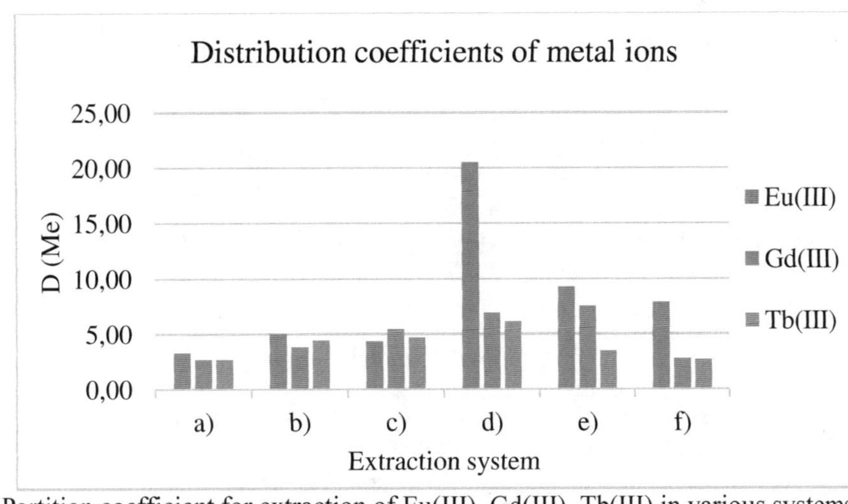

Fig. 1. Partition coefficient for extraction of Eu(III), Gd(III), Tb(III) in various systems a) PEG-1500 (16.3 wt.%) - $NaNO_3$ (36 wt.%) -H_2O; b) PEG-1500 (16.3 wt.%) - $NaNO_3$ (36 wt.%) - H_2O + TOMAN; c) PEG-1500 (16.3 wt.%) - $NaNO_3$ (36 wt.%) - H_2O + TBP; d) PEG-1500 (16.3 wt.%) -

$NaNO_3$ (36 wt.%) - H_2O + TBP:D2EHPA (1:4); f) PEG-1500 (16.3 wt.%) - $NaNO_3$ (36 wt.%) - H_2O + TBP:D2EHPA (2:3); e) PEG-1500 (16.3 wt.%) - $NaNO_3$ (36 wt.%) - H_2O + TBP:D2EHPA (4:1).

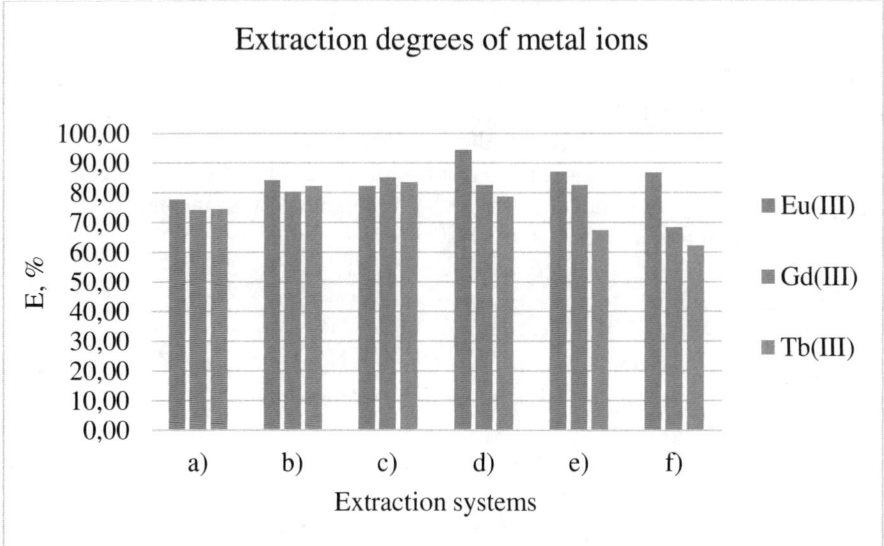

Fig. 2. Extraction degree of Eu (III), Gd (III), Tb (III) in different systems a) PEG-1500 (16.3 wt.%) - $NaNO_3$ (36 wt.%) -H_2O; b) PEG-1500 (16.3 wt.%) - $NaNO_3$ (36 wt.%) - H_2O + TOMAN; c) PEG-1500 (16.3 wt.%) - $NaNO_3$ (36 wt.%) - H_2O + TBP; d) PEG-1500 (16.3 wt.%) - $NaNO_3$ (36 wt.%) - H_2O + TBP:D2EHPA (1:4); f) PEG-1500 (16.3 wt.%) - $NaNO_3$ (36 wt.%) - H_2O + TBP:D2EHPA (2:3); e) PEG-1500 (16.3 wt.%) - $NaNO_3$ (36 wt.%) - H_2O + TBP:D2EHPA (4:1).

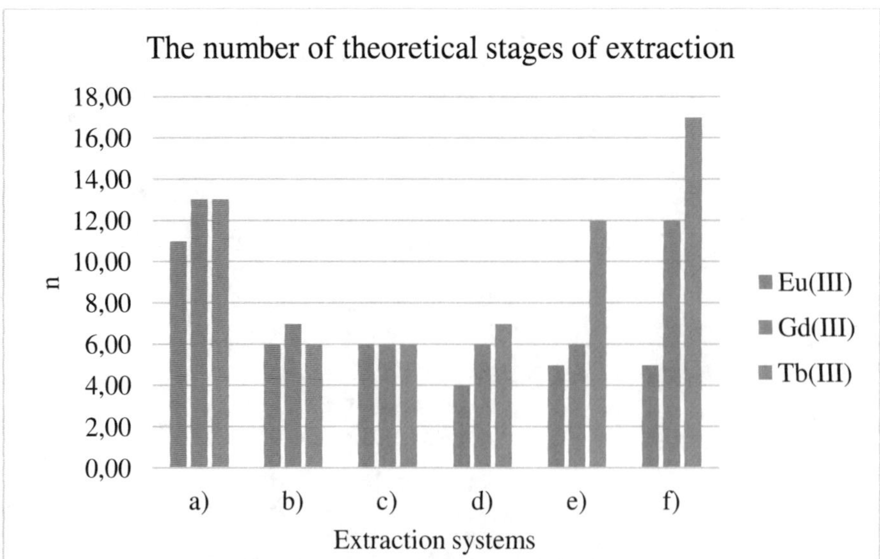

Fig. 3. The number of theoretical stages of extraction of Eu (III), Gd (III), Tb (III) in various systems a) PEG-1500 (16.3 wt.%) - $NaNO_3$ (36 wt.%) -H_2O; b) PEG-1500 (16.3 wt.%) - $NaNO_3$ (36 wt.%) - H_2O + TOMAN; c) PEG-1500 (16.3 wt.%) - $NaNO_3$ (36 wt.%) - H_2O + TBP; d) PEG-1500 (16.3 wt.%) - $NaNO_3$ (36 wt.%) - H_2O + TBP:D2EHPA (1:4); f) PEG-1500 (16.3 wt.%) - $NaNO_3$ (36 wt.%) - H_2O + TBP:D2EHPA (2:3); e) PEG-1500 (16.3 wt.%) - $NaNO_3$ (36 wt.%) - H_2O + TBP:D2EHPA (4:1).

4. Conclusions

The extraction system based on PEG-1500 (16.3 wt.%) - $NaNO_3$ (36 wt.%) - H_2O with the addition of TBP and D2EHPA in ratios of 1:4, 2:3, and 4:1 is more efficient for the extraction of Eu(III), Gd(III) and Tb(III) compared to systems without the addition of this mixture of extractants or with the addition of TOMAN or TBP. It was found that in the system with the addition of TBP and D2EHPA in a ratio of 1:4, Eu(III) is extracted with high efficiency.

Acknowledgments

This work was supported by IGIC RAS state assignment.

References

[1] Gupta C K and Krishnamurthy N 2016 Extractive metallurgy of rare earths *CRC Press* 522

[2] Omodara L, Pitkaaho S, Turpeinen E M, Saavalainen P, Oravisjärvi K and Keiski R L 2019 Recycling and substitution of light rare earth elements, cerium, lanthanum, neodymium, and praseodymium from end-of-life applications - A review *J. Clean. Prod.* **236** 1 117573

[3] Belova V V, Voshkin A A, Egorova N S and Kholkin A I 2012 Solvent extraction of rare earth metals from nitrate solutions with di(2,4,4-trimethylpentyl)phosphinate of methyltrioctylammonium *J. Mol. Liq.* **172** 144-6

[4] Binnemans K and Jones PT 2014 Perspectives for the recovery of rare earths from end-of-life fluorescent lamps *J. Rare Earths* **32** 3 195-200.

[5] Tunsu C, Ekberg C and Retegan T 2014 Effect of solution composition on the optimum redox potential for chalcopyrite leaching in sulfuric acid solutions *Hydrometallurgy* **91** 144-9

[6] De Carolis R, Fontana D, Pietrantonio M, Pucciarmati S and Torelli G N 2015 A hydrometallurgical process for recovering rare earths and metals from spent fluorescent lamps *Environ. Eng. Manag. J.* **14** 1603-10

[7] Takahashi T, Takano A, Saitoh T, Nagano N, Hirai S and Shimakage K 2001 Separation and recovery of rare earth elements from phosphor sludge in processing plant of waste fluorescent lamp by pneumatic classification and sulfuric acidic leaching *Shigen to Sozai* **117** 579-85

[8] Поляков Е Г 2018 *Металлургия редкоземельных металлов* ред. Нечаев А В и Смирнов А В (Москва: Металлургиздат ЗАО)

[9] Mohammadi M, Forsberg K, Kloo L, De La Cruz J M and Rasmuson A 2015 Separation of Nd(III), Dy(III) and Y(III) by solvent extraction using D2EHPA and EHEHPA *Hydrometallurgy* **156** 215-24

[10] Zakhodyaeva Y A, Voshkin A A, Belova V V and Khol'kin A I 2012 Extraction of monocarboxylic acids by trioctylmethylammonium di(2-ethylhexyl)phosphate *Theor. Found. Chem. Eng.* **46** 413-8

[11] Степанов С И 2004 *Экстракция редких металлов солями четвертичных аммониевых оснований* ред. Степанов СИ и Чекмарев АМ (Москва: ИздАТ)

[12] Пяртман А К, Копырин А А, Пузиков Е А и Богатов К Б 1996 Экстракция нитратов редкоземельных металлов смесями три-н-бутилфосфата и нитрата триметилалкиламмония из водно-солевых растворов *ЖНХ* **41** 2 347-51

[13] Belova V V, Voshkin A A, Egorova N S and Khol'kin A I 2010 Extraction of rare earth metals from nitrate solutions with a binary extractant based on Cyanex 272 *Russ. J. Inorg. Chem.* **5** 629–33

[14] Fedorova M I, Zinov'eva I V, Zakhodyaeva Y A and Voshkin A A 2020 Extraction of Fe(III), Zn(II), and Mn(II) Using a System with a Green Solvent for Trioctylmethylammonium Thiocyanate *Theor. Found. Chem. Eng.* **54** 313-8

[15] Zakhodyaeva Y A, Izyumova K V, Solov'eva M S and Voshkin A A 2017 Extraction separation of the components of leach liquors of batteries *Theor. Found. Chem. Eng.* **51** 5 883-7
[16] Anastas P and Warner J 1998 *Green Chemistry: Theory and Practice* (Oxford: Oxford University Press)
[17] Kostanyan A E, Voshkin A A and Kodin N V 2011 Pulsed cyclic device for liquid countercurrent chromatography *Theor. Found. Chem. Eng.* **45** 779
[18] Kostanian A E and Voshkin A A 2007 Analysis of new counter-current chromatography operating modes *J. Chromatogr. A* **1151** 126-30
[19] Kostanyan A E and Voshkin A A 2009 Pulsation cyclic liquid-liquid chromatography *Theor. Found. Chem. Eng.* **43** 729-33
[20] Abbott A, Capper G, Davies D, Rasheed R and Tambyrajah V 2003 Novel solvent properties of choline chloride/urea mixtures *Chem. Commun.* 70-1
[21] Liu F, Porvali A, Wang J L, Wang H, Peng C, Wilson B P and Lundstrom M 2020 Recovery and separation of rare earths and boron from spent Nd-Fe-B magnets *Miner. Eng.* **145** 106097
[22] Voshkin A A, Zakhodyaeva Y A, Zinov'eva I V and Shkinev V M 2018 Interphase Distribution of Aromatic Acids in the Polyethylene Glycol–Sodium Sulfate–Water System *Theor. Found. Chem. Eng.* **52** 890-3
[23] Shkinev V M, Zakhodyaeva Y A, Dzhenloda R Kh, Mokhodoeva O B and Voshkin A A 2017 Synthesis of magnetic iron oxide nanoparticles at the interface of the polyethylene glycol–ammonium sulfate–water extraction system *Mendeleev Commun.* **27** 485-6
[24] Zakhodyacva Y A, Solov'ev V O, Zinov'eva I V, Rudakov D G, Timoshenko A V and Voshkin A A 2019 Interphase Distribution of Thiophene, Toluene, and o-Xylene in the Hexane–Polymer–Water Extraction System *Theor. Found. Chem. Eng.* **53** 4 550-5
[25] Belova V V,Voshkin A A, Kholkin A I and Payrtman A K 2009 Solvent extraction of some lanthanides from chloride and nitrate solutions by binary extractants *Hydrometallurgy* **97** 198–203
[26] Voshkin A A, Shkinev V M and Zakhodyaeva Y A 2017 A new extraction method for the preparation of zinc oxide nanoparticles in aqueous two-phase systems *Russ. J. Phys. Chem. A* **91** 226–8
[27] Belova V V, Voshkin A A and Khol'kin A I 2011 Liquid-liquid distribution of metals in systems with trioctylmethylammonium dinonyl naphthalenesulfonate *Theor. Found. Chem. Eng.* **45** 764
[28] Fedorova M I, Zakhodyaeva Y A and Voshkin A A 2020 Interphase Distribution of Fe(III) and Zn(II) in Cloride Systems with Aliquat 336 in Polypropylene Glycol 425 *Theor. Found. Chem. Eng.* **54** 425-30
[29] Voshkin A A, Belova V V and Khol'kin A I 2003 Extraction of iron(III) by binary extractants based on quaternary ammonium bases and organic acids *Russ. J. Inorg. Chem.* **48** 4 608-13
[30] Zakhodyaeva Y A, Zinovyeva I V, Tokar E S and Voshkin A A 2019 Complex extraction of metals in an aqueous two-phase system based on poly(ethylene oxide) 1500 and sodium nitrate *Molecules* **24** 4078
[31] Zinov'eva I V, Zakhodyaeva Y A and Voshkin A A 2020 Data on the extraction of benzoic, salicylic and sulfosalicylic acids from dilute solutions using PEG-based aqueous two-phase systems *Data in Brief* **28** 105033
[32] Zakhodyaeva Y A, Rudakov D G, Solov'ev V O, Voshkin A A and Timoshenko A V 2019 Liquid–Liquid Equilibrium in an Extraction System Based on Polyvinylpyrrolidone-3500 and Sodium Nitrate *Theor. Found. Chem. En.* **53** 2 159-65
[33] Nayl A 2010 Extraction and separation of Co(II) and Ni(II) from acidic sulfate solutions using Aliquat 336 *J. Hazard. Mater.* **173** 223-30
[34] Zakhodyaeva Y A, Rudakov D G, Solov`ev V O, Voshkin A A and Timoshenko A V 2019 Liquid-liquid Equilibrium of Aqueous Two-Phase System Composed of Poly(Ethylene Oxide) 1500 and Sodium Nitrate *J. Chem. Eng. Data.* **64** 3 1250-5

[35] Дошарова Д Т и Сальникова Е В 2016 Влияние смеси экстрагентов трибутилфосфата и ди-(2-этилгексил)фосфорной кислоты на извлечение неодима из кислых сред *Изв. вузов. Прикладная химия и биотехнология.* **6** 4 86-91

[36] Valadares A, Valadares C F, Rodrigues de Lemos L, Mageste A B and Rodrigues G D 2018 Separation of cobalt and nickel in leach solutions of spent nickel-metal hydride batteries using aqueous two-phase systems (ATPS) *Hydrometallurgy* **181** 180-8

Vanadium(IV) ions extraction in the aqueous two-phase system based on poly(ethylene glycol)

A V Levina*, M I Fedorova

Kurnakov Institute of General and Inorganic Chemistry of the Russian Academy of Sciences, Leninsky prospect, 31, 119991, Moscow, Russia

*Corresponding author: lev@igic.ras.ru

Abstract. Vanadium has found wide applications in various industries: metallurgy, electronics, batteries, and others. This element, being dispersed, does not form its minerals and, therefore, is mined along the way. An alternative approach of vanadium extraction is to separate it from heavy oil, by flushing the oil with strong acids solutions and further hydrometallurgical processing. The existing methods of oil dehydration solutions processing for the vanadium extraction are not perfect - they do not meet the "green" chemistry principles. Thus, we investigated V(IV) ions extraction in the aqueous two-phase system (ATPS) based on poly(ethylene glycol) 1500 (16.3wt%) – $NaNO_3$ (36wt%) – H_2O. The dependences of V(IV) extraction efficiency on phases time contact, and dependences V(IV) distribution coefficients on salt phase acidity have been obtained. This system is shown to be a prospective solution for the vanadium(IV) recovery from acidic waste steams problem as it is possible to extract vanadium with an efficiency of more than 81.6% per one extraction step (the distribution coefficient of vanadium (IV) was 4.84).

Keywords: liquid extraction, aqueous two-phase systems, “green” chemistry, vanadium, vanadyl-ion

1. Introduction

Vanadium, due to the peculiarities of its electronic structure, has a wide range of compounds, exhibiting valency II, III, IV and V [1]. These compounds have found applications in such areas as metallurgy [2], electronics [3], chemical current sources production [4], etc. Since the demand for this metal is growing [5], and vanadium reserves in ores are limited, it is necessary to develop new ways of vanadium extraction. One of the most perspective solutions for this problem is vanadium extraction from high-viscosity heavy oil [6]. Such kind of fuels contained a large number of heavy organic compounds containing heteroatoms (porphyrins, pyridines, pyrroles, mercaptans, thioethers, arenothiophenes, etc.), capable of binding with V (IV) ions due to the formation of coordination and covalent bonds [7].

At the first stage of processing this type of raw material, demetallization is carried out using strong acids solutions (sulfuric, hydrochloric) to destroy bonds with complexing agents of organic origin. Further, vanadium, in the form of vanadyl-ion, is extracted from an aqueous solution using liquid extraction [8]. Since in acidic water solutions (pH<3) vanadium is in the cationic form VO^{2+}, the most effective extractants for its recovery will be cation-exchange ones.

The most widely used cation-exchange extractants for metals extraction are derivatives of phosphoric (bis(2-ethylhexyl) phosphoric acid and its esters) or phosphinic (bis(2,4,4-trimethylpentyl)phosphinic acid and its esters) acids. Voshkin et al. had studied Ferrum (III) ions distribution in a system based on water – toluene with using binary extragent based on organophosphorus acids [9]. It was shown that the extraction, in this case, occurs both due to the formation of anionic complexes of iron (III) and due to the formation of complexes with an organic acid. Egorova et al. studied the extraction of lanthanides [10], as well as the uranyl ion [11], using extractants based on phosphinic acid derivatives. Belova et al. studied the ability of an extractant based on Cyanex 272 [12], based on a phosphoric acid derivative [13], based on phosphinic acid derivatives [14,15], to extract several rare earth metals, as well as uranium and thorium. The works of these scientists showed a high extraction ability of organophosphorus extractants (practically quantitative extraction of individual metals can be achieved).

Voshkin et al. had studied (Fe(III), La(III), Nd(III), Yb(III), Y(III), Pd(II)) metals extraction in high-frequency vibrating mini-columns. It has been shown that the use of these installations makes it possible to efficiently extract and separate metals. [16].

Thus, for the extraction of V (IV), mono-2-ethylhexyl ester of 2-ethylhexylphosphoric acid [17], and di (2-ethylhexyl) phosphoric acid [18], diluted in organic solvent (kerosene), were highly efficient. With an excess of the extractant 50 times relative to the metal concentration, the quantitative extraction of vanadyl ion is possible [18].

Even though the extraction of metals, including vanadium (IV), in water-organic solvent systems is well studied and the efficiency of using these systems for the metal extraction is noticeable, such systems do not meet the principles of "green" chemistry due to the use of toxic and fire- and explosive organic solvents. Two-phase water systems based on water-soluble polymers and inorganic salts can be an effective replacement for classical extraction systems. [19-20]. So, a system based on poly(ethylene glycol) and polypropylene glycol has shown its effectiveness for the transition metal ions without complexing agents [21-23], and with the addition of traditional extragents: Aliquat 336 [24], trioctylmethylammonium thiocyanate [25], potassium thiocyanate [26]. Systems based on water-soluble polymers, in particular, based on poly(ethylene glycol), have shown their superiority compared to classical extraction systems in the extraction of organic substances [27-28]. In addition to extraction, ATPSs based on water-soluble polymers are used in the synthesis of nanoparticles, which is described in the works of Shkinev [29] and Voshkin [30].

To date, it is known about a relatively small number of works devoted to the vanadium ions extraction [31-32]. The main disadvantage of these systems is that to achieve a quantitative or close to quantitative recovery, it is necessary to introduce complexing agents, for example, potassium thiocyanate, which are mainly toxic.

Thus, this work aimed to study the extraction capacity of an aqueous two-phase system based on poly(ethylene glycol) 1500 and sodium nitrate by the V(IV) ions extraction example.

2. Experimental

The tools were used in this work: 50 ml measuring beakers, measuring pipettes (from 10µ to 5ml), 20 ml separation funnels, 15 ml graduated plastic tubes, drop tube, analytical Balance (AND HR-100AZ), Magnetic stirrer (IKA C-MAG HS4) with stirrer bar, centrifuge (ELMI CM-6MT), desalinated filter paper, Enviro-Genie thermostatically controlled shaker (Scientific Industries, Inc.), spectrophotometer Cary-60 (Agilent Tech.), quartz cuvettes (l=10mm).

The initial solution of vanadyl sulfate was obtained by dissolving an accurate sample of $VOSO_4 \cdot 3H_2O$, «chemical grade», weighed on an analytical balance, in a 0.05 M sulfuric acid solution.

The extraction system was made using poly(ethylene glycol) 1500 (Acros Organics) and sodium nitrate «chemical grade». The system based on poly(ethylene glycol) 1500 (16.3wt%) – $NaNO_3$ (36wt%) – H_2O was used in all experiments. To prevent the transition of VO^{2+} to another form (V_2O_4, HV_2O^{5-}) during extraction processes, nitric acid was introduced into the system to pH = 1.1, except for the cases in which the determination dependences of metal distribution coefficients on the salt phase pH values. The acidity of the medium was measured using pH-meter Starter 5000 (OHAUS) with a combination electrode calibrated against buffer solutions (pH = 1.68, 4.01, 7.00, 10.01) at a temperature of 25°C.

The V(IV) extraction was carried out at a temperature of 25°C in graduated plastic tubes in an Enviro-Genie thermostatically controlled shaker (Scientific Industries, Inc.) at a rotation speed of 30 rpm until the thermodynamic equilibrium establishment (30 min).

The V(IV) ion concentration in the initial solution and in aqueous phases after extraction was determined spectrophotometrically using 4-(2-pyridylazo)resorcinol, which forms complexes with V(IV), that absorb in the visible spectrum at 536 nm. The optical density values were determined using a Cary-60 device (Agilent Tech.) in quartz cuvettes l = 10 mm.

Efficiency assessment was carried out using such characteristics as the distribution coefficient (D) and the degree of recovery (E,%), calculated using formulas (1) and (2):

$$D = \frac{C_{t.ph.}}{C_{b.ph.}}, (1)$$

$$E = \frac{C_{t.ph.} \cdot V_{t.ph.}}{C_{t.ph.} \cdot V_{t.ph.} + C_{b.ph.} \cdot V_{b.ph.}}, (2)$$

where $C_{t.ph.}$ – metal concentration in the top (polymer) phase, $C_{b.ph.}$ – metal concentration in the bottom (salt) phase, $V_{t.ph.}$ – top (polymer) phase volume, $V_{b.ph.}$ – bottom (salt) phase volume.

The presented experimental data are the result of a series of experiments and processed by methods of mathematical statistics. All the data, obtained in these experiments are new and unique.

3. Results and discussion

To determine the time required to reach equilibrium, we carried out a study on the variation phases time contact during the V(IV) ions extraction in the poly(ethylene glycol) 1500 (16.3wt%) - NaNO3 (36wt%) - H2O system. Figure 1 shows the dependence V(IV) ions extraction degree on the phases time contact. Figure 1 shows that the degree of VO^{2+} recovery remains constant in the studied time interval. The fact that the vanadium(IV) distribution coefficients have not change over the entire studied period of time has a significant meaning as it allows scaling up the extraction process in the design of technological plants.

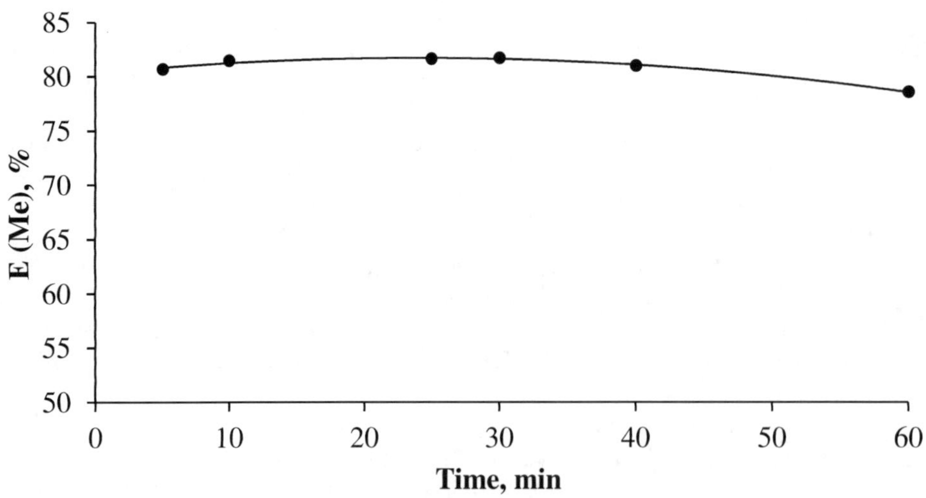

Figure 1. The dependence V(IV) ions extraction efficiency on the phases time contact in the system poly(ethylene glycol) 1500 (16.3wt%) - NaNO3 (36wt%) - H2O, phases ratio – 1:1, $C(V(IV))_{init.}$ = 0.01 mol L^{-1}, $pH_{salt\ ph.}$=1.1.

For comparison, Simonova et al. [33] investigated the vanadium (IV) extraction in the thiocyanate complexes form in two-phase systems based on PEG1500/alcohols. It is shown that in these systems the vanadium (IV) extraction by 93-98% takes place. However, in this work thiocyanate, which is known to be harmful and toxic, is used as a complex-forming agent.

To determine the V(IV) distribution coefficients dependence on the salt phase acidity, we carried out several experiments on varying pH by changing the initial amount of nitric acid. Figure 2 shows that the acidity of the salt phase does not affect the extraction of vanadyl-ions; therefore, the proton does not participate in the extracted compound formation. The extraction of vanadium (IV) in the region of higher pH proceeds with the formation of insoluble compounds in the salt phase.

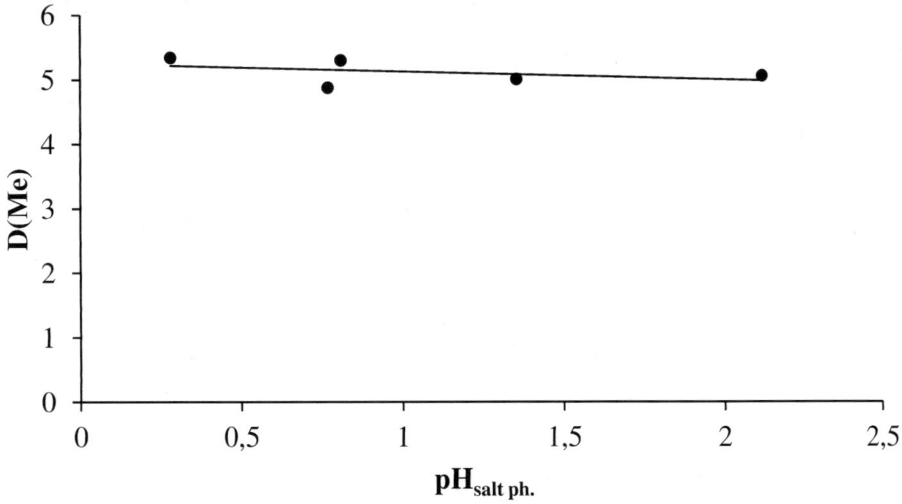

Figure 2. The dependence V(IV) ions distribution coefficients on the salt phases acidity in the system poly(ethylene glycol) 1500 (16.3wt%) - NaNO3 (36wt%) - H2O, phases ratio – 1:1, $C(V(IV))_{init.}$ = 0.01 mol L^{-1}. Thus, the efficiency of a two-phase aqueous system based on polyethylene glycol and sodium nitrate for the extraction of V (IV) ions from acidic aqueous solutions has been shown.

The data obtained in [31-32] suggested that the best vanadium recovery observed at pH=2, which correlates with the data obtained by us.

4. Conclusion

In this work, it was proposed to use an extraction system based on poly(ethylene glycol) 1500 (16.3wt%) - NaNO3 (36wt%) - H2O for the V(IV) from acidic aqueous solutions. It is shown that the phases time contact as well as the salt phase acidity, practically does not affect the extraction of vanadyl-ions. Therefore, this system is shown to be a prospective solution for the vanadium(IV) recovery from acidic waste steams problem.

References

[1] Maurya M R 2019 Probing the synthetic protocols and coordination chemistry of oxido-, dioxido-, oxidoperoxido-vanadium and related complexes of higher nuclearity *Coord. Chem. Rev.* **383** 43–81

[2] Yin B, Maresca F and Curtin W A 2020 Vanadium is an optimal element for strengthening in both fcc and bcc high-entropy alloys *Acta Mater.* **188** 486-91

[3] Li J, Li J, Zhang Y, Li J, Yang Y, Zhao H, Zheng C, Li J, Huang J, Li F, Tang T and Yao J 2020 All-optical switchable terahertz spin-photonic devices based on vanadium dioxide integrated metasurfaces *Opt. Commun.* **460** 124986

[4] Jing M, Qin Y, Shi Q, Liu J and Yan C 2017 A new insight into electrode processes of vanadium redox flow battery by thermo-electro-chemistry method *Journal of Energy Storage* **14** 163-7
[5] Gilligan R and Nikoloski A N 2020 The extraction of vanadium from titanomagnetites and other sources *Miner. Eng.* **146** 106106
[6] Zuliani J E, Miyata T, Mizoguchi T, Feng J, Kirk D W and Jia C Q 2016 Characterization of vanadium in oil sands fluid petroleum coke using electron microscopy *Fuel* **178** 24-8
[7] Li Y, Shang H, Zhang Q, Elabyouki M and Zhang W 2020 Theoretical study of the structure and properties of Ni/V porphyrins under microwave electric field: A DFT study *Fuel* **278** 118305
[8] Cai Z, Feng Y, Li H and Zhou Y 2013 Selective separation and extraction of vanadium(IV) and manganese(II) from co-leaching solution of roasted stone coal and pyrolusite via solvent extraction *Ind. Eng. Chem. Res.* **52** 13768-76.
[9] Voshkin A A, Belova V V and Khol'kin A I 2003 Extraction of iron(III) by binary extractants based on quaternary ammonium bases and organic acids *Russ. J. Inorg. Chem.* **48** 4 608-13
[10] Egorova N S, Belova V V, Voshkin A A, Zhilov V I and Khol'kin A I 2005 Extraction of lanthanide chlorides by binary extractants based on phosphinic acid derivatives *Russ. J. Inorg. Chem.* **50** 1781-4
[11] Egorova N S, Belova V V, Voshkin A A, Khol'kin A I, Pyartman A K and Keskinov V A 2008 Extraction of Uranyl Nitrate with a Binary Extractant Based on Di(2,4,4-trimethylpentyl)phosphinic Acid *Theor. Found. Chem. Eng.* **42** 5 708-13
[12] Belova V V, Voshkin A A, Egorova N S and Khol'kin A I 2010 Extraction of rare earth metals from nitrate solutions with a binary extractant based on Cyanex 272 *Russ. J. Inorg. Chem.* **55** 4 629-33
[13] Belova V V, Voshkin A A, Egorova N S and Kholkin A I 2012 Solvent extraction of rare earth metals from nitrate solutions with di(2,4,4-trimethylpentyl)phosphinate of methyltrioctylammonium *J. Mol. Liq.* **172** 144-6
[14] Belova V V, Egorova N S, Voshkin A A and Khol'kin A I 2015 Extraction of rare earth metals, uranium, and thorium from nitrate solutions by binary extractants *Theor. Found. Chem. Eng.* **49** 4 545-9
[15] Belova V V, Voshkin A A, Kholkin A I and Payrtman A K 2009 Solvent extraction of some lanthanides from chloride and nitrate solutions by binary extractants *Hydrometallurgy* **97** 3-4 198–203
[16] Voshkin A A, Kodin N V, Kondakov D F and Kostanyan A E 2010 Study of metal extraction in high-frequency vibration mini-columns *Russ. J. Inorg. Chem.* **55** 5 794-7
[17] Li X, Wei C, Wu J, Li M, Deng Z, Li C and Xu H 2012 Co-extraction and selective stripping of vanadium (IV) and molybdenum (VI) from sulphuric acid solution using 2-ethylhexyl phosphonic acid mono-2-ethylhexyl ester *Sep. Purif. Technol.* **86** 64-9
[18] Zhang Y, Zhang T-a, Lv G, Zhang G, Liu Y and Zhang W 2016 Synergistic extraction of vanadium(IV) in sulfuric acid media using a mixture of D2EHPA and EHEHPA *Hydrometallurgy* **166** 87-93
[19] Zakhodyaeva Y A, Rudakov D G, Solov'ev V O, Voshkin A A and Timoshenko A V 2019 Liquid-Liquid Equilibrium in an Extraction System Based on Polyvinylpyrrolidone-3500 and Sodium Nitrate *Theor. Found. Chem. Eng.* **53** 2 159-65
[20] Zakhodyaeva Y A, Rudakov D G, Solov'ev V O, Voshkin A A and Timoshenko A V 2019 Liquid-Liquid Equilibrium of Aqueous Two-Phase System Composed of Poly(Ethylene Oxide) 1500 and Sodium Nitrate *J. Chem. Eng. Data* **64** 3 1250-5
[21] Zakhodyaeva Y A, Zinov'eva I V, Tokar E S and Voshkin A A 2019 Complex Extraction of Metals in an Aqueous Two-Phase System Based on Poly(Ethylene Oxide) 1500 and Sodium Nitrate *Molecules* **24** 22 4078
[22] Zakhodyaeva Y A, Zinov'eva I V and Voshkin A A 2019 Extraction of Iron(III) Chloride Complexes Using the Polypropylene Glycol 425-NaCl-H2O *Theor. Found. Chem. Eng.* **53** 5 735-40

[23] Gradov O M, Zakhodyaeva Y A, Zinov'eva I V and Voshkin A A 2019 Some Features of the Ultrasonic Liquid Extraction of Metal Ions *Molecules* **24** 19 3549
[24] Fedorova M I, Zakhodyaeva Y A and Voshkin A A 2020 Interphase Distribution of Fe(III) and Zn(II) in Chloride Systems with Aliquat 336 in Polypropylene Glycol 425 2020 *Theor. Found. Chem. Eng.* **54** 3 425-30
[25] Fedorova M I, Zinov'eva I V, Zakhodyaeva Y A and Voshkin A A 2020 Extraction of Fe(III), Zn(II), and Mn(II) Using a System with a Green Solvent for Trioctylmethylammonium Thiocyanate 2020 *Theor. Found. Chem. Eng.* **54** 2 313-8
[26] Zakhodyaeva Y A, Izyumova K V, Solov'eva M S and Voshkin A A 2017 Extraction separation of the components of leach liquors of batteries *Theor. Found. Chem. Eng.* **51** 5 883-7
[27] Zinov'eva I V, Zakhodyaeva Y A and Voshkin A A 2020 Data on the extraction of benzoic, salicylic and sulfosalicylic acids from dilute solutions using PEG-based aqueous two-phase systems *Data in Brief* **28** 105033
[28] Zakhodyaeva Y A, Solov'ev V O, Zinov'eva I V, Rudakov D G, Timoshenko A V and Voshkin AA 2019 Interphase Distribution of Thiophene, Toluene, and o-Xylene in the Hexane-Polymer-Water Extraction System *Theor. Found. Chem. Eng.* **53** 4 550-5
[29] Shkinev V M, Zakhodyaeva Y A, Dzhenloda R K, Mokhodoeva O B and Voshkin A A 2017 Synthesis of magnetic iron oxide nanoparticles at the interface of the polyethylene glycol-ammonium sulfate-water extraction system *Mendeleev Commun.* **27** 5 485-6
[30] Voshkin A A, Shkinev V M and Zakhodyaeva Y A 2017 A New Extraction Method for the Preparation of Zinc Oxide Nanoparticles in Aqueous Two-Phase Systems *Russ. J. Phys. Chem. A* **91** 2 226-8
[31] Pan Y, Sun X, Zhang Y Separation of vanadium and molybdenum from aqueous solution using PEG2000 + sodium sulfate + water aqueous two-phase system 2019 *SN Applied Sciences* **1** 1461
[32] Sun T, Pan Y, Sun X, Zhang Y Recovery of vanadium using an aqueous two-phase system consisting of poly (ethylene glycol) 2000 and sodium sulfate 2019 *Hydrometallurgy* **189** 105135
[33] Simonova T N and Dubrovina V A Separation of Vanadium(IV) from Vanadium(V) and Its Determination in the Presence of Thiocyanate Ions Using TwoPhase Aqueous Systems 2014 *J. Anal.. Chem.* **69** 9 845-50

Acknowledges
The reported study was funded by RFBR, project number 20-33-80015.

Implementation of the extraction method for the separation of Fe^{3+} and Ni^{2+} ions on cascade of mixing-settling extractors

Milevskii N.A., Klychevskikh Y.A., Solov'ev V.O., Zinov'eva I.V., Fedorova M.I.

Kurnakov Institute of General and Inorganic Chemistry, Russian Academy of Sciences, Russia, 117901, Moscow, Leninskii prosp., 31

milevskynikita@gmail.com

Abstract. Extraction processes are one of the most efficient methods of purification and separation of compounds. However, laboratory-developed methods commonly are not a complete technique that can be applied to the real manufacturing process with its extraction equipment. The extraction method for the separation of Fe^{3+} and Ni^{2+} ions using hydrochloric acid solution on a cascade of mixing-settling extractors has been determined. The extraction scheme chosen provides high efficiency of a cascade by using liquid pseudomembranes (LPM). Besides its extraction scheme simplicity, the LPM method is more economically beneficial compared to classical extraction methods. The process includes stages of Fe^{3+} ions extraction into the feed, followed by reextraction into the water and the feed purification for reusing it later on. The consumption of solvents and the stirring speed for the stable and efficient cascade work have been optimized. The described process is a complete technological solution for recycling processed battery waste.

Keywords: extraction, separation, reextraction, liquid pseudomembranes, tributylphosphate, Ni-MH batteries

1. Introduction

Compared to other separation methods, extraction processes have a number of significant advantages. To date plenty of fundamental studies demonstrating possibilities for applying extraction to such processes as a removal of organic compounds [1-3] and mineral acids [4] in industrial waste or radiochemical [5, 6] and rare earth metal [7, 8] productions have been carried out. At the moment there are some main directions of a modern extraction technologies development. Voshkin et al. have conducted researches on the new extraction systems, e.g., aqueous two-phase systems (ATPS) [9-11] that are able to reduce an environmental impact of metal ions separation processes [12-14]. New, more efficient extractants and their extraction mechanisms [15, 16] are being studied as well. There are a continuing development and perfecting of extraction equipment corresponding to the modern production processes requirements [17].

Extraction processes provide a solution to the important problem of electronic [18] and chemical current sources waste [19] separation and recycling. A growing amount of nickel metal hydride (NiMH) battery waste has attracted attention of many researches as it contains highly valuable metals such as Al,

Ni, Co, Fe, Zn [20, 21]. In this case classical separation methods using tributylphosphate (TBP) as an extractant [23, 24] are in favor since construction of mass-produced extraction equipment is made for more hydrodynamically stable systems that can be easily stripped.

At the moment different methods and technologies for extraction processes efficiency increase gain more popularity, e.g., Gradov and colleagues' [25-27] studies show an expediency of ultrasound usage for extraction intensification. However, application of new methods often requires an acquisition of equipment with a new construction [28], which is both technologically and financially undesirable. Relatively new extraction method using LPM [29] lacks these disadvantages since it can be realized on a cascade of mixer-settlers that can be easily acquired commercially and have an affordable price.

The LPM method falls into the range of contactless interphase mass transfer processes and currently is being actively developed as an efficient technic for extraction, separation and purification of compounds [30]. Mass exchange between two or more liquid phases takes place as components transfer from one phase into another via an exchange medium. The benefits of the LPM method are the usage of a finite amount of the exchange medium and increased extraction effectiveness in comparison to classical extraction schemes for mixer-settler cascades.

Therefore, the aim of this study is to model and carry out the LPM method of separation of the solution containing Fe^{3+} and Ni^{2+} ions in amounts close to their concentrations in the leaching solution of Ni-MH batteries [31].

2. Experimental

2.1 Research materials

The standardized solution was prepared by dissolving the precise weights of $FeCl_3 \cdot 6\,H_2O$, $NiCl_2 \cdot 6\,H_2O$ with the qualification of "chemical grade" in distilled water suspended on an analytical balance (AND HR-100AZ) following by addition of the precise volume of hydrochloric acid with the qualification of "chemical grade".

The extractant was prepared by mixing 20 vol% of 1-octanol, 30 vol% of kerosene and 50 vol% of TBP with the qualification of "chemical grade". Distilled water acts as the reextractant.

2.2 Research method

Modelling of **a** and **b** extraction schemes which are a combination of the LPM method and the countercurrent flow extraction has been carried out using following formulas (1), (2), (3) derivation of which is thoroughly reviewed in the paper [32]:

$$c_i = \frac{1}{D_i \cdot \frac{V}{O}}, (1)$$

$$\varphi_i = \frac{1 - c_i^n}{1 - c_i^{2 \cdot n}}, (2)$$

$$C_{ir} = C_i - \varphi_i \cdot C_i, (3)$$

where V and O are the volumetric flow rates of the aqueous and organic phases respectively, D_i is the equilibrium distribution coefficient for the *i*th component, c_i is the extraction factor of the *i*th component, φ_i is the recovery of the *i*th component in the extractant, C_{ir} is the concentration of the *i*th component in the raffinate in the extraction and stripping, C_i is the concentration of the *i*th component in aqueous phase, n is the number of extractors.

Modelling of the LPM method extraction scheme **c** has been carried out using formulas (4), (5) derivation of which is thoroughly reviewed in Kostanyans' study [30]:

$$F_i = \frac{v_i}{w \cdot m_i}, (4)$$

$$x_1'' = x_1' \cdot \frac{1 - F_1 \cdot F_2}{\left[\frac{1 + F_1}{(1 + F_2) \cdot F_1}\right]^n - F_1 \cdot F_2}, (5)$$

where F_1 and F_2 are the mass-exchange factors in the extraction and stripping, respectively, v_1, v_2 and w are the volumetric flow rates of the phases, m_i is the equilibrium distribution coefficient for the *i*th component, x_i'' is the concentration of the *i*th component in the raffinate or the reextractant.

Effectiveness evaluation was conducted using mole fraction (χ) calculated by the formula (6):

$$\chi_1 = \frac{x_1''}{x_1'' + x_2''}, (6)$$

A cascade of LEU-12-1.0 mixer-settlers produced by GK "Rusredmet" was used as an extraction apparatus. Each extractor has a volume of 300 ml. Figure 1 shows flow sheets of **a**, **b** and **c** extraction schemes.

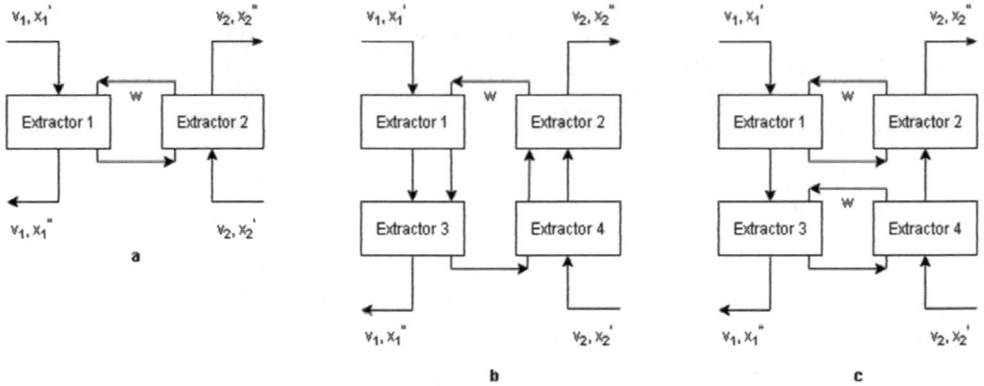

Figure 1. Flow sheets of **a**, **b** and **c** extraction schemes.

Aqueous phases flows were realized by peristaltic pumps with YZ2515x nozzles produced by Shenchen Pump, volumetric flow rates of the phases are of 0.5 liters per hour. Rotation frequency of turbine stirrers has been issued at 700 rpm (the power supply unit operating voltage is 8V).

Raffinate and reextractant samples have been collected three times after achieving hydrodynamical equilibrium, the obtained values have been averaged.

Metal ions concentrations in the feed solutions and in the aqueous phases after extraction were determined spectrophotometrically. For Fe(III) determination sulfosalicylic acid was used, the absorption maximum was at 420 nm. Nickel forms a complex with 4-(2-pyridylazo)resorcinol which has the absorption maximum at 492 nm. Optical density values were determined using Cary-60 (Agilent Tech.) spectrophotometer and quartz cuvettes l = 10 mm.

The presented experimental data is the result of a series of experiments and was processed by methods of mathematical statistics.

3. Results and discussion

The obtained data demonstrates efficient separation of hydrochloric acid solution containing nickel and iron chlorides using different extraction schemes. The results of the modelling correspond with the results of the experiment and confirm the LPM method to be more efficient than classical extraction scheme.

Table 1 and table 2 display the raffinate and reextractant compositions respectively obtained by calculation and by experiment. The experimental data validates the theoretical calculations of all extraction schemes which confirms the efficiency of calculation methods usage for technological processes designing.

Table 1 shows the results demonstrating a decrease of the mole fraction of iron ions in the raffinate from 0.256 to 0.122 in the transition from scheme **a** to scheme **b** due to the addition of two mixer-settlers to the scheme **b**. However, a decrease of the mole fraction from 0.122 to 0.082 can be observed in the transition from scheme **b** to scheme **c** despite the same number of extractors used which demonstrates the efficiency of the LPM method.

Table 1. Raffinate composition.

	$C(Ni^{2+})$, mol/l		$C(Fe^{3+})$, mol/l		$\chi(Fe^{3+})$
	Calculation	Experiment	Calculation	Experiment	
Scheme a	0.174	0.1686	0.0524	0.058	0.256
Scheme b	0.168	0.1589	0.0193	0.022	0.122
Scheme c	0.197	0.1958	0.017	0.0176	**0.082**

Table 2 also demonstrates an increase of the mole fraction of iron ions $\chi(Fe^{3+})$ from 0.859 to 0.966 with the LPM method usage.

Table 2. Reextractant composition.

	$C(Ni^{2+})$, mol/l		$C(Fe^{3+})$, mol/l		$\chi(Fe^{3+})$
	Calculation	Experiment	Calculation	Experiment	
Scheme a	0.022	0.018	0.124	0.11	0.859
Scheme b	0.0337	0.0297	0.179	0.152	0.837
Scheme c	0.01	0.011	0.31	0.314	**0.966**

4. Conclusion

In this study quantitative characteristics of iron and nickel ions extraction from the standardized solution using different extraction schemes were obtained. The efficiency of the LPM method compared to the countercurrent flow extraction is demonstrated both experimentally and theoretically. The benefits of the LPM method are that it can be easily carried out and does not require additional equipment or sophistication of technological process unlike countercurrent flow extraction. Moreover, the extractant in the LPM method is not removed from extraction cascade which is cost-effective. Therefore, the LPM method can be used as a complete technological scheme, for example for separation of the components of leach solutions.

References

[1] Sulaiman R, Adeyemi I, Abraham S R, Hasan S W and AlNashef I M 2019 Liquid-liquid extraction of chlorophenols from wastewater using hydrophobic ionic liquids *J. Mol. Liq.* **294** 111680

[2] Zakhodyayeva Y A, Voshkin A A, Belova V V, Kostanyan A E and Khol'kin A I 2010 Extraction of lactic acid from technological (concentrated) solutions *Theor. Found. of Chem. Eng.* **44** 782–3

[3] Zakhodyaeva Y A, Solov'ev V O, Zinov'eva I V, Rudakov D G, Timoshenko A V and Voshkin A A 2019 *Theor. Found. of Chem. Eng.* **53** 550–5

[4] Belova V V, Kulichenkov S A, Voshkin A A, Khol'kin A I, Kuvaeva Z I and Soldatov V S 2007 Extraction of mineral acids with methyltrioctylammonium dinonylnaphthalenesulfonate *Russ. J. Inorg. Chem.* **52** 3 460–4

[5] Belova V V, Egorova N S, Voshkin A A and Khol'kin A I 2015 Extraction of rare earth metals, uranium, and thorium from nitrate solutions by binary extractants *Theor. Found. of Chem. Eng.* **49** 4 545–9

[6] Egorova N S, Belova V V, Voshkin A A, Khol'kin A I, Pyartman A K and Keskinov V A 2008 Extraction of Uranyl Nitrate with a Binary Extractant Based on Di(2,4,4-trimethylpentyl)phosphinic Acid *Theor. Found. of Chem. Eng.* **42** 708–13

[7] Belova V V, Voshkin A A, Khol'kin A I and Payrtman A K 2009 Solvent extraction of some lanthanides from chloride and nitrate solutions by binary extractants *Hydrometallurgy* **97** 3–4 198–203

[8] Belova V V, Voshkin A A, Egorova N S and Khol'kin A I 2012 Solvent extraction of rare earth metals from nitrate solutions with di(2,4,4-trimethylpentyl)phosphinate of methyltrioctylammonium *J. Mol. Liq.* **172** 144–6

[9] Voshkin A A, Zakhodyaeva Y A, Zinov'eva I V and Shkinev V M 2018 Interphase Distribution of Aromatic Acids in the Polyethylene Glycol–Sodium Sulfate–Water System. *Theor. Found.*

Chem. Eng. **52** 890–3

[10] Zakhodyaeva Y A, Rudakov D G, Solov'ev V O, Voshkin A A and Timoshenko A V 2019 Liquid-Liquid Equilibrium of Aqueous Two-Phase System Composed of Poly(Ethylene Oxide) 1500 and Sodium Nitrate *J. Chem. Eng. Data* **64** 3 1250–5

[11] Zakhodyaeva Y A, Rudakov D G, Solov'ev V O, Voshkin A A and Timoshenko A V 2019 Liquid–Liquid Equilibrium in an Extraction System Based on Polyvinylpyrrolidone-3500 and Sodium Nitrate *Theor Found Chem Eng* **53** 159–65

[12] Fedorova M I, Zinov'eva I V, Zakhodyaeva Y A and Voshkin A A 2020 Extraction of Fe(III), Zn(II), and Mn(II) Using a System with a Green Solvent for Trioctylmethylammonium Thiocyanate *Theor. Found. Chem. Eng.* **54** 2 313–8

[13] Zakhodyaeva Y A, Zinovyeva I V, Tokar E S and Voshkin A A 2019 Complex extraction of metals in an aqueous two-phase system based on poly(ethylene oxide) 1500 and sodium nitrate *Molecules* **24** 4078

[14] Zakhodyaeva Y A, Zinov'eva I V and Voshkin A A 2019 Extraction of Iron(III) Chloride Complexes Using the Polypropylene Glycol 425–NaCl–H_2O System *Theor. Found. Chem. Eng.* **53** 735–40

[15] Guor-Tzo Wei, Zusing Yang and Chao-Jung Chen 2003 Room Temperature Ionic Liquid as a Novel Medium for Liquid/liquid Extraction of Metal Ions *Anal. Chim. Acta* **488** 183–92

[16] Belova V V, Voshkin A A, Egorova N S and Khol'kin A I 2010 Extraction of rare earth metals from nitrate solutions with a binary extractant based on Cyanex 272 *Russ. J. Inorg. Chem.* **55** 4 629–33

[17] Voshkin A A, Kodin N V and Kostanyan A E 2010 Column miniextractors with a vibrating nozzle *Theor. Found. Chem. Eng.* **44** 616–8

[18] Khaliq A, Rhamdhani M A, Brooks G and Masood S 2014 Metal Extraction Processes for Electronic Waste and Existing Industrial Routes: A Review and Australian Perspective *Resources* **3** 152–79.

[19] Zakhodyaeva Y A, Izyumova K V, Solov'eva M S and Voshkin A A 2017 Extraction Separation of the Components of Leach Liquors of Batteries *Theor. Found. Chem. Eng.* **51** 5 883–7

[20] Zhang L and Xu Z 2016 A Review of Current Progress of Recycling Technologies for Metals from Waste Electrical and Electronic Equipment *J. Clean. Prod.* **127** 19-36.

[21] Carla Lupi and Daniela Pilone 2002 Ni–MH spent batteries: a raw material to produce Ni–Co alloys *Waste Manage.* **22** 871–4

[22] Daniela da Silveira Leite, Pablo Luis Gutierrez Carvalho, Leandro Rodrigues de Lemos, Aparecida Barbosa Mageste and Guilherme Dias Rodrigues 2017 Hydrometallurgical separation of copper and cobalt from lithium-ion batteries using aqueous two-phase systems *Hydrometallurgy* **169** 245–52

[23] Xintao Yi, Guangsheng Huo and Wei Tang 2020 Removal of Fe(III) from Ni-Co-Fe chloride solutions using solvent extraction with TBP *Hydrometallurgy* **192**

[24] Chiraz Z, Rafik T, Mourad B and Mahmoud D 2010 Liquid–liquid extraction and transport across supported liquid membrane of phenol using tributyl phosphate *J. Membr. Sci.* **360** 334–40

[25] Gradov O M, Zakhodyaeva Y A and Voshkin A A 2018 Breakup of immiscible liquids at the interface using high-power acoustic pulses *Chem Eng Process* **131** 125–30

[26] Gradov O M, Voshkin A A and Zakhodyaeva Y A 2017 Estimating the parameters of ultrasonically induced mass transfer and flow of liquids in the pseudomembrane method *Chem Eng Process* **118** 54–61

[27] Gradov O M, Zakhodyaeva Y A, Zinov'eva I V and Voshkin A A 2019 Some Features of the Ultrasonic Liquid Extraction of Metal Ions *Molecules* **24** 3549

[28] Voshkin A A, Kodin N V, Kondakov D F and Kostanyan A E 2010 Study of metal extraction in high-frequency vibration mini-columns *Russ. J. Inorg. Chem.* **55** 794–7

[29] Zhong-Mao Gu 1990 A new liquid membrane technology — electrostatic pseudo liquid membrane *J. Membr. Sci.* **52** 77-88

[30] Kostanyan A E, Egorova N S, Voshkin A A and Zonov S S 2008 Extraction of uranyl, ytterbium, and lanthanum nitrates in a three-compartment multiphase extractor *Theor. Found. Chem. Eng.* **42** 718–23

[31] Larsson K, Ekberg C and Ødegaard-Jensen A 2013 Dissolution and characterization of HEV

NiMH batteries *Waste Manage.* **33**(3) 689–98
[32] Robert E. Treybal 1951, Liquid Extraction, McGraw Hill Book Company Inc.

Acknowledgments
This work was supported by IGIC RAS state assignment.

Extraction of pyridine using systems based on water-soluble polymers

V O Solovev[1,2], S V Soloveva[1,2] and N A Milevsky[2]
1 MIREA – Russian Technological University, Moscow, Russian Federation
2 Kurnakov Institute of General and Inorganic Chemistry, Moscow, Russian Federation

*Corresponding author: svo@igic.ras.ru

Abstarct. In the process of using hydrocarbon fractions containing a large amount of nitrogenous compounds, nitrogen oxides are released into the atmospheric air, which have a negative impact on the environment and human health. The traditional cleaning method is treatment with a 25% sulfuric acid solution and subsequent hydrotreating. However, this process becomes disadvantageous due to its inability to achieve ultra-low concentrations of nitrogen-containing compounds (<10 ppm). Extraction using non-toxic and environmentally friendly water-soluble polymers is a promising alternative compared to traditional methods.This work presents the dependence of the interphase distribution of pyridine on the composition of extraction systems based on water-soluble polymers. According to the results of the study, it was found that polyethylene glycol-400, polypropylene glycol-425 and methyl ether of polyethylene glycol-350 exhibit effective extraction properties in relation to pyridine and extract it by 90.95%, 90.33% and 87.82% in one extraction stage, respectively. It was also found that the use of two-phase aqueous systems based on water-soluble polymers in the process of extracting pyridine is promising.

Keywords: green chemistry, liquid–liquid extraction, extraction efficiency, interphase distribution, pyridine, water soluble polymers, light hydrocarbon fractions.

1. Introduction

Light hydrocarbon fractions are the most important product of the oil refining industry. Also, in the process of using such fractions, nitrogen oxides are released into the atmospheric air, negatively affecting the environment and human health [1]. The quality of the yield of light fractions is subject to stringent requirements for the content of nitrogen compounds, since at the stage of processing petroleum feedstock they significantly reduce the activity of catalysts, cause tar formation and darkening of petroleum products [2]. All nitrogen-containing compounds in the oil are divided into three groups, the largest of which is the group of basic heterocycles, the share of such compounds in oils is up to 60% of the total nitrogen content [3]. The traditional method for removing nitrogen-containing compounds is treatment with a 25% sulfuric acid solution and subsequent hydrogenation [4]. However, obtaining fractions of the desired quality with a low content of nitrogen compounds (<10 ppm) becomes economically unprofitable [5]. As a consequence of this, recently, alternative methods for the isolation and concentration of nitrogen-containing compounds have been developed, such as oxidation [6], sorption [7], and extraction [8]. Extraction is a priority in comparison with other separation methods since in most cases it proceeds under normal conditions (T = 25^{0}C, atmospheric pressure), does not require complex instrumentation and does not change the chemical structure of compounds and, therefore, does not affect the quality of liquid fuel [9, 10]. Liquid-liquid extraction is applicable in the processes of isolation, separation, and concentration of a number of carboxylic acids [11, 12], non-ferrous [13, 14], and rare earth metals [15-17]. Extraction is applicable in other areas of technology such as chromatography [18, 19], pseudo-liquid membranes [20], and supercritical fluids [21]. Traditional organic solvents used as extractants in the processes of removing nitrogen compounds from fuels, for example, dimethylformamide [22], are toxic and fire hazardous. Systems based on water-soluble polymers are used as an alternative [23]. The advantage of these systems is their non-toxicity and environmental safety [24]. The applicability of these systems in the extraction of carboxylic acids has been described by Zinovieva et al. [25]. The authors of the article [25] studied the dependences of the interphase distribution of benzoic, salicylic, and sulfosalicylic acids on the composition of two-phase aqueous systems, temperature, and pH values. Many researchers have studied the extraction of metals, so the article [26] studied the extraction of iron (III) chloride in a two-phase aqueous system polypropylene glycol 425 - sodium chloride - water. The authors of the article [27] in their work present an ecologically safe aqueous two-phase system based on polyethylene oxide for the complex extraction of Ni (II), Co (II), Fe (III), Mn (II), Zn (II), Cu (II) and Al (III) from nitrate solutions, in this article we studied the effect of various parameters of the system on the extraction behavior, such as the acidity of the medium, temperature, initial metal concentration and contact time of the phases. Oscherel, Gradov, et al. Studied the effect of ultrasonic radiation on the extraction of $Fe3^{+}$ ions from chloride and nitrate solutions in systems based on water-soluble polymers [28]. Previously, we studied the interphase distribution of thiophene in systems based on water-soluble polymers, and it was found that the system based on polyethylene glycol 400 exhibits effective extraction properties with respect to thiophene. [29] Also, systems based on water-soluble polymers are used in the processes of removing nitrogen-containing compounds from petroleum products, so Jiang et al. [30] investigated the dependence of the extraction of basic and non-basic nitrogen-containing compounds on the composition of systems based on water-soluble polymers. Zhu et al. [31] proved the effectiveness of the use of deep eutectic solvents based on water-soluble polymers in the processes of fuel purification from nitrogen-containing compounds.

As a result, in the processes of purification of light hydrocarbon fractions, extraction systems that correspond to the principles of "green" chemistry are of particular relevance, and the purpose of this work was to study the interphase distribution of pyridine in two-phase systems polymer - n-hexane - water.

2. Experimental details

2.1. Reagents

The following reagents were used in experimental studies: pyridine (CHIMMED), n-hexane (CHIMMED), polyethylene glycol 400 (Clariant), polypropylene glycol 425 (Acros Organics), polyethylene glycol methyl ether 350 (Acros Organics). All reagents were used without additional purification.

2.2 Research methods

An initial solution simulating light hydrocarbon fractions, a ready-made pyridine solution with a concentration of 0.05% (wt.) in n-hexane.

To study the process of pyridine extraction, we used graduated test tubes and separating funnels with ground stoppers. Extraction of solutions simulating hydrocarbon fractions, solutions of an extractant based on a water-soluble polymer. Next, the resulting mixture was stirred for 1 hour in a thermostated shaker Enviro-Genie (Scientific Industries,Inc.) at a temperature of 25°C and at a speed of 45 rpm to achieve thermodynamic equilibrium of the system. Then the mixture was centrifuged for 10 min at 2500 rpm (centrifuge CM-6MT, SIA ELMI), after which the volumes of the phases were measured. The concentration of pyridine in the organic phase was determined spectrophotometrically (spectrophotometer Cary-60, Agilent) at a wavelength of 251 nm (for this, the electronic absorption spectrum in the ultraviolet region was captured, fig. 1) in quartz cells with an optical path length of 1 mm relative to a solution in which pyridine is dissolved.

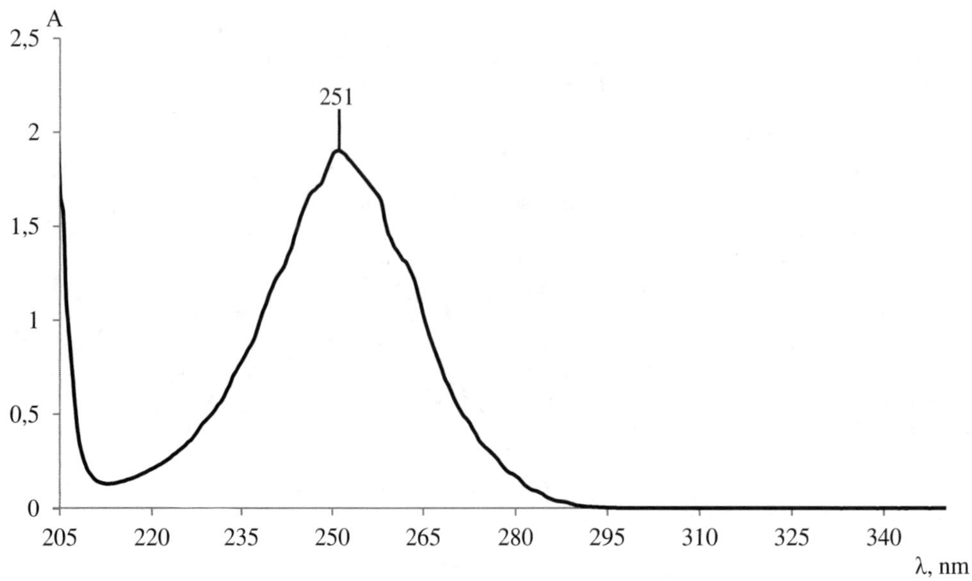

Figure: 1. Electronic absorption spectrum of pyridine

For the quantitative determination of pyridine, a calibration straight line (Fig. 2) was constructed at five points, at the corresponding wavelength of 251 nm and R^2 was 0.9994.

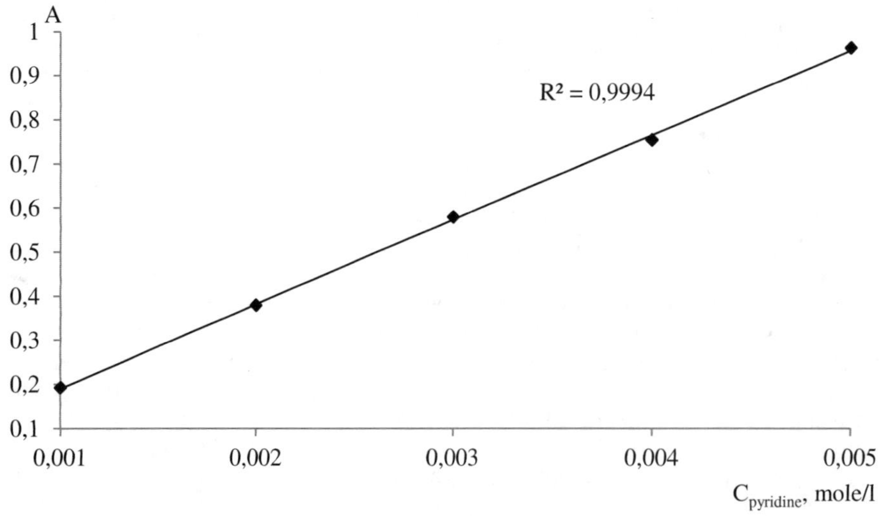

Figure: 2. Calibration straight line to determine the quantitative characteristics of pyridine extraction in the polymer - n-hexane - water system.

3. Results and discussion

In the course of the experimental work, the interfacial distribution of pyridine was studied in the systems polyethylene glycol 400 - n-hexane - water, polypropylene glycol 425 - n-hexane - water, and polyethylene glycol methyl ether 350 - n-hexane - water.

In fig. 3 shows the results of the dependence of the degree of extraction of pyridine from n-hexane on the composition of extraction systems based on water-soluble polymers.

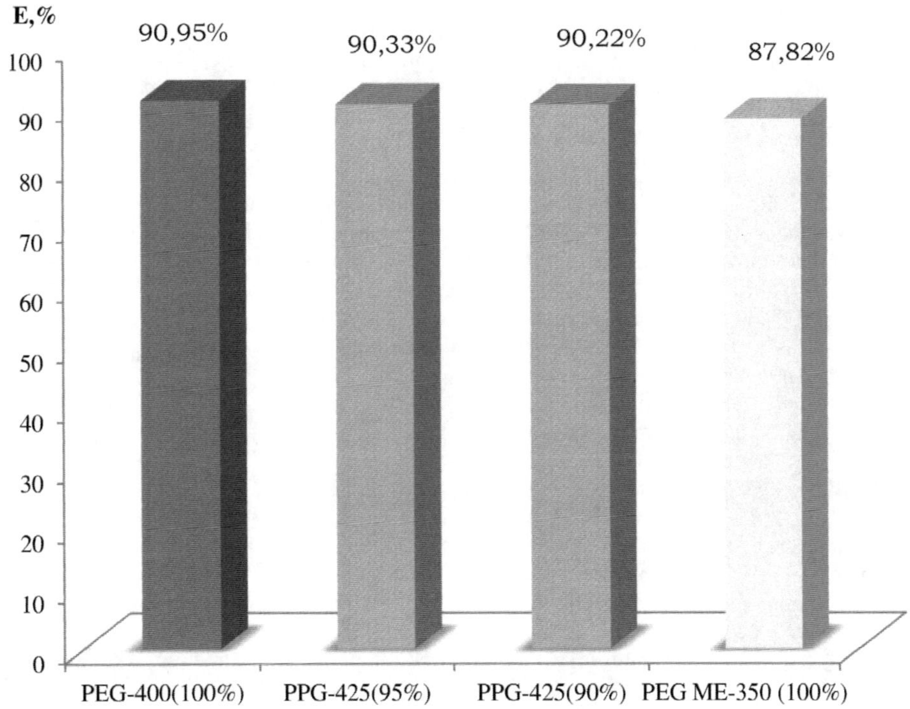

Figure: 3. Dependence of the degree of extraction of pyridine on the composition of extraction systems based on water-soluble polymers.

As can be seen, extraction systems based on polyethylene glycol 400, polypropylene glycol 425, and polyethylene glycol methyl ether 350 have high selectivity with respect to pyridine, extracting it up to 90.95 %, 90.33% and 87.82% in one extraction stage, respectively.. It should be noted that the use of 100% polypropylene glycol 425 is excluded since this leads to the formation of a homogeneous system and the process of extracting pyridine from n-hexane, in this case, becomes impossible. Therefore, the use of an aqueous solution of polypropylene glycol 425 makes it possible to implement the pyridine extraction process. Also from fig. 3 it can be seen that systems containing 90% and 95% have practically the same efficiency, extracting pyridine from n-hexane by 90.22% and 90.33% for one degree of extraction, respectively.

4. Conclusions

Within the framework of this work, a series of systematic studies of the dependence of the degree of extraction of pyridine from n-hexane on the composition of extraction systems based on water-soluble polymers was carried out. It was found that polyethylene glycol 400, polypropylene glycol 425 and polyethylene glycol methyl ether 350 exhibit effective extraction abilities with respect to pyridine and extract it by 90.95%, 90.33% and 87.82% in one extraction stage, respectively. Based on the results of this study, there is a need for further research to improve these systems in order to increase the efficiency of extracting nitrogen compounds from light hydrocarbon fractions.

References

[1] Mathidala S and Ogunlaja A S 2019 Selective removal of pyridine in fuel by imprinted polymer (poly 4-vinyl aniline-co-DVB) as adsorbent *Pet. Sci. Technol.* **37** 14 1691-703.

[2] Wen J, Lin H, Han X, Zheng Y and Chu W 2017 Physicochemical studies of adsorptive denitrogenation by oxidized activated carbons *Ind Eng Chem Res.* **56** 5033–41.

[3] Manovyan A K 2011 The technology of primary processing of oil and natural gas: A textbook for high schools. 2 nd ed. M.: p. 568. (in Russ.).

[4] Hu S, Luo G, Shima T, Luo Y and Hou Z 2017 Hydrodenitrogenation of pyridines and quinolines at a multinuclear titanium hydride framework *Nature communications* **8** 1 1-8.

[5] Bauserman J W, Mushrush G W, Hardy D R, Willauer H D, Laskoski M and Keller T M 2014 The removal of organic nitrogen compounds from unstable diesel fuels *Petrol. Sci. Technol.* **32** 638–645.

[6] de Souza W F, Guimarães I R, Guerreiro M C and Oliveira L C A 2009 Catalytic oxidation of sulfur and nitrogen compounds from diesel fuel *Appl. Catal. A* **360** 205–20

[7] Khan N A and Jhung S H 2019 Phytic acid-encapsulated MIL- 101(Cr): Remarkable adsorbent for the removal of both neutral indole and basic quinoline from model liquid fuel *Chem. Eng.* **375** 121948.

[8] Zakhodyaeva Y A, Voshkin A A, Belova V V and Khol'kin A I 2011 Extraction of monocarboxylic acids with binary extracting agents based on amines and quaternary ammonium bases *Theor. Found. Chem. Eng.* **45** 5 739.

[9] Zakhodyaeva Y A, Rudakov D G, Solov'ev V O, Voshkin A A and Timoshenko A V 2019 Liquid–liquid equilibrium of aqueous two-phase system composed of poly (ethylene oxide) 1500 and sodium nitrate *J. Chem. Eng. Data* **64** 3 1250-5.

[10] Zakhodyaeva Y A, Solov'ev V O, Zinov'eva I V, Rudakov D G, Timoshenko A V and Voshkin A A 2019 Interphase distribution of thiophene, toluene, and o-xylene in the hexane–polymer–water extraction system *Theor. Found. Chem. Eng.* **53** 4 550-5.

[11] Voshkin A A, Zakhodyaeva Y A, Zinov'eva I V and Shkinev V M 2018 Interphase distribution of aromatic acids in the polyethylene glycol–sodium sulfate–water system *Theor. Found. Chem. Eng.* **52** 5 890-3.

[12] Zinov'eva I V, Zakhodyaeva Y A and Voshkin A A 2019 Extraction of Monocarboxylic Acids f rom Diluted Solutions with Polyethylene Glycol *Theor. Found. Chem. Eng.* **53** 5 871-4.

[13] Fedorova M I, Zinov'eva I V, Zakhodyaeva Y A and Voshkin A A 2020 Extraction of Fe (III), Zn (II), and Mn (II) Using a System with a Green Solvent for Trioctylmethylammonium Thiocyanate *Theor. Found. Chem. Eng.* **54** 313-8.

[14] Voshkin A A, Belova V V and Zakhodyaeva Y A 2018 Iron extraction with di (2-ethylhexyl) dithiophosphoric acid and a binary extractant based on it *Russ. J. Inorg. Chem.* **63** 3 387-390.

[15] Belova V V, Voshkin A A, Egorova N S and Kholkin A I 2012 Solvent extraction of rare earth metals from nitrate solutions with di (2, 4, 4-trimethylpentyl) phosphinate of methyltrioctylammonium *Journal of Molecular Liquids.* **172** 144-6.

[16] Belova V V, Egorova N S, Voshkin A A and Khol'kin A I 2015 Extraction of rare earth metals, uranium, and thorium from nitrate solutions by binary extractants. *Theor. Found. Chem. Eng.* **49** 4 545-9.

[17] Belova V V, Voshkin A A, Kholkin A I and Payrtman A K 2009 Solvent extraction of some lanthanides from chloride and nitrate solutions by binary extractants *Hydrometallurgy* **97** 3-4 198-203.

[18] Kostanyan A E and Voshkin A A 2011 Analysis of cyclic liquid chromatography *Theor. Found. Chem. Eng.* **45** 1 68-74.

[19] Kostanyan A E, Voshkin A A and Kodin N V 2011 Pulsed cyclic device for liquid countercurrent chromatography *Theor. Found. Chem. Eng.* **45** 5 779.

[20] Gradov O M, Voshkin A A and Zakhodyaeva Y A 2017 Estimating the parameters of ultrasonically induced mass transfer and flow of liquids in the pseudomembrane method *Chem. Eng. Process: Process Intensification* **118** 54-61.

[21] Kostenko M O, Pokrovskiy O I, Zakhodyaeva Y A, Voshkin A A and Lunin V V 2020 Unusual effect of flow rate on retention in analytical supercritical fluid chromatography exemplified by polyethylene glycol separation *J. Chromatogr. A* **1610** 460513.

[22] Ali M C, Yang Q, Fine A A, Jin W, Zhang Z, Xing H, et al. 2016 Efficient removal of both basic and non-basic nitrogen compounds from fuels by deep eutectic solvents. *Green Chem.* **18** 157–64.

[23] Zakhodyaeva Y A, Rudakov D G, Solov'ev V O, Voshkin A A and Timoshenko A V 2019 Liquid–Liquid Equilibrium in an Extraction System Based on Polyvinylpyrrolidone-3500 and Sodium Nitrate *Theor. Found. Chem. Eng.* **53** 2 159-65.

[24] Zinov'eva I V, Zakhodyaeva Y A and Voshkin A A 2019 Interphase Distribution of Caffeine and Coumarin in Extraction Systems with Polyethylene Glycol and Sodium Sulfate *Theor. Found. Chem. Eng.* **53** 6 996-1000.

[25] Zinov'eva I V, Zakhodyaeva Y A and Voshkin A A 2020 Data on the extraction of benzoic, salicylic and sulfosalicylic acids from dilute solutions using PEG-based aqueous two-phase systems *Data in brief* **28** 105033.
[26] Zakhodyaeva Y A, Zinov'eva I V and Voshkin A A 2019 Extraction of iron (III) chloride complexes using the polypropylene glycol 425–NaCl–H 2 O system *Theor. Found. Chem. Eng.* **53** 5 735-40.
[27] Zakhodyaeva Y A, Zinov'eva I V, Tokar E S and Voshkin A A 2019 Complex extraction of metals in an aqueous two-phase system based on poly (ethylene oxide) 1500 and sodium nitrate *Molecules* **24** 22 4078.
[28] Gradov O M, Zakhodyaeva Y A, Zinov'eva I V and Voshkin A A 2019 Some Features of the Ultrasonic Liquid Extraction of Metal Ions *Molecules* **24** 19 3549.
[29] Solov'ev V O, Zakhodyaeva Y A and Voshkin A A 2020 On influence of polymer additives, sodium nitrate and 1-methyl-2-pyrrolidone on the extraction of thiophene in the n-hexane - water system *Theor. Found. Chem. Eng* **54** 5 623-8.
[30] Jiang X, Zhu S, Gao J, Yu Y, Xiong C, Li C and Yang W 2019 Extractive removal of both basic and non-basic nitrogens from fuel oil by dicarboxyl-modified polyethylene glycol: Performance and mechanism *Fuel* **254** 115626.
[31] Zhu S, Xu J, Cheng H, Gao J, Jiang X, Li C and Yang W 2019 Poly (ethylene glycol) Diacid-Based Deep Eutectic Solvent with Excellent Denitrogenation Performance and Distinctive Extractive Behavior *Energy & Fuels* **33** 10 10380-8.

Acknowledgement
The reported study was funded by RFBR according to the research project № 18-29-06070

IOP Conf. Series: Materials Science and Engineering 1212 (2022) 012016 doi:10.1088/1757-899X/1212/1/012016

Synthesis of composite nanostructure natural ZrO_2 and magnetite particles (Fe_3O_4@ZrO_2) and study of its lead ion adsorption efficiency

C Irawan[1], M D Putra[1], I F Nata[1], M F Refki[1], R Hidayat[1], A Triantoro[2], Abbas[2]

[1] Department of Chemical Engineering, Faculty of Engineering, Universitas Lambung Mangkurat, Jl. A. Yani Km. 36, Banjarbaru, South Kalimantan, Indonesia 70714
[2] Department of Mining Engineering, Faculty of Engineering, Universitas Lambung Mangkurat, Jl. A. Yani Km. 36, Banjarbaru, South Kalimantan, Indonesia 70714

Corresponding author: cirawan@ulm.ac.id

Abstract. The potential of zircon minerals in Indonesia, especially in Central Kalimantan, has not been adequately explored and developed into valuable materials with high technical and economic value and environmentally friendly. This research has the potential to be processed and formed into advanced materials, seeing its high potential as an excellent adsorbent for anions/cations in water treatment and industrial wastewater. This research aims to develop raw zircon minerals into zircon oxides, which will later be composited with magnetic nanoparticles. The zircon mineral processing is carried out using hydrothermal methods. It is known that the physical and mechanical characteristics are suitable to be developed by having good reusability and durability as advanced materials. The adsorbent characterizations of FTIR, SEM, and XRF analysis showed that the Fe_3O_4@ZrO_2 had many different functional groups and a high specific surface area for adsorption processes. The Fe_3O_4@ZrO_2 showed high adsorption uptake capacity and selectivity for the lead in the Sasirangan textiles wastewater. Therefore, the Fe_3O_4@ZrO_2 have the potential to be used as an adsorbent in water and wastewater treatment.

1. Introduction

In Indonesia, zircon minerals are often found on land, oceans, coastlines, and rivers. These zircon minerals are found together with calciterite, electro minerals such as gold and silver, and others minerals like ilmenite, magnetic, monalacite, xenotime, pyrite, sulfide and silica. Economically, zircon minerals are found in granules, which are breakdown of acid and metamorphic igneous rocks that are then transported, concentrated and sedimentated together with zircon grains including calciterite (the main mineral of tin), rutile, ilminite, pyrite, magnetic, gold and iron oxide. Zircon minerals have great potential to be used and developed in various applications, for example: fuel cells, electro and energy catalysis, protective coatings for optical glass and filters, nano-electronics, microelectronics, ceramic biomaterials, thermo-luminescence UV dosimeters, piezoelectric manufacturing, electro-optic, dielectric, nano-composites and nanocapping materials and other industrial applications [1, 2].

Currently, zirconium and its oxides are widely used in water treatment and purification applications using various methods to synthesize ZrO_2-based nanostructures with different physicochemical properties [3-5]. Ehrlich et al. synthesized zirconium dioxide nanostructures with a hydrothermal template of chitin and had good potential from these zirconium-chitin composites for different applications. However, one of the factors limiting the use of pure chitin is its relatively low mechanical strength [6]. Ciesielczyk et al. synthesized micrometric $ZrO_2@SiO_2$ binary oxide through the sol-gel method [7]. The results showed that the adsorption and release capabilities of the synthesized material were significantly influenced by the number of surface functional groups and the porosity of the material. In another recent study, $MgO@ZrO_2$ oxide material was synthesized via the sol-gel method, which showed good adsorption due to the high surface area of the synthesized material. Wang et al., conducted research on the synthesis of a nano-zero-iron-reduced-zero graphite oxide composite and hydrated zirconium-oxide nanocomposite using a graphene oxide (GO) template due to its superior mechanical properties and high surface area [1, 8]. ZrO_2 nanosheets have a higher surface area, more active sites and a porous structure for better adsorption applications.

In addition, magnetic nanoparticles have attracted great interest research in recent years as adsorbents to remove and recover pollutants from water because the use of magnets provides a very simple and effective method of separating particles from water. But because of the possible oxidation of Fe^{2+} in the structure of magnetite in air, the limited types of pollutants that have magnetite, and a relatively low adsorption capacity, the modification of the surface of magnetite for use as magnetic adsorbent for attraction is investigated intensively [9-11]. This research conducted similarly with our previous work using the one-step synthesis hydrothermal and application of magnetic adsorbents which have the advantage of not being easily oxidized, magnetic composite nanoparticles with biomass fibers and chitosan to remove and recover anions from Municipal Water Work water sources, artificial and industrial wastewater with relatively high concentrations [12-14]. The results obtained indicate that the resulting material has a high adsorption performance for pollutants (anions and cations in terms of adsorption capacity, selectivity, and kinetics. However, the mechanism is still poorly understood and its potential applications in natural water and industrial wastewater with relatively low anion concentrations. It is hoped that the zircon oxide composite with magnetic nanoparticles ($Fe_3O_4@ZrO_2$) using in this research which is applied to drinking water treatment can provide a real contribution to the treatment of household or industrial wastewater containing other ions and dissolved species.

2. Materials and Methods

2.1. The synthesis of zircon mineral with magnetic nanoparticles composites

The synthesis of zircon mineral composites and magnetic nanoparticles was carried out by means of a one-pot solvothermal reaction (Irawan et al. 2018b). Initially anhydrous sodium acetate (1.6 g) and FeCl3.6H2O (0.8 g) were dissolved in ethylene glycol (24 mL), the mixture was heated at 60 oC while stirring (100 rpm) until the solution was homogeneous and colored orange. The zircon mineral (1 g) is added to the mixture and left for 20 minutes, then poured into a solvothermal reactor. The reaction lasts for 6 hours at 200 °C. After the reaction has been achieved, it is cooled down at room temperature. The resulting composites are washed with distilled water until the solution is clean from the rest of the other chemicals. Subsequent washing with ethanol (50%) for 3 times. To simplify the washing process, magnets are used to separate the composite from the mixture. The resulting composites are stored in distilled water for the next stage of use, the purpose of which is storage in distilled water is to avoid the oxidation process with air. On the other hand, to modify the amino functional group on the composite surface was added 1,6-hexanediamine to the mixture (7 mL) before entering the reactor.

2.2. The batch adsorption experiment

The adsorption process by batch experiment was carried out by taking a sample of 50 mL of Sasirangan textile-contaning lead wastewater which was placed in a 100 mL glass bottle and the pH value was adjusted to 1 M NaOH or 1 M HCl. A certain amount of adsorbent is put into the glass bottles. The mixture is then placed in a shaker with a stirring speed of 100 rpm, for a certain contact time at room temperature and then samples are taken for analysis. The adsorption experiments were done in triplicate samples and the average value was taken. The data from the experiment were fitting and figure out by Sigma Plot® version 10 of software.

3. Results and Discussion

3.1. Characterization of Fe_3O_4@ZrO_2 composites

The structure morphology properties and chemical composition properties of Zircon mineral and Fe_3O_4@ZrO_2 composites have been characterized using scanning electron microscopy (SEM, JOEL JSM-6500F) with energy-dispersive X-ray spectroscopy (EDAX) and and elemental analysis by Energy Dispersive X-ray Fluorescence (XRF, PANalytical's MiniPal 4 energy-dispersive EDXRF bench-top spectrometer) to conclude the identity and quantities of the elements in samples.

Typical SEM images of a representative sample show the smooth surface of the zircon mineral (**Fig. 1a**). The Fe_3O_4@ZrO_2 composites revealed the round-shaped magnetic nanoparticle along zircon surface with diameter size in the range of 20–30 nm when $FeCl_3.6H_2O$ was included in the reaction solution. In addition, due to the presence of magnetic nanoparticles that covered space layer of zircon, the color of composites is dark-brown (**Fig. 1b**).

Figure 1. SEM images of (a) Zircon mineral and (b) Fe_3O_4@ZrO_2 composites

Table 1 Elemental analysis of zircon mineral and Fe_3O_4@ZrO_2 composites measured by XRF

Sample	Chemical composition (Percentage of mass, %)				
	SiO_2	Fe_2O_3	ZrO2	HfO_2	Other
Zircon mineral	28.21	10.80	58.95	1.28	0.76
Fe_3O_4@ZrO_2 composites	0.5	66.26	32.0	0.69	0.55

The XRF data provide evidence of the chemical composition of each the adsorbent in **Table 1**, respectively. It illustrates that the major component are Si, Fe, and Zr. it can only determine the total amount of chemical composition in each sample. These data show that the zircon mineral dominated by Zr, Si then Fe, while the Fe_3O_4@ZrO_2 composites dominated by Fe, Zr then Si that seem similar to zircon but the amount of element increased when combined each other. Therefore, these materials of the samples are identical. It illustrates that XRF cannot differentiate between the chemical compositions in the samples. Still, it can only determine the total amount of Si, Fe and Zr as a major component in the samples.

3.2. Batch Adsorption Studies

The effect of contact time on the adsorption of lead onto Fe_3O_4@ZrO_2 composites as adsorbent was examined by conducting an experiment at room temperature, initial lead concentration of ca. 60 mg/L, adsorbent dose of 0.5 g/L, and shaking rate of 100 rpm. Fig. 1a shows that during the first 60 minutes, lead uptake capacity increased significantly from 0 mg/g to 108 mg/g, which was probably due to the diffusion taking place into the pores and or adsorption onto the surface of the adsorbent. Initially, the lead concentration gradient was relatively high diffusing into and through the pore of the adsorbent. Afterward, the lead uptake capacity slightly increased to lead uptake capacity of 112 mg/g and being constant until reached equilibrium condition within 6 hours. The equilibrium adsorption experiments were conducted in 6 hours of reaction time.

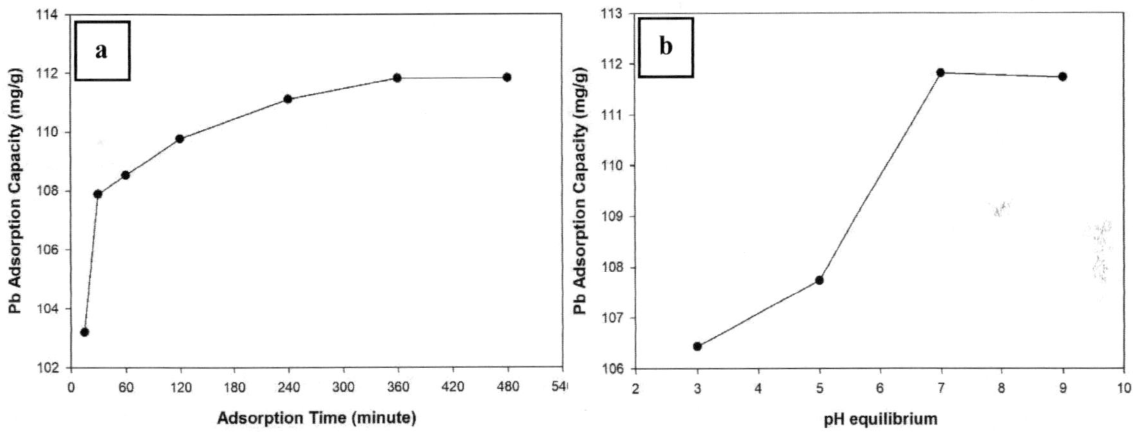

Figure 2. Pb adsorption capacity of (a) The effect of contact time (pH_e of 7.0±0.2) (b) The effect of equilibrium pH (6 hours), at room temperature, stirring rates of 100 rpm, adsorbent dose of ca. 0.5 g/L, respectively.

Several investigators reported that lead adsorption process usually pH-dependent and found optimum pH range of lead removal on the different adsorbents such as activated carbon [15], nano-zirconium silicate [16], Fe_3O_4@BDC@AGPA nanocomposites [17] and magnetic nanoparticle coated montmorillonite [14]. Fig. 2a shows that pH affected lead adsorption density over a wide pH range of 3.0 - 9.0. The uptake adsorption capacity is relatively low at lower pH. However, increasing equilibrium pH, a significant enhancement in adsorption uptake capacity, is recorded. The optimum pH for Fe_3O_4@ZrO_2 composites as adsorbent was found to be 7.0±0.2, with the uptake capacity of about 110 mg/L Pb(II). At lower pH, Pb(II) uptake decreased because the adsorbent's surface area was more protonated and competitive adsorption occurred between H^+ protons and free Pb(II) ions and their hydroxides towards the fixation sites. Therefore, H^+ ions react with anionic functional groups on the surface of Fe_3O_4@ZrO_2 composites and results on the reduction of the number of binding sites available for the adsorption of Pb(II). The amount adsorbed increased as pH increased from 3.0 to 9.0.

This increase may be due to the presence of negative charge on the adsorbent's surface that may be responsible for metal binding. However, as the pH is lowered, the hydrogen ions compete with the metal ions for the sorption sites in the sorbent; the overall surface charge on the adsorbent becomes positive, and hinds the binding of positively charged metal ions. At pH 9.0, a decrease in Pb(II) adsorption was observed due to the formation of soluble hydroxyl complexes. It could also increase concentration of hydroxyl ion (OH^-) which intermolecular competitiveness with lead ion for the Fe_3O_4@ZrO_2 composites surface site adsorption and depressed the lead uptake capacity. The excess of hydroxyl ions surrounding the adsorbent's binding sites was made the adsorption process unfavourable and leading to precipitated was observed.

Moreover, the force acting between the neutral charge of lead ($Pb(OH)_2$) and the surface is the weaker van der Waals interaction due to the attractive force between molecular activities with one another or with neutral molecules. The lowest of molecular interaction and these interactions appear because molecules have an unsymmetrical electronic distribution that creates small dipoles that can induce complementary dipoles in the neighboring surface, which depend on the molecule's orientation respect to the surface of the adsorbent.

4. Conclusions

The Fe_3O_4@ZrO_2 was successfully synthesized using a one-pot hydrothermal process. It showed high selectivity adsorption uptake capacity up to 110 mg/g for the lead in the Sasirangan textiles wastewater at room temperature, pH_e of 7.0±0.2, stirring rates of 100 rpm, adsorbent dose of ca. 0.5 g/L. The adsorption mechanism of lead adsorption by Fe_3O_4@ZrO_2 was a combination of electrostatic attraction and Van der Waals interaction, whereas at pH above 7 was precipitated. Therefore, the Fe_3O_4@ZrO_2 has the potential to be used as an adsorbent in water and wastewater treatment.

References

1. Fletcher, A., *2 - SOURCES OF ZIRCONIA*, in *Zirconia (Third Edition)*. 1992, Elsevier. p. 17-18.
2. Keerthana, L., C. Sakthivel, and I. Prabha, *MgO-ZrO2 mixed nanocomposites: fabrication methods and applications.* Materials Today Sustainability, 2019. **3-4**: p. 100007.
3. Cui, H., et al., *Strong adsorption of arsenic species by amorphous zirconium oxide nanoparticles.* Journal of Industrial and Engineering Chemistry, 2012. **18**(4): p. 1418-1427.
4. Hang, C., et al., *As(III) and As(V) Adsorption by Hydrous Zirconium Oxide Nanoparticles Synthesized by a Hydrothermal Process Followed with Heat Treatment.* Industrial & Engineering Chemistry Research, 2012. **51**(1): p. 353-361.
5. Luo, J., et al., *Zirconia (ZrO2) Embedded in Carbon Nanowires via Electrospinning for Efficient Arsenic Removal from Water Combined with DFT Studies.* ACS Applied Materials & Interfaces, 2016. **8**(29): p. 18912-18921.
6. Ehrlich, H., et al., *Extreme Biomimetics: formation of zirconium dioxide nanophase using chitinous scaffolds under hydrothermal conditions.* Journal of Materials Chemistry B, 2013. **1**(38): p. 5092-5099.
7. Ciesielczyk, F., et al., *The development of zirconia/silica hybrids for the adsorption and controlled release of active pharmaceutical ingredients.* Colloids and Surfaces A: Physicochemical and Engineering Aspects, 2018. **545**: p. 39-50.
8. Wang, C., et al., *Removal of As(III) and As(V) from aqueous solutions using nanoscale zero valent iron-reduced graphite oxide modified composites.* Journal of Hazardous Materials, 2014. **268**: p. 124-131.
9. Khani, R., S. Sobhani, and M.H. Beyki, *Highly selective and efficient removal of lead with magnetic nano-adsorbent: Multivariate optimization, isotherm and thermodynamic studies.* Journal of Colloid and Interface Science, 2016. **466**: p. 198-205.
10. Rajabi, H.R., et al., *Surface modified magnetic nanoparticles as efficient and green sorbents: Synthesis, characterization, and application for the removal of anionic dye.* Journal of Magnetism and Magnetic Materials, 2015. **394**: p. 7-13.

11. Sarkar, A., S.K. Biswas, and P. Pramanik, *Design of a new nanostructure comprising mesoporous ZrO2 shell and magnetite core (Fe3O4@mZrO2) and study of its phosphate ion separation efficiency.* Journal of Materials Chemistry, 2010. **20**(21): p. 4417-4424.
12. Irawan, C., et al., *Biopolymer of Chitosan from Fish Scales as Natural Coagulant for Groundwater Treatment.* IOP Conference Series: Earth and Environmental Science, 2018. **175**: p. 012028.
13. Irawan, C., et al., *Biosorption of Lead (II)–containing Sasirangan Textile Wastewater using Nanocomposites of Eleocharis dulcis Fibers with Iron (III) Nanoparticles as Adsorbent.* MATEC Web Conf., 2018. **156**.
14. Irawan, C., I.F. Nata, and C.-K. Lee, *Removal of Pb(II) and As(V) using magnetic nanoparticles coated montmorillonite via one-pot solvothermal reaction as adsorbent.* Journal of Environmental Chemical Engineering, 2019. **7**(2): p. 103000.
15. Mouni, L., et al., *Adsorption of Pb(II) from aqueous solutions using activated carbon developed f rom Apricot stone.* Desalination, 2011. **276**(1): p. 148-153.
16. Mahmoud, M.E., G.M. Nabil, and S.M.E. Mahmoud, *High performance nano-zirconium silicate adsorbent for efficient removal of copper (II), cadmium (II) and lead (II).* Journal of Environmental Chemical Engineering, 2015. **3**(2): p. 1320-1328.
17. Alsuhybani, M., et al., *Highly efficient removal of Pb(II) from aqueous systems using a new nanocomposite: Adsorption, isotherm, kinetic and mechanism studies.* Journal of Molecular Liquids, 2020. **301**: p. 112393.

Acknowledgment
A grant of Universitas Lambung Mangkurat financially supported this work, The Ministry of Education and Culture of Indonesia (Project No: 212.319/UN8.2/PL/2020), the fund of Research for Lecturer Grant.

IOP Conf. Series: Materials Science and Engineering 1212 (2022) 012017 doi:10.1088/1757-899X/1212/1/012017

The effect of plastic type on the yield and quality of lignite and plastic waste pyrolysis products

H Wijayanti, A Tuhuloula, L Subekti, M Aisyah, I D Pitaloka, and Siti Nurmajidah

Department of Chemical Engineering, Faculty of Engineering, Lambung Mangkurat University, South Kalimantan, Indonesia

hesti.wijayanti@ulm.ac.id

Abstract. Lignite is a low rank coal which has great potential in South Kalimantan. However, it has not been used optimally due to its low quality. One of the ways to improve it is pyrolysis. Pyrolysis is the thermal decomposition of organic material in the absence of oxygen which will produce three products (char, tar and gas). Pyrolysis can make lignite into liquid fuel (pyrolysis liquid), but it still requires improvement due to the need for pure hydrogen donors. Plastic waste has a higher hydrogen/carbon ratio than coal. This material can be used as an additive in the pyrolysis process because it is rich in hydrogen. The samples of plastic waste used were ppolyethylene (PE), ppolypropylene (PP), and polystyrene (PS). Samples of lignite and plastic (plastic composition was 25 wt%) were used for every experiment, and pyrolysis was carried out with a holding time of 60 minutes at 500°C. The pyrolysis liquid obtained is then analyzed for its yields and properties (density, kinematic viscosity, heating value). The most feasible result for fuel alternative was obtained with the addition of PE.

1. Introduction

Indonesia has abundant potency of coal. The data from ESDM in 2018 [1] mentioned that the coal deposit in Indonesia is approximately 26.2 billion tons. The deposit in Kalimantan island leads as the highest with the amount 14.9 billion tons, followed by Sumatera and Sulawesi with the total amount of 11.2 and 0.12 billion tons, respectively. However, coal from Kalimantan island including to lignite and sub-bituminous coal [2] which is considered as a low grade (low rank) coal due to its low energy content and high moisture [3].

Some previous research have been conducted to optimize the utilization of lignite as an efficient energy source namely gasification, hydothermal liquefaction [4] and pyrolysis [5]. Pyrolysis is an alternative method for clean and efficient utilization of lignite. This technology breaks the sample (such as lignite) structure into smaller species [6] similar to the technique for treating petroleum products and its derivatives [7].In pyrolysis, biomass is decomposed into liquid, solid (char) and gas in the absence of oxygen. However, the oxygen content of lignite is considerably high, so it is become a drawback in the pyrolysis process, since oxygen contributes in lowering calorific value of its product [8]. The addition of hydrogen can improve the quality of pyrolysis product. However, pure gas hydrogen is still expensive.

Plastic waste is still a big issue for the environment. Not only it is unable to be decomposed by nature, but it causes air pollution if burned [9]. Three main types most used of plastics waste are made of polyethylene, polypropylene and polystyrene [10]. Baofeng et al. [4] found that the composition of carbon and hydrogen in plastics is very high, even higher than that of in coal. They also observed that there is an interaction between plastics and lignite during the pyrolysis process. It is proposed that plastic waste will take part as the hydrogen "donor" during the lignite pyrolysis process.

Zhou et al [11] observed that pyrolysis of coal with polyethylene (PE) and polypropylene (PP) addition at various conditions result in the improvement of the liquid yields and the coal conversion. Moreover, Ozsin and Putun [12] discovered the pyrolysis of walnut shell and peach stones with polystyrene decreased activation energy value.

The research about pyrolysis lignite mixed with plastics is interesting to be conducted, in order to increase the quality of its pyrolysis product and at the same time also contribute to solve plastic waste problem issue. In this work, pyrolysis lignite (low rank coal) with three common types of plastics, those are polyethylene, polypropylene and polystyrene, is conducted. The effect of plastic type on pyrolysis product yields and also its quality are investigated. Furthermore, this study could contribute to develop lowrank coal as an efficient energy source.

2. Materials and Method

Lignite was obtained from Satui Region at Tanah Bumbu district, South Kalimantan. It was ground and sieved to the size of 0.5-1 mm to homogenize the sample. The lignite powder was dried in an oven. Whereas polyethylene (PE), polypropylene (PP) and polystyrene (PS) plastics were pelletized to the size of 2-4 mm. Then, these samples were characterized using proximate analysis. The result of proximate analysis for the feedstocks is presented in Table 1.

Table 1. The Proximate Analysis of Lignite and Plastics

Sample	% Weight			
	Moisture	Ash	Volatile Matter	Fixed Carbon
Lignite	17.10	6.51	59.15	17.24
PE	0.34	3.37	95.61	0.68
PP	0.30	5.48	93.40	0.82
PS	0.24	0.48	96.18	3.10

Pyrolysis reaction was investigated in a stainless steel tubular reactor with 16 cm of diameter and 25 cm of height. For every run, 500 grams of lignite was used. This sample was heated up to 500°C at the heating rate of 8°C/min and held at that temperature for 1 hour. The pyrolysis temperature (500°C) was chosen as observed by Kojic et al.[13] that at that temperature showed the positive synergetic effect. The vapor that produced from the reactor was flown through the connecting pipe and condensed through the condenser then the result was collected as pyrolysis oil in the flask. When the reaction was completed, the reactor was cooled down to room temperature so it could be opened to collect the char (solid product). The yields of product were obtained from the weight percentage of liquid, gas and char products to the total weight of sample used for every reaction. Furthermore, the pyrolysis oil was separated from its aqueous phase (contains much water) to get the organic phase that would be analyzed to measure its physical properties such as: heating value, viscosity and density. Furthermore, the pyrolysis of lignite and polyethylene (L/PE), polypropylene (L/PP) and polystyrene (L/PS) plastics were conducted as the same method as pyrolysis of lignite. The weight ratio of lignite:plastics used was 3:1.

3. Results and Discussion

3.1. Effect of plastic type on pyrolysis products yields

The yield of product is highly influenced by temperature, reactor configuration, type of biomass, catalyst and heating rate [14]. The pyrolysis of lignite will produce products in the form of liquid pyrolysis (tar and water), gas and char. The addition of plastic waste to lignite pyrolysis is expected to affect the yield of the resulting liquid pyrolysis. The yield char, gas and pyrolysis liquid of each type of plastic addition can be seen in Figure 1.

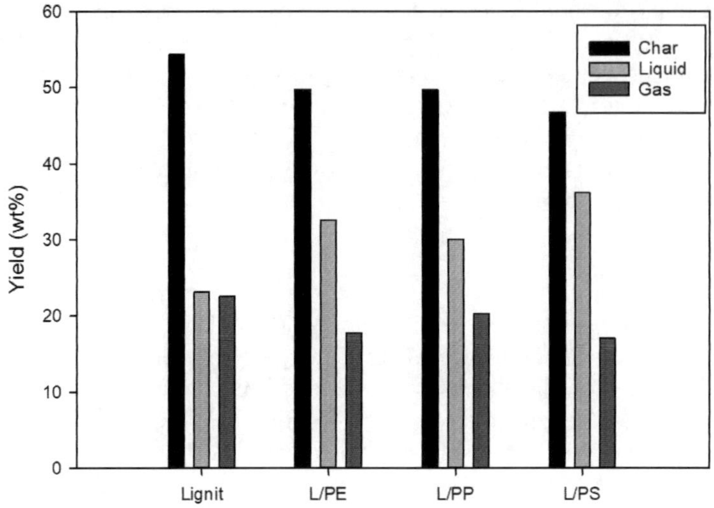

Figure 1. Effect of plastic type on the yields of pyrolysis products

From Figure 1, the different type of plastic used results different trend of products yield compared to the result of lignite pyrolysis itself. In general, the highest char product was resulted from lignite pyrolysis (55.61%), while the lowest was produced from L/PS at 46.74%. On the other hand, L/PS resulted the highest liquid yield at 36.92% compared to L/PE, L/PP and lignite pyrolysis while the smallest resulted from L/PP. As observed by Pinto et al [10], that found pyrolysis of PS resulted the highest liquid yield compared to PE and PP. As observed by Kiran et al. [15] the PS pyrolysis yielded higher liquid than that of PE and PS. L/PP pyrolysis performed the highest gas product at 32.96%. In contrast, the gas resulted from L/PS pyrolysis was the lowest. These results imply that the addition of plastic PE, PS and PP in lignite pyrolysis will increase the production of liquid and gas, but lowering char products. These in accordance to Qian's (2014)[16] research on product yield with the addition of PE and PP at various compositions, the most products formed are tar and gas due to the high hydrocarbon content. Furthermore, the pyrolysis liquid product consists of two different phases, namely the organic phase (tar) and the aqueous phase (water). The organic phase of the liquid pyrolysis product has the potential to be used as fuel. The yields of the organic phase and the aqueous phase between the type of plastic waste used (PP, PE, PS) during lignite pyrolysis is presented in Figure 2.

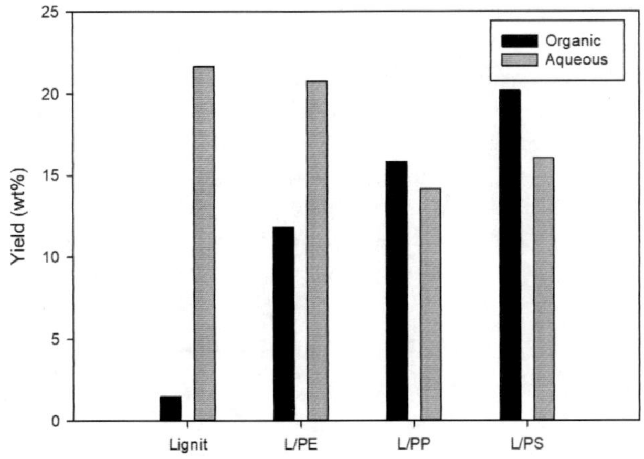

Figure 2. Effect of plastic type on the yields of organic and aqueous phase in pyrolysis liquid

As can be seen in Figure 2, the addition of three types of plastics shows significant differences in results compared to lignite alone. The order of the organic phase yield from the lowest to the highest is lignite, L/PP, L/PE and L/PS. In lignite pyrolysis, as the temperature increases, free radicals will be generated when the linking bonds between the basic structural units of lignite are broken so that hydrocarbon gases can be released from the system[17] . This causes the yield of gas and organic phase increase, while char yield decline [18-21]. In addition, Qian et al. [16] stated that free radicals from lignite participate in plastic decomposition reactions, and plastics act as a hydrogen donor for free radicals from lignite. In the co-pyrolysis process, the less active hydrogen from lignite is combined with the hydrogen from plastic, and the combination of these two can form a molecule with the same size of liquid pyrolysis [18]. Thus, the organic phase yields increased. Sinag et al.[22] observed a similar trend with the addition of LDPE plastic to Mustafa Kemal Pasüa (MKP) lignite compared to the absence of LDPE. Moreover, lignite pyrolysis itself without the addition of plastic will require a higher pyrolysis temperature [22].

3.2. Effect of plastic type on pyrolysis liquid properties

Analysis of physical properties is carried out to determine the viscosity, density and also heating value of the organic phase. The results of the analysis and also the properties of some types of other fuel can be seen in Table 2.

Table 2. Properties of Pyrolysis Liquid (Organic Phase) Compared with Fuels

	High Heating Value (cal/g)	Kinematic Viscosity at 20°C (mm^2/s)	Density at 20°C (kg/m^3)
L/PE	10935.859	9.0478	825
L/PS	9823.637	1.3543	878
L/PP	10338.097	9.0258	804
Crude oil[23]	11059.000	5 – 20	700-1000
Gasoline[23]	10605.000	0.5	720-760
Diesel[23]	10844.000	2 – 4	780-860
Biodiesel[23]	10079.000	4 – 6	860-900
Jet Fuels (kerosene based) [23]	10461.000	8	760-810

Based on the data in Table 2, the heating value of the organic phase resulting from pyrolysis shows a value that is not much different from other fuels. Bhattacharya et al. [24] reported the co-pyrolysis of PE mixed with pine wood caused the reduction of oxygen and water content that increase high heating rates. L/PE has the highest heating value. This is because the H element in PE is the highest at 14.3% [25] compared to other types of plastics. L/PE and L/PP meet the specifications as fuel oil because generally fuel oil has a heating value of 10079-10844 cal / g. Meanwhile, L/PS still has a heating value below the fuel oil specification. The kinematic viscosity of the organic phase obtained in the three types of plastics in this study is still between the viscosity values for fuel. When viewed from the viscosity value, L/PE and L/PP are close to the characteristics of kerosene-based jet fuels. However, the L/PS viscosity is almost not comparable with other types of oil. As for the density parameters L/PE and L/PP are included in the diesel oil range.

4. Conclusion

The different type of plastic used results different trend of products yield compared to the result of lignite pyrolysis itself. These results imply that the addition of plastic PE, PS and PP in lignite pyrolysis will increase the production of liquid and gas, but lower char products. Moreover, the addition of three types of plastics shows significant differences in results compared to lignite alone. The order of the organic phase yield from the lowest to the highest are: lignite, L/PP, L/PE and L/PS. The properties in the term of heating value, viscosity and density of the organic phase from pyrolysis of lignite and PE and PP in this study are almost similar to other fuels, however, the heating value and density of lignite and PS are not comparable with others.

References

[1] ESDM. 2018. Cadangan Batubara Indonesia Sebesar 26 Miliar Ton. *https://www.esdm.go.id/id/media-center/arsip-berita/cadangan-batubara-indonesia-sebesar-26-miliar-ton (in Bahasa)* Accessed on 25 July 2018
[2] Speight J 2005 *Handbook of Coal Analysis. John Wiley & Sons, Inc., Publication* 222
[3] Shi Y, Li S, Hu H 2012. *Journal of Analytical and Applied Pyrolysis* **95** 75.
[4] Baofeng W, Yaru H, Jinjun Z 2015. *China Petroleum Processing and Petrochemical Technology* **17** 24
[5] Chunmei Q, Min Z, Jianghong W, Puhai Y, Xu Y 2014. *International Journal of Mining Science and Technology* **24** 137
[6] Pinto F, Gulyurtlu I, Goncalves M, Cabrita I, Costa P 1997. *Proceedings of Recycle 97 Switzerland* **2** 256
[7] Shell Group Companies 1983. *The Petroleum Handbook* 6th ed. 235
[8] Bridgwater A V 2012. *Biomass and Bioenergy* **38** 68
[9] Tipayarom D, Oanh N T K 2007. *ScienseAsia* **33(3)** 339
[10] Pinto F, Costa P, Gulyurtlu I, Cabrita I 1999. Journal of Analytical and Applied Pyrolysis **51** 39
[11] Zhou L, Luo T, Huang Q 2009. *Energy Conversion and Management* **50** 705
[12] Ozsin G, Putun A E 2017. *Energy Conversion and Management* **149** 675
[13] Kojic I, Bechtel A, Kittinger F, Stevanovic N, Obradovic M, Stojanovic K 2018. *Journal of Serbian Chemical Society* **83** 1
[14] Putun E, Uzun B B, Putun A E 2006. *Bioresources Technology* **97** 701
[15] Kiran N, Ekinci E, Snape C E 2000. *Resources, Conservation and Recycling* **29** 273
[16] Qian C, Zhou M, Wei J, Ye P, Yang X 2014. *International Journal of Mining Science and Technology* **24** 137
[17] Qiu N S, Li H L, Jin Z J, Zhu Y K 2007. *International Journal of Coal Geology* **69(3)** 220
[18] Juntgen H 1984. *Fuel* 63(6) 731
[19] Sharypov V I, Beregovtsova N G, Kuznetsov B N, Cebolla B L, Collura S 2007. *Pyrolysis* **8(2)** 257
[20] Sakurovs R 2003. *Fuel* **82** 1911
[21] Fei J X, Zhang J, Wang F C, Wang J 2012. *Journal of Analytical and Applied Pyrolysis* **95** 61
[22] Sınağ A, Sungur M, Canel M 2006. Energy and Fuels **20** 109
[23] Demirel Y 2012. *Energy Production, Coversion, Storage, Conservation, and Coupling*. Lincoln NE 68588-0643 USA.
[24] Bhattacharya P, Steele P H, Hassan M, Mitchell B, Ingram L, Pittman U 2009. *Fuel* **88(7)**
[25] [25] Li D, Li W, Li B 2001. *Journal of Fuel Chemistry Technology* **29** 19

Acknowledgments

The authors wish to thank the Lambung Mangkurat University for the "Dosen Wajib Meneliti" Grant Year 2020.

A modified shrinking core model for microwave-assisted leaching of aluminum from peat clay

Hairullah, A Mirwan*, M D Putra, B H Ilmanto, H S H Putri, M B Ulum, M R Haka

Chemical Engineering Department, Faculty of Engineering, Lambung Mangkurat University, Banjarbaru, South Kalimantan, 71714, Indonesia

*Corresponding author's email: agusmirwan@ulm.ac.id

Abstract. Aluminum oxide in peat clay has the potential to be used as a catalyst, coagulant, and adsorbent for the water treatment process. The usefulness of aluminum oxide in peat clay is enhanced by the leaching process. Aluminum was leached from peat clay in a variety of microwave power, HCl concentrations, and particle size. The effect of the microwave leaching parameters on the aluminum leaching rate was observed. The shrinking core (SC) model used in microwave-assisted leaching was assumed a pseudo steady state with chemical reactions. Effective diffusivity (D_e), mass transfer coefficient (k_c), and reaction rate constants (k) are used as fitting parameters. The best fitting parameters D_e, k_c, and k obtained 0.0049 cm^2/s, 2.49 cm/s, and 10.5 cm/s, respectively. The comparison of experimental data and model calculations shown that the SC model can describe experimental data well for all microwave-assisted leaching conditions. Precious information on the results of this research can be given for the goal of the scaling-up and design of the microwave assisted leaching process.

1. Introduction

Peat clay used in this study is located at a depth of about 1.5 to 3.0 meters from surface soil. Peat clays are found in Indonesia, especially in South Kalimantan [1]. Mirwan et al. [2] reported that peat clay has high mineral content of silicon oxide (SiO_2), iron oxide (Fe_2O_3), titanium oxide (TiO_2), calcium oxide (CaO), and aluminum oxide (Al_2O_3), and other oxides in significant amounts small. The content of aluminum oxide (Al_2O_3) in peat clay is around 7.20% which by calcination process can increase the percentage of its content to 17.90%. Aluminum oxide has the potential to be used as a catalyst, coagulant, and adsorbent for the water treatment process [1,2]. Therefore, increasing the leaching process of aluminum oxide from peat clay should be of interest to industries related to the various potentials of aluminum oxide compounds.

Microwave-assisted leaching (MAL) has been used to increase the level of leaching process for mineral and secondary resources, especially copper ore, nickel ore, sewage sludge, fly ash, and industrial dust [5]. The MAL method has several advantages over conventional methods, especially in the fields of hydrometallurgy, food, pharmacological medicine, and natural medicine [6]–[8]. Aluminum leaching from peat clay using acidic solvents with the microwave-assisted has been considered as a possible field of MAL because of shorter processing times, high recovery values, and environmentally friendly.

Although the leaching process equipment with microwaves requires high maintenance costs than conventional process equipment [9], energy costs are reduced to one-tenth of conventional processes and can increase selectivity so as to reduce the energy and solvents used in the leaching process [10]. To improve the efficiency of MAL, there are several kinetic models that have been postulated for effective component diffusion. In engineering, mathematical modeling can simplify process control, optimize, design a process, and provide correct information about equipment scale up procedures. Moreover, the kinetic law provides better information about leaching behavior [11].

he proposed mathematical model is based on the integration of the balanced mass balance differential. These models are based on the mass transfer mechanism which is reviewed on a macro scale. The shrinking core (SC) model illustrates that the reaction occurs first in the outer shell of the particle and then the reaction area moves to areas that don't react more deeply. The reaction area occurs at ash layer assumed to be inert [12]. Macro-scale shrinking core models have been used in solid-liquid reactions with conventional heating and successfully applied to experimental data for aluminum leaching from water treatment sludge and peat clay [3], [13]. Aluminum leaching from peat clay has been modeled by Mirwan et al. [3] however, the model is only applied to temperature condition of the process. Aluminum was located in the solid core particles of shrinking peat soils as extracted solutes, and assumed that the structure of the particles changed, the kinetic mechanism of first-order leaching, and linear equilibrium at the solid-liquid interface. In their model, effective diffusivity, mass transfer coefficient, and reaction rate constants are used as fitting parameters.

In this work, microwave-assisted leaching aluminum from peat clay is carried out at various microwave power, acid concentration, and particle size. The novelty of this work is the application of the SC model to aluminum leaching from peat clay using microwave-assisted to describe the aluminum leaching process based on experimental data

2. Materials and Method

2.1. Materials
The peat clay was collected from Peat Village, South Kalimantan, Indonesia and approximately 3.0 meters in the depths from the surface of the earth. Peat clays are cleaned of small logs and other impurities, and dried by sunlight directly. Grinding and sieving are carried out to reduce particle size (0.0074 and 0.0149 cm) and calcined at 700 °C for 120 min.

2.2. Microwave aluminum leaching
The microwave-assisted leaching devices are arranged in advance as shown in Figure 1. 5 grams of sieved peat clay and 1 M HCl solution that has been made are inserted into a three-neck flask. The microwave power is set at 100 W and 60 min the operating time. The stirring motor with 300 rpm is turned on. The sample solution is taken using a syringe in the span of 5, 10, 15, 20, 25, 30, 40, 50 and 60 minutes.

Sample solution was stored in a glass sample bottle. The stages of the leaching process are repeated for 2 M and 4 M HCl solutions, microwave power of 80 W, and particle size of 0.0074 and 0.0149 cm. The aluminum content was determined using an inductively coupled plasmacluster optical emission spectrometer (ICP-OES, 9060-D Teledyne Leeman Labs, USA). Each analysis was repeated three times to obtain an average value.

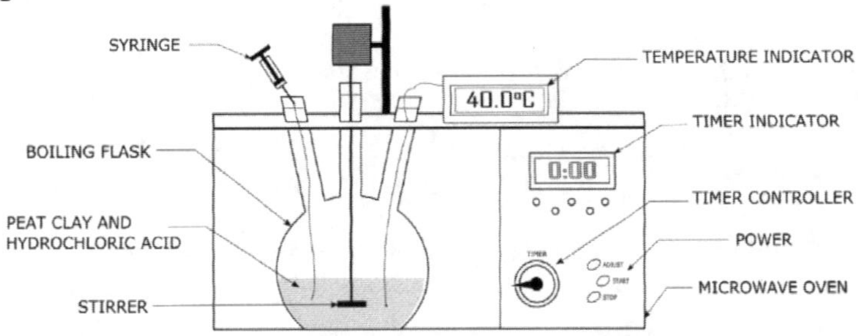

Figure 1. The microwave leaching experiment equipment

2.3. Mathematical model of shrinking core (SC)

The shrinking core (SC) model was adopted and modified from Mirwan et al. [13] with several to assume. Peat clay has a shape spherical particle. Fixed temperature during the microwave-assisted leaching process. Other compounds found in peat clay do not have a significant effect on process kinetics. The particles are perfectly mixed and the spherical shape of the particles can be maintained; although it has an effect on some cases. Based on assumptions and considerations that the microwave leaching process occurs in a stirred with reaction process, the mass balance equation is described as:

$$\frac{dC_{Al}}{dt} = \frac{k_c}{R}\frac{(1-\varepsilon)}{\varepsilon}\left(C_{A(R)} - C_{Al}\right) \tag{1}$$

Changed the concentration of solute (aluminum) in the solid phase (peat clay) are related to the mass transfer coefficient of the solute:

$$\frac{d\overline{q}}{dt} = \frac{4\pi R^2 k_c \rho_p}{3m}\left(C_{A(R)} - C_{Al}\right) \tag{2}$$

Diffusion equation outside the particle:

$$-\frac{d}{dr}\left(-D_e 4\pi r^2 \frac{dC_{A(R)}}{dr}\right) = 0 \tag{3}$$

The initial conditions in the equation (4) and boundary conditions in the equation (5 and 6):

$$r_c = R \text{ and } \overline{q} = q_u \text{ at } t = 0 \tag{4}$$

$$r = r_c \rightarrow D_e \frac{dC_{A(R)}}{dr} = kC_{A(R)} \tag{5}$$

$$-D_e \frac{dC_{A(R)}}{dr} = k_c\left(C_{A(R)} - C_{Al}\right) \tag{6}$$

The average value of the solute concentration in the solid phase:

$$\frac{\overline{q}}{q_u} = \left(\frac{r_c}{R}\right)^3 \tag{7}$$

Equation (3) was integrated to produce an equation (8):

$$C_{A(R)} = -\frac{K_1}{r} + K_2 \tag{8}$$

Equation (5) was substituted into the equation (8) to obtain equation (9):

$$K_1 = \frac{kK_2 r_c^2}{\left(D_e + kr_c\right)} \tag{9}$$

Equation (8) was substituted into the equation (6) so that the equation is obtained (10)

$$K_2 = \frac{k_c C_{Al}\left(D_e + kr_c\right)R^2}{D_e\left(kr_c^2 + R^2 k_c\right) + k_c k\left(r_c R^2 - r_c^2 R\right)} \tag{10}$$

Equation (10) was substituted into the equation (9). These equations are substituted into the equation (8) to obtain equation (11):

$$C_{A(R)} = \frac{kk_cC_{Al}R^2r_c^{\ 2} + k_cC_{Al}(D_e + kr_c)R^2}{D_e(kr_c^{\ 2} + R^2k_c) + k_ck(r_cR^2 - r_c^{\ 2}R)} \quad (11)$$

3. Results and Discussion

3.1. Effect of microwave power

Figure 2 shown that leaching recovery increases with increasing microwave power. Mass transfer coefficient was decreased, but the rate of formation reaction increases with increasing microwave power (Table 1). At the same time, mass transfer resistance increases with decreasing mass transfer coefficient. The shrinking core model was illustrated experimental data and SC model good fit with small error value.

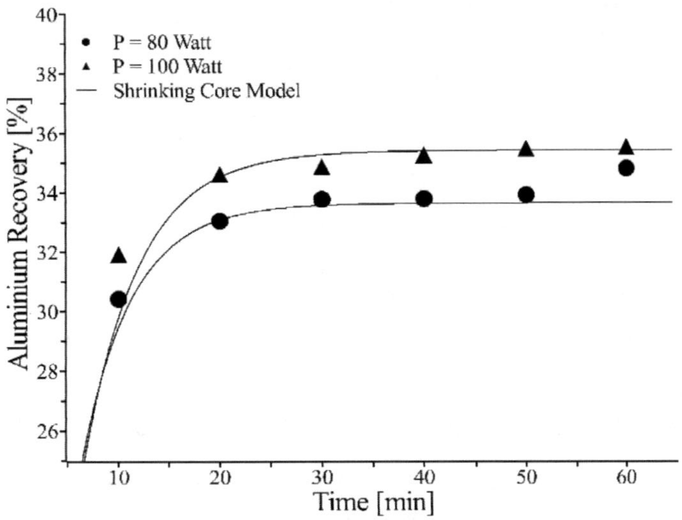

Figure 2. Effect of microwave power (80 and 100 W) on aluminum recovery by SC models at 0.0074 cm particle size and 1 M HCl concentration.

Table 1. Values of fitting parameters in SC mode

		SC model		
		D_e (cm^2/s)	k (cm/s)	k_c (cm/s)
Microwave power	80			
P (Watt)		0.002708	1.28	2.756
	100	0.00139	3.58	0.99
HCl concentration	1			
C_{HCl} (M)		0.00137	6.85	0.929
	2	0.003369	3.8	2.03
	4	0.009308	3.6	4.119
Particle size d_p (cm)	0.0074	0.01516	6.2	6.289
	0.0149	0.00358	1.2	0.909

3.2. Effect of HCl concentration

Figure 3 shown that the effect of HCl concentrations (1, 2, and 4 M) on recovery aluminum of microwave leaching. Recovery aluminum leaching increases with an increase in acid concentration.

Increasing acid concentration was caused an increase in the amount of acid mass per unit volume to enter the particle, thereby increasing inter-molecular interactions between acid and solutes, and increasing solubility of solutes. For this condition, mass transfer is influenced by increasing acid concentration, mass transfer coefficient and effective diffusivity. According to Figure 3, the SC model more good fit illustrates experimental data because it has a relatively small error value.

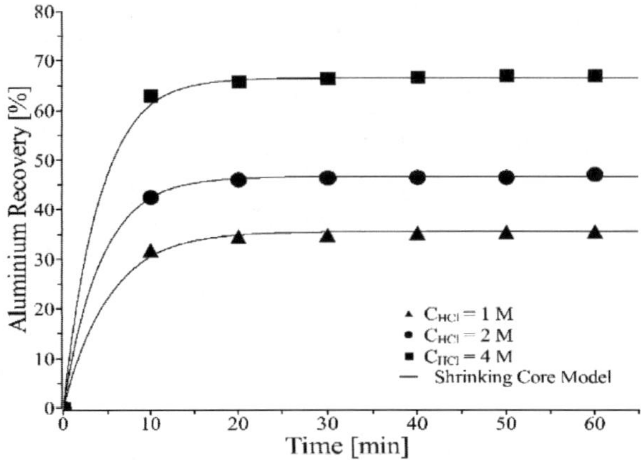

Figure 3. Effect of HCl concentrations (1, 2, and 4 M) on aluminum recovery by SC model at 0.0074 cm particle size and 100 W of microwave power

3.3. Effect of particle size

The effect of particle size (0.0074 and 0.0149 cm) on recovery leaching for SC model is shown in Figure 4. A reduction in recovery leaching for an increase in particle size is observed. The milling process increases surface area, reducing mass transfer resistance, making solutes more accessible to solvents, and consequently, increasing recovery aluminum leaching. Mirwan et al. [1] obtained similar results for aluminum leaching using HCl solvents under conventional heating. Smaller particle sizes can make large contact surface areas thereby increasing recovery leaching. The SC model can describe experimental data well at larger particle sizes of 0.0149 cm.

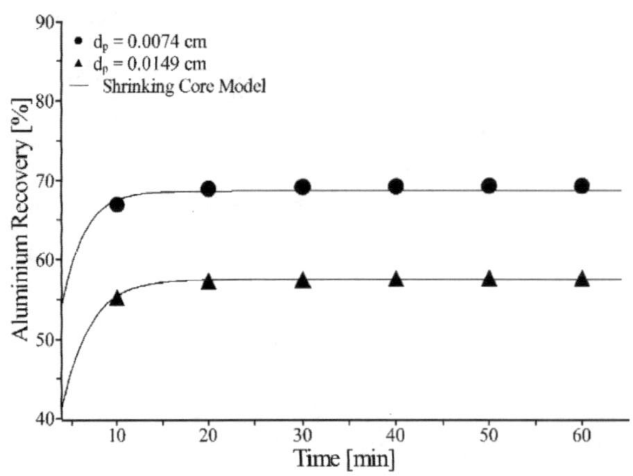

Figure 4. Effect of particle size (0.0074 and 0.0149 cm) on aluminum recovery by SC model at 4 M HCl concentration, 40 °C temperature, and 100 W microwave power

4. Nomenclature

$\bar{q}$	average value of q (g/L)	k_c	liquid phase mass transfer coefficient (cm/s)
C_{Al}	solvent concentration (g/L)	K_1	first partition constant (g/cm^3)
r	radial coordinates	K_2	second partition constant (g/cm^3)
$C_{A(R)}$	aluminum leaching concentration (g/L)	m	solid mass (g)
D_e	effective diffusivity (cm^2/s)	π	circle ratio (3,14159)
R	particle radius (cm)	ε	empty fraction (-)
k	reaction rate constants (cm/s)	ρ_p	solid density (g/cm^3)
t	time (s)		

5. Conclusions

Aluminum from peat clay was leached using microwave assisted leaching to observe the effects of microwave power, HCl concentration, and particle size. Aluminium recovery increase with in increasing microwave power, HCl concentration, and particle size. The shrinking core model can illustrate experimental data well for all extraction conditions except for particle size. The best fit of parameters is 0.0049 cm^2/s, 10.5 cm/s and 2.49 cm/s for D_e, k, and k_c, respectively.

References

[1] Mirwan A, Susianto, Altway A, Handogo R 2018 Kinetic model for identifying the rate controlling step of the aluminum leaching from peat clay *J. Teknol* **80** 2 37–44.

[2] Mirwan A, Putra M D, Lestari R A 2019 Potential peat clay for minerals source utilized as adsorbent and catalyst *MATEC Web of Conferences* **280** 04003.

[3] Mirwan A, Susianto, Altway A, Handogo R 2020 Temperature-dependent kinetics of aluminum leaching from peat clay *MJFAS* **16** 2 248–251.

[4] Wang J, Zhang Y, Huang J, Liu T 2018 Kinetic and mechanism study of vanadium acid leaching from black shale using microwave heating method *JOM* **70** 6 1031–1036.

[5] Ma Z, Liu Y, Zhou J, Liu M, Liu Z 2019 Recovery of vanadium and molybdenum from spent petrochemical catalyst by microwave-assisted leaching *Int. J. Miner. Metall. Mater.* **26** 1 33–40.

[6] Chen Y, Zhou J, Zhang L, Peng J, Li S, Yin S, Yang K, Lin Y 2018 Microwave-assisted and regular leaching of germanium from the germanium-rich lignite ash *Green Process. Synth.* **7** 6 538–545.

[7] Lovrić V, Putnik P, Bursać Kovačević D, Jukić M, Dragović-Uzelac V 2017 Effect of microwave-assisted extraction on the phenolic compounds and antioxidant capacity of blackthorn flowers *Food Technol. Biotechnol.* **55** 2 243–250.

[8] Umama Y, Venkatajah G, Shourabh R, Kumar R, Verma A, Kumar A, Gayoor M K 2019 Topic-the scenario of pharmaceuticals and development of microwave as; sisted extraction technique *World J. Pharm. Pharm. Sci.* **8** 7 1260–1271.

[9] Kapoore R V, Butler T O, Pandhal J, Vaidyanathan S 2018 Microwave-assisted extraction for microalgae: from biofuels to biorefinery *Biology* **7** 1 18.

[10] Guo L, Lan J, Du Y, Zhang T C, Du D 2020 Microwave-enhanced selective leaching of arsenic from copper smelting flue dusts *J. Hazard. Mater.* **386** 121964.

[11] Alara O R, Abdurahman N H 2019 Microwave-assisted extraction of phenolics from hibiscus sabdariffa calyces: kinetic modelling and process intensification *Ind. Crops Prod.* **137** 528–535.

[12] Wanta K C, Perdana I, Petrus H. T 2016 Evaluation of shrinking core model in leaching process of Pomalaa nickel laterite using citric acid as leachant at atmospheric conditions *IOP Conference Series: Materials Science and Engineering* **162** 1 012018.

[13] Mirwan A, Susianto, Altway A, Handogo R 2017 A modified shrinking core model for leaching of aluminum from sludge solid waste of drinking water treatment *Int. J. Technol* **8** 1 19.

Acknowledgments

The authors are thankful for the financial support from the Ministry of Research and Technology/National Research and Innovation Agency, Republic of Indonesia No. B/87/E3/RA.00/2020 (Master's Thesis Research Grant 2020).

Reviewers' comments 1

C1: Research contribution is quite significant
R1: We are very grateful to the Reviewer for his comments.

C2: More study is needed regarding the aluminum leaching model approach
R2: We are grateful to Reviewer for their comments which helped improve the revised version of this manuscript. We learn more about the microwave-assisted aluminum leach model approach

Reviewers' comments 2

The paper is quit interesting which discuss about modified shrinking core model for microwave-assisted leaching of aluminum from peat clay. However, some improvements need to address as follows:
C1: Please refer to the IOP template how to write author name etc.
R1: We gratefully acknowledge reviewers for their comments. We have adjusted the standard of IOP template as suggested in http://cms.iopscience.iop.org/alfresco/d/d/workspace/SpacesStore/f67538ae-18b2-11e4-831a-29411a5deefe/WordGuidelines.zip

C2: The abstract is too short, I suggest to add in the abstract research background, aims, method, results, and the implication of the benefits.
R1: We thank reviewers for their comments. We've added the abstract section as requested.

Determination of rice husk activated Carbon capacity in adsorption of Cu Metal from Sasirangan liquid waste based on isotherm model

I Syauqiah[1,2], D Nurandini[1,2], N P Prihatini[2,3], and Jamiyaturrasidah[3]

[1] Department of Chemical Engineering, Faculty of Engineering, Lambung Mangkurat University, South Kalimantan, Indonesia
[2] Wetland Based Materials Research Group, Lambung Mangkurat University, South Kalimantan, Indonesia
[3] Department of Environmental Engineering, Faculty of Engineering, Lambung Mangkurat University, South Kalimantan, Indonesia

E-mail: desi.nurandini@ulm.ac.id

Abstract. The process of manufacturing Sasirangan - a traditional fabric of South Kalimantan - has an impact that affects environmental pollution, namely the dyeing process of the fabric. The synthetic dyes used contain heavy metals and one of those toxic metals is copper (Cu). This study aims to determine the adsorption capacity of rice husk activated carbon adsorbent by adjusting the adsorption pattern based on isotherm models as the treatment to sasirangan liquid waste. The method consists of three stages: preparation of adsorbent by carbonization process, chemical and physical activation, then continued by adsorption process of Cu metal with carbon from rice husks with variations of adsorbent dose (2, 4, and 6 grams). This treatment was conducted by batch process. In this reseach, the adsorption capacity of rice husk adsorbent towards heavy metal Cu in sasirangan liquid waste was determined from the equilibrium state with the Langmuir isotherm equation and Freundlich isotherm equation. Based on isothermal studies of adsorption data, the correlation coefficient values obtained from the isotherm model approaches are: for dose of 2 grams adsorbent, Langmuir R^2 = 0.9991 and Freundlich R^2 = 0.9981; for dose of 4 grams adsorbent, Langmuir R^2 = 0.9992 and Freundlich R^2 = 0.9989; for dose of 6 grams adsorbent, Langmuir R^2 = 0.9990 and Freundlich R^2 = 0.9986. The results of investigation indicate that adsorption data correlated well with Langmuir isotherm model.

Keywords: adsorption, Cu metal, sasirangan liquid waste, isotherm model

1. Introduction

Sasirangan is a traditional fabric from South Kalimantan which is produced by the Banjar tribe on a home industry scale. An important part of making sasirangan fabrics is making patern by dyeing the finished fabric using synthetic dyes which are relatively stable and firmly attached [1]. The synthetic dyes used contain heavy metals, such as copper (Cu) [2]. Heavy metal contamination in ecosystems related to their accumulation in living tissues throughout the food chain as non-biodegradable pollutants potentially causing significant health problems [3]. If the heavy metal content exceeds the threshold, then the liquid waste of Sasirangan textile industry has a high potential for contamination that it does not meet the requirements to be discharged into the environment before being treated first [4].

Traditional processing methods commonly used for metal-containing wastes are chemical precipitation and filtration, electrochemical treatment, chemical oxidation or reduction, reverse osmosis, solvent extraction, ion exchange, and evaporation. However there are several disadvantages of these methods, such as high cost, imperfect metal removal, low selectivity, high energy requirements and by-products a poisonous slurry that is difficult to remove [5]. Nowadays, adsorption methods have been widely used to remove toxic metals from wastewater. The adsorbents used are prepared from agricultural byproducts, which are cheap and widely available [6]. Agricultural wastes such as mangosteen peel [7], sawdust [8], sunflower stalks [9], peanut shells [10], and wheat husks [11] have been used to remove metal ions from wastewater.

The adsorption process is an effective technology for heavy metals removal in Sasirangan liquid waste. Adsorption is kind of attachment process of adsorbate on the surface of an adsorbent [12]. One type of adsorbent used is activated carbon made from rice husks. Currently, the existence of rice husks in Indonesia has not received proper attention. Its utilization is still only for simple purposes such as rubbing ash and as a plant medium. In certain areas, rice husks are even discarded and considered less useful. Rice husks can be used as adsorbents because they contain cellulose which can adsorb metal cations from the solution medium. Thus, rice husks have the potential to be utilized as heavy metal adsorbents from liquid medium [13]. Research about the potential of rice husk activated carbon as an adsorbent has been carried out. It was reported that the maximum adsorption capacity of rice husk activated carbon towards paraquat is 317.7 mg/g.

The relationship between the amount of adsorbed adsorbate and the concentration of adsorbate in solution at equilibrium and constant temperature can be expressed by the adsorption isotherm [14]. The amount of substance adsorbed on the surface of the adsorbent is a balanced process, because the rate of adsorption is accompanied by desorption. At the beginning of the reaction, the adsorption event is more dominant than the desorption event, so that adsorption takes place quickly. At certain times the adsorption event tends to be slow, and conversely the rate of desorption tends to increase. The time when the adsorption rate is the same as the desorption rate is often referred to as the equilibrium state [15].

Adsorption isotherm model is applied to effectively use the adsorption process in waste treatment for environmental protection purposes [16]. Adsorption isotherms is generally a curve presenting the phenomenon controlling the mobility and the retention of a substance from an aqueous porous medium or aquatic environment to a solid-phase at constant temperature and pH [17, 18]. The ratio between the amount of adsorbate and the residual in solution (adsorption equilibrium) is determined when the adsorbate-containing phase and the adsorbate concentration in the bulk solution have been contacted with the adsorbent for sufficient time [19, 20]. In this study, the adsorption capacity of rice husk activated carbon adsorbent was defined by observing the adsorption pattern at the equilibrium with Langmuir isotherm and Freundlich isotherm models.

2. Materials and Method

2.1. Matfierials

The materials used in this research were sasirangan liquid waste obtained from Banjarmasin, chloric acid (HCl), nitric acid (HNO_3), and distilled water.

2.2. Method

This study consists of three stages.The first is preparation of adsorbent by carbonization process. The second stage is chemical activation with HCl solution and physical activation by burning processing in the the furnace. Meanwhile, the third stage was testing the ability of activated carbon from rice husks to adsorb Cu metal with variables of adsorbent doses (2, 4 and 6 gr). The treatment was conducted by batch process using a rotary shaker. The sample volume of Sasirangan liquid waste used was 200 ml.

Then the experiment was conducted to determine the ability of activated carbon of rice husk as an adorbent.The test results were analyzed descriptively to describe the decrease in heavy metal Cu concentrations at various adsorbent doses. The reduction of Cu content is determined by Equation 1 as follows:

$$E = \frac{Co - Ce}{Co} \text{ x } 100\% \quad (1)$$

Where:

E = Decrease percentage of Cu concentation (%)
C_0 = Initial concentration of Cu (mg/L)
C_e = Final concentration of Cu (mg/L)

The relationship between the amount of adsorbed adsorbate and the concentration of adsorbate in solution at equilibrium and constant temperature can be expressed by the adsorption isotherm. The Langmuir isotherm is expressed in Equation 2 as follows[21]:

$$Q = \frac{b.K.Ce}{1+K.Ce} \quad (2)$$

Where :

Q = the amount of adsorbate per adsorbent mass (mg/g)
b = maximum capacity of adsorbent (mg/g)
K = adsorption equilibrium constant (L/mg)
Ce = solute concentration at equilibrium (mg/g)

Meanwhile, the mathematical expression of Freundlich isotherm is defined in Equation 3 as follows [22]:

$$x/m = k\, C^{1/n} \quad (3)$$

Where:

x = amount of solute adsorbed (mg)
m = adsorbent mass (mg)
C = concentration of the same adsorbent (mg/g)
k = adsorbent constant
n = adsorbent constant

3. Result and Discussion

The adsorption capacity of rice husk activated carbon can be determined by observing the adsorption pattern of adsorbent interacted with sasirangan liquid waste, with variations in its dose. In this study, the adsorption capacity of the adsorbent towards heavy metal Cu in Sasirangan liquid waste was defined by adsorption isotherm model. Two common isotherm models were established in equilibrium state: the Langmuir and Freundlich isotherm models. The applicability of the isotherm equation was judged by the determination of correlation coefficients, R^2 [23]. The adsorption isotherm has the ability to describe the relationship between the adsorbate molecule and adsorbent surface well and highlight distribution adsorbate molecules between the liquid and the solid phase [24]. Following are the results of capacity calculations with the Langmuir isotherm and Freundlich isotherm equations as can be seen in Table 1.

Table 1. The adsorption capacity of rice husk activated carbon with by isotherm equation

Contact time (min)	Adsorbent dose (gr)	C_0 (mg/l)	Ce (mg/l)	x/m	Ce/(x/m)	Log Ce	Log (x/m)	R^2	
								Langmuir	Freundlich
30	2	0.047	0.028	0.002	14.737	-1.553	-2.721		
60	2	0.047	0.029	0.002	16.111	-1.538	-2.745	0.9991	0.9993
120	2	0.047	0.027	0.002	13.500	-1.569	-2.699		
30	4	0.047	0.019	0.001	13.571	-1.721	-2.854		
60	4	0.047	0.018	0.001	12.414	-1.745	-2.839	0.9991	0.9985
120	4	0.047	0.013	0.002	7.647	-1.886	-2.770		
30	6	0.047	0.021	0.001	24.231	-1.678	-3.062		
60	6	0.047	0.022	0.001	26.400	-1.658	-3.079	0.9990	0.9987
120	6	0.047	0.017	0.001	17.000	-1.770	-3.000		

Where:
C_0 = initial concentration of Cu (ppm)
Ce = final concentration of Cu (ppm)

Recently, linear regression analysis has been one of the most widely applied tools to determine the best fitting adsorption models because it quantifies the distribution of adsorbate, analyzes the adsorption system, and verifies the consistency of hypothesis of the adsorption isotherm model [25]. The comparison of linear plot between Langmuir and Freundlich Isotherm model of 2 grams adsorbent is presented in Figure 1.

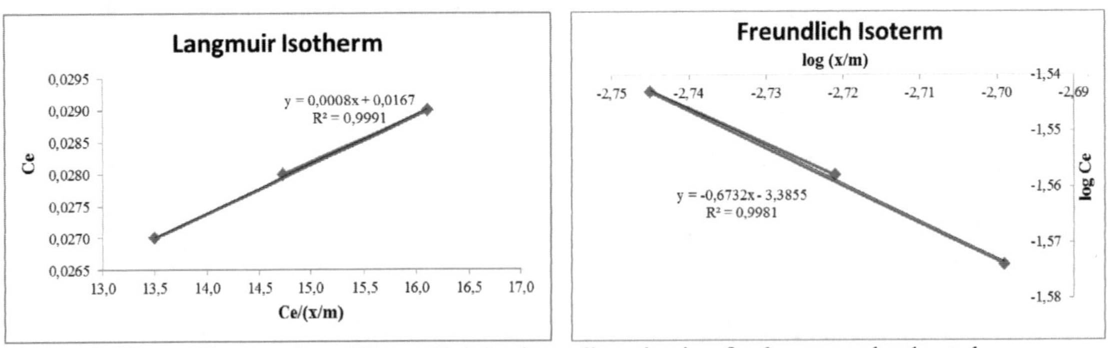

Figure 1. Chart of Langmuir and Freundlich isotherm linearization for 2 grams adsorbent dose

The linearity pattern formed from the Langmuir isotherm equation has the value of correlation coefficient $R^2 = 0.9991$ and for the Freundlich isotherm, $R^2 = 0.9981$. Then the adsorption process of heavy metal Cu in sasirangan liquid waste with activated carbon from rice husk follows the Langmuir isotherm model, because its R^2 value is more approached to linearity.

The comparison between concentrations of Cu adsorbed by rice husk activated carbon with a dose of 4 grams is presented in Figure 2.

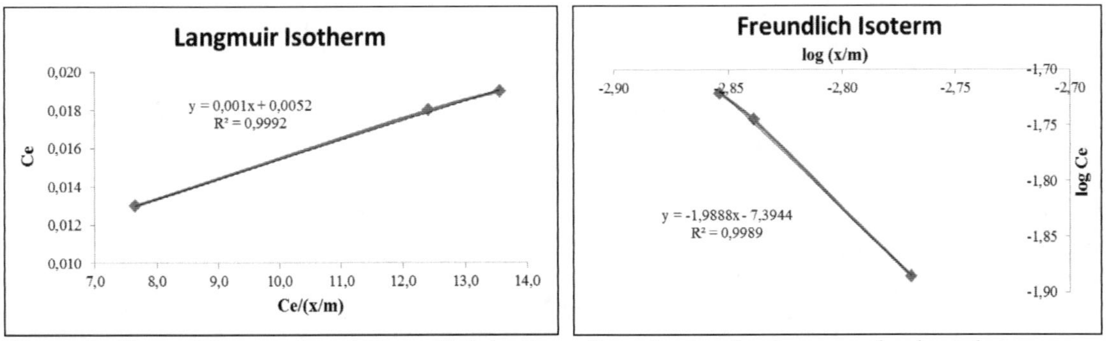

Figure 2. Chart of Langmuir and Freundlich isotherm linearization for 4 grams adsorbent dose

Figure 2 shows the difference in the linearity pattern between two isotherm models. The linearity pattern formed from the Langmuir isotherm equation has the value of R^2 = 0.9992 and for the Freundlich isotherm, R^2 = 0.9989. Then the adsorption process of heavy metal Cu in sasirangan liquid waste with activated carbon from rice husks with activation follows the Langmuir isotherm model, because of its higher approachment to linearity.

The comparison between concentrations of Cu adsorbed by rice husk activated carbon with a dose of 6 grams is presented in Figure 3.

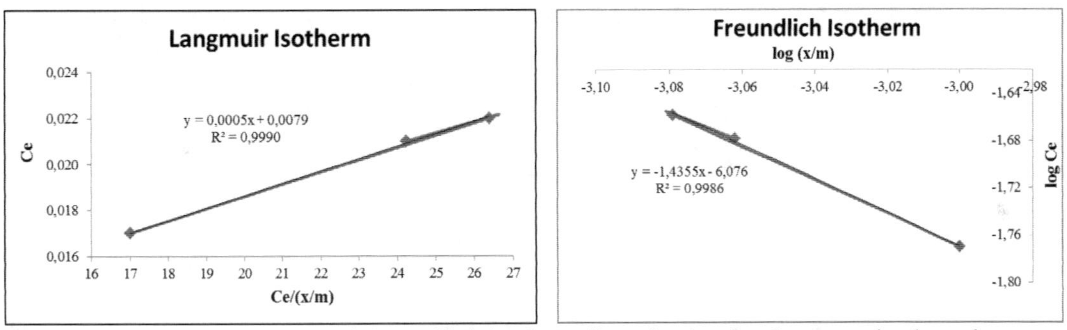

Figure 3. Chart of Langmuir and Freundlich isotherm linearization for 6 grams adsorbent dose

Figure 3 displays the difference in the linear plot between two isotherm patterns. The linearity pattern that is formed from the Langmuir isotherm equation shows the value of R^2 = 0.9990 and for the Freundlich isotherm, R^2 = 0.9986. It is onsidered that the adsorption process of heavy metal Cu in sasirangan liquid waste with activated carbon from rice husks with activation follows the Langmuir isotherm model.

The adsorption capacity by the adsorbent is number of groups that can be exchanged in the adsorbent. The adsorption exchange capacity of an adsorbent is the number of ions that can be exchanged for each gram of dry adsorbent or the number of ions that can be exchanged per 1 ml of wet adsorbent. Determination of the adsorption capacity of the rice husk adsorbent can be determined through an adsorption isotherm model approach, by observing the adsorption pattern of the two adsorbents interacted with sasirangan liquid waste with optimum weight concentration and contact time. Adsorption isotherm is the relationship between the amount of substance adsorbed by the weight of the adsorbent with the concentration of solute at a certain temperature. Adsorption equilibrium occurs when the amount of adsorbate absorbed by the adsorbent (on the surface phase) and the adsorbate remaining in solution (in the bulk phase) remains relatively constant with respect to the shaking time [26].

The adsorption isotherm is used to characterize the equation between the amount of adsorbate accumulated in the adsorbent and the concentration of the adsorbate solution. Based on previous research that has been conducted in determining the most appropriate correlation for balance data in the adsorption system design, there are two isotherm models used, namely Langmuir and Freundlich [23]. Based on the calculation for the adsorption pattern of rice husk activated carbon adsorbent towards heavy metal Cu in Sasirangan liquid waste using the Langmuir and Freundlich isotherm model approach, the value of the isotherm model is obtained. The most suitable correlation value for adsorption equilibrium data is the R^2 value which is closer to linearity. From the three doses of rice husk adsorbent, the R^2 value for the 2 grams dose is R^2 Langmuir = 0.9991 and R^2 Freundlich = 0.9993, for the 4 grams dose the value of R^2 Langmuir = 0.9991 and R^2 Freundlich = 0.9985 and for the adsorbent dose 6 grams of R^2 Langmuir = 0.9990 and R^2 Freundlich = 0.9987. The linearity of the curve that is formed indicates that the adsorption pattern follows the Langmuir isotherm model. This shows that on the surface of the rice husk activated carbon adsorbent, adsorption process occurs in the outer layer.

Adsorption isotherm describes the relationship between the adsorbed substances in a certain amount of adsorbent weight in equilibrium units [27]. Freundlich isotherm model is an empirical equation describing the nonideal and reversible adsorption. The model can be applied to multilayer adsorption, with non-uniform distribution of adsorption heat and affinities over the heterogeneous surface [28]. Meanwhile, Langmuir isotherm is an adsorption model where on the adsorbent surface there are homogeneous active sites which are proportional to the surface area. Each active site can only adsorb one adsorbate molecule, so that the adsorption will only be limited to the formation of a single layer [22]. If every active site has adsorbed the adsorbate, the adsorbent can no longer adsorb. Chemical adsorption occurs because of the interaction between the active site of the adsorbent and the adsorbate which involves chemical bonds, chemical interactions only occur in the single absorption layer on the surface of the adsorbent cell wall. The applicability of the Langmuir model isotherm in this study means that the adsorption process occurs in a mono-layer with each molecule having the same enthalpy and activation energy. The results also show that no interaction and transmigration of dyes on neighboring surface areas [23].

4. Conclusion

This study demonstrated that rice husk activated carbon could be used as an effective adsorbent for the treatment of Sasirangan liquid waste containing heavy metal Cu. Isothermal study indicated that Langmuir adsorption isotherm model is better fits than the Freundlich model based on the values of corresponding linear correlation coefficients (R^2). The highest regression values of 0.9992 obtained from 4 grams of adsorbent dose. It is therefore in conclusion that different models should be used to fit adsorption experimental data to evaluate the best model to explain the adsorption.

References

[1] Rossi B S, Paryanti, Ristianingsih Y, and Tuhuloula A B 2014 Penurunan Konsentrasi Logam Pb^{2+} dan Cd^{2+} pada Limbah Cair Industri Sasirangan dengan Metode Fitoremediasi *J. Teknologi Agro-Industri* **1** No. 1

[2] Ronaldo Imelda H S and Wahyuni N 2013 Adsorpsi Ion Logam Cu (II) Menggunakan Biomassa Alga Coklat yang Terenkapsulasi Aqua-gel Silika *JKK* **2** 148-152

[3] Ho Y S 2003 Removal of Copper Ions from Aqueous Solution by Tree Fern *Water Res.* **37** 2323-2330

[4] Irawati U, Utami U B L and Muslima H 2011 Pengolahan Limbah Cair Sasirangan Menggunakan Filter Arang Aktif Cangkang Kelapa Sawit Berlapiskan Kitosan Setelah Koagulasi dengan $FeSO_4$ *Journal of Sains and Applied Chemistry* **5** 34-44

[5] Volesky B 2001 Detoxification of Metal-Bearing Effluents: Biosorption for The Next Century *Hydrometallurgy* **59** 203-216

[6] Khan N, Yahayaa E M, Pakir M F, Latiffa M, Abustana I, Bellob O S and Ahmad M A 2011 Adsorptive Removal of Cu(II) Using Activated Carbon Prepared from Rice Husk by $ZnCl_2$ Activation and Subsequent Gasification with CO_2 IJET-IJENS *IJET-IJENS* **11** 01

[7] Ahmad M A and Alrozi R 2010 Optimization of Preparation Conditions for Mangosteen Peel-

Based Activated Carbons for The Removal of Remazol Brilliant Blue R Using Response Surface Methodology *Chem. Eng. J.* **165** 883-890
[8] Vaishya R C and Prasad S C 1991 Adsorption of Copper (II) on Sawdust *Indian J. Environ. Prot.* **11** 284-289
[9] Sun G and Shi W 1998 Sunflower Stalks as Adsorbents for The Removal of Metal Ions from Wastewater *Ind. Eng. Chem. Res.* **37** 1324-1328
[10] Shukla S R and Pai R S 2005 Adsorption of Cu(II), Ni(II) and Zn(II) on Dye Loaded Groundnut Shells and Sawdust *Sep. Purif. Technol.* **43** 1-8
[11] Basci N, Kocadagistan E and Kocadagistan B 2004 Biosorption of Copper (II) from Aqueous Solutions by Wheat Shell *Desalination* **164** 135-140
[12] Hadiwidodo M 2008 Penggunaan Abu Sekam Padi Sebagai Adsorpsi dalam Pengolahan Air Limbah yang Menggandung Logam Cu *J.Teknik* **29** 0852-1697
[13] Nurhasni, Hendrawati, and Nubzah S 2014 Sekam Padi untuk Menyerap Ion Logam Tembaga dan Timbal dalam Air Limbah *Valensi* **4** 36-44
[14] Bregas S T S 2005 Isoterm Kesetimbangan Adsorpsi Timbal pada Abu Sekam Padi *Ekuilibrium* **4** 100-105
[15] Swiatkowski A 1998 *Adsorption and its Aplication in Industry and Environmental Protection Studies in Surface Science and Catalysis* (Netherland: Elsevier)
[16] Kamau J and Kamau G 2017 Modeling of Experimental Adsorption Isotherm Data for Chlorothalonil by Nairobi River Sediment *Mod Chem Appl* **5** 203
[17] Limousin G, Gaudet J P, Charlet L and Szenknect S 2007 Sorption Isotherms: A Review on Physical Bases, Modeling, and Measurement *Appl Geochem* **22** 249-275
[18] Allen S J, Mckay G andPorter J F 2004 Adsorption Isotherm Models for Basic Dye Adsorption by Peat in Single and Binary Component Systems *J Colloid Interface Sci* **280** 322-333
[19] Kumar K V and Sivanesan S 2007 Sorption Isotherm for Safranin onto Rice Husk: Comparison of Linear and Non-Linear Methods *Dyes Pigments* **72** 130-133
[20] Ghiaci M, Abbaspur A, Kia R and Seyedeyn F 2004 Equilibrium Isotherm Studies for The Sorption of Benzene, Toluene, and Phenol onto Organo-Zeolites and As-Synthesized MCM-41 *Sep Purif Technol* **40** 217-229
[21] Wijayanti I E and Kurniawati E A 2019 Studi Kinetika Adsorpsi Isoterm Persamaan Langmuir dan Freundlich pada Abu Gosok sebagai Adsorben *EduChemia* **4** 2
[22] Oscik J 1994 *Adsorption* (New York: Ellis Horwood)
[23] Krishni R R, Foo K Y, and Hameed B H 2013 Adsorptive Removal of Methylene Blue Using The Natural Adsorbent-Banana Leaves *J. Desalination and Water Treatment* **52** 31-33
[24] K.Y. Foo, B.H. Hameed, Insights into the modeling of adsorption isotherm systems, Chem. Eng. J. 156 (2010) 2–10.
[25] Ayawei N, Ebelegi A N and Wankasi D 2017 Modelling and Interpretation of Adsorption Isotherms *Journal of Chemistry* **11**
[26] Zhang Y, Yu F, Cheng W, Wang J and Ma J 2017 *Journal of Chemistry* **2017** 1936829
[27] Purwaningsih 2009 Adsorpsi Multi Logam Ag, Pb, Cr, Cu dan Ni pada Hibrida Etilendiamino-Silika dari Abu Sekam Padi *J. Penelitian Saintek* **14** 56-76
[28] Freundlich H 1906 U¨ber die adsorption in lo¨sungen (Adsorption in solution) *Z. Phys. Chem.* **57** 384–470

Acknowledgment

The authors would like to express gratitude for The PNBP Research Grant from University of Lambung Mangkurat.

IOP Conf. Series: Materials Science and Engineering 1212 (2022) 012020 doi:10.1088/1757-899X/1212/1/012020

Study of caffeine and coumarin extraction kinetics in an aqueous two-phase system based on polyethylene glycol 1500

I V Zinov'eva

Kurnak ov Institute of General and Inorganic Chemistry of the Russian Academy of Sciences, 31 Leninsky Prospect, Moscow, 119991, Russia

iz@igic.ras.ru

Abstract. Caffeine and coumarin are organic compounds of plant origin, which have biological activity and have found wide application in medicine, pharmaceutical, perfumery and food industries. Recovery of caffeine and coumarin from aqueous solutions using liquid- liquid extraction is the most effective method. In the present work the kinetic dependences of caffeine and coumarin in the aqueous two-phase system (ATPS) of PEG 1500 - Na_2SO_4 - H_2O were experimentally obtained. It was established that equilibrium in the system is achieved after 17 minutes for caffeine and after 12 minutes for coumarin. The effect of ultrasound on the extraction rate of caffeine and coumarin in an aqueous two-phase system was also studied.

Keywords: extraction, aqueous two-phase system, caffeine, coumarin, ultrasound

1. Introduction

The recovery of bioactive compounds from aqueous solutions is an urgent issue that requires special attention. To extract organic substances from aqueous solutions, filtration [1], sorption [2], extraction [3-5], chromatography [6], etc. [7-9] are mainly used. Today, numerous works are aimed at the development of environmentally friendly, effective methods for the extraction of organic compounds that are not inferior to classical extraction systems based on toxic and harmful organic solvents [10-12]. The aqueous two-phase system is of great interest as an alternative extraction system, which is eco-friendly, easy to utilize, efficient and cost-effective [13-17]. Yazdabadi A et al. established the regularities of caffeine extraction in the system based on tetrabutilphosphonium bromide and sorbitol [18]. de Araujo Sampaio D et al. conducted the study on the caffeine extraction in a system based on polyethylene glycol 400 and sodium sulfate at different temperatures, which showed that caffeine is mainly distributed in the polymer phase [19]. All these research results show the prospects of ATPS application for the extraction of organic compounds. However, little information is presented in the literature on the kinetics of extraction of substances in ATPS, which is necessary for the calculation of mass transfer characteristics and the design of extraction equipment [20,21].

Recently, mass transfer processes, including liquid-liquid extraction, are undergoing greater modernization, and combining with the method of ultrasonic intensification [22-26]. However, attempts to intensify, using ultrasound, the processes of metal ion extraction are accompanied by a decrease in the quantitative characteristics of extraction. Gradov et al. carried out an experimental study of the interphase distribution dependence of Fe (III) ion using an aqueous two-phase system on the time of phase contact with the ultrasonic exposure use and in its absence [27]. It was found that the use of ultrasound irradiation has the effect of slowing down the extraction of this metal ion into the PEG phase. Hu Y et al. studied the effect of ultrasound on the lactic acid extraction from the fermentation solution. Lactic acid was extracted into extractant after 10-15 minutes with ultrasound exposure, which is several times faster than without ultrasound [28].

Caffeine and coumarin, widely used in the pharmaceutical, food and perfume industries, were chosen as model objects. In this work, the caffeine and coumarin extraction kinetics in an aqueous two-phase system based on polyethylene glycol 1500 and sodium sulfate with constant interphase area was studied for the first time. The time to achieve equilibrium in the distribution of the studied substances between the polymer and salt phases was determined. The kinetic dependencies of caffeine and coumarin distribution in ATPS under the influence of ultrasound were obtained for the first time.

2. Experimental details

2.1. Reagents

Polyethylene glycol with a molecular weight of 1500 was purchased from Fluka (Shanghai, China). Na_2SO_4 was used as phase-forming salt and was purchased from Chimmed (Moscow, Russia). Caffeine and coumarin were used from Sigma-Aldrich (St. Louis, MO, USA) (99% purity). All solutions were prepared using distilled water purified in a UPVA-5 unit for the production of analytical grade water (Livam, Belgorod, Russia). The structural formulas of the research objects are shown in Figure 1.

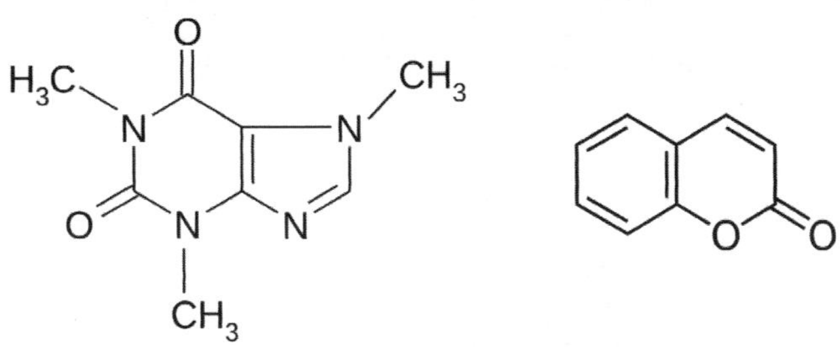

Figure 1. The structural formulas of caffeine and coumarin.

2.2. Experiment method

To carry out the extraction experiment, a glass cylinder with an interphase area value of 5.31 cm^2 was used when mixing equal volumes (15 ml) of polymer and salt phases. The recoverable substance of a given concentration was prepared by dissolution in the salt phase. Both phases were mixed using a top-driven stirrer with a stirring speed of 100 rpm. We used an ultrasonic generator with a frequency of 35 kHz and a maximum power of 110 W. Figure 2 shows the installation for an extraction experiment.

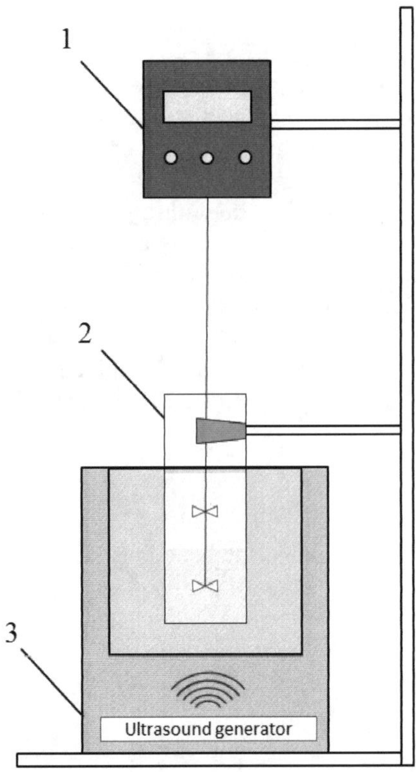

Figure 2. Pilot installation: 1 – top-drive stirrer; 2 – ATPS extraction system; 3 – ultrasound bath

The concentration of recoverable compounds was detected in the polymer phase by spectrophotometry in the ultraviolet region of the spectrum (λ = 270 nm) using a fiber optic probe with an optical path length of 10 mm. Measurement of optical density was carried out on the Cary-60 spectrophotometer (Agilent Tech., USA). Determination of the optical density of caffeine and coumarin in experiments without ultrasound was carried out online in a cylinder every 10 seconds. With ultrasound, measurement of optical density carried out every minute at the initial site of dependence before the establishment of constant values of optical density, then the measurement frequency was reduced to 5 minutes. The spectrophotometer was controlled and the results were processed using the Cary WinUV software. All presented experimental data were the result of a series of experiments and processed by the methods of mathematical statistics.

3. Results and Discussion

3.1. Caffeine and coumarin extraction kinetics in PEG 1500 - Na_2SO_4 - H_2O system with and without the use of ultrasound

The dependence of interphase distribution of caffeine (initial concentration of 0.00225 mol/L) and coumarin (initial concentration of 0.002 mol/L) in an aqueous two-phase system (PEG 1500 (15 wt%) – Na_2SO_4 (9 wt %) – water) on time was experimentally investigated. Figures 3 and 4 illustrate the data obtained without ultrasound. The equilibrium in the system is achieved in 17 and 12 minutes for caffeine and coumarin, respectively, which is probably due to the low interphase tension of the ATPS. Comparing the results with the experiments obtained by Kaplanow I et al. where the mass transfer of lysozyme and bromelain was studied in the PEG system 4000/sodium citrate, the equilibrium of caffeine and coumarin in the system proposed here is achieved faster than the one of proteins [29]. The rapid transition of caffeine and coumarin can be justified by the use of polyethylene glycol with a lower molecular weight. The results obtained can be used to calculate the mass transfer coefficient and extraction rate for the design of liquid-liquid processes [30,31].

Figures 5 and 6 show the kinetic dependencies for caffeine and coumarin with ultrasound exposure. Comparing the results obtained, it can be seen that the time to reach equilibrium between caffeine and coumarin remains unchanged. It can be noted that in the case of caffeine and coumarin extraction in ATPS it is not necessary to use additional effects, because the equilibrium in the system is reached quickly enough, unlike the results in other works [32].

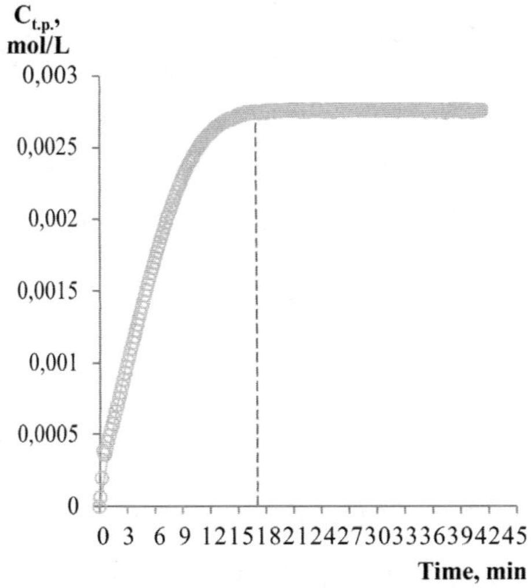

Figure 3. Concentration profile of caffeine in the polymer phase over the time of extraction in the PEG 1500 - Na_2SO_4 - H_2O system without ultrasound.

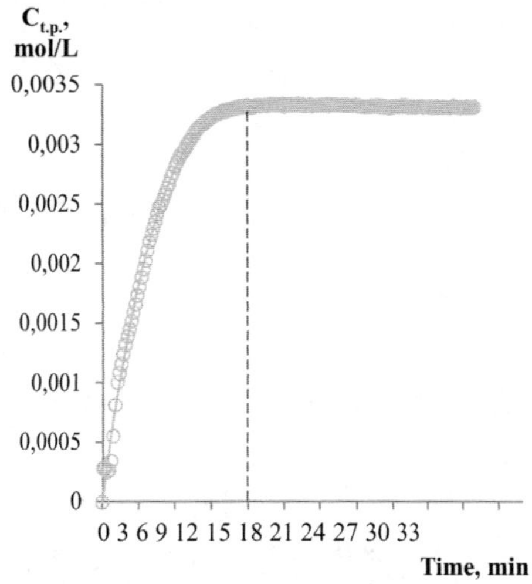

Figure 4. Concentration profile of coumarin in the polymer phase over the time of extraction in the PEG 1500 - Na_2SO_4 - H_2O system without ultrasound.

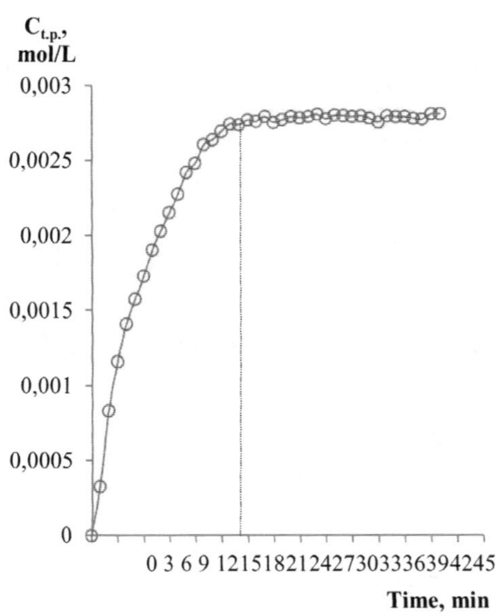

Figure 5. Concentration profile of caffeine in the polymer phase over the time of extraction in the PEG 1500 - Na_2SO_4 - H_2O system with ultrasound.

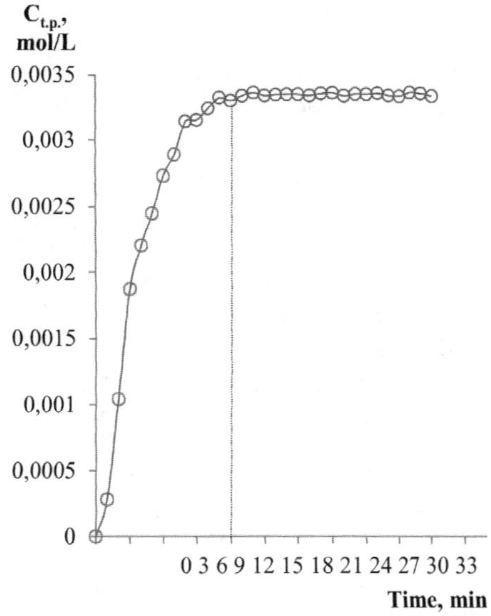

Figure 6. Concentration profile of coumarin in the polymer phase over the time of extraction in the PEG 1500 - Na_2SO_4 - H_2O system with ultrasound.

Conclusions

In this work, the kinetics of extraction of caffeine and coumarin in an aqueous two-phase system based on polyethylene glycol 1500 and sodium sulfate was studied. It was established the time after which equilibrium is reached in the system for caffeine and coumarin, equal to 17 and 12 minutes, respectively. The effect of ultrasound on the extraction of the studied compounds in aqueous two-phase system was evaluated.

References

[1] López-Garzón C S and Straathof A J J 2014 Recovery of carboxylic acids produced by fermentation *Biotechnol. Adv.* **32** 5 873–904.

[2] Tsai W T, Hsien K J and Hsu H C 2009 Adsorption of organic compounds from aqueous solution onto the synthesized zeolite *J. Hazard. Mater.* **166** 2-3 635–41.

[3] Voshkin A A, Zakhodyaeva Yu A, Zinov'eva I V and Shkinev V M 2018 Interphase distribution of aromatic acids in the polyethylene glycol-sodium sulfate-water system *Theor. Found. Chem. Eng.* **52** 5 890-3.

[4] Kholkin A I, Belova V V, Zakhodyaeva Y A and Voshkin A A 2013 Solvent extraction of weak acids in binary extractant systems *Sep. Sci. Technol.* **48** 9 1417–25.

[5] Belova V V, Kulichenkov S A,Voshkin A A, Khol'kin A I, Kuvaeva Z I and Soldatov V S 2007 Extraction of mineral acids with methyltrioctylammonium dinonylnaphthalenesulfonate *Russ. J. Inorg. Chem.* **52** 3 460-4.

[6] Zakhodyaeva Yu A and Voshkin A A 2013 Extraction and separation of carboxylic acids by liquid-liquid extraction and chromatography *Theor. Found. Chem. Eng.* **47** 4 461-6.

[7] Chuo S C, Abd-Talib N, Mohd-Setapar S H et al. 2018 Reverse micelle extraction of antibiotics using an eco-friendly sophorolipids biosurfactant *Sci. Rep.* **8** 477.

[8] Lee H D, Lee M Y, Hwang Y S, Cho Y H, Kim H W and Park H B 2017 Separation and purification of lactic acid from fermentation broth using membrane-integrated separation processes *Ind. Eng. Chem. Res.* **56** 29 8301–10.

[9] Kostenko M O, Pokrovskiy O I, Zakhodyaeva Yu A, Voshkin A A and Lunin V V 2019 Unusual effect of flow rate on retention in analytical supercritical fluid chromatography exemplified by polyethylene glycol separation *J. Chrom. A* **1610** 460513.

[10] Zakhodyaeva Yu A, Rudakov D G, Solov'ev V O, Voshkin A A and Timoshenko A V 2019 Liquid-liquid equilibrium in an extraction system based on polyvinylpyrrolidone-3500 and sodium nitrate *Theor. Found. Chem. Eng.* **53** 2 159–65.

[11] Da Cruz Silva K, Abreu C S, Vieira A W, Mageste A B, Rodrigues G D, de Lemos L R 2020 Aqueous two-phase systems formed by different phase-forming components: Equilibrium diagrams and dye partitioning study *Fluid Phase Equilib.* **520** 112664

[12] Zakhodyaeva Yu A, Rudakov D G, Solov'ev V O, Voshkin A A and Timoshenko A V 2019 Liquid-liquid equilibrium of aqueous two-phase system composed of poly(ethyleneoxide) 1500 and sodium nitrate *J. Chem. Eng. Data* **64** 3 1250–55.

[13] Zakhodyaeva Y A, Zinov'eva I V, Tokar E S and Voshkin A A 2019 Complex extraction of metals in an aqueous two-phase system based on poly(ethylene oxide) 1500 and sodium nitrate *Molecules* **24** 22 4078.

[14] Fedorova M I, Zinov'eva I V, Zakhodyaeva Yu A and Voshkin A A 2020 Extraction of Fe(III), Zn(II), and Mn(II) using a system with a green solvent for trioctylmethylammonium thiocyanate *Theor. Found. Chem. Eng.* **54** 2 313–8.

[15] Zinov'eva I V, Zakhodyaeva Yu A and Voshkin A A 2019 Extraction of monocarboxylic acids from diluted solutions with polyethylene glycol *Theor. Found. Chem. Eng.* **53** 5 871–4.
[16] Zinov'eva I V, Zakhodyaeva Yu A and Voshkin A A 2020 Data on the extraction of benzoic, salicylic and sulfosalicylic acids from dilute solutions using PEG-based aqueous two-phase systems *Data in Brief* **28** 105033.
[17] Zinov'eva I V, Zakhodyaeva Yu A and Voshkin A A 2019 Interphase distribution of caffeine and coumarin in extraction systems with polyethylene glycol and sodium sulfate. *Theor. Found. Chem. Eng.* **53** 996–1000.
[18] Yazdabadi A, Shahriari S and Salehifar M 2019 Extraction of caffeine using aqueous two-phase systems containing ionic liquid and sorbitol *Fluid Phase Equilib.* **502** 112287.
[19] de Araujo Sampaio D, Mafra L I, Yamamoto C I, de Andrade E F, de Souza M O, Mafra M R and de Castilhos F 2016 Aqueous two-phase (polyethylene glycol+sodium sulfate) system for caffeine extraction: Equilibrium diagrams and partitioning study *J. Chem. Thermodyn.* **98** 86–94.
[20] Pandey N K, Augustine E, Murali R, Desigan N, Kamachi Mudali U and Joshi J B 2016 Kinetics of extraction of nitric acid into binary mixture of tri-n-butyl phosphate and normal paraffin hydrocarbon *Chem. Eng. Res. Des.* **111** 492–503.
[21] Wang Y, Wang Y, Zhou H, Li F and Sun X 2017 Extraction kinetics of mixed rare earth elements with bifunctional ionic liquid using a constant interfacial area cell *RSC Adv.* **7** 39556-563.
[22] Gradov O M, Zakhodyaeva Yu A and Voshkin A A 2018 Breakup of immiscible liquids at the interface using high-power acoustic pulses *Chem. Eng. Process.: Process Intensification*, **131** 125–30.
[23] Gradov O M, Voshkin A A and Zakhodyaeva Yu A 2017 Estimating the parameters of ultrasonically induced mass transfer and flow of liquids in the pseudomembrane method. *Chem. Eng. Process.: Process Intensification* **118** 54–61.
[24] Voshkin A A and Gradov O M 2017 Parametric splitting and transfer of liquid cuts for the intensification of mass exchange in a cylindrical volume *Theor. Found. Chem. Eng.* **51** 3 274–81.
[25] Gradov O M, Voshkin A A and Zakhodyaeva Yu A 2017 Analysis of the possible applications of the acoustic flow effect for the breakup and transfer of liquid substances in a cylindrical volume *Theor. Found. Chem. Eng.* **51** 5 876–882.
[26] Zhou L, Jiang B, Zhang T, Li S 2019 Ultrasound-assisted aqueous two-phase extraction of resveratrol from the enzymatic hydrolysates of Polygonum cuspidatum *Food Biosci.* **31** 100442.
[27] Gradov O M, Zakhodyaeva Yu A, Zinov'eva I V and Voshkin A A 2019 Some features of the ultrasonic liquid extraction of metal ions *Molecules* **24** 19 3549.
[28] Hu Y, Kwan T H, Daoud W A and Lin C S K 2017 Continuous ultrasonic-mediated solvent extraction of lactic acid from fermentation broths *J. Clean. Prod.* **145** 142–50.
[29] Kaplanow I, Goerzgen F, Merz J et al. 2019 Mass transfer of proteins in aqueous two-phase systems *Sci. Rep.* **9** 3692.
[30] Kostanian A E and Voshkin A A 2007 Analysis of new counter-current chromatography operating modes *J. Chromatog. A* **1151** 1-2 126–30.
[31] Kostanyan A E and Voshkin A A 2011 Analysis of cyclic liquid chromatography. *Theor. Found. Chem. Eng* **45** 1 68–74.
[32] Kulaguin Chicaroux A and Zeiner T 2019 Theoretical and experimental investigation of mass transfer in aqueous two-phase systems based on linear and branched polymers *Fluid Phase Equilib.* **479** 106–13.

Acknowledgments

This work was supported by IGIC RAS state assignment.

Application of ionic liquid based on Aliquat 336 and D2EHPA in the extraction of transition metals

M I Fedorova*, A V Levina

Kurnakov Institute of General and Inorganic Chemistry of the Russian Academy of Sciences, Leninsky prospect, 31, 119991, Moscow, Russia

*Corresponding author: prusachenkova.maria@gmail.com

Abstract. To date, there are a number of methods for selective extraction of transition metal ions based on liquid-liquid extraction. One of the most interesting methods of metal ions extraction is liquid-liquid extraction with application of ionic liquids based on quaternary ammonium bases and organic acid residues, in particular, organophosphorus, diluted with organic solvent (toluene, xylene, kerosene, etc.). However, using of organic solvents does not correspond to the modern tendencies of harmful effects on the environment reducing. Thus, we propose to use Aliquat 336 and D2EHPA-based ionic liquid for extraction of transition metal ions in ecologically safe aqueous two-phase system based on polypropylene glycol 425 and sodium chloride. Di(2-ethylhexyl)phosphate trioctylmethylammonium has been shown to yield more than 80% Fe(III) and Y(III) ions.

Key words: liquid-liquid extraction, aqueous two-phase systems, ionic liquid, "green" chemistry

1. Introduction

Ionic liquids (IL) based on quaternary ammonium bases and organic acid residues are widely used in the extraction of organic [1] and inorganic substances [2]. Zakhodyaeva et al. [3-5] were studied the distribution of monocarboxylic acids from aqueous phases to organic phase (toluene) with using of ionic liquids that include a quaternary ammonium base (QAB) (trioctylmethylammonium) and an organophosphorus residue (di(2-ethylhexyl)phosphate, di(2-ethylhexyl)dithiophosphate or dinonylnaphthalene sulfonate). In this case, the extraction of organic acids occurs due to the non-covalent interaction between the proton formed by the organophosphorus acid and the anion of the acid residue of the extracted monobasic acid [1]: $HX_{(aq)}+R_4NA_{(org)}\leftrightarrow R_4NX\cdot HA_{(org)}$. It should be noted that the extraction of monocarboxylic acids by these ionic liquids is characterized by high distribution coefficients, for example, for caproic acid it is 68, for butanoic acid - 34 [2]. Li et al. [6] were studied cresol distribution with using IL based on imidazolium cation and rhodanid anion. It was shown that the extraction efficiency could reach 98% under optimal conditions: temperature 298.15 K, mass ratio IL: model solution = 0.2 and phase contact time – 30 minutes. Sulaiman et al. [7] were studied the range of chlorophenols distribution in system based on water – IL (based on bis(trifluoromethylsulfonyl)imide). The synthesized ILs proved to be effective for the whole range of chlorophenols - the extraction rate was more than 80% in one extraction stage. Skoronski et al. [8] studied the effect of ILs based on phosphonium cation and organophosphoric acid residue on phenol extraction. At volume ratio of phases water - IL, which is 135, practically quantitative extraction of both 2,4-dichlorophenol and phenol is achieved.

The use of ILs for extraction has found extensive applying in the extraction of inorganic substances: acids [2], rare earth metals [10-12], nonferrous [13,14] and alkali [15] metals. Belovoi et al. [16-19] were studied the distribution of rare-earth metals (REM) using ionic liquids based on phosphinic or phosphoric acids and a methyltrioctylammonium cation. REMs are able to form complex anions (like other metals of side groups) due to coordination interaction with residues of organic and inorganic acids. Pavón et al. [20] had described the REM extraction mechanism with using IL based on organic base and phosphoric acid. It can be presented by following way:

$$3RNH_3^+A^-_{(org.)} + Me^{3+}_{(aq.)} + 3Cl^-_{(aq.)} \leftrightarrow MeA_3 \cdot 3RNH_3^+Cl^-_{(org.)}.$$

Voshkin et al. [21-22] had described iron (III) extraction mechanism with using IL based on quaternary ammonium base and organic acid. In general, it can be presented by following equation:

$$12FeCl^-_{4\,(aq.)} + 9R_4NA_{(org.)} \leftrightarrow 9(R_4N)FeCl_{4\,(org.)} + Fe_3A_{9\,(org.)} + 12Cl^-_{(aq.)}.$$

One of metal ionic liquids advantages is easier re-extraction than in case of using only organophosphoric acids, due to the complex compound with ammonium base formation. Thus, the use of ionic liquids makes it possible to combine cation-exchange and anion-exchange extraction.

However, organic solvents application does not meet the principles of "green chemistry": they are toxic and flammable. Therefore, it is necessary to find an environmentally friendly replacement for them. One of the most developing directions is the creation of two-phase water systems based on water-soluble polymers [23] and their use for the extraction of metal ions both without the introduction of complexing agents [24] and with the introduction of such agents [25-27], these systems may be an alternative to organic solvents.

Thus, the goal of this work was to study the effect of an ionic liquid based on a quaternary ammonium base and organophosphorus acid on the extraction of a number of transition metals in an aqueous two-phase system.

2. Experimental details

2.1. Reagents

The tools and materials were used in this work: 50 ml measuring beakers, measuring pipettes (from 100μ to 5ml), 20 ml separation funnels, 15 ml graduated plastic tubes, drop tube, analytical Balance (AND HR-100AZ), Magnetic stirrer (IKA C-MAG HS4) with stirrer bar, centrifuge (ELMI CM-6MT), desalinated filter paper, Enviro-Genie thermostatically controlled shaker (Scientific Industries, Inc.). spectrophotometer Cary-60 device (Agilent Tech.), quartz cuvettes (*l=10*MM).

The initial solutions of metal chlorides or nitrates were prepared by dissolving precise weights of $FeCl_3 \cdot 6H_2O$, $NiCl_2 \cdot 6H_2O$, $CoCl_2 \cdot 6H_2O$, $MnCl_2 \cdot 4H_2O$, and $Y(NO_3)_3 \cdot 6H_2O$ with the qualification of "chemical grade" in distilled water suspended on an analytical balance (AND HR-100AZ).

An ionic liquid based on trioctylmethylammonium chloride (Aliquat 336) and di(2-ethylhexyl)phosphoric acid (D2EHPA) was used as an extragent. This extragent was made by mixing of equal amounts of Aliquat 336 and D2EHPA during 20 minutes followed by stripping with alkali of organic phase for chloride-ions removing.

2.1.1. Research methods.

The extraction of metal ions was carried out using an aqueous two-phase system based on polypropylen glycol 425 (30 wt.%) – NaCl (8 wt.%) - H_2O with an initial metal concentration of 0.01mol L^{-1}. The ratio of polymer and salt phases was 1:3.

The extraction of metal ions (Fe (III), Ni (II), Co (II), Mn (II), Y(III)) was carried out at a temperature of 25°C in graduated plastic tubes in a thermostatic Enviro-Genie shaker (Scientific Industries, Inc.) at a rotation speed of 30 rpm to establish thermodynamic equilibrium (30 min).

The evaluation of extraction efficiency was carried out using characteristics such as distribution coefficient (D) and extraction degree (E, %) calculated using formulas (1) and (2):

$$D = \frac{C_{t.ph.}}{C_{b.ph.}}, (1)$$

$$E = \frac{C_{t.ph.} \cdot V_{t.ph.}}{C_{t.ph.} \cdot V_{t.ph.} + C_{b.ph.} \cdot V_{b.ph.}}, (2)$$

where $C_{t.ph.}$ – metal concentration in top (polymer) phase, $C_{b.ph.}$ – metal concentration in bottom (salt) phase, $V_{t.ph.}$ – upper (polymer) phase volume, $V_{b.ph.}$ – bottom (salt) phase volume.

The concentration of metal ions in the initial solutions and in the aqueous phases after extraction was determined spectrophotometrically using 4-(2-pyridylazo)resorcinol, which forms complexes with metals that absorb in the visible spectrum at the following wavelengths: Ni (492 nm), Co (508 nm), Mn (496 nm). To determine Fe (III), sulfosalicylic acid was used, with a maximum absorption of complex at a wavelength of 420 nm. The complexometric titration using EDTA and xylenol orange at pH = 6 was used to determine the Y(III) ions concentration in the initial solutions and in the aqueous phases after extraction. The optical density values were determined using a Cary-60 device (Agilent Tech.) In quartz cuvettes l = 10 mm.

The presented experimental data are the result of a series of experiments and processed by methods of mathematical statistics

3. Results and Discussion

The investigation on a wide range of metals (Fe(III), Ni(II), Co(II), Mn(II), Y(III)) interphase distribution in system based on polypropylen glycol 425 (30 wt.%) – NaCl (8 wt.%) - H_2O with trioctylmethylammonium di(2-ethylhexyl)phosphate ionic liquid has been carried out. It should be noted, that there is a possibility of precipitate formation (tri-phase system consisted of two liquids and one solid phases) for individual metals (Fe(III), Y(III)) that can be dissolve by hydrochloric acid or extra amounts of ionic liquid (IL) addition. Thus, for the Fe(III), Y(III) ions the conditions under which the formed system is two-phase (consisting of two liquid phases) have been determined.

Quantitative characteristics of metal ions extraction are shown in Table 1. This table contains also data about conditions (hydrochloric acid and trioctylmethylammonium di(2-ethylhexyl)phosphate concentrations) required for heterogeneous system formation.

Table 1. The quantitative characteristics of metal ions extraction in ATPS with ionic liquid addition

	C(IL), mol L^{-1}	C(HCl), mol L^{-1}	$C_{b.ph.}$, mol L^{-1}	$C_{t.ph.}$, mol L^{-1}
Fe(III)	0.1	1.25	0.00037	0.023
Ni(II)	0.1	0	0.012	0.00
Co(II)	0.1	0	0.010	0.00
Mn(II)	0.1	0	0.07	0.00076
Y(III)	0.4	1.07	0.0003	0.019

It can be seen from figure 1 and figure 2 that obtained extragent is effective for following metals: Fe(III) and Y(III) as they are capable for complex compounds formation with the D2EHPA residue and QAB cation, which were shown in works by Voshkin [21-22], Fedorova [26], Tait [27], Mishra [28]. At the same time for Ni(II), Co(II) and Mn(II) ions this extractant was not effective, which was also shown in Tait [27]. This may be due to the fact that these metals are unable to form complex compounds with organophosphoric acid residual and QAB cation in similar media. Also di(2-ethylhexyl)phosphate trioctylmethylammonium extract Y(III) ions. This may be due to the fact that the radius of yttrium ion is small and inclined to complex formation [28]. Thus, with using this extragent it is possible almost quantitative extraction of these metals, and metal ions effective separation.

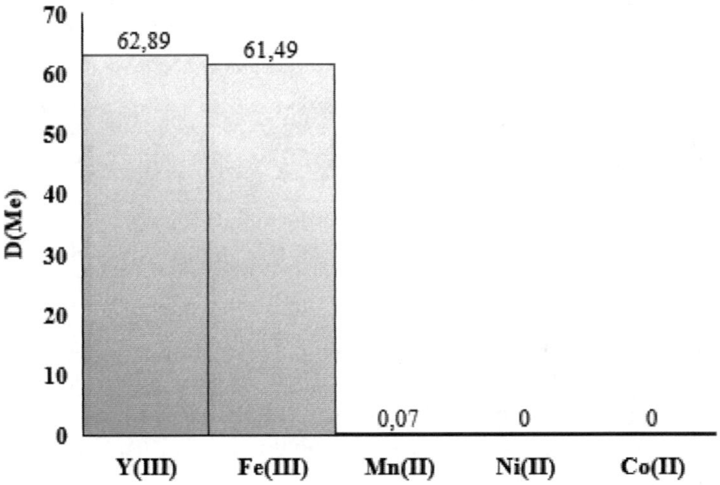

Figure 1. The distribution coefficients of metals in ATPS with IL addition

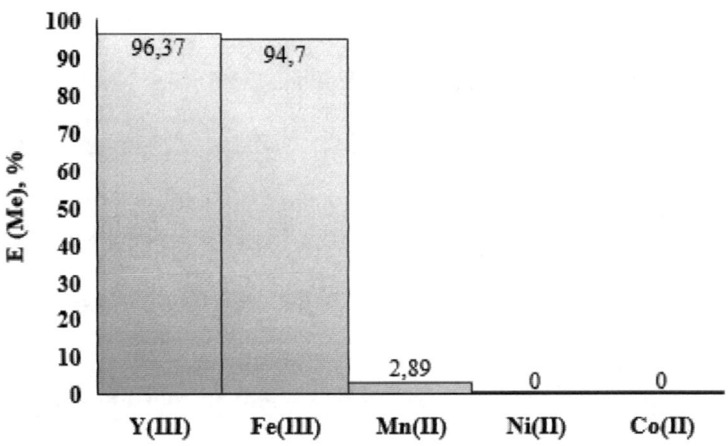

Figure 2. The extraction efficiency of metals in ATPS with IL addition

4. Conclusions

In the course of the study, quantitative characteristics of the extraction of transition metal ions in an aqueous two-phase system based on a water-soluble polymer with the introduction of trioctylmethylammonium di(2-ethylhexyl)phosphate were obtained. It was found that this extractant selectively extracts ions of metals such as Fe(III) and Y(III). Thus, it is possible to use trioctylmethylammonium di(2-ethylhexyl)phosphate at certain stages of the implementation of technological schemes, for example, in the separation of metals from leaching solutions using an environmentally friendly extraction system.

References

[1] Kholkin A I, Belova V V, Zakhodyaeva Y A, Voshkin A A 2013 Solvent Extraction of Weak

Acids in Binary Extractant Systems *Sep. Sci. Technol.* **48** 9 1417–25

[2] Belova V V, Kulichenkov S A, Voshkin A A, Khol'kin A I, Kuvaeva Z I, Soldatov V S 2007 Extraction of mineral acids with methyltrioctylammonium dinonylnaphthalenesulfonate *Russ. J. Inorg. Chem.* **52** 3 460–4

[3] Zakhodyaeva Yu A, Voshkin A A, Belova V V 2013 Extraction of acetic acid by binary extractants *Russ. J. Inorg. Chem.* **58** 4 481–5

[4] Zakhodyaeva Yu A, Voshkin A A, Belova V V, Khol'kin A I 2012 Extraction of monocarboxylic acids by trioctylmethylammonium di(2-ethylhexyl)phosphate *Theor. Found. Chem. Eng.* **46** 4 413–8

[5] Zakhodyaeva Yu A, Voshkin A A, Belova V V, Khol'kin A I 2011 Extraction of Monocarboxylic Acids with Binary Extracting Agents Based on Amines and Quaternary Ammonium Bases *Theor. Found. Chem. Eng.* **45** 5 739–43

[6] Li A, Xu X, Zhang L, Gao J, Xu D, Wang Y 2020 Separation of cresol from coal tar by imidazolium-based ionic liquid [Emim][SCN]: Interaction exploration and extraction experiment *Fuel* **264** 116908

[7] Sulaiman R, Adeyemi I, Abraham S R, Hasan S W, AlNashef I M 2019 Liquid-liquid extraction of chlorophenols from wastewater using hydrophobic ionic liquids *J. Mol. Liq.* **294** 111680

[8] Skoronski E, Fernandes M, Malaret F J, Hallett J P 2020 Use of phosphonium ionic liquids for highly efficient extraction of phenolic compounds from water *Sep. Purif. Technol.* **248** 117069

[9] Egorova N S, Belova V V, Voshkin A A, Zhilov V I, Khol'kin A I 2005 Extraction of lanthanide chlorides by binary extractants based on phosphinic acid derivatives *Russ. J. Inorg. Chem.* **50** 11 1781–4

[10] Rout A, Venkatesan K A 2020 Synergic extraction of europium(III) in hydrophobic ammonium ionic liquid containing neutral and acidic extractants *J. Mol. Liq.* **312** 113377

[11] Rama Swami K, Prathibha T, Venkatesan K A, Antony M P 2019 Separation of Am(III) from Eu(III) present in 3 M nitric acid medium using completely incinerable binary extractants *J. Mol. Liq.* **279** 427–33

[12] Tran T T, Azra N, Iqbal M, Lee M S 2020 Synthesis of succinimide based ionic liquids and comparison of extraction behavior of Co(II) and Ni(II) with bi-functional ionic liquids synthesized by Aliquat336 and organophosphorus acids *Sep. Purif. Technol.* **238** 116496

[13] Belova V V, Voshkin A A, Khol'kin A I 2011 Liquid-liquid distribution of metals in systems with trioctylmethylammonium dinonyl naphthalenesulfonate *Theor. Found. Chem. Eng.* **45** 5 764–8

[14] Shi C, Li H, Liu B, Qin Y, Song G 2020 Solvent extraction of lithium from aqueous solution using an ammonium ionic liquid extractants *J. Mol. Liq.* **304** 112756

[15] Belova V V, Voshkin A A, Egorova N S, Khol'kin A I 2010 Extraction of rare earth metals from nitrate solutions with a binary extractant based on Cyanex 272 *Russ. J. Inorg. Chem.* **55** 4 629–33

[16] Belova V V, Voshkin A A, Khol'kin A I, Payrtman A K 2009 Solvent extraction of some lanthanides from chloride and nitrate solutions by binary extractants *Hydrometallurgy* **97** 3-4 198–203

[17] Belova V V, Egorova N S, Voshkin A A, Khol'kin A I 2015 Extraction of rare earth metals, uranium, and thorium from nitrate solutions by binary extractants *Theor. Found. Chem. Eng.* **49** 4 545–9

[18] Belova V V, Voshkin A A, Egorova N S, Khol'kin A I 2012 Solvent extraction of rare earth metals from nitrate solutions with di(2,4,4-trimethylpentyl)phosphinate of methyltrioctylammonium *J. Mol. Liq.* **172** 144 - 6

[19] Pavón S, Fortuny A, Coll M T , Sastre A M 2018 Rare earths separation from fluorescent lamp wastes using ionic liquids as extractant agents *Waste Management* **82** 241–8

[20] Voshkin A A, Belova V V, Khol'kin A I 2003 Extraction of iron(III) by binary extractants based on quaternary ammonium bases and organic acids *Russ. J. Inorg. Chem.* **48** 4 608–13

[21] Voshkin A A, Belova V V, Zakhodyaeva Y A 2018 Iron Extraction with Di(2-Ethylhexyl)dithiophosphoric Acid and a Binary Extractant Based on It *Russ. J. Inorg. Chem.* **63** 3 387–90
[22] Zakhodyaeva Yu A, Rudakov D G, Solov'ev V O, Voshkin A A, Timoshenko A V 2019 Liquid-Liquid Equilibrium of Aqueous Two-Phase System Composed of Poly(Ethylene Oxide) 1500 and Sodium Nitrate *J. Chem. Eng. Data* **64** 3 1250 – 5
[23] Fedorova M I, Levina A V, Zakhodyaeva Y A, Voshkin A A 2020 Interphase Distribution of V(IV) in the Polyethylene Glycol 1500–Sodium Nitrate–Water System *Theor. Found. Chem. Eng.* **54** 4 604–9
[24] Zakhodyaeva Yu A, Izyumova K V, Solov'eva M S, Voshkin A A 2017 Extraction Separation of the Components of Leach Liquors of Batteries *Theor. Found. Chem. Eng.* **51** 5 883–7
[25] Fedorova M I, Zakhodyaeva Yu A, Voshkin A A 2020 Interphase Distribution of Fe(III) and Zn(II) in Chloride Systems with Aliquat 336 in Polypropylene Glycol 425 *Theor. Found. Chem. Eng.* **54** 3 425–30
[26] Fedorova M I, Zinov'eva I V, Zakhodyaeva Y A, Voshkin A A 2020 Extraction of Fe(III), Zn(II), and Mn(II) Using a System with a Green Solvent for Trioctylmethylammonium Thiocyanate *Theor. Found. Chem. Eng.* **54** 2 313–8
[27] Tait B K The extraction of some base metal ions by Cyanex 30L, Cyanex 302, and their binary extractant mixtures with Aliquat 336 1992 *Solvent Extr. Ion Exch.* **10** 5 799–809
[28] Mishra B B, Devi N 2018 Solvent extraction and separation of europium (III) using a phosphonium ionic liquid and an organophosphorus extractant-A comparative study *J. Mol. Liq.* **271** 389 – 96

Acknowledge
The reported study was funded by RFBR and Moscow city Government according to the research project № 19-33-70011.

Interphase distribution of a number of polybasic carboxylic acids in aqueous two-phase systems based on polyethylene glycol-1500.

S V Solov'eva[1,2], V O Solov'ev[1,2], I V Zinov'eva[2]
1 MIREA – Russian Technological University, Moscow, Russian Federation
2 Kurnakov Institute of General and Inorganic Chemistry of the Russian Academy of Sciences, Moscow, Russia

*Corresponding author: svo@igic.ras.ru

Abstract. In this work, we studied the interphase distribution of a number of polybasic carboxylic acids, in particular, malic, succinic, glutaric, citric, ascorbic and tartaric in two-phase aqueous systems polyethylene glycol-1500 (PEG-1500) - sodium sulfate - water and PEG-1500 - ammonium sulfate - water. Using the spectrophotometric method of analysis, quantitative characteristics (interfacial distribution and degree of recovery) of the extraction of carboxylic acids were determined. In conclusion, it was found that the considered two-phase aqueous systems are promising in the process of extraction of polybasic carboxylic acids. In particular, the system polyethylene glycol 1500 - sodium sulfate - water has effective extraction properties for succinic malic and ascorbic acids. On the other hand, depending on the composition of the extraction systems, the quantitative characteristics for citric, glutaric and tartaric acids practically do not differ.

Key words: green chemistry, liquid extraction, aqueous two-phase systems, extraction degree, interphase distribution, polyethylene glycol 1500, sodium sulfate, ammonium sulfate, carboxylic acids.

1. Introduction

The development and implementation of «green» extraction systems to solve problems associated with the purification, concentration and separation of mixtures is one of the rapidly developing areas in chemical technology [1, 2]. Thus Egorova et al investigated the extraction of lanthanides using binary extractants [3]. In work Zakhodyaeva the extraction separation of non-ferrous metals from leaching solutions of active elements of batteries was studied [4]. In investigate Voshkin studied the extraction of iron (III) ions using binary extractants [5]. Aqueous two-phase systems are environmentally friendly, non-toxic, cheap and easy to use, in comparison with classical extraction systems [6, 7]. Voshkin et al. шnvestigated the interphase distribution of aromatic acids in a two-phase aqueous system polyethylene glycol – sodium sulfate – water [8]. The authors of the work [9] studied the extraction of thiophene, o-xylene, and toluene in hexane – polymer – water systems. In work Zinov'eva the interfacial distribution of caffeine and coumarin in a system based on polyethylene glycol and sodium sulfate was considered [10]. Zakhodyaeva et al. developed phase diagrams of two-phase aqueous systems polyethylene glycol 1500 – nitrate nitrate – water and polyvinylpyrrolidone 3500 – sodium nitrate – water at three different temperatures [11, 12]. Such systems have found their application in extraction technologies for the purification and separation of organic and inorganic substances [13, 14]. Zinovyeva et al. studied the extraction of benzoic, salicylic, and sulfosalicylic acids using two-phase aqueous systems based on polyethylene glycol [15]. In work Chen, the authors investigated the selective separation of neodymium (III) and samarium (III) in two-phase systems [16]. The authors of the work [17] studied the extraction of metal ions in two-phase aqueous systems based on polyethylene glycol and sodium sulfate. Tong et al. investigated the extraction of platinum from hydrochloric acid solutions using two-phase systems [18]. Also, in the processes of acid extraction, both binary extractants [19-21] and neutral [22, 23] are used. In addition to traditional liquid-liquid extraction, there are also liquid-liquid chromatographic separation of substances [24, 25], as well as supercritical fluid technologies [26, 27].

These systems can also be used in the extraction of carboxylic acids [13, 15]. Polybasic carboxylic acids, such as malic, succinic, glutaric, citric, ascorbic and tartaric acids are widely used in various fields of various industries, such as food, pharmaceutical, cosmetic, etc., as food additives, organic fertilizers, in the preparation of medicines, in the synthesis of polyamides, in the production of 1,5-pentanediol. Polybasic carboxylic acids are found in natural or man-made products, for example, in water-acid effluents of caprolactam production [28, 29]. So Jazz et al. investigated the extraction of carboxylic acids from organic solvents using supercritical fluid technologies [30]. As a result, there is a need to allocate carboxylic acids of technogenic water runoff.

The aim of this work is to study the interphase distribution of a number of carboxylic acids, in particular malic, succinic, glutaric, citric, ascorbic and tartaric acids, in two-phase aqueous systems PEG-1500 - sodium sulfate - water and PEG-1500 - ammonium sulfate - water during the extraction of a number polybasic carboxylic acids.

2. Experimantal details

2.1. Reagents

The following reagents were used in experimental studies: polyethylene glycol 1500 (PEG-1500) (Fluka), sodium sulphate («c.p.»), ammonium sulphate («c.p.»), malic acid (CHEMMED, «p»), succinic acid (CHEMMED, «p»), glutaric acid (CHEMMED, «p»), citric acid (CHEMMED, «c.p.»), ascorbic acid (CHEMMED, «c.p.»), tartaric acid (CHEMMED, «p.a.»). All reagents were used without further purification.

2.2. Research methods

The initial solutions of polybasic carboxylic acids were prepared by dissolving precise attachments suspended on analytical scales in distilled water.

To study the process of extraction of carboxylic acids in aqueous two-phase systems, graduated test tubes and dividing funnels with sanded stoppers were used. For this, exact weighed portions of a water-soluble polymer and a phase-forming salt, weighed on an analytical balance (OHAUS Explorer), were placed in a test tube and dissolved in aqueous solutions of carboxylic acids, then the resulting mixtures were mixed for 30 minutes. at a rotation speed of 45 rpm in an Enviro-Genie thermostatic shaker (Scientific Industries, Inc.) at 25 °C to achieve thermodynamic equilibrium. The mixture was then centrifuged for 5 minutes. at 2500 rpm (centrifuge CM-6MT, SIA ELM), after which phase volumes were measured.

Quantitative characteristics of the extraction process were determined by spectrophotometric analysis (Cary-60, Agilent spectrophotometer). For the quantitative determination of malic, succinic, glutaric, citric, ascorbic and tartaric acids, calibration lines were constructed at five points, with wavelengths of 205, 201, 203, 206, 244 and 210 nm, respectively, and R^2 was 0.9998, 0.9996, 0.9986, 0.9989, 0.9999, 0.9999, respectively. The wavelength data for each of the acids were determined experimentally using a spectrophotometric analysis method.

IOP Conf. Series: Materials Science and Engineering 1212 (2022) 012022 doi:10.1088/1757-899X/1212/1/012022

3. Results and discussion

During the experimental work, the interphase distribution of malic, succinic, glutaric, citric, ascorbic and tartaric acids in aqueous two-phase systems PEG – 1500 – sodium sulfate-water and PEG – 1500 – ammonium sulfate-water was studied.

Fig. 1-2 shows experimental results of the distribution coefficient of the considered polybasic carboxylic acids in aqueous two-phase systems PEG-1500-sodium sulfate-water and PEG-1500-ammonium sulfate-water. These systems also exhibit effective extraction properties with respect to multi-base carboxylic acids in comparison with traditional organic extractants. According to these figures, the distribution coefficient for citric, ascorbic, glutaric and tartaric acids practically does not differ depending on the composition of the aqueous two-phase system. However, the PEG-1500-sodium sulfate-aqueous system shows more pronounced extraction properties in relation to succinic and malic acids, compared to the PEG-1500 – ammonium sulfate – aqueous system. As a result, there is a need for further research to increase the efficiency of these systems in the extraction of polybasic carboxylic acids. determine the equilibrium time a series of experiments on varying the phase contact time from 3 to 60 minutes Figure 1 were carried out. As can be seen from Figure 1 a slight increase in the Ce(III) distribution coefficient over the studied time interval is observed. In case of La(III) an increase in the distribution coefficient is observed up to 20 minutes and then it decreases, which may be due to the emulsification effect.

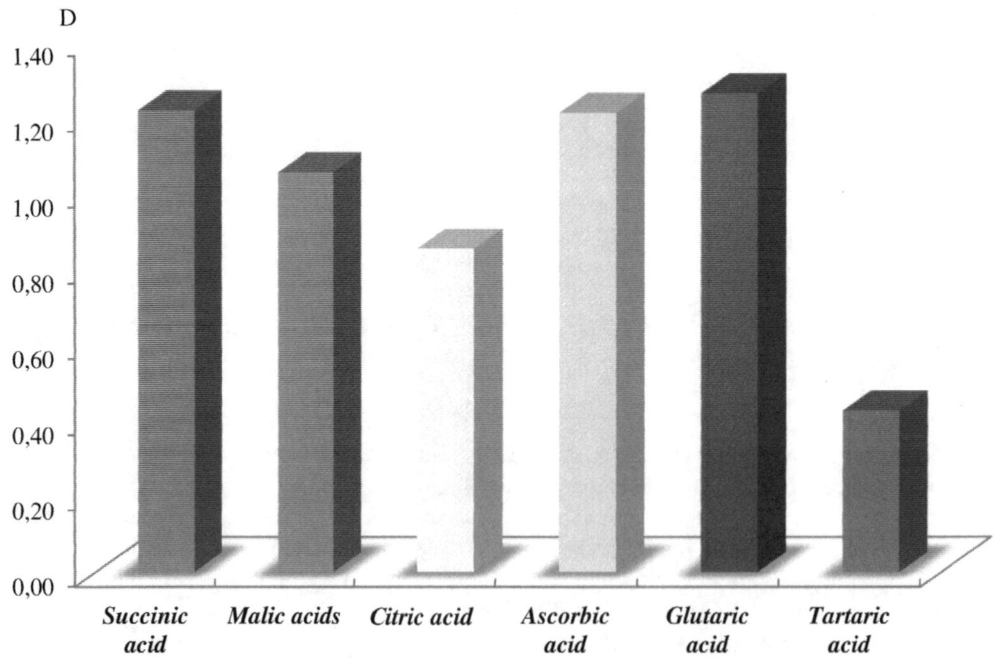

Figure 1. Distribution coefficients polybasic carboxylic acids in aqueous two-phase system PEG-1500-sodium sulfate-water, where *D* is the distribution coefficient.

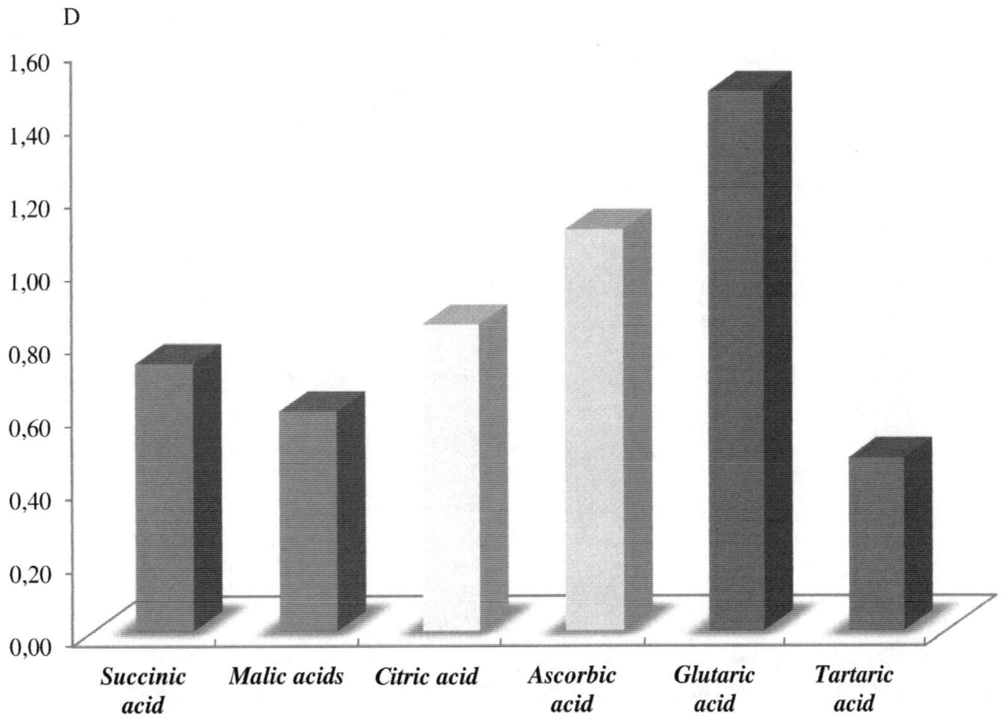

Figure 2. Distribution coefficients polybasic carboxylic acids in aqueous two-phase system PEG-1500-ammonium sulfate-water, where D is the distribution coefficient.

Comparing the results obtained in figure 1 and 2 show that the PEG-1500 - sodium sulfate - water system has more pronounced extraction properties in relation to succinic malic and ascorbic acids. On the other hand, depending on the composition of the extraction system, the quantitative characteristics for citric, glutaric and tartaric acids practically do not differ.

4. Conclusions

A series of systematic studies of the interfacial distribution of a number of polybasic carboxylic acids in aqueous two-phase systems of different composition was carried out. During the study, it was found that the systems PEG-1500-sodium sulfate-water and PEG-1500-ammonium sulfate-water exhibit effective extraction properties in relation to the considered polybasic carboxylic acids in comparison with traditional organic extractants. The use of two-phase aqueous systems in the extraction of polybasic carboxylic acids is promising in view of their non-toxicity, fire safety, and low cost in comparison with traditional organic extractants. Research will also continue to improve these systems in order to increase the efficiency of extraction of polybasic carboxylic acids.

References

[1] Belova V V, Voshkin A A, Egorova N S and Khol'kin A I 2010 Extraction of rare earth metals from nitrate solutions with a binary extractant based on Cyanex 272 *Russ. J. Inorg. Chem.* **55** 4 629-33.

[2] Królikowski M and Lipińska A 2019 Separation of thiophene, or benzothiophene from model fuel using glycols. Liquid–liquid phase equilibria and oxidative desulfurization study *Fluid Phase Equilib.* **482** 11-23.

[3] Egorova N S, Belova V V, Voshkin A A, Zhilov V I and Khol'kin A I 2005 Extraction of

lanthanide chlorides by binary extractants based on phosphinic acid derivatives *Zh. Neorg. Khim.* **50** 11 1902-5.

[4] Zakhodyaeva Y A, Izyumova K V, Solov'eva M S and Voshkin A A 2017 Extraction separation of the components of leach liquors of batteries *Theor. Found. Chem. En.* ***51*** 5 883-7.

[5] Voshkin A A, Belova V V and Khol'kin A I 2003 Extraction of iron (III) by binary extractants based on quaternary ammonium bases and organic acids *Russ. J. Inorg. Chem.***48** 4 608-13.

[6] Li F T, Wu B, Liu R H, Wang X J, Chen L J and Zhao D S 2015 An inexpensive N-methyl-2-pyrrolidone-based ionic liquid as efficient extractant and catalyst for desulfurization of dibenzothiophene *Chem. Eng. J.* **274** 192-9.

[7] Li F T, Kou C G, Sun Z M, Hao Y J, Liu R H and Zhao D S 2012 Deep extractive and oxidative desulfurization of dibenzothiophene with $C_5H_9NO \cdot SnCl_2$ coordinated ionic liquid *J. Hazard. Mater* **205** 164-70.

[8] Voshkin A A, Zakhodyaeva Y A, Zinov'eva I V and Shkinev V M 2018 Interphase distribution of aromatic acids in the polyethylene glycol–sodium sulfate–water system *Theor. Found. Chem. En.* ***52*** 5 890-3.

[9] Zakhodyaeva Y A, Solov'ev V O, Zinov'eva I V, Rudakov D G, Timoshenko A V and Voshkin A A 2019 Interphase distribution of thiophene, toluene, and o-xylene in the hexane–polymer–water extraction system *Theor. Found. Chem. En.* **53** 4 550-5.

[10] Zinov'eva I V, Zakhodyaeva Y A and Voshkin A A 2019 Interphase Distribution of Caffeine and Coumarin in Extraction Systems with Polyethylene Glycol and Sodium Sulfate Theor. Found. Chem. En. 53 6 996-1000.

[11] Zakhodyaeva Y A, Rudakov D G, Solov'ev V O, Voshkin A A and Timoshenko A V 2019 Liquid–liquid equilibrium of aqueous two-phase system composed of poly (ethylene oxide) 1500 and sodium nitrate J. Chem. Eng. Data 64 3 1250-5.

[12] Zakhodyaeva Y A, Rudakov D G, Solov'ev V O, Voshkin A A and Timoshenko A V 2019 Liquid–Liquid Equilibrium in an Extraction System Based on Polyvinylpyrrolidone-3500 and Sodium Nitrate Theor. Found. Chem. En. 53 2 159-65.

[13] Zinov'eva I V, Zakhodyaeva Y A and Voshkin A A 2019 Extraction of Monocarboxylic Acids from Diluted Solutions with Polyethylene Glycol Theor. Found. Chem. En. 53 5 871-4.

[14] Fedorova M I, Zinov'eva I V, Zakhodyaeva Y A and Voshkin A A 2020 Extraction of Fe (III), Zn (II), and Mn (II) Using a System with a Green Solvent for Trioctylmethylammonium Thiocyanate Theor. Found. Chem. En. 54 313-8.

[15] Zinov'eva I V, Zakhodyaeva Y A and Voshkin A A 2020 Data on the extraction of benzoic, salicylic and sulfosalicylic acids from dilute solutions using PEG-based aqueous two-phase systems Data in brief 28 105033.

[16] Chen Y, Wang H, Pei Y, Ren J and Wang J 2015 pH-controlled selective separation of neodymium (III) and samarium (III) from transition metals with carboxyl-functionalized ionic liquids ACS Sustainable Chem. Eng. 3 12 3167-74.

[17] Shibukawa M, Nakayama N, Hayashi T, Shibuya D, Endo Y and Kawamura S 2001 Extraction behaviour of metal ions in aqueous polyethylene glycol–sodium sulphate two-phase systems in the presence of iodide and thiocyanate ions Anal. Chim. Acta 427 2 293-300.

[18] Tong Y, Wang C, Huang Y and Yang Y 2015 Extraction and stripping of platinum from hydrochloric acid medium by mixed imidazolium ionic liquids Ind. Eng. Chem. Res. 54 2 705-11.

[19] Zakhodyaeva Y A, Voshkin A A, Belova V V and Khol'kin A I 2011 Extraction of monocarboxylic acids with binary extracting agents based on amines and quaternary ammonium bases Theor. Found. Chem. Eng. 45 5 739-43.

[20] Kholkin A I, Belova V V, Zakhodyaeva Y A and Voshkin A A 2013 Solvent extraction of weak acids in binary extractant systems Separation Science and Technology 48 9 1417-25.

[21] Zakhodyaeva Y A, Voshkin A A and Belova V V 2013 Extraction of acetic acid by binary extractants Russian Journal of Inorganic Chemistry 58 4 481-485.

[22] Zakhodyayeva Y A, Voshkin A A, Belova V V, Kostanyan A E and Khol'kin A I 2010 Extraction of lactic acid from technological (concentrated) solutions Theor. Found. Chem. Eng. 44 5 782-5.

[23] Belova, V. V., Zakhodyaeva, Y. A., & Voshkin, A. A. (2017). Extraction of carboxylic acids with neutral extractants Theor. Found. Chem. Eng. 51 5 786-94.

[24] Kostanyan A E and Voshkin A A 2009 Pulsation cyclic liquid-liquid chromatography Theor. Found. Chem. Eng. 43 5 729-33.

[25] Zakhodyaeva Y A and Voshkin A A 2013 Extraction and separation of carboxylic acids by liquid-liquid extraction and chromatography Theor. Found. Chem. Eng. 474 461-6.

[26] Kostenko M O, Pokrovskiy O I, Zakhodyaeva Y A, Voshkin A A and Lunin V V 2020 Unusual effect of flow rate on retention in analytical supercritical fluid chromatography exemplified by polyethylene glycol separation Journal of Chromatography A 1610 460513.

[27] Yousefi M, Rahimi-Nasrabadi M, Pourmortazavi S M, Wysokowski M, Jesionowski T, Ehrlich H and Mirsadeghi S 2019 Supercritical fluid extraction of essential oils TrAC Trends in Analytical Chemistry 118 182-93.

[28] Lopez-Garzon C S and Straathof A J 2014 Recovery of carboxylic acids produced by fermentation: Biotechnol. Adv 32 5 873-904.

[29] Levanova S V, Gerasimenko V I, Glazko I L, Sokolov A B, Sumarchenkova I A and Kanayev A V 2006 Sintez slozhnykh efirov iz zhidkikh otkhodov proizvodstva kaprolaktama Russ. J. Gen. Chem. 3 37-42.

[30] Djas M and Henczka M 2018 Reactive extraction of carboxylic acids using organic solvents and supercritical fluids: a review Sep. Purif. Technol 201 106-19.

Acknowledgement
The reported study was funded by RFBR according to the research project № 18-29-06070

The interphase distribution of light REE Ce(III) and La(III) in system based on PEG-1500 – $NaNO_3$ – H_2O with the quaternary ammonium base addition

A V Levina[*], A Ya Fedorov, M I Fedorova

Kurnakov Institute of General and Inorganic Chemistry of the Russian Academy of Sciences, Moscow, Russia

*Corresponding author: levinaarina95@gmail.com

Abstract. The problem of e-waste processing and recovery of valuable metals from such waste for the second use is attracting more and more scientists' attention. Liquid extraction as one of hydrometallurgy steps is a traditional method for the metal recovery. However, application of solvent extraction is not meet the green chemistry principles due to organic solvents. Aqueous two-phase systems based on water-soluble polymers are promising alternative for hazardous organic solvents. In this work the dependencies of Ce(III) and La(III) distribution coefficients from process time and the initial quaternary ammonium base concentration have been achieved. Also, based on the Ce(III) and La(III) extraction isotherms it has been shown that the metals initial concentrations are highly affects the distribution coefficients of studied metals. The possibility of aqueous two-phase system application as a solvent for quaternary ammonium salt for light REE (Ce(III) and La(III)) extraction from water solution has been shown.

Key words: liquid extraction, aqueous two-phase systems, rare earths elements extraction, waste processing, «green» chemistry

1. Introduction

The increase in demand, and, as a consequence, the consumption of electronic devices around the world is constantly growing [1-3]. In that case, the amount of e-waste containing valuable components, including rare-earth metals (Ce, La, etc.) is increasing [4,5]. The problem of processing such kind of waste and extracting valuable metals from them, for their reuse, is a key one, since the storage of chemical current sources entails environmental pollution by heavy metals [6]. To implement this task, one of the promising stages of the hydrometallurgical process - liquid extraction can be use [7-10]. For example, the possibility of rare earth elements [11-13], some actinides [14] and iron ions [15,16] extraction with binary organic extragents has been shown. However, the application of traditional approaches is connected with organic solvents that make extraction processes expensive and environmental unsafe [17]. Also, using strong acids in extraction and re-extraction processes is required complicated waste utilization processes after the recovery of valuable components [18].

The promising alternative for the hazardous organic solvents can be aqueous two-phase systems based on water-soluble polymers that are biodegradable, fire-resistant and safe [19-21]. Moreover, such systems can extract the range of metals from aqueous media with high effectivity and selectivity [22-25].

The aim of this work was the investigation of quaternary ammonium base salt effect on the light rare-earth elements (Ce(III) and La(III) extraction in aqueous two-phase system based on polyethylene glycol 1500. The influence of the phase contact time on the distribution coefficients of Ce(III) and La(III) has been determined. Also, effects of the initial concentration of metals and quaternary ammonium base salt on the Ce(III) and La(III) distribution coefficients in aqueous two-phase system has been investigate.

2. Experimental details

2.1. Reagents
The initial solutions of metal nitrates were obtained by dissolving the sample weights of $La(NO_3)_3$*$6H_2O$ and $Ce(NO_3)_3$*6H2O of the qualification "chemical grade" suspended on an analytical balance (AND HR-100AZ) in distilled water. Trioctylmethylammonium nitrate (TOMAN) was prepared as follows: 1.6 mol L^{-1} sodium nitrate solution was mixed with Aliquat 336 in a 1:1 ratio [26]. Stirring was performed until the chloride ions were completely replaced by nitrate ions. The mixing time was 20 minutes

2.1.1. Research methods
The metal ions extraction was carried out using a two-phase aqueous system polyethylene glycol 1500 (16.3%) - $NaNO_3$ (36%) - H_2O with an initial metal concentration of 0.01mol L^{-1} (except for the cases when the investigations on the effects of the initial concentrations of metals on the distribution coefficients were carried out). The nitric acid concentration in the salt phase was equal 0.01mol L^{-1} (for the hydrolyze avoiding).

The metal ions La (III) and Ce (III) extraction was carried out at a temperature of 25°C in graduated plastic tubes in an Enviro-Genie thermostatically controlled shaker (Scientific Industries, Inc.) at a rotation speed of 30 rpm until the thermodynamic equilibrium was established (20 min).

The complexometric titration using EDTA and xylenol orange at pH = 6 was used to determine the metal ions concentration in the initial solutions and in the aqueous phases after extraction.

The presented experimental data are the result of a series of experiments and processed by methods of mathematical statistics

3. Results and discussion
To determine the equilibrium time a series of experiments on varying the phase contact time from 3 to 60 minutes Figure 1 were carried out. As can be seen from Figure 1 a slight increase in the Ce(III) distribution coefficient over the studied time interval is observed. In case of La(III) an increase in the distribution coefficient is observed up to 20 minutes and then it decreases, which may be due to the emulsification effect.

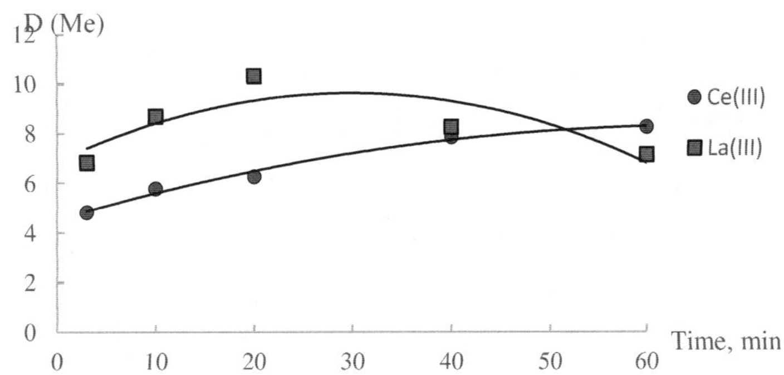

Figure 1. The dependence Ce(III) and La(III) distribution coefficients on the phase contact time in system based on PEG-1500 (16.3%) – $NaNO_3$ (36%) – H_2O, $[Me]_{init.}$=0.01mol L^{-1}, $[TOMAN]_{t.p.}$=0.05mol L^{-1}.

To determine the effect of the metals initial concentration on the Ce(III) and La(III) distribution coefficients, the extraction isotherms of both metals were obtained Figure 2. Based on the Figure 2, it is clear that the extraction isotherms of both metals have a similar form - they increase over the entire studied interval from [Me]0.001 to 0.01mol L^{-1}.

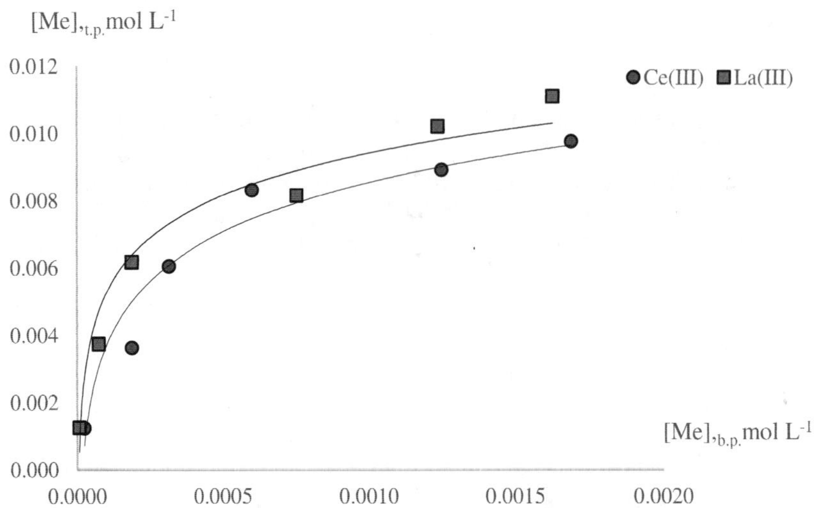

Figure 2. Ce(III) and La(III) extraction isotherms in system based on PEG-1500 (16.3%) – $NaNO_3$ (36%) – H_2O, $[TOMAN]_{t.p.}$=0.05mol L^{-1}, where $[Me]_{,t.p.}$mol L^{-1} – metal concentration in the equivalent top phase, $[Me]_{,b.p.}$mol L^{-1} – metal concentration in the equivalent bottom phase.

The Figure 3 shows the dependence of the Ce(III) and La(III) distribution coefficients on the TOMAN initial concentration in the top phase. It can be clearly seen the exponential nature of the graph – from the [TOMAN]=0.07 mol L^{-1} there is a sharp increase in the Ce(III) and La(III) distribution coefficients.

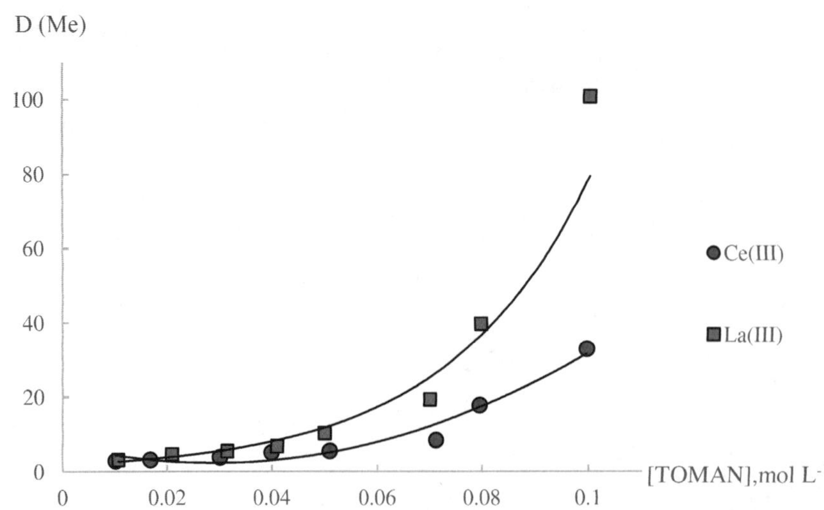

Figure 3. The dependences of Ce(III) and La(III) distribution coefficients on the TOMAN initial concentration.

4. Conclusions

As a result of this work Ce(III) and La(III) extraction isotherms at temperature of 25°C in the system based on PEG-1500 – $NaNO_3$ – H_2O with the TOMAN addition has been obtained. The nonlinear nature of these dependences indicates that the distribution coefficients of Ce(III) and La(III) strongly depends on the initial metals concentrations. Obtained kinetic dependences of Ce(III) and La(III) distribution coefficients shows that in case of Ce(III) the dependence slightly increases through the studied period, and in case of La(III) this dependence has a maximum (at 20min) and then decreasing (possibly because of emulsification process). The obtained dependences of the distribution coefficients of Ce (III) and La (III), depending on the initial concentration of TOMAN, show that with a seven-fold excess of TOMAN in the upper phase, the distribution coefficient of La (III) equals 100. This work shows the possibility applying aqueous two-phase systems as an alternative solvent for traditional organic solvent.

References

1. Zhang L, Lijuan L, Rui H, Shi D, Peng X, Ji L, Song X 2020Lithium recovery from effluent of spent lithium battery recycling process using solvent extraction *J. Hazard. Mater.* **398** 122840
2. Nowakowski P, Pamula T 2020 Application of deep learning object classifier to improve e- waste collection planning *Waste Manage.* **109** 1-9
3. Ismail H, Hanafiah M M 2020A review of sustainable e-waste generation and management: Present and future perspectives *J. Environ. Manage.* **264** 110495
4. Kumari A, Panda R, Lee J, Thriveni T, Jha M, Pathak D 2019 Extraction of rare earth metals (REMs) from chloride medium by organometallic complexation using D2EHPA *Sep. Purif. Technol.* **227** 115680
5. Zhang J, Li X, Song D, Miao Y, Song J, Zhang L 2018Effective regeneration of anode material recycled from scrapped Li-ion batteries *J. Power Sources* **390** 38-44
6. Li W, Achal V 2020 Environmental and health impacts due to e-waste disposal in China–A review *Sci. Total Environ.* **737** 139745
7. Granata G, Pagnanelli F, Moscardini E, Takacova Z, Havlik T, Toro L 2012 Simultaneous. Recycling of nickel metal hydride, lithium ion and primary lithium batteries: Accomplishment of European Guidelines by optimizing mechanical pre-treatment and solvent extraction operations *J. Power Sources* **212** 205-11.
8. Fedorova M I, Zinov`eva I V, Zakhodyaeva Y A, Voshkin A A 2020 Extraction of Fe(III), Zn(III), and Mn(II) using a system with a Green Solvent for trioctylmethylammonium thiocyanate *Theor. Found. Chem. Eng.* **54** 2 313-8
9. Takahashi V C I, Botelho Junior A B, Espinosa D C R, Tenorio J A S 2020 Enhancing cobalt recovery from Li-ion batteries using grinding treatment prior to the leaching and solvent extraction process *J. Environ. Chem. Eng.* **8** 103801
10. Shuya L, Yang C, Xuefeng C, Wei S, Yaqing W, Yue Y 2020 Separation of lithium and transition metals from leachate of spent lithium-ion batteries by solvent extraction method with Versatic 10 *Sep. Purif. Technol.* **250** 117258
11. Belova V V, Voshkin A A, Egorova N S, Kholkin A I 2012 Solvent extraction of rare earth metals from nitrate solutions with di(2,4,4-trimethylpentyl)phosphinate of methyltrioctylammonium *J.Mol. Liq.* **172** 144-6
12. Belova V V, Voshkin A A, Egorova N S, Khol`kin A I 2010 Extraction of rare earth metals from nitrate solutions with a binary extractant based on Cyanex 272 *Rus. J. Inorg. Chem.* **55** 4 629-33
13. Belova V V, Voshkin A A, Egorova N S, Khol`kin A I 2015 Extraction of rare earth metals, uranium, and thorium from nitrate solutions by binary extragents *Theor. Found. Chem. Eng.* **49** 4 545-9

14. Egorova N S, Belova V V, Voshkin A A, (...) Pyartman A K, Keskinov V A 2008 Extraction of uranyl nitrate with a binary extractant based on di(2,4,4-trimethylpentyl)phosphinic acid *Theor. Found. Chem. Eng.* **42** 5 708-13
15. Voshkin A A, Belova V V, Khol`kin A I 2003 Extraction of iron(II) by binary extractants based on quaternary ammonium bases and organic acid *Rus. J. Inorg. Chem.* **48** 4 608-13
16. Voshkin A A, Belova V V, Zakhodyaeva Y A 2018 Iron extraction with Di(2-ethylhexyl)dithiophosphoric acid and a binary extractant based on it *Rus. J. Inorg. Chem.* **63** 3 387-90
17. Sun Y, Zhu M, Yao Y, Wang H, Tong B, Zhao Z 2020 A novel approach for the selective extraction of Li^+ from the leaching solution of spent lithium-ion batteries using benzo-15-crown-5 ether as extractant *Sep. Purif. Technol* **237** 116325
18. Batchu N, Sonu C, Lee M 2013 Synergistic solvent extraction of manganese(II) with a mixture of Cyanex 272 and Cyanex 301 from chloride solutions *Hydrometallurgy* **140** 89
19. Zakhodyaeva Y A, Rudakov D G, Solov`ev V O, Voshkin A A, Timoshenko A V 2019 Liquid-liquid equilibrium in an extraction system based on polyvinylpyrrolidone-3500 and sodium nitrate *Theor. Found. Chem. Eng.* **53** 2 159-65
20. Zakhodyaeva Y A, Rudakov D G, Solov`ev V O, Voshkin A A, Timoshenko A V 2019 Liquid-liquid extraction equilibrium of aqueous two-phase system composed of poly(ethylene oxide)1500 and sodium nitrate *J. Chem. Eng. Data* **63** 3 1250-5
21. Oan Y, Sun X, Qi M, Qin R, Che X, Zhang Y 2020 A clean and efficient method for separation of vanadium and molybdenum by aqueous two-phase systems *J. Mol. Liq.* **313** 113540
22. Zakhodyaeva Y A, Zinov'eva I V, Tokar E S, Voshkin A A 2019 Complex Extraction of Metals in an Aqueous Two-Phase System Based on Poly(Ethylene Oxide) 1500 and Sodium Nitrate *Molecules* **22** 22 4078
23. Zakhodyaeva Y A, Izuymova K V, Solov`eva M S, Voshkin A A 2017 Extraction separation of the components of leach liquors of batteries *Theor. Found. Chem. Eng.* **51** 11 883-7
24. Voshkin A A, Zakhodyaeva Y A , Zinov'eva I V, Shkinev V M 2018 Interphase distribution of aromatic acids in the polyethylene glycol – sodium sulfate – water system *Theor. Found. Chem. Eng.* **52** 5 890-3
25. Zakhodyaeva Y A, Zinov'eva I V, Voshkin A A 2019 Extraction of Iron(III) Chloride Complexes Using the Polypropylene Glycol 425–NaCl–H2O *Theor. Found. Chem. Eng.* **53** 5 735-40
26. Nayl A 2010 Extraction and separation of Co(II) and Ni(II) from acidic sulfate solutions using Aliquat 336 *J. Hazard. Mater.* **173** 223

Acknowledgement
The reported study was funded by RFBR according to the research project № 18-29-24170

The possibility of using DES based on polypropylene glycol 425 and tetrabuthylammonium bromide in the extraction processes of transition metals

A Y Fedorov*, **A V Levina**

Kurnakov Institute of General and Inorganic Chemistry of the Russian Academy of Sciences, Leninsky prospect, 31, 119991, Moscow, Russia

*Corresponding author: alfedwork@gmail.com

Abstract. Traditionally, the method of liquid extraction is used to extract metals from aqueous. This work is devoted to the combination of perspective alternative for hazardous solvents (aqueous two-phase systems based on water-soluble polymers) and the novel deep eutectic solvents in the non-ferrous metals extraction processes. In this work, the synthesis of deep eutectic solvent based on a water-soluble polymer (PPG-425) and tetrabutylammonium bromide (TBAB) by stirring for 10 minutes at 80°C has been shown. The obtained results showed not only the possibility of using DES in the metal extraction process, but the selectivity to the Fe(III) and Zn(III), the distribution coefficients were 71.64 and 25.17 respectively. The metal concentrations were determined spectrophotometrically using 4-(2-pyridylazo)resorcinol. This work shows the perspectives of using DESs in the metal extraction processes.

Key words: liquid-liquid extraction, aqueous two-phase systems, deep eutectic solvents, «green» chemistry

1. Introduction

In recent years, much attention has been paid to the transition from traditional methods of extraction and separation of various metals from discharged lithium-ion batteries to methods that take into account the principles of «green» chemistry [1]. The traditional methods involved the using of various solvents and substances that were harmful to the environment and health [2]. At the beginning of the twenty-first century, deep eutectic solvents (DESs) were developed, they became promising alternative for hazardous organic solvents. DESs are systems formed from a eutectic mixture of Lewis and Bronsted acids and bases, which may contain various anionic and/or cationic substances. One of the DESs components should act as a hydrogen bond donor, and the other as an acceptor, as a result of which they form a eutectic mixture with a melting point much lower than the melting temperature of any of the individual components [3]. At present DESs have found applications in different fields of chemistry. They are used in biotechnological processes, for example, for the separation of RNA [4], for the electrodeposition of metals [5], for the extraction or metal compounds from aqueous media [6], in analytical chemistry [7], for the synthesis of iron [8] or zinc [9] nanoparticles.

In addition to the principles of «green» chemistry, researchers began to worry about the problems of the limited non-renewable resources of the Earth, including various metals, the mining and processing of which is a source of environmental pollution. Electronic waste is a huge problem, because if it is not recycled, then the heavy metals that are contained in them are released into the environment. Their processing involves the use of strong acids at the leaching stage, and the following extraction of metals from the resulting leach solutions in water-organic solvent systems, which does not comply with the principles of «green» chemistry and are toxic, flammable and explosive [10,11]. One of the most common methods for the extraction of organic and inorganic substances is liquid extraction by the systems based on water and organic solvent. At present, hybrid methods of liquid-liquid chromatography are known, which are the most effective in comparison with traditional extraction using mixing and settling apparatus. Counter-current chromatography [12,13,14] is of the greatest interest, since its use implies not only an analytical aspect, but also a preparative one, that is, application in order to extract target components, processes of separation of rare earth and other metals have been implemented. In addition, this method is easily scalable, that is, it is applicable not only in laboratory conditions, but also on an industrial scale [15,16].

The use of extraction systems based on water-soluble polymers is a more environmentally friendly method. Traditionally, for the extraction of V(IV) [17], Hg(II), Co(II), Zn(II) [18] and Tl(I) [19] ions, it is carried out using toxic and flammable organic solvents, but this can avoid if using aqueous two-phase systems. In addition, they allow efficient extraction of metals such as Fe (III), Ni (II), La (III), Ce (III). These systems are an environmentally friendly replacement for organic solvents. However, with all the obvious advantages of ATPS, there are a number of weakness. So, for example, in [20] was shown that wide range of metals is extracted without selectivity, which necessitates their additional separation. Another problem is the initial medium from which metals are extracted. For example, it has been shown to extract such metals as Fe(III), Co(II), Cu(II) [21, 22], Co(II) and Ni(II) [23], Fe(III) [24] in ATPs from sulfate media, which requires additional introduction into the system of complexing agents such as potassium thiocyanate or 1-nitroso-2-naphthol, which entails the problem of waste disposal after extraction. In addition, attention is paid to liquid-liquid equilibria of aqueous two-phase systems, which can subsequently be used for extraction processes [25, 26]. Research also concerns the extraction of Fe(III), Co(II), Cu(II) and Ni(II) from chloride media [27].

It should be noted that in [28] was shown the extraction of metals using a quaternary ammonium base (QAB) (Aliquat 336) in an aqueous two-phase system with high efficiency, which is caused by the formation of metal complexes with Aliquat 336. However, despite the high metals distribution coefficient, QABs are toxic and do not satisfy the requirements of «green» chemistry.

In [29] was studied the extraction of a number of alkali and transition metal chlorides from aqueous media using DESs based on decanoic acid and lidocaine in various ratios (2:1, 3:1, 4:1). The study showed that most metal cations are extracted with high efficiency.

In this regard, the aim of this work was synthesis DES based on polypropylene glycol 425 (PPG425) and tetrabutylammonium bromide (TBAB) and its application for extraction the range of metals: Fe(III), Zn(II), Cu(II), Ni(II), Co(II), Mn(II).

2. Experimental

The tools were used in this work: 50 ml measuring beakers, measuring pipettes (from 10μ to 5ml), 20 ml separation funnels, 15 ml graduated plastic tubes, drop tube, analytical Balance (AND HR-100AZ), Magnetic stirrer with heating plate (IKA C-MAG HS4) with stirrer bar, centrifuge (ELMI CM-6MT), desalinated filter paper, Enviro-Genie thermostatically controlled shaker (Scientific Industries, Inc.), spectrophotometer Cary-60 device (Agilent Tech.), quartz cuvettes (*l=10*мм)

The initial solutions of metal chlorides or nitrates were prepared by dissolving precise weights of $FeCl_3 \cdot 6H_2O$, $ZnCl_2$, $NiCl_2 \cdot 6H_2O$, $CoCl_2 \cdot 6H_2O$, $MnCl_2 \cdot 4H_2O$, $CuCl_2 \cdot 2H_2O$ with the qualification of "chemical grade" in distilled water suspended on an analytical balance (AND HR-100AZ).

DES based on PPG-425 and TBAB in (2:1) mole ratio were used as the extractant. The molar ratio was chosen based on literatures data, in which similar DES were successfully synthesized and applied [30,31]. This extractant was prepared at a temperature of 80°C with constant mixing for 10 minutes. By the method of IR spectroscopy was confirmed the formation of hydrogen bonds between PPG-425 and TBAB, which indicates the DES formation.

The extraction of metal ions was carried out using an aqueous two-phase system based on DES(33 wt.%) - NaCl(8 wt.%) - H_2O with an initial metal concentration of 0.01mol L^{-1} and a hydrochloric acid concentration in the salt phase of 0.03mol L^{-1}, for suppression of the hydrolysis reaction. The ratio of polymer and salt phases was 1:3.

The extraction of metal ions (Fe(III), Zn(II), Cu(II), Ni(II), Co(II), Mn(II)) was carried out at a temperature of 25°C in graduated plastic tubes in a thermostatic Enviro-Genie shaker (Scientific Industries, Inc.) at a rotation speed of 30 rpm to establish thermodynamic equilibrium (20 min).

The concentration of metal ions in the initial solutions and in the aqueous phases after extraction was determined spectrophotometrically using 4-(2-pyridylazo)resorcinol, which forms complexes with metals that absorb in the visible spectrum at the following wavelengths: Zn (490 nm), Ni (492 nm), Co (508 nm), Mn (496 nm), Cu (494 nm). To determine Fe (III), sulfosalicylic acid was used, with a maximum absorption of complex at a wavelength of 420 nm. The concentration of metal ions in the organic phases was determined by the difference between the concentrations in the initial solution and in the aqueous phase after extraction, taking into account the ratio of phase volumes. The optical density values were determined using a Cary-60 device (Agilent Tech.) in quartz cuvettes l = 10 mm.

The presented experimental data are unique and are the result of a series of experiments and processed by methods of mathematical statistics.

3. Results and discussion

We studied the interphase distribution of a wide range of metal ions (Fe(III), Zn(II), Cu(II), Ni(II), Co(II), Mn(II)) in the system based on DES(33 wt.%) - NaCl(8 wt.%) - H_2O. Figures 1 and figures 2 shows the values of the distribution coefficients and the extraction efficiency for the listed metal ions.

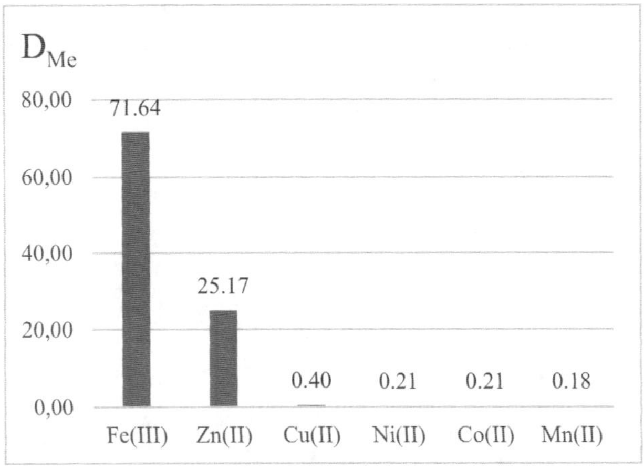

Fig. 1. The values of the distribution coefficients for the metal ions extracted in an aqueous two- phase system based on DES(PPG-425 - TBAB)-NaCl- H2O

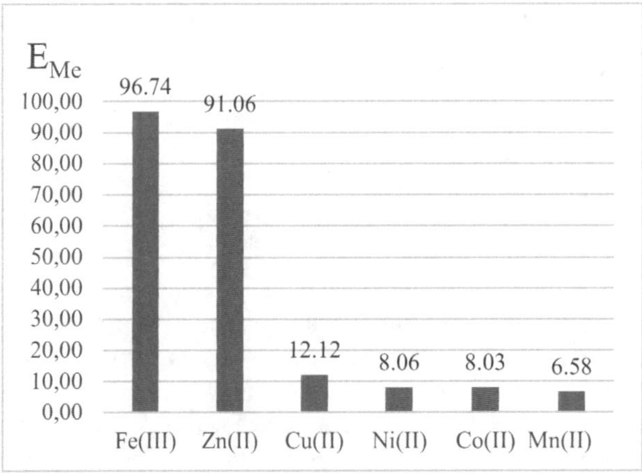

Fig. 2. The values of the extraction efficiency for the metal ions extracted in an aqueous two-phase system based on DES(PPG-425 - TBAB)-NaCl- H2O

Based on the data from figures 1 it is obvious that only Fe(III) and Zn(II) are well extracted, the distribution coefficients (D) are 71.64 and 25.17, respectively, for the rest metals distribution coefficients do not exceed 0.4. It is possible that the high values of the distribution coefficients for iron and zinc are associated with the fact that they easily forming chloride complexes, which is not typical for other metals.

4. Conclusions

The study showed the effectiveness of using DES based on PPG-425 and TBAB in the extraction of a number of metal ions by the system DES(33 wt.%) - NaCl(8 wt.%) - H_2O. It was found that using DES it is possible to achieve not only high extraction efficiency of Fe(III) and Zn(II), but also selective extraction of these metals in comparison with a number of studied metals. Thus, it is possible to use DES based on PPG-425 and TBAB (2:1) mole ratio for the selective extraction of Fe(III) and Zn(II) from aqueous media.

References

[1] Tran M, Rodrigues F, Kato K, Babu G and Ajayan P 2019 Deep eutectic solvents for cathode recucling of Li-ion batteries *Nat. energy* **4** 339-45

[2] Anastas P and Warner J 1998 Green Chemistry: Theory and Practice *Oxford University Press, New York* 30

[3] Abbott A, Capper G, Davies D, Rasheed R and Tambyrajah V 2003 Novel solvent properties of choline chloride/urea mixtures *Chem. Commun.* 70–71

[4] Zhang H, Wang Y, Zhou Y, Xu K, Li N, Wen Q and Yang Q 2017 Aqueous biphasic systems containing PEG-based deep eutectic solvents for high-performance partitioning of RNA *Talanta* **170** 266–74

[5] Smith E, Abbott A and Ryder K 2014 Deep Eutectic Solvents (DESs) and Their Applications / *Chem. Rev.* **114** 11060−82

[6] Ola P and Matsumoto M 2018 Use of deep eutectic solvent as extractant for separation of Fe (III) and Mn (II) from aqueous solution *Sep. Sci. Technol.* **54** 5 759-65

[7] Shishov A, Bulatov A, Locatelli M, Carradori S and Andruch V 2017 Application of deep eutectic

solvents in analytical chemistry. A review *Microchem. J.* **135** 33-8
[8] Shkinev V M, Zakhodyaeva Y A, Dzhenloda R K, Mokhodoeva O B and Voshkin A A 2017 Synthesis of magnetic iron oxide nanoparticles at the interface of the polyethylene glycol-ammonium sulfate-water extraction system *Mendeleev Commun.* **5** 485-6
[9] Voshkin A A, Shkinev V M and Zakhodyaeva Y A 2017 A New Extraction Method for the Preparation of Zinc Oxide Nanoparticles in Aqueous Two-Phase Systems *Russ. J. Phys. Chem. A* **91** 226–8
[10] Zeng X, Li J and Singh N 2014 Recycling of Spent Lithium-Ion Battery: A Critical Review *Crit. Rev. Env. Sci. Technol* **44** 1129–65
[11] Zhao Y, Yuan X, Jiang L, Wen J, Wang H, Guan R, Zhang J and Zeng G 2020 Regeneration and reutilization of cathode materials from spent lithium-ion batteries *Chem. Eng. J.* **383** 123089
[12] Kostanyan A E, Voshkin A A and Kodin N V 2011 Controlled-cycle pulsed liquid–liquid chromatography. A modified version of Craig's counter-current distribution *J. Chromatogr. A*,**1218** 36 6135-43
[13] Kostanian A E and Voshkin A A 2007 Analysis of new counter-current chromatography operating modes *J. Chromatogr. A* **1151** 126-30
[14] Hewitson P, Sutherland I, Kostanyan A E, Voshkin A A and Ignatova S 2013 Intermittent counter-current extraction-Equilibrium cell model, scaling and an improved bobbin design *J. Chromatogr. A* **1303** 18-27
[15] Kostanyan A E and Voshkin A A 2011 Analysis of cyclic liquid chromatography *Theor. Found. Chem. Eng.* **45** 68-74
[16] Kostanyan A E and Voshkin A A 2009 Support-free pulsed liquid-liquid chromatography *J. Chromatogr. A* **1216** 7761-6
[17] Fedorova M I, Levina A V, Zakhodyaeva Y A, Voshkin A A 2020 Interphase Distribution of V(IV) in the Polyethylene Glycol 1500–Sodium Nitrate–Water System *Theor. Found. Chem. Eng.* **54** 4 604–9
[18] Hamta A and Dehghani M 2017 Application of polyethylene glycol based aqueous two-phase systems for extraction of heavy metals *J. Mol. Liq.* **231** 20-4
[19] Huang Y, Chen D, Kong L, Su M and Chen Y 2020 Aqueous two-phase systems (polyethylene glycol + ammonia sulfate) for thallium extraction: Optimization of extraction efficiency, structural characterization, and mechanism exploration *Sep. Purif. Technol.* **235** 115740
[20] Zakhodyaeva Yu A, Zinovyeva I V, Tokar E S and Voshkin A A 2019 Complex extraction of metals in an aqueous two-phase system based on poly(ethylene oxide) 1500 and sodium nitrate *Molecules* **24** 4078
[21] Zakhodyaeva Yu A, Izyumova K V, Solov'eva M S and Voshkin A A 2017 Extraction separation of the components of leach liquors of batteries *Theor. Found. Chem. Eng.* **51** 5 883
[22] Voshkin A A, Belova V V and Khol'kin A I 2003 Extraction of iron(III) by binary extractants based on quaternary ammonium bases and organic acids *Russ. J. Inorg. Chem.* **48** 4 608-13
[23] Valadares A, Valadares C, De Lemos L, Mageste A and Rodrigues G 2018 Separation of cobalt and nickel in leach solutions of spent nickel-metal hydride batteries using aqueous two-phase systems (ATPS) *Hydrometallurgy* **181** 180
[24] Belova V V, Voshkin A A and Khol'kin A I 2011 Liquid-Liquid Distribution of Metals in Systems with Trioctylmethylammonium Dinonyl Naphthalenesulfonate *Theor. Found. Chem. En.* **45** 764
[25] Zakhodyaeva Y A, Zinov'eva I V and Voshkin A A 2019 Extraction of Iron(III) Chloride Complexes Using the Polypropylene Glycol 425–NaCl–H2O System *Theor. Found. Chem. En.* **53** 5 735-40
[26] Zakhodyaeva Y A, Rudakov D G, Solov'ev V O, Voshkin A A and Timoshenko A V 2019 Liquid–Liquid Equilibrium in an Extraction System Based on Polyvinylpyrrolidone-3500 and Sodium Nitrate *Theor. Found. Chem. En.* **53** 2 159-65
[27] Zakhodyaeva Y A, Rudakov D G, Solov'ev V O, Voshkin A A and Timoshenko A V 2019

Liquid–Liquid Equilibrium of Aqueous Two-Phase System Composed of Poly(Ethylene Oxide) 1500 and Sodium Nitrate *J. Chem. Eng. Data.* **64** 3 1250-5

[28] Fedorova M I, Zinov'eva I V, Zakhodyaeva Y A and Voshkin A A 2020 Extraction of Fe(III), Zn(II), and Mn(II) Using a System with a Green Solvent for Trioctylmethylammonium Thiocyanate *Theor. Found. Chem. En.* **54** 2 313-18

[29] Osch D, Parmentier D, Dietz C, van den Bruinhorst A, Tuinier R and Kroon M 2016 Removal of Alkali and Transition Metal Ions from Water with Hydrophobic Deep Eutectic Solvents *Chem. Commun.* **52** 11987-90

[30] Jibril B, Mjalli F, Naser J and Gano Z 2014 New tetrapropylammonium bromide-based deep eutectic solvents: Synthesis and characterizations *J. Mol. Liq.* **199** 462-9

[31] Singh A, Walvekar A, Khalid M, Wong W Y and Gupta T C S M 2017 Thermophysical properties of glycerol and polyethylene glycol (PEG 600) based DES *J. Mol. Liq.* **252** 439-44

Acknowledgments
This work is supported by the Russian Science Foundation under grant 20-13-00387.

Thiophene extraction in PEG - hexan - water based systems

V O Solov'ev[1,2], S V Solov'eva[1,2], N A Milevsky[2]
1 MIREA – Russian Technological University, Moscow, Russian Federation
2 Kurnakov Institute of General and Inorganic Chemistry of the Russian Academy of Sciences, Moscow, Russia

*Corresponding author: svo@igic.ras.ru

Abstract. The effect of individual and multicomponent additives of sodium chloride, iron chloride, sodium nitrate and 1-methyl-2-pyrrolidone on the interfacial distribution of thiophene in the systems polyethylene glycol 1500 (PEG-1500) - hexane - water and polyethylene glycol 8000 (PEG-8000) - hexane - water has been studied in this investigation. It was found that in most cases, the introduction of a small amount of individual and joint additives increases the extraction of thiophene from the organic phase to the aqueous.

Key words: liquid – liquid equilibrium, interphase distribution, thiophene, water soluble polymers, inorganic salts, n-hexane, light hydrocarbon fractions.

1. Introduction

Light hydrocarbon fractions obtained during oil refining contain a large number of sulfur-containing compounds [1]. Sulfur oxides are released into the atmospheric air, which negatively affect the environment [2] and human health [3] during the combustion of such fractions. Recently, increasingly stringent environmental requirements for sulfur content have been imposed on the quality of light fractions [4]. The traditional method for removing sulfur compounds from light hydrocarbon fractions is hydrotreating [2], sorption [5], oxidation [3], and extraction [6, 7]. Liquid-liquid extraction is a priority in comparison with other traditional methods of separating multicomponent mixtures, for example, rectification, which is energy-intensive [8, 9]. For example, extraction is used to isolate, separate, and concentrate both acids [10] using binary extractants [11, 12] and metals [13]. Extraction is also used in other studies, for example, in chromatography [14,15] or in supercritical fluid technologies [16, 17]. Recently, extraction methods of purification of light hydrocarbon fractions from sulfur-containing compounds have become especially relevant [18]. In the process of desulfurization of light hydrocarbon fractions, traditional extractants such as DMSO, DMF, acetonitrile, sulfolane and others are toxic, fire hazardous and environmentally unsafe [19]. Alternatively, extraction systems based on water-soluble polymers are used [20], so Voshkin et al. Studied the extraction of benzoic, salicylic, and sulfosalicylic acids in a two-phase aqueous system PEG-1500 - sodium sulfate - water [21], Zakhodyaeva et al. Studied the liquid-liquid equilibrium of the two-phase system PEG-1500 - sodium nitrate - water at three different temperatures and plotted phase diagrams [22], these systems are also used in the extraction of metals [23, 24], and also Gradov et al. Investigated the distribution of iron (III) ions in two-phase aqueous systems in the presence of ultrasound [25]. Such systems are a priority in comparison with classical water – organic solvent extraction systems, since these systems are non-toxic, fire-safe, cheap and easy to use [26], for example, these systems were used by Zinoviev et al. To study the distribution of caffeine and coumarin [27]; they also studied the extraction of monocarboxylic acids from dilute solutions using extraction systems based on polyethylene glycols with different molecular weights [28]. Zakhodyaeva et al. Studied the liquid-liquid equilibrium of a non-toxic and fire-safe two-phase water system, consisting of polyvinylpyrrolidone 3500 and sodium nitrate, and also mentioned that this system is promising in the process of extracting polycyclic sulfur-containing compounds from light hydrocarbon fractions [29]. In work [30] Zakhodyaeva et al. Investigated the extraction separation of non-ferrous metals from leaching solutions of active elements of batteries using a two-phase water system PEG-1500 - ammonium sulfate - water.

The aim of this work was to study the effect of individual and multicomponent additives in the systems PEG-1500 - hexane - water and PEG-8000 - hexane - water on the degree of extraction of thiophene from n-hexane in the polymer phase.

2. Experimantal details

2.1. Reagents

The following reagents were used in experimental studies: thiophene (Biochem Chemopharma, ≥99.0%), n-hexane (CHEMMED, «c.p.»), polyethylene glycol 1500 (Fluka), polyethylene glycol 8000 (Fluka), sodium chloride (CHEMMED, «p»), iron (III) chloride (CHEMMED, «c.p.»), sodium nitrate (CHEMMED, «c.p.»), 1-methyl-2-pyrrolidone (Acros Organics, 99%).

2.1.1. Research methods

The initial solution simulating light hydrocarbon fractions was prepared by dissolving thiophene with a concentration of 0.05% (wt) in n-hexane. The extractant solution was prepared by dissolving an exact suspension of a hydrophilic polymer and an injectable additive suspended on an analytical balance (OHAUS Explorer) in distilled water.

To study the thiophene extraction process, graduated tubes and dividing funnels with ground stoppers were used. The extraction was carried out by mixing equal volumes of an aqueous solution of the extractant and a solution of thiophene in n-hexane, then the resulting mixture was mixed in an Enviro-Genie temperature-controlled shaker (Scientific Industries, Inc.) at a temperature of 25 °C at a rotation speed of 45 rpm for 1 hour for achieving thermodynamic equilibrium of the system. The mixture was then centrifuged for 10 min at 2500 rpm (centrifuge CM-6MT, SIA ELMI), after which the phase volumes were measured. The concentration of thiophene in the organic phase was determined spectrophotometrically (spectrophotometer Cary-60, Agilent) at a wavelength of 228 nm. The thiophene concentration in the aqueous phase was determined by the difference between its concentration in the initial solution and in the organic phase after extraction.

3. Results and discussion

During the experimental work, the influence of individual and joint additives on the degree of interphase distribution of thiophene in the PEG-1500 – hexane – water and PEG-8000 – hexane – water systems was studied.

Previously, we experimentally found that the introduction of individual additives of more than 10% sodium nitrate, 5% 1-methyl-2-pyrrolidone, 2% ferric chloride in the hexane - water system are optimal and a further increase in the concentration of reagents does not lead to an increase in the degree of extraction of thiophene. Consequently, the concentrations of substances in multicomponent additives are the same as individual ones.

In fig. 1-2 show the results of the dependence of the extraction efficiency on the content of various additives in the polymer-hexane-water systems.

As can be seen from the presented fig. 1, multicomponent additives (2% iron (III) chloride + 10% sodium nitrate and 10% sodium nitrate + 5% 1-methyl-2-pyrrolidone) exhibit effective extraction properties in comparison with individual ones. This is due to the combined effect of the additives on the extraction of thiophene from hexane. Also their rice. 1 that the addition of 10% sodium nitrate to the PEG-1500 - hexane - water system results in 25% recovery of thiophene from hexane.

The obtained experimental data (Fig. 2) show that the introduction of multicomponent additives has practically no effect on the efficiency of thiophene extraction in comparison with the PEG-8000 - hexane - water system. It is also seen from the obtained data that the introduction of 10% sodium nitrate leads to almost the same value of extraction efficiency as in the case of the PEG-1500 – hexane – water system.

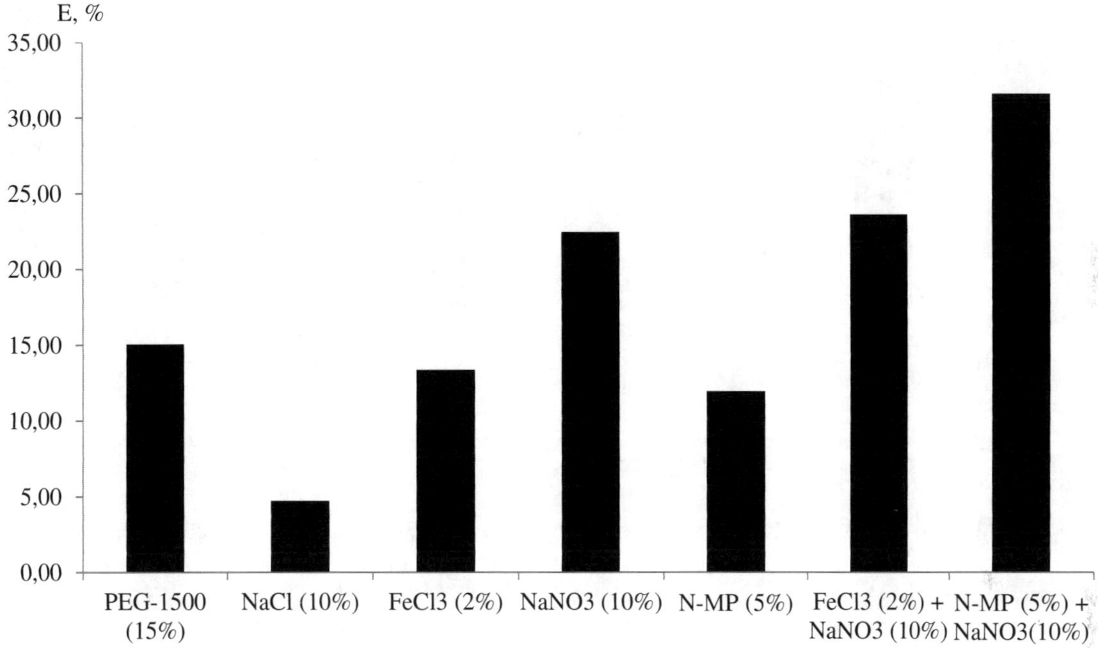

Figure 1. The dependence of the degree of thiophene extraction on the content of various additives in the PEG-1500 – hexane – water system.

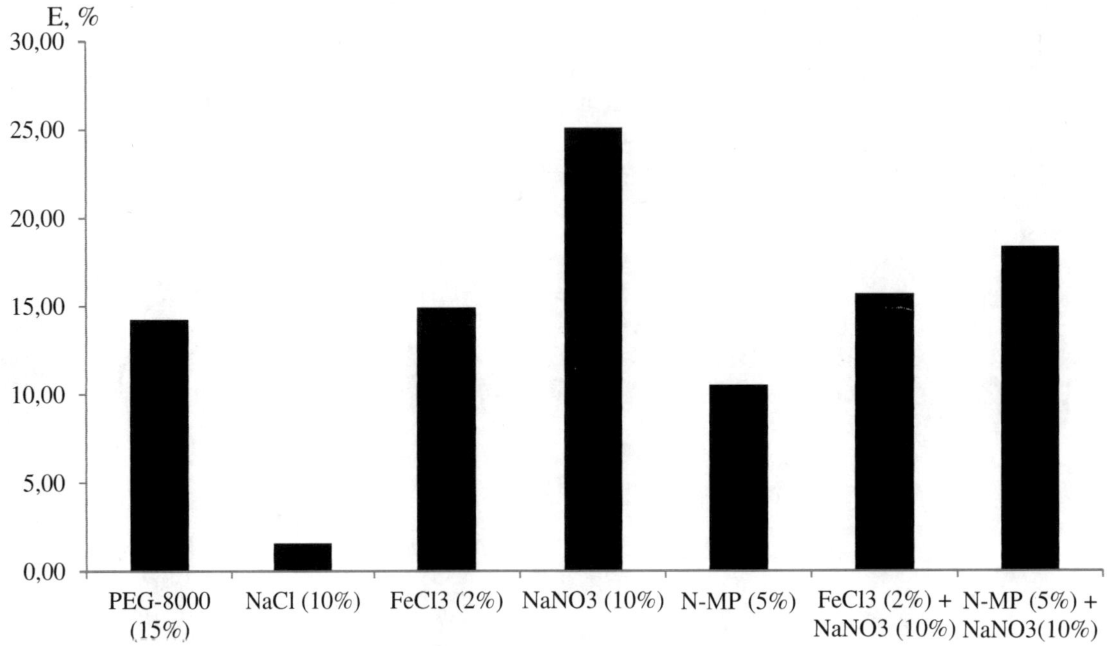

Figure 2. The dependence of the degree of thiophene extraction on the content of various additives in the PEG-8000 – hexane – water system.

Comparing Fig. 1 and 2, it can be seen that with the addition of 2% ferric chloride, 10% sodium nitrate and 5% 1-methyl-2-pyrrolidone to the extraction system, the efficiency of thiophene extraction is practically the same for PEG-1500 - hexane - water and PEG-8000 - hexane - water. It is also seen that sodium chloride as an additive does not exhibit effective extraction properties and practically does not extract thiophene from the organic phase. However, the addition of multicomponent additives to the PEG-1500 - hexane - water system exhibits more effective properties in the process of extracting thiophene from hexane, compared to their addition to the PEG-8000 - hexane - water system.

4. Conclusions

The study conducted a series of systematic studies of the dependence of the degree of thiophene extraction on the content of various additives, in particular, sodium chloride, iron (III) chloride, sodium nitrate and 1-methyl-2-pyrrolidone in the PEG-1500 – hexane – water and PEG-8000 – hexane – water systems. It was found that individual and multicomponent introduction of additives into systems in most cases increases the degree of thiophene extraction from n-hexane to the polymer phase. As a result, there is a need for further research to increase the efficiency of these systems in the processes of purification of light hydrocarbon fractions from sulfur containing compounds.

Acknowledgement

The reported study was funded by RFBR according to the research project № 18-29-06070

References

[1] Królikowski M and Lipińska A 2019 Separation of thiophene, or benzothiophene from model fuel using glycols. Liquid–liquid phase equilibria and oxidative desulfurization study *Fluid Phase Equilib.* **482** 11-23.

[2] Królikowski M 2019 Liquid–liquid extraction of sulfur compounds from heptane with tricyanomethanide based ionic liquids *J. Chem. Thermodyn.* **131** 460-70.

[3] Li F T, Kou C G, Sun Z M, Hao Y J, Liu R H and Zhao D S 2012 Deep extractive and oxidative desulfurization of dibenzothiophene with $C_5H_9NO\cdot SnCl_2$ coordinated ionic liquid *J. Hazard. Mater.* **205** 164-70.

[4] Saha B and Sengupta S 2017 Extraction of thiophenic sulfur compounds from model fuel using a water-based solvent *Energy Fuels* **31** 1 996-1004.

[5] Zhang X F, Wang Z, Feng Y, Zhong Y, Liao J, Wang Y and Yao J 2018 Adsorptive desulfurization from the model fuels by functionalized UiO-66 (Zr) *Fuel* **234** 256-62.

[6] Zakhodyaeva Y A, Solov'ev V O, Zinov'eva I V, Rudakov D G, Timoshenko A V and Voshkin A A 2019 Interphase distribution of thiophene, toluene, and o-xylene in the hexane–polymer–water extraction system *Theor. Found. Chem. Eng.* **53** 4 550-5.

[7] Belova V V, Zakhodyaeva Y A and Voshkin A A. 2017 Extraction of carboxylic acids with neutral extractants Theor. Found. Chem. Eng. **51** 5 786-94.

[8] Zakhodyaeva Y A, Voshkin A A, Belova V V and Khol'kin A I 2011 Extraction of monocarboxylic acids with binary extracting agents based on amines and quaternary ammonium bases *Theor. Found. Chem. Eng.* **45** 5 739-43.

[9] Zakhodyaeva Y A and Voshkin A A 2013 Extraction and separation of carboxylic acids by liquid-liquid extraction and chromatography *Theor. Found. Chem. Eng.* **474** 461-6.

[10] Khol'kin A I, Zakhodyaeva Y A, Voshkin A A and Belova V V 2013 Interphase distribution of weak acids in systems with binary extractants *Theor. Found. Chem. Eng.* **47** 4 453-60.

[11] Zakhodyaeva Y A, Voshkin A A, Belova V V and Khol'kin A I 2012 Extraction of monocarboxylic acids by trioctylmethylammonium di (2-ethylhexyl) phosphate *Theor. Found. Chem. Eng.* **46** 4 413-18.

[12] Kholkin A I, Belova V V, Zakhodyaeva Y A and Voshkin A A 2013 Solvent extraction of weak acids in binary extractant systems *Separation Science and Technology* **48** 9 1417-25.

[13] Voshkin A A, Belova V V and Zakhodyaeva Y A 2018 Iron extraction with di (2-ethylhexyl) dithiophosphoric acid and a binary extractant based on it *Russ. J. Inorg. Chem.* **63** 3 387-90.

[14] Kostanyan A E and Voshkin A A 2011 Analysis of cyclic liquid chromatography *Theor. Found. Chem. Eng.* **45** 1 68-74.

[15] Kostanyan A E and Voshkin A A 2009 Support-free pulsed liquid–liquid chromatography *Journal of Chromatography A* **1216** 45 7761-66.

[16] Yousefi M, Rahimi-Nasrabadi M, Pourmortazavi S M, Wysokowski M, Jesionowski T, Ehrlich H and Mirsadeghi S 2019 Supercritical fluid extraction of essential oils *TrAC Trends in Analytical Chemistry* **118** 182-93.

[17] Cossey H L, Guigard S E, Underwood E, Stiver W H, McMillan J and Bhattacharya S 2019 Supercritical fluid extraction of bitumen using chemically modified carbon dioxide *The Journal of Supercritical Fluids* **154** 104599.

[18] Chen Y, Song H, Meng H, Lu Y, Li C, Lei Z and Chen B 2017 Polyethylene glycol oligomers as green and efficient extractant for extractive catalytic oxidative desulfurization of diesel *Fuel Process. Technol.* **158** 20-5.

[19] Li F T, Wu B, Liu R H, Wang X J, Chen L J and Zhao D S 2015 An inexpensive N-methyl-2-pyrrolidone-based ionic liquid as efficient extractant and catalyst for desulfurization of dibenzothiophene *Chem. Eng. J.* **274** 192-9.

[20] Voshkin A A, Shkinev V M and Zakhodyaeva Y A 2017 A new extraction method for the preparation of zinc oxide nanoparticles in aqueous two-phase systems *Russ. J. Phys. Chem. A* **91** 2 226-8.

[21] Voshkin A A, Zakhodyaeva Y A, Zinov'eva I V and Shkinev V M 2018 Interphase distribution of aromatic acids in the polyethylene glycol–sodium sulfate–water system *Theor. Found. Chem. Eng.* **52** 5 890-3.

[22] Zakhodyaeva Y A, Rudakov D G, Solov'ev V O, Voshkin A A and Timoshenko A V 2019 Liquid–liquid equilibrium of aqueous two-phase system composed of poly (ethylene oxide) 1500 and sodium nitrate *J. Chem. Eng. Data* **64** 3 1250-5

[23] Zakhodyaeva Y A, Zinov'eva I V, Tokar E S and Voshkin, A. A 2019 Complex extraction of metals in an aqueous two-phase system based on poly (ethylene oxide) 1500 and sodium nitrate *Molecules* **24** 22 4078.

[24] Fedorova M I, Zinov'eva I V, Zakhodyaeva Y A and Voshkin A A 2020 Extraction of Fe (III), Zn (II), and Mn (II) Using a System with a Green Solvent for Trioctylmethylammonium Thiocyanate *Theor. Found. Chem. Eng.* **54** 313-8.

[25] Gradov O M, Zakhodyaeva Y A, Zinov'eva I V and Voshkin A A 2019 Some Features of the Ultrasonic Liquid Extraction of Metal Ions *Molecules* **24** 19 3549.

[26] Zinov'eva I V, Zakhodyaeva Y A and Voshkin A A 2020 Data on the extraction of benzoic, salicylic and sulfosalicylic acids from dilute solutions using PEG-based aqueous two-phase systems *Data in brief* **28** 105033.

[27] Zinov'eva I V, Zakhodyaeva Y A and Voshkin A A 2019 Interphase Distribution of Caffeine and Coumarin in Extraction Systems with Polyethylene Glycol and Sodium Sulfate *Theor. Found. Chem. Eng.* **53** 6 996-1000.

[28] Zinov'eva I V, Zakhodyaeva Y A and Voshkin A A 2019 Extraction of Monocarboxylic Acids from Diluted Solutions with Polyethylene Glycol *Theor. Found. Chem. Eng.* **53** 5 871-4.

[29] Zakhodyaeva Y A, Rudakov D G, Solov'ev V O, Voshkin A. A and Timoshenko A. V 2019 Liquid–Liquid Equilibrium in an Extraction System Based on Polyvinylpyrrolidone-3500 and Sodium Nitrate *Theor. Found. Chem. Eng.* **53** 2 159-65.

[30] Zakhodyaeva Y A, Izyumova K V, Solov'eva M S and Voshkin A A 2017 Extraction separation of the components of leach liquors of batteries *Theor. Found. Chem. Eng.* **51** 5 883 7.

The functionalization of pyrolyzed palm empty fruit bunches-based membranes adsorbent by fourier-transform infrared spectroscopy

N Hidayah[1], V D Putri[2], M Elma[3,5a], Mahmud[4], I Syauqiah[3], A Amenia[4,5], D G L Putra[4,5] and H R Akbar[4,5], A Rahma[5]

[1]. System Information Department, Sari Mulia University, Jl. Pramuka No. 2A, Banjarmasin, South Kalimantan 70725, Indonesia

[2]. Pharmacy Department, Sari Mulia University, Jl. Pramuka No. 2A, Banjarmasin, South Kalimantan 70725, Indonesia

[3]. Chemical Engineering Department, Lambung Mangkurat University, Jl. A. Yani KM 36, Banjarbaru, South Kalimantan 70714, Indonesia

[4]. Environmental Engineering Department, Lambung Mangkurat University, Jl. A. Yani KM 36, Banjarbaru, South Kalimantan 70714, Indonesia

[5]. Materials and Membranes Reasearch Group (M^2ReG). Lambung Mangkurat University, Jl. A. Yani KM 36, Banjarbaru, South Kalimantan, Indonesia 70714

Email : [a]melma@ulm.ac.id

Abstract. Membranes adsorbent are successfully prepared derived from palm empty fruit bunches (PEFB) which pyrolyzed by furnace as physical activation. The PEFB membrane adsorbent was activated to develop porous structures and surface area which able to be applied for gas separation. The aims of this study are to fabricated the pyrolyzed PEFB-based membrane adsorbent with different loading of PEFB mass to identify the surface organic functional groups of the PEFB membrane adsorbent. Fabrication of this membrane adsorbent was conducted into three steps, i.e. (1) pre-treated PEFB materials; (2) pyrolyzed the PEFB adsorbent at 500°C; and (3) PEFB membrane adsorbent fabrication by mixed both of PVA and PEG polymers into PEFB adsorbent with varied mass (15-17.5 grams). The functionalization of this membrane adsorbents was analysed by Fourier Transform Infra-Red (FTIR) spectra. The result shows the three variations of the PEFB membrane adsorbents present the surface oxygen, functional group. The effect of PEFB mass loading to the carbon pores formation of PEFB membrane adsorbent was exhibited by the escalating of C-H and C-O groups. The membrane adsorbent by adding 17.5 grams of PEFB mass indicating the highest peak of hydroxyl C-O at wavenumber 1070 cm^{-1}. It demonstrates that membrane adsorbent with high PEFB mass loading and physic activation by pyrolyzing is great to tailoring the membrane adsorbent structure properties which capable to be applied for gas separation, especially for biogas upgrading.

1. Introduction

In the last decade, the clean and renewable energy has attracted among the industry and researcher. Biogas is one of the alternatives to alter fossil fuels thought as energy conservation. It is a mixture of gases mainly composed of methane (CH_4) and carbon dioxide (CO_2) of 60% and 30%, respectively [1]. The methane fraction of biogas has a huge consideration for household utilization, generated electricity or vehicles. Thus, an effective method for upgrading biogas is needed. Nowadays, the upgrading biogas mainly made using adsorption process [2]. However, this method also has limitations, including high operating costs and problems with regeneration [3].

There has been a considerable growth in the research and development of organic and inorganic membrane materials [4-6] and for the separation of gas mixtures [7, 8]. Numerous novel membrane materials have been synthesized and evaluated. A recent study was reported by Iovane et al. that exhibited the polymer membrane has capable to remove CO_2 over 60% and successfully upgrading the methane concentration of > 80% [9]. However, polymer membranes have also a limitation due to the requirement of a multi-stage separation process to allow the membrane operate under low pressures, and need extra treatment to extend the membrane long term [10]. In order that such a concept has already been shown to improve the separation properties of polymers membrane by incorporated adsorbent into a polymeric matrix membrane. In the present study, the zeolite adsorbents have been incorporated into a polymeric matrix to form a heterogeneous membrane for separating CH_4 and CO_2 [11]. It founds presence of zeolite adsorbent was shown suitable pore size and able to improve the separation performance of polymer polydimethylsiloxane membrane. Unfortunately, zeolite is not economic materials. Affordable materials are needed to address this limitation.

Activated carbons from biomass waste such as Oil palm empty fruit bunches (PEFB) was continuing to attracted attention for biogas upgrading as they are reversible CO_2 adsorbents and present high adsorption capacity per mass unit. PEFB is rich by cellulose acetate, lignin and carbon content which favourable for membrane production [12]. The employing of PEFB has become attention to surmount the environmental issue and reused them as economical products. In the present study, Raja [12] has been reported the employing PEFB activated carbon as a membrane adsorbent derived from cellulose acetate modified by zeolite for water treatment. The previous study [13] has reported that activated carbon was fabricated from PEFB was carried out via physical carbonization by pyrolyzed. Regarding Sopiah, et al. [13] the carbonization process may improve the pore structure of absorbent with small density by opening the pores are closed with tar and amorphous carbon. Furthermore, the reasearch conducted by Elma, et al. [14] also explaining incorporation of a high concentration carbon into membrane matrix may be affected to functional group content in its membrane. It is due to carbon loading was given to enhancing the condensation reaction effect [15]. In this work, we focused on fabricated the pyrolyzed PEFB-based membrane adsorbent with different loading of PEFB mass to identify the surface organic functional groups of the PEFB membrane adsorbent.

2. Methodology

2.1 Chemical and Materials

The chemicals and materials were employed to fabricate the membrane adsorbent are isopropyl alcohol (2-propanol, Merck), ammonium chloride (NH_4Cl, Merck), polyethylene glycol 400 (PEG 400), polyvinyl alcohol (PVA), filter paper, deionized water (DI water Merck), and palm empty fruit bunches (PEFB) collected from PT. Citra Hasnur Terpadu, Sungai Puting Village, South Kalimantan-Indoneisa.

2.2 Experimental Methods

2.2.1 Fabrication and Characterization of Membrane Adsorbent

The palm empty fruit bunches (PEFB) were collected from oil palm plantation that has been washed by freshwater and dried for 7 days under the sunlight. After a week, the PEFB were physically activated using pyrolysis apparatus at 500°C for 30 minutes. The activated PEFB subsequently was mashed and sieved with a size of 200-400 mesh to obtain the uniform size. Furthermore, the activated charcoal is

IOP Conf. Series: Materials Science and Engineering 1212 (2022) 012026 doi:10.1088/1757-899X/1212/1/012026

weighed as much as 12.5 grams, 15 grams, and 17.5 grams which are then mixed with 35 ml 2-propanol and the mixture is put in a 100 ml measuring cup, stirred at 600 rpm for 10 minutes [16]. Added 3,5 grams of NH_4Cl as a cationic surfactant which has been dissolved in 300 ml of distilled water. The mixture is then stirred with a magnetic stirrer for 1 hour to form nano-sized membrane pores [16]. Separate between solid and solution, then in variations of 12.5, 2.82 grams of PVA and 4.17 ml of PEG is added to the solid, and add 5 ml of remaining solution so that the membrane can adhere well. In the 15 grams variation, 3.4 grams of PVA and 5 ml of PEG were added, and 8 ml of the remaining solution gradually until the membrane mixture was solid and well adhered. Then added 5.09 grams of PVA, 7.52 ml of PEG and 8 ml of the remaining solution for a variation of 17.5 grams. The membrane is then stirred evenly until it is shaped like a paste and sticks together perfectly. The membrane mixture is then printed by stainless steels cast with a diameter of 7 cm, then the membrane is dried in the sun for about one day. After the upper surface of the membrane is slightly dry, the membrane is then pressed using a hydraulic jack with a pressure of 100 psi. The pressed membrane is then dried again for about 2 days and followed by drying using a furnace at 105°C for 30 minutes.

Characterization of the membrane adsorbent was carried out using the FTIR test to identify the content of functional groups of organic compounds based on the resulting spectra according to the peaks formed by a functional group. The spectra were recorded from 3000 to 500 cm^{-1}. Then, the results of the FTIR data were analysed by Fityk program to determines the quantity of peak area by gaussian models.

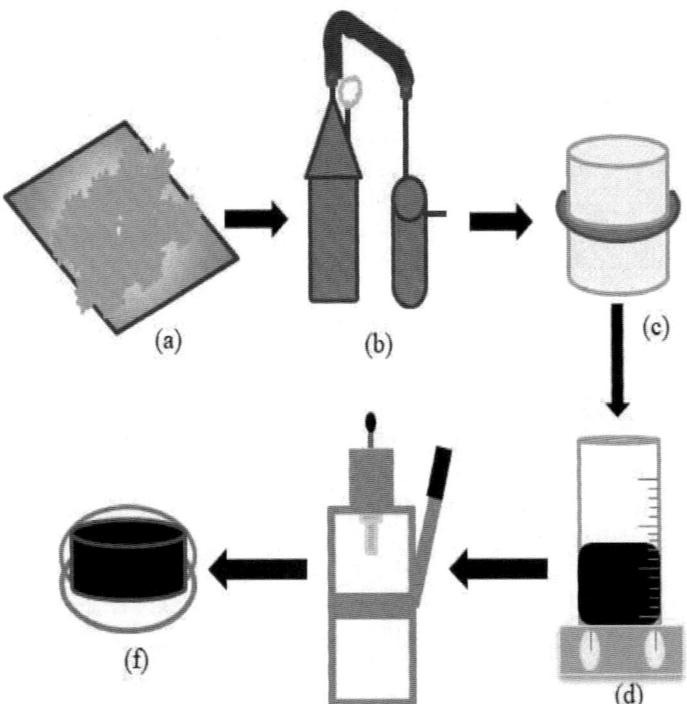

Where:
a. Palm empty fruit bunches (PEFB) are dried in the sun
b. Pyrolysis Apparatus
c. 200 & 400 mesh sieve size
d. Hot plate
e. Hydraulic jack
f. Adsorber membrane

Figure 1. The schematic fabrication of membrane adsorbent

3. Result and Discussion

3.1 Preparation of PEFB Membrane Adsorbent

Oil palm biomass has been great importance as various parts such as trunks, fronds, leaves, shells, and empty fruit bunches (EFB) for extensively studied as adsorbents. Palm empty fruit bunches (PEFB) is one of the largest amounts of solid wastes produced from an oil palm plantation is about 22-24% of the total fresh fruit bunch weight [17]. In the industrial sector, PEFB has an economic enough value, due to has millions of organic fibres that can be used in various fields, especially as raw material for producing such as activated charcoal, briquette raw materials, membranes adsorbent and many other benefits [18]. Due to high organic content in PEFB, makes it perfect for processing membrane adsorbent which can be applied for biogas upgrading. The complete chemical composition of the crude PEFB is shown in **Table 1**.

Table 1. Chemical Composition of Palm Empty Fruit Bunches Biomass

Chemical composition	(%) of Raw PEFB		
	[17]	[19]	[20]
Lignin	22.60	15.70	17.84
Pentosan	25.90	-	
A-Cellulose	45.80	-	50.49
Holocellulose	71.88	-	80.09
Ash	1.6	-	
Pectin	12.85	-	

Table 1 shows the summarized of the chemical composition of raw PEFB biomass. The highest chemical composition contains in PEFB are cellulose and holocellulose. In fundamental, lignocellulose substances contain three chief structural components: hemicellulose, cellulose and lignin. That three main parts have high molecular weights and contribute much mass, while the extractives are of small molecular size and are available in little quantity [20]. The hemicellulose chains have short branches and are amorphous. Amorphous morphology is partially soluble or swellable in water as shown in **Table 2**.

Table 2. Physical Properties of Palm Empty Fruit Bunces Biomass

PEFB Physic Properties	(%) of PEFB	References
Solubility in aqueous sodium hydroxide 1%	19.50-31.17	[19]
Solubility in cold water (30°C)	13.89-14.79	[17]
Solubility in hot water (100°C)	2.50-14.79	[17]
Solubility in alcohol-benzene	5.00	[19]

Hemicellulose is mostly soluble in alkali and as such is more easily hydrolysed. Cellulose is formed by joining the anhydro-glucose units into glucose chains and act as a linear polymer chain. Cellulose is insoluble in largely solvents and has low accessibility to enzymatic hydrolysis and acid as shown in **Table 2**. Lignin is a natural polymer or aromatic compounds. It acts as a cementing agent matrix of cellulose fibres in the structures of plants along with hemicellulose. Their functions are to provide structural strength, provide sealing of water conducting system that links roots with leaves, and protect plants against degradation. This matrix comprises a variety of functional groups, such as hydroxyl, methoxyl, and carbonyl, which impart a high polarity to the lignin macromolecule [20]. PEFB biomass is now considered to be one of the most promising non-wood lignocellulosic raw materials as adsorbents.

Generally, there was three main steps process to fabricates the PEFB membrane adsorbent in this work, i.e. (1) pre-treated PEFB materials by washed the raw PEFB using freshwater and dried it under sunlight for 7 days. indirectly these compounds must be removed or first so that they do not directly affect the surface of the absorbent membrane and increase effectiveness activated carbon. (2) Activated PEFB adsorbent using pyrolyzed at 500 °C for 30 min in activated carbon reservation. (3) Finally, PEFB adsorbents were sieved at 200-400 mesh to obtain the uniform sizes of an adsorbent. The study has reported a physical activation process gives a similar morphological compared to adsorbent from chemical-physical activation [21]. Larger pore size has found than activated carbon resulting from chemical activation [22].

The functional groups or carbon bonds can be seen based on the results of the FTIR analysis. In the Hidayu et al. (2013) work states that crude oil palm empty bunches have a complex and very clear spectrum, this is indicated by the adsorption peak at 3302 cm^{-1} which is associated with the O-H stretch functional group and this indicates the presence of bound hydroxide compounds in crude PEFB. The distribution of the peak point of the wave is also found at a wavelength of 1739 cm^{-1} which is indicated as the functional group C = O. The wavelength at 1216 cm^{-1} and 1032 cm^{-1} of PEFB membrane adsorbent also indicates the C-O stretch functional group [15]. Broadly speaking, crude oil palm empty bunches cannot be applied directly in the manufacture of adsorbent membranes, the presence of several compounds and functional groups contained in oil palm empty bunches will of course indirectly affect the surface area and effectiveness of the membrane. So that the compound or carbon content contained in it must be removed first using carbonization and physical activation, by the pyrolysis process at a temperature of 500°C as shown in **Figure 2**. The temperature of 500°C was chosen because most of the volatile material of PEFB evaporates at 410°C - 520°C. The small number of impurities are disappeared and become easier for further processing [23]. In addition, this physical activation is also required so that the pore structure of the membrane more stable and offers high absorption capacity [24].

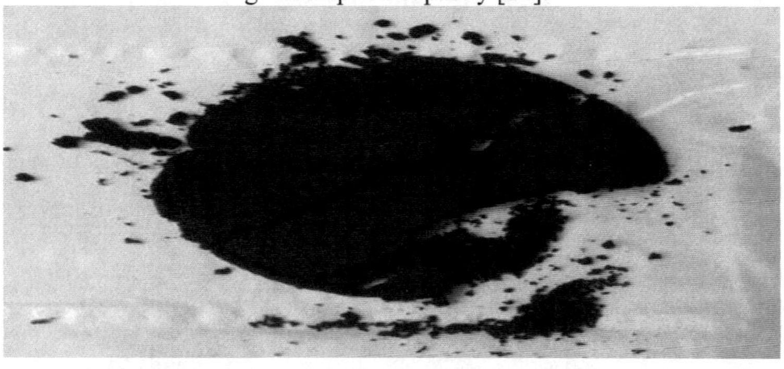

Figure 2. Membrane adsorbent derived from PEFB which pyrolyzed at 500 °C by loading of 17.5 grams biomass

The PEFB membrane adsorbent has been cast using stainless steels mold, before the samples were analyzed using the Fourier Transform Infrared (FTIR) analysis to see the functional groups of all the sample. The conducted of PEFB membrane adsorbent after the cast has a diameter and thickness of 7 cm and 0.8 cm, respectively for biomass loading 17.5 grams (**Figure 2**). The multiple PEFB membrane adsorbent physic shape results are shorted in **Table 3**. The entire upper and lower surface of each membrane is smooth and flat and does not leave residual grains as can be seen in **Figure 2**.

Table 3. PEFB Membrane Adsorbent Physic Shape Result Measurement

PEFB loading mass (grams)	Diameter (cm)	Thickness
12.5	7	0.4
15	7	0.6
17.5	7	0.8

3.2 Functionalization of PEFB Membrane Adsorbent

This study discusses the pyrolyzed PEFB-based membrane adsorbent with multiple biomass addiction is given in **Figure 3.** It shows FTIR spectra by adding 12.5, 15 and 17.5 grams of PEFB biomass at wavelength 3000-500 cm^{-1}. The band found in the region of 1067-1076 cm^{-1} indicates to the C-O stretching vibration as a function of secondary alcohol groups (**Figure 3**). In line with the result reported by Nandiyanto, et al. [25]. The deconvolution of C-O group was presented in **Table 4**. Regarding the result, the highest PEFB loading mass of 17.5 grams also demonstrates the highest peak which almost 2-3 times broader than PEFB loading mass of 12.5 and 15 grams. Meanwhile, the band of 1236-1243 cm^{-1} were identified as aromatic ether groups with functional groups alkyl O–H stretch [25]. The broader of O-H group would imply increasing the water absorbing on the surface. This result same in line with a similar study by [26], has investigated coconut shell based activated carbon. Interestingly, the C-H mode only appears by PEFB loading mass of 12.5 and 17.5 grams at the wavelength region of 1440 and 1439 cm^{-1}. The C-H stretching mode was indicating the function of cellulose groups [27]. Sharpening bands of C = C-C which a whole wave is associated with the aromatic ring stretch which observed at ~1580 cm^{-1}.

Table 4. Deconvolution of PEFB Membrane Adsorbent Functionalize using Fityk Program

PEFB loading mass (grams)	C-O groups (unit area)	C=C-C group (unit area)
12.5	2.30	0.25
15	1.89	0.18
17.5	4.58	0.71

Based on this work the difference biomass weight loading has a significant effect on the functional groups produced, especially C-O and aromatic groups (C=C-C) [28]. Moreover, the pyrolyzed process greatly affects the deoxygenation process with lignocellulosic material to reduce the aliphatic structure in the C-O aromatic structure [29]. The pyrolyzed excellently affected based on **Figure 3**. The activation temperature is an important factor on the formation quality of activated carbon, so it has been impacted by the absorption capacity of carbon in ionic levels [30]. The physical activations carried out on each of the adsorbent membrane variations still leave compounds that are dominated by carbon elements which are covalently bonded, and of course, this is very helpful in expanding the membrane surface so that the pores are open and with thus has a great absorption of other substances, especially in the gas phase [13]. On the outer surface OH increase rapidly, this results in line with previous research [31, 32].

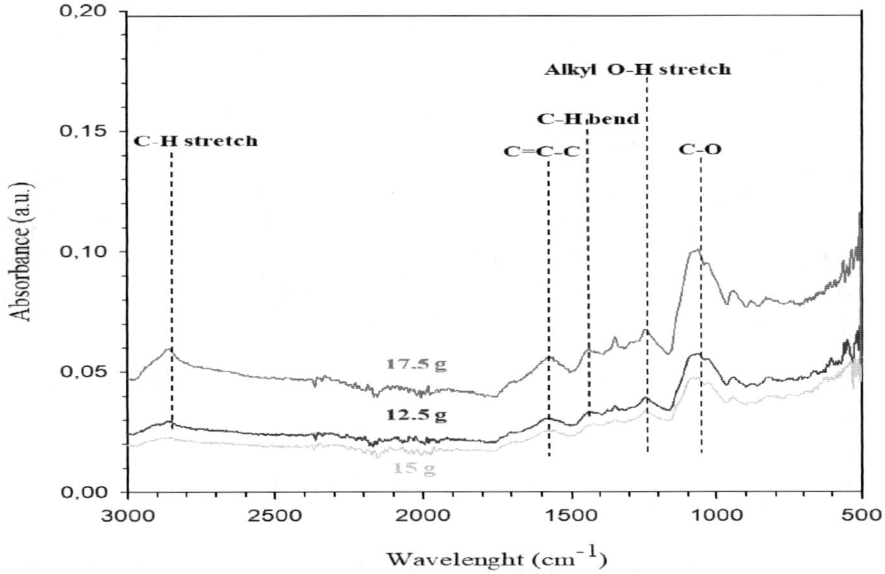

Figure 3. FTIR spectra of membrane adsorbent derived from PEFB which pyrolyzed at 500 °C by loading of varied PEFB mass

4. Conclusion

Membranes adsorbent are successfully fabrication by employing palm empty fruit bunches (PEFB) biomass waste. The PEFB membranes adsorbent were activated by physical activation to the developed porous structure and high surface area. The functionalization of all biomass weight variation on the membrane adsorbent identifying the C-O stretch functional groups dominates in all sample. Most of the O-H (water) content in the membrane perfectly evaporates during the activation process. So that what remains is only other carbon elements that are needed in the process of forming membrane pores. Furthermore, physical activation leads the membrane adsorbent to carbonization and enlarging the surface area of the membrane to apply for gas purification.

References

[1] M. A. Lourenço, C. Nunes, J. R. Gomes, J. Pires, M. L. Pinto, and P. Ferreira, "Pyrolyzed chitosan- based materials for CO2/CH4 separation," *Chemical Engineering Journal,* vol. 362, pp. 364-374, 2019.

[2] C. A. Grande, D. Morence, A. M. Bouzga, and K. A. Andreassen, "Silica gel as a selective adsorbent for biogas drying and upgrading," *Industrial & Engineering Chemistry Research,* 2020.

[3] C. G. Joseph, M. Wan, K. S. Quek, and S. Kogularama, "Parametric and adsorption kinetic studies of reactive black 5 removal from textile simulated wastewater using oil palm (Elais guineensis) empty fruit bunch," *Journal of Applied Sciences,* vol. 15, no. 8, pp. 1103-1111, 2015.

[4] A. Rahma, M. Elma, E. L. A. Rampun, A. E. Pratiwi, A. Rakhman, and Fitriani, "Rapid Thermal Processing and Long Term Stability of Interlayer-free Silica-P123 Membranes for Wetland Saline Water Desalination," *Advanced Research in Fluid Mechanics and Thermal Sciences,* vol. 71, no. 2, pp. 1-9, July 2020 2020, Art no. 1.

[5] A. Rahma, M. Elma, A. E. Pratiwi, and E. L. Rampun, "Performance of interlayer-free pectin template silica membranes for brackish water desalination," *Membrane Technology,* vol. 2020, no. 6, pp. 7-11, 2020.

[6] M. Elma *et al.*, "Carbon templated strategies of mesoporous silica applied for water desalination: A review," *Journal of Water Process Engineering,* vol. 38, p. 101520, 2020/12/01/ 2020.

[7] F. M. Baena-Moreno, E. le Saché, L. Pastor-Pérez, and T. Reina, "Membrane-based technologies for biogas upgrading: a review," *Environmental Chemistry Letters,* pp. 1-10, 2020.

[8] C. Liang, G. Sha, and S. Guo, "Carbon membrane for gas separation derived from coal tar pitch," *Carbon,* vol. 37, no. 9, pp. 1391-1397, 1999.

[9] P. Iovane, F. Nanna, Y. Ding, B. Bikson, and A. Molino, "Experimental test with polymeric membrane for the biogas purification from CO2 and H2S," *Fuel,* vol. 135, pp. 352-358, 2014/11/01/ 2014.

[10] K. Khulbe and T. Matsuura, "Removal of heavy metals and pollutants by membrane adsorption techniques," *Applied water science,* vol. 8, no. 1, p. 19, 2018.

[11] J. M. Duval, B. Folkers, M. H. V. Mulder, G. Desgrandchamps, and C. A. Smolders, "Adsorbent filled membranes for gas separation. Part 1. Improvement of the gas separation properties of polymeric membranes by incorporation of microporous adsorbents," *Journal of Membrane Science,* vol. 80, no. 1, pp. 189-198, 1993/06/02/ 1993.

[12] P. M. Raja, "PEMBUATAN MEMBRAN SELULOSA ASETAT DARI TANDAN KOSONG KELAPA SAWIT TERMODIFIKASI MIKRO ZEOLIT ALAM UNTUK FILTRASI AIR SUNGAI," *Jurnal Agro Estate,* vol. 1, no. 1, pp. 27-33, 2017.

[13] N. Sopiah, D. Prasetyo, and D. B. Aviantara, "Pengaruh Aktivasi Karbon Aktif dari Tandan Kosong Kelapa Sawit terhadap Adsorpsi Kadmium Terlarut," *Jurnal Riset Teknologi Pencegahan Pencemaran Industri,* vol. 8, no. 2, pp. 55-66, 2017.

[14] M. Elma, D. K. Wang, C. Yacou, and J. C. D. d. Costa, "Interlayer-Free P123 Carbonised Template Silica Membranes for Desalination with Reduced Salt Concentration Polarisation," *Journal of Membrane Science,* vol. 475, pp. 376-383, 2015.

[15] G. Q. Lu *et al.*, "Inorganic membranes for hydrogen production and purification: A critical review and perspective," *Journal of Colloid and Interface Science,* vol. 314, no. 2, pp. 589-603, 2007/10/15/ 2007.

[16] V. Rachmawati and A. Damayanti, "Pengolahan Limbah Cair Industri Pewarnaan Jeans Menggunakan Membran Silika Nanofiltrasi Aliran Cross Flow untuk Menurunkan Warna dan Kekeruhan," *Jurnal Teknik ITS,* vol. 2, no. 2, pp. D113-D117, 2013.

[17] M. R. L. L. Gaol, R. Sitorus, S. Yanthi, I. Surya, and R. Manurung, "Pembuatan selulosa asetat dari α-selulosa tandan kosong kelapa sawit," *Jurnal Teknik Kimia USU,* vol. 2, no. 3, pp. 33-39, 2013.

[18] J. M. L. Thoe, N. Surugau, and H. L. H. Chong, "Application of Oil Palm Empty Fruit Bunch as Adsorbent: A Review," *Transactions on Science and Technology,* vol. 6, no. 1, pp. 9-26, 2019.

[19] B. Wirjosentono, P. Guritno, and H. Ismail, "Oil palm empty fruit bunch filled polypropylene composites," *International Journal of Polymeric Materials,* vol. 53, no. 4, pp. 295-306, 2004.

[20] T. Ahmad, M. Rafatullah, A. Ghazali, O. Sulaiman, and R. Hashim, "Oil palm biomass–Based adsorbents for the removal of water pollutants—A review," *Journal of Environmental Science and Health, Part C,* vol. 29, no. 3, pp. 177-222, 2011.

[21] R. H. Khuluk, "Pembuatan dan karakterisasi karbon aktif dari tempurung kelapa (Cocous nucifera L.) sebagai adsorben zat warna metilen biru," 2016.

[22] W. Rahmalia, F. Yulistira, J. Ningrum, M. Qurbaniah, and M. Ismadi, "Pemanfaatan potensi tandan kosong kelapa sawit (Elais guineensis Jacq) sebagai bahan dasar c-aktif untuk adsorpsi logam

perak dalam larutan," *Pontianak: Jurusan Kimia, Fakultas Matematika dan Ilmu Pengetahuan Alam Universitas Tanjungpura,* 2009.

[23] R. K. Liew *et al.*, "Oil palm waste: an abundant and promising feedstock for microwave pyrolysis conversion into good quality biochar with potential multi-applications," *Process Safety and Environmental Protection,* vol. 115, pp. 57-69, 2018.

[24] A. Geng, "Conversion of Oil Palm Empty Fruit Bunch to Biofuels," 2013.

[25] A. B. D. Nandiyanto, R. Oktiani, and R. Ragadhita, "How to read and interpret FTIR spectroscope of organic material," *Indonesian Journal of Science and Technology,* vol. 4, no. 1, pp. 97-118, 2019.

[26] I. Tan, M. Abdullah, L. Lim, and T. Yeo, "Surface modification and characterization of coconut shell-based activated carbon subjected to acidic and alkaline treatments," *Journal of Applied Science & Process Engineering,* vol. 4, no. 2, pp. 186-194, 2017.

[27] P. M. Abdul *et al.*, "Effects of changes in chemical and structural characteristic of ammonia fibre expansion (AFEX) pretreated oil palm empty fruit bunch fibre on enzymatic saccharification and fermentability for biohydrogen," *Bioresource Technology,* vol. 211, pp. 200-208, 2016.

[28] M. Bakhtiar, N. Sari, A. B. Yaacob, M. Yunus, and K. B. Ismail, "Characterization of oil palm Empty Fruit Bunch (EFB) biochar activated with potassium hydroxide under different pyrolysis temperature," *Journal of Engineering Science and Technology,* vol. 14, no. 5, pp. 2792-2807, 2019.

[29] S. Yavari, A. Malakahmad, N. B. Sapari, and S. Yavari, "Synthesis optimization of oil palm empty fruit bunch and rice husk biochars for removal of imazapic and imazapyr herbicides," *Journal of environmental management,* vol. 193, pp. 201-210, 2017.

[30] R. Idrus, B. P. Lapanporo, and Y. S. Putra, "Pengaruh suhu aktivasi terhadap kualitas karbon aktif berbahan dasar tempurung kelapa," *Prisma Fisika,* vol. 1, no. 1, 2013.

[31] C.-H. Ooi, W.-K. Cheah, Y.-L. Sim, S.-Y. Pung, and F.-Y. Yeoh, "Conversion and characterization of activated carbon fiber derived from palm empty fruit bunch waste and its kinetic study on urea adsorption," *Journal of environmental management,* vol. 197, pp. 199-205, 2017.

[32] D. Prahas, Y. Kartika, N. Indraswati, and S. Ismadji, "Activated carbon from jackfruit peel waste by H3PO4 chemical activation: Pore structure and surface chemistry characterization," *Chemical Engineering Journal,* vol. 140, no. 1-3, pp. 32-42, 2008.

[33] N. H. A. Rani, N. F. Mohamad, S. Matali, and S. A. S. A. Kadir, "Preparation and characterization of activated carbon made from oil palm empty fruit bunch," *Key Engineering Materials,* 2014.

Acknowledgement

Authors thank to Materials and Membranes Research Group (M^2ReG) for the facilities provided. Hidayah thanks to the Research Collaboration Between Universities Grant 2020-2021. Muthia thanks also to Applied Research Universities Grant 2019-2020, Thesis Magister Grant 2020, World Class Research Grant 2021-2022 the Directorate of Research and Community Service, the Ministry of Research and Technology/the National Agency for Research and The Innovation Republic of Indonesia and Lambung Mangkurat University Research Grant 2020.

Membranes adsorber from oil palm empty fruit branches (OPEFB): preparation and fabrication

N Hidayah[1], P D Darsono, M Elma[3,5,a], Mahmud[4], I Syauqiah[3], H R Akbar[4,5], A Amenia[4,5], D G L Putra[4,5], N Huda[4,5], and A Rahma[5]

[1]System Information Department, Sari Mulia University, Jl Pramuka No. 2A, Banjarmasin, South Kalimantan 70725, Indonesia
[2]Pharmacyst Department, Sari Mulia University, Jl Pramuka No. 2A, Banjarmasin, South Kalimantan 70725, Indonesia
[3]Chemical Engineering, Faculty of Engineering, Lambung Mangkurat University, Jl. A. Yani Km 36, Banjarbaru 70714, South Kalimantan, Indonesia
[4]Environmental Engineering, Faculty of Engineering, Lambung Mangkurat University, Jl. A. Yani Km 36, Banjarbaru 70714, South Kalimantan, Indonesia
[5]Materials and Membranes Research Group (M^2ReG), Lambung Mangkurat University, Jl. A. Yani Km 36 Banjabaru 70714, South Kalimantan, Indonesia

Email : amelma@ulm.ac.id

Abstract. Oil Palm empty fruit branches (OPEFB) are solid waste that are numerous produced from palm oil mills. OPEFB is economically and potentially used as membrane adsorber material due to has good thermal stability, chemical resistance and biodegradability. The objectives of this work is to preparate and fabricate the OPEFB membrane adsorber which is activated by physical activation. The OPEFB has been cleaned and dried, subsequently heated at 500 °C for 30 min via pyrolysis. The activated OPEFB was sieved using 200-400 mesh and followed by the addition of 2-propanol, NH_4Cl, polyvinyl alcohol (PVA) and polyethylene glycol (PEG) to become a mixture. The activated OPEFB ratio were varied in the mixtures to obtain the best composition in order to produce a good membrane adsorber texture for casting. FTIR shows on wavenumber at 1082 cm^{-1} indicates that there is O-H stretching functional groups and bands at 943 cm^{-1} correspons to C=O functional group. It is concluded that the membrane mixtures can be employed as membrane adsorber due to carbon content which creates strong matrix applied for gas separation.

1. Introduction

Indonesia is the largest palm oil producer in the world, especially in South Kalimantan-Indonesia [1, 2]. Production of fresh palm fruit oil branches is increased as much as the demand every year. Nevertheless, the increase in fresh palm fruit oil branches also brings out increasing solid and wastewater production [3]. Oil palm empty fruit branches (OPEFB) is one of the solid waste produced by palm oil mills is about 23% of the total fresh palm fruit oil [4]. Whereas, OPEFB still containing high hemicellulose (37.76%), cellulose (66.07%), fiber (72.67%), and lignin [5]. Cellulose is a carbon compound that consists of much glucose which is bound by beta 1.4 glycoside bonds and easily decomposes into sample C by cellulosic organisms compounds. Meanwhile, lignin is a component of OPEFB which is relatively difficult to deteriorate. This compound is a structural polymer associated with cellulose and hemicellulose. When observed from its structure, OPEFB has economic value and potential to be used for gas adsorption. It due to the OPEFB causes a polar character in the adsorber [6].

Cellulose derived OPEFB is a renewable natural polymer that is superfluous available and has unique characteristics such as good thermal stability, chemical resistance, and biodegradability [2, 7]. Furthermore, the fibrous structure on the nanometer-scale of cellulose also has several advantages i.e. high surface area, controllable porous structure, and outstanding mechanical properties [8]. Cellulose has the desired structure which is a challenge due to its strong hydrogen bonds. This leads to very poor solubility in several solvents [9]. This material is appropriate and favorable to tailored the membrane adsorber [10].

In generally, OPEFB was used to preparation of activated carbon due to its porous material containing carbon. Moreover, activated carbon also offering a high surface area [11-13]. Activated carbon can adsorb particularly gases and chemical compounds depend on its adsorption properties by the size, pore-volume, and surface area [14]. The adsorption capacity of activated carbon ranges from 25 to 100%. Activated carbon can be made through a physical or chemical activation process [15]. Adsorption is a physical phenomenon that occurs when gas or liquid molecules are in contact with a solid surface and some of the molecules condense on the solid surface. One of the adsorption technologies for biogas purification can use adsorbents such as activated carbon for CO_2 adsorption [16]. Some purification technologies were developed with various types including water scrubbing, chemical adsorption, and membrane purification technology [17]

Membrane technology has come a long way in recent decades [18-21]. Membrane technology has several advantages both technically and economically [22-24]. Thus, it is often used in the environmental, health, and industrial fields [25]. The membrane functions as a thin, selective barrier between the two phases by only passing certain components and holding the other components [26-29]. Recently, adsorbent membranes have emerged as a popular technology [30]. When compared to conventional separation methods such as chemical precipitation, ion exchange, and adsorption column, the adsorbent membrane technology has unique advantages. These advantages are high removal efficiency, normal productivity, low-pressure drop, and good stability [31]. Therefore, this study was focused to designing the membrane adsorbent structure properties derived from OPFEB witch suitable to apply for gas purification.

2. Method

2.1. Chemicals and Materials

Preparation of OPEFB membrane adsorber by employing of the chemicals and materials such as oil palm empty fruit branches (OPEFB) was taken from PT. Kharisma Alam Persada South Kalimantan Indonesia, isopropyl alcohol (2-propanol, Merck), ammonium chloride (NH_4Cl, Merck), polyethylene glycol 400 (PEG 400), polyvinyl alcohol (PVA), filter paper, deionized water (DI water Merck).

2.2. Pre-treatment of OPEFB

Oil palm empty fruit branches (OPEFB) must be firstly mashed until it becomes fibrous. Then dry the OPEFB under the sun for ± 7 days and stored in a closed container before use.

2.3. Preparation of OPEFB Membrane Adsorber

After Oil palm empty fruit branches (OPEFB) have been dried under the sun for ± 7 days. Afterwards, the pyrolysis process was carried out with a temperature of 500°C for 30 minutes, then the membrane was sieved to get a size of 200 mesh to 400 mesh. After OPEFB has been carbonized, it is ready to be used in the fabrication membrane processed.

2.4. Fabrication of OPEFB Membrane Adsorber

The preparation of the adsorber membrane dope solution in this research was carried out by trial and error method to obtain the best adsorber membrane. Weigh the OPEFB activated carbon to a predetermined size, then mix with 35 ml 2-propanol, then centrifuged the media for 10 minutes at a speed of 600 rpm. Filter the media that has been centrifuged, then add 3.5 gr of NH_4Cl that has been dissolved with 300 ml of DI water and stirred for 1 h to form pores on the membrane. Once in the stirrer, the media is filtered. Then, add 3.4 gr of PVA and 5 ml of PEG so that the media can be produced. Schematic set up fabrication of membrane adsorber OPEFB can be seen in **Figure 1**, as follow.

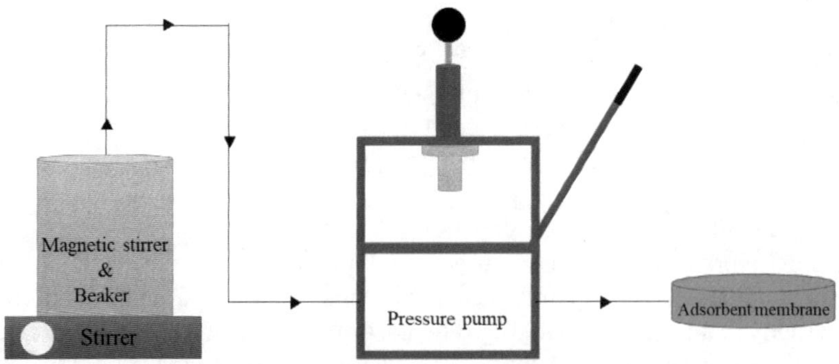

Figure 1. Schematic set up fabrication of membrane adsorber OPEFB

3. Results and Discussion

3.1. Effect of Composition and Method Fabrication Membrane Adsorber OPEFB

A comparison of the composition fabrication of membrane adsorber OPEFB can be seen in **Table 1,** as follow. **Table 1** shows that the polymerization process in the fabrication of membrane adsorber OPEFB is influenced by the polymer concentration and the treatment of the fabrication membrane. In experiment 7th to 10th, the OPEFB membrane adsorber were produced good physic texture as shows in **Figure 2**.

Table 1. Comparison of the composition fabrication of membrane adsorber OPEFB

Trial	Composition		Method	Texture result
	OPEFB (g)	PEG (mL)		
1	10	5	Dried under the sun until the membrane is dry and heated by ovenat 70 °C for an hour	Moist
2	10	5	Heated by oven at 80 °C for 24 h	Moist
3	10	5	Dried under the sun for 4 h and heated by a furnace at 105 °C for 30 min	Too sticky
4	10	5	Dried under the sun for 5 to 6 days and heated by a furnace at 105 °C for 30 min	Too thick
5	10	5	Dried under the sun for 5 to 6 days and heated by a furnace at 105 °C for 30 min	Moist
6	5	3-3.2	Dried under the sun for 2 to 3 days	Crushed
7	15	0.5	Dried under the sun for ± 2 days and heated by a furnace at 105°C for 1 h	Good but too thick
8	12.5	5	Dried under the sun for ± 2 days and heated by a furnace at 105°C for 1 h	Good but easliy crack
9	5	5	Dried under the sun for ± 2 days and heated by a furnace	Good but the edge is fragile

Trial	Composition		Method	Texture result
	OPEFB (g)	PEG (mL)		
10	17.5	5	at 105°C for 1 h Dried under the sun for ± 2 days and heated by a furnace at 105°C for 1 h	Good membrane casting

After the membrane was made according to the membrane fabrication procedure. Then the membrane was fabricated and dried in the sun until the membrane dried. After the membrane was heated by oven at 70°C for 1 h and result in the membrane become fragile. After the membrane fabrication process, is carried out according to the procedure. The fabrication process is carryout out in the oven at 80°C for ± 24 h. The process is too long for heated the membrane causes the upper of the membrane to dry faster so that the reversing process is required. This causes the center of the membrane to be imperfect and still moist.

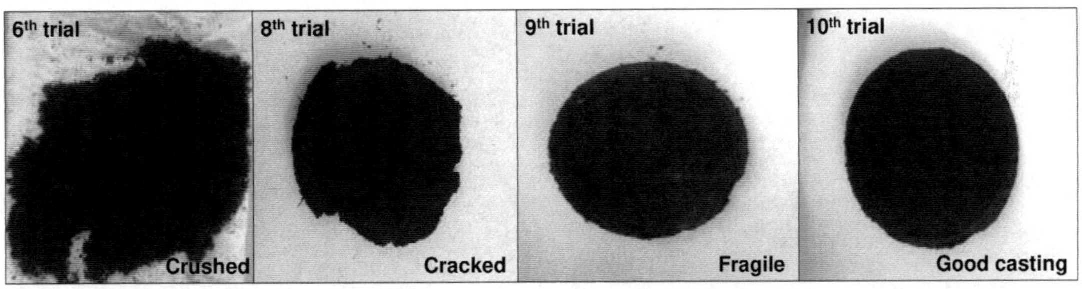

Figure 2. The OPEFB membrane adsorber obtained by added different weight of biomass at 6 to 10 trials

In the next Experiment, the OPEFB activated carbon was firstly soaked in DI water for ± 1 h, then filtered. After the membrane was fabricated, the membrane was heated by an oven at 40°C for 10 h and followed by drying under the sun for ± 4 h. After the membrane dry, then the membrane was heated by a furnace at 105 °C for 30 minutes. This process makes the membrane adsorber OPEFB is less than perfect because the bottom of the membrane is sticky to cast. Then, the procedure for fabrication membrane adsorber OPEFB was the same as in the previous work. However, the membrane is not in the oven, however, the membrane is dried under the sun for 5 to 6 days then it dried by a furnace at 105°C for 30 minutes. As a result, the membrane in the this experiment was too thick, so the weight composition of OPEFB must be reduced.

In the next experiment, 10 g of OPEFB were used. After all the ingredients are mixed on the media, the result is still not as desired. As a result, the membrane is still too wet, this is presumably because of the PEG was added too much, in this research used 5 ml PEG. 6th experiment, using only 5 gr of OPEFB and divided into 2 spaces, each weight is 2.5 g. The membrane fabrication procedure is still the same as the previous procedure. PEG was measured as much as 3 ml and 3.5 ml respectively and then put into each activated carbon OPEFB weight 2.5 g. Without the addition of NH_4Cl and DI water, the membrane was immediately fabricated and press with a pressure of 80 bar. Then the membrane is dried under the sun for ± 2 to 3 days. After the membrane is dried and then the membrane is removed from the casting. The result, the membrane was still too fragile, this is presumably because the PEG added was still lacking.

In the 7th experiment, activated carbon OPEFB was washed with DI water and filtered. The procedure is still the same as the previous adsorber membrane preparation. However, only 0.5 ml of PEG were added in this step. The media was stirred using a mixer, then added the remaining solution of NH_4Cl and 5 ml of DI water. Then the membrane is fabricated and pressed with a pressure of 100 bar. The membrane was dried under the sun for ± 2 days. After drying, the membrane dried by a furnace at 105°C for 1 h. In this research, its procedures a good membrane, and the membrane surface look flat. Therefore, the 8th experiment to 10th experiment was carried out by varying the weight of activated carbon OPEFB, namely 12.5 g, 15 g, and 17.5 g with the same procedure as the 7th experiment to fabricated a good membrane.

3.2. Characterization of Membrane Adsorber OPEFB
The FTIR spectra of membrane adsorber OPEFB were given in **Figure 3,** based on these figures, activated carbon OPEFB showed at 1082 cm^{-1} which attributed to O-H stretching functional group. The bands at 943 cm^{-1} corresponded to C=O functional group. Rani, et al. [32], conducted a study compare FTIR spectra result of the raw material OPEFB, activated carbon OPEFB and commercial activated carbon. Based on the research, showed the raw material OPEFB has a complicated and clear spectrum than activated carbon OPEFB and commercial activated carbon.

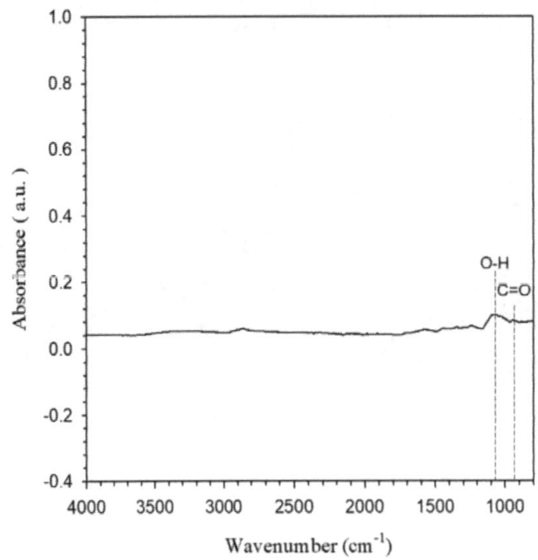

Figure 3. FTIR spectra of membrane adsorber OPEFB

Based on research Yacob [33], the functional groups from raw material were released as volatile materials when the heat was supplied to the sample during the carbonization and activation process. This proved that the activation process has taken place successfully. It could be seen from **Figure 3** that the curves of membrane adsorber OPEFB. The membrane adsorber OPEFB that has been fabricated can be seen from **Figure 4,** as follow. The color of the membrane is black due to the physical activation process. Membrane adsorber OPEFB has a diameter of 7.2 cm and the thickness of the membrane is 0.6 cm.

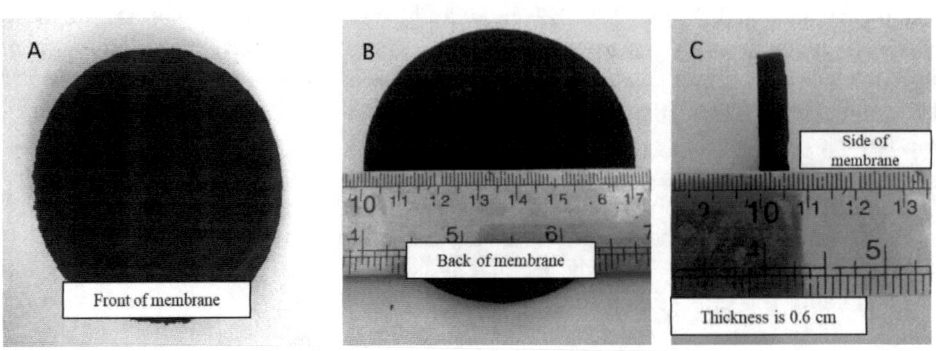

Figure 4. (A) Front of image membrane adsorber OPEFB, (B) Back of image membrane adsorber OPEFB and (C) Thickness of image membrane adsorber OPEFB

Morphology of membrane adsorber OPEFB observe used mini microscope with 1000x magnification to observe the surface and cross-section of the membrane. **Figure 5**, there exist very small pores available on the surface of membrane adsorber OPEFB. This is maybe due to the temperature given during the physical activation of activated carbon OPEFB. According to Rani, et al. [32], physical activation with a temperature of 756°C and activation time 77 minutes, can be making larger pores on the membrane.

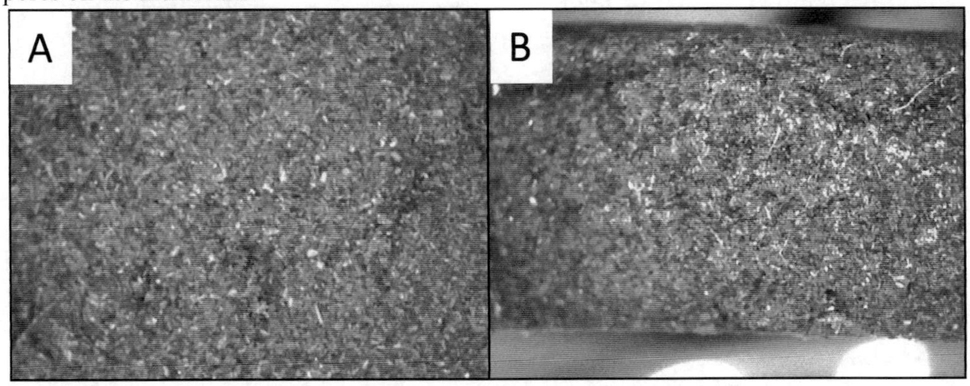

Figure 5. (A) The surface section and (B) cross section of the microscope image of membrane adsorber OPEFB with 1000x magnification

Oil palm empty fruit branches (OPEFB) have fibers belonging to the type of lignocellulose so that the OPEFB has a complex structure [1]. On figure 3 and figure 4 can be seen that the OPEFB fibers are rod-shaped. The OPEFB structure can be divided into two types, namely fibers with a rough surface and porous surfaces containing silica. It is according to the content of OPEFB. The OPEFB is covered by lignin, hemicellulose, and another component that binds cellulose [1].

4. Conclusions

Membrane adsorber from palm empty fruit branches have been successfully fabricated by varied composition and activated via physical activation. The best composition of OPEFB membrane adsorber is OPEFB:2-propanol:PEG:NH_4Cl: PVA of 17.5 g: 35 ml: 5 ml: 3.5 g: 3.4 g. The membrane has diameters of 7.2 cm and a thickness of 0.6 cm. Morphology of the membrane seen using a mini microscope showed that the membrane has very small pores and the fibers are rod-shaped. The FTIR spectra of membrane absorbent OPEFB showed that at 1082 cm^{-1} which attributed to O-H stretching functional group. The bands at 943 cm^{-1} corresponded to C=O functional group. In conclusion, the membrane can be used as membrane adsorber due to containing rich by carbon which affordable for gas separation

References

[1] A. Gazaly, I. N. Karima, and I. F. Nata, "Utilization of Palm Fibre Biomass for Fabrication of Nanoparticle Magnetic Biocomposite as Adsorbent for Sasirangan Wastewater Treatment. *(SNIKSDA National, Conference 2016)*, 2020.

[2] T. Ahmad, M. Rafatullah, A. Ghazali, O. Sulaiman, and R. Hashim, "Oil palm biomass–Based a dsorbents for the removal of water pollutants—A review," *Journal of Environmental Science and Health, Part C,* vol. 29, no. 3, pp. 177-222, 2011.

[3] I. G. Wenten, K. Khoiruddin, P. T. Aryanti, A. V. Victoria, and G. Tanukusuma, "Membrane-based zero-sludge palm oil mill plant," *Reviews in Chemical Engineering,* vol. 36, no. 2, pp. 237-263, 2020.

[4] Y. H. Teow, S. A. Tajudin, K. C. Ho, and A. W. Mohammad, "Synthesis and characterization of graphene shell composite from oil palm frond juice for the treatment of dye-containing wastewater," *Journal of Water Process Engineering,* vol. 35, p. 101185, 2020.

[5] V. Viena, B. Bahagia, and Z. Afrizal, "Production of Activated Carbon from Paml Shell and Application for Adsorption Fe, Mn and pH of Ground Water," *Jurnal Serambi Engineering,* vol. 5, no. 1, 2020.

[6] M. Maslahat, R. P. Hutagaol, and S. Lestari, "Potentially of Empty Palm Fruit Branches (EPFB) Bioadsorbent for Phenol Waste Recovery," *Journalof Natural Science,* vol. 2, no. 2, pp. 155-168, 2017.

[7] D. Atila, D. Keskin, and A. Tezcaner, "Cellulose acetate based 3-dimensional electrospun scaffolds for skin tissue engineering applications," *Carbohydrate polymers,* vol. 133, pp. 251-261, 2015.

[8] H. Lee, M. Nishino, D. Sohn, J. S. Lee, and I. S. Kim, "Control of the morphology of cellulose acetate nanofibers via electrospinning," *Cellulose,* vol. 25, no. 5, pp. 2829-2837, 2018.

[9] M. R. L. L. Gaol, R. Sitorus, S. Yanthi, I. Surya, and R. Manurung, "Synthesis of Cellulose Acetate from α-cellulose of Empty Palm Fruit Bunches," *Journal of Chemical Engineering USU,* vol. 2, no. 3, pp. 33-39, 2013.

[10] S. Xu, J. Zhang, A. He, J. Li, H. Zhang, and C. C. Han, "Electrospinning of native cellulose from nonvolatile solvent system," *Polymer,* vol. 49, no. 12, pp. 2911-2917, 2008.

[11] C. Yang, Z. Du, J. Jin, J. Chen, and J. Mi, "Epoxide-functionalized tetraethylenepentamine encapsulated into porous copolymer spheres for CO2 capture with superior stability," *Applied Energy,* vol. 260, p. 114265, 2020.

[12] F. Yalcinkaya, E. Boyraz, J. Maryska, and K. Kucerova, "A review on membrane technology and chemical surface modification for the oily wastewater treatment," *Materials,* vol. 13, no. 2, p. 493, 2020.

[13] G. Dognani *et al.*, "Effective chromium removal from water by polyaniline-coated electrospun adsorbent membrane," *Chemical Engineering Journal,* vol. 372, pp. 341-351, 2019/09/15/ 2019.

[14] C. A. Grande, R. Blom, A. Möller, and J. Möllmer, "High-pressure separation of CH4/CO2 using activated carbon," *Chemical Engineering Science,* vol. 89, pp. 10-20, 2013.

[15] W. A. Wibowo, P. Paryanto, R. A. Lutfiani, and R. M. Putra, "Purification of Biogas from H_2S Gas Using Activated Carbon From Mangrove Fruit," *Jurnal of Innovation Chemical Egineering,* vol. 5, no. 1, 2020.

[16] Y. Wang and R. T. Yang, "Chemical Liquid Deposition Modified 4A Zeolite as a Size-Selective Adsorbent for Methane Upgrading, CO2 Capture and Air Separation," *ACS Sustainable Chemistry & Engineering,* vol. 7, no. 3, pp. 3301-3308, 2019/02/04 2019.

[17] P. Iriani and A. Heryadi, "Purification of Biogas Through Activated Carbon Coloum Adsorbent," *Sigma-Mu,* vol. 6, no. 2, pp. 36-42, 2014.

[18] A. Rahma, M. Elma, E. L. A. Rampun, A. E. Pratiwi, A. Rakhman, and Fitriani, "Rapid Thermal Processing and Long Term Stability of Interlayer-free Silica-P123 Membranes for Wetland Saline Water Desalination," *Advanced Research in Fluid Mechanics and Thermal Sciences,* vol. 71, no. 2, pp. 1-9, July 2020 2020, Art no. 1.

[19] A. Rahma, M. Elma, A. E. Pratiwi, and E. L. Rampun, "Performance of interlayer-free pectin template silica membranes for brackish water desalination," *Membrane Technology,* vol. 2020, no. 6, pp. 7-11, 2020.

[20] M. Elma *et al.*, "Carbon templated strategies of mesoporous silica applied for water desalination: A review," *Journal of Water Process Engineering,* vol. 38, p. 101520, 2020/12/01/ 2020.

[21] M. Elma, A. Rahma, A. E. Pratiwi, and E. L. A. Rampun, "Coagulation as pretreatment for membrane-based wetland saline water desalination," *Asia-Pacific Journal of Chemical Engineering,* vol. n/a, no. n/a, p. e2461, 2020.

[22] R. A. Lestari *et al.*, "Organo Silica Membranes for Wetland Saline Water Desalination: Effect of membranes calcination temperatures," *E3S Web Conf.,* vol. 148, p. 07006, 2020.

[23] M. Elma, A. E. Pratiwi, A. Rahma, E. L. A. Rampun, and N. Handayani, "The Performance of Membranes Interlayer-Free Silica-Pectin Templated for Seawater Desalination via Pervaporation Operated at High Temperature of Feed Solution," *Materials Science Forum,* vol. 981, pp. 349-355, 2020.

[24] A. Rahma, M. Elma, M. Mahmud, C. Irawan, A. E. Pratiwi, and E. L. A. Rampun, "Removal of natural organic matter for wetland saline water desalination by coagulation-pervaporation," *Jurnal Kimia Sains dan Aplikasi,* vol. 22, no. 3, pp. 85-92, 2019.

[25] A. Rafique, K. M. Zia, M. Zuber, S. Tabasum, and S. Rehman, "Chitosan functionalized poly (vinyl alcohol) for prospects biomedical and industrial applications: A review," *International journal of biological macromolecules,* vol. 87, pp. 141-154, 2016.

[26] A. E. Pratiwi, M. Elma, A. Rahma, E. L. A. Rampun, and G. S. Saputro, "Deconvolution of pectin carbonised template silica thin-film: synthesis and characterisation," *Membrane Technology,* vol. 2019, no. 9, pp. 5-8, 2019/09/01/ 2019.

[27] M. Elma *et al.*, "Fabrication of interlayer-free silica-based membranes – effect of low calcination temperature using an organo-catalyst," *Membrane Technology,* vol. 2019, no. 2, pp. 6-10, 2019/02/01/ 2019.

[28] D. K. Wang, M. Elma, J. Motuzas, W.-C. Hou, F. Xie, and X. Zhang, "Rational design and synthesis of molecular-sieving, photocatalytic, hollow fiber membranes for advanced water t reatment applications," *Journal of Membrane Science,* vol. 524, pp. 163-173, 2017/02/15/ 2017.

[29] D. K. Wang *et al.*, "Physicochemical and photocatalytic properties of carbonaceous char and titania composite hollow fibers for wastewater treatment," *Carbon,* vol. 109, pp. 182-191, 2016/11/01/ 2016.

[30] V. K. Gupta, S. Agarwal, and T. A. Saleh, "Synthesis and characterization of alumina-coated carbon nanotubes and their application for lead removal," *Journal of hazardous materials,* vol. 185, no. 1, pp. 17-23, 2011.

[31] S. Mehdipour, V. Vatanpour, and H.-R. Kariminia, "Influence of ion interaction on lead removal by a polyamide nanofiltration membrane," *Desalination,* vol. 362, pp. 84-92, 2015.

[32] N. H. A. Rani, N. F. Mohamad, S. Matali, and S. A. S. A. Kadir, "Preparation and characterization of activated carbon made from oil palm empty fruit bunch," *Key Engineering Materials,* 2014.

[33] A. Yacob, *The Malaysian Journal of Analytical Science.,* vol. 12, 2008.

Acknowledgement

Authors thank to Materials and Membranes Research Group (M^2ReG) for the facilities provided. Hidayah thanks to the Research Collaboration Between Universities Grant 2020-2021. Muthia thanks also to Applied Research Universities Grant 2019-2020, Thesis Magister Grant 2020, World Class Research Grant 2021-2022 the Directorate of Research and Community Service, the Ministry of Research and Technology/the National Agency for Research and Innovation Republic of Indonesia and Lambung Mangkurat University Research Grant 2020.

Analysis of rock mass behavior with Empirical and Numerical method for the construction of diversion tunnel, Laos

D J W Mboussa, S Sun

School of Earth Science and Engineering, Hohai University, Nanjing 210098, China

Email : weaverjosias@gmail.com

Abstract. Tunneling construction in the mountain area is a challenge for engineers and geotechnicians because of instability due to the presence of discontinuities. The objective of this paper is the modeling of surrounding rock masses for the stability of the diversion tunnel to predict the behavior of rock masses during the excavation process for the Nam Phoun hydropower station project in Laos. Field investigation and laboratories test was realized; Empirical methods as Rock mass designation and Geological Strength Index were performed, rock masses were classified in three categories (RM-1, RM-2, and RM-3); in situ stresses were obtained from existing equations, numerical modeling was performed by the 2D plane strain finite element code Phase2 developed by Rocscience, using Generalized Hoek-Brown criterion for each type of rock masses. The results of numerical modeling show the strength zones of stresses and deformations around the tunnel and predict the instabilities around the tunnel during excavations processes. Thus, for all rock's masses, it will be necessary to consider an analysis for the supports design before the excavation's process. The findings of this study allow a clearer understanding of the importance to assess a predictive analysis of slope stability during the feasibility phase of a project by engineers to have an idea of instabilities and its significant in preventing the impact on the cost of the project.

1. Introduction

The construction of structures in mountain environments has always been a major challenge for engineers and geotechnicians. In most cases, the excavated ground presents risks of failure related to the behavior of the rock masses due to their heterogeneity and anisotropy. In the case of the construction of deep or shallow tunnels, road, a dam in a massif with discontinuities, great attention should be paid to the stress redistribution, strain, displacement and behavior of the rock mass, which is control by the number of discontinuities set and their characteristics (orientation, appearance, and spacing), the characteristics of rock's matrix (e.g. resistance, deformability, weathered) and groundwater condition; In most of the case, the detail on these parameters are absent [1]. But, taking into account these factors during the design of the structure by engineers is very important; This makes it possible to evaluate the risks, the avoid any impact on the investment cost of the project. However, with the progress in the rock engineering area, some empirical methods and numerical methods have been developed by many researchers. The empirical methods allow obtaining the data on engineering properties of rock mass such as shear strength properties, in situ stress, deformation modulus, the general state of discontinuities, hydrological condition of the rock mass, the quality of rock mass and others parameters controlling the behavior of rock mass. But in the case of tunneling designing, empirical methods present some limitations in the evaluation of the performance of support systems, stress distribution and the deformation around the tunnel [2]; In the aim to solve this problem, the numerical method is used by engineers. These methods have many advantages in the evaluation of rock mass behavior and it provides quick modeling of the studied problem and the lowest financial cost.

The purpose of this study is the modeling of surrounding rock masses for the stability of diversion tunnel in the aim to predict the behavior of rock masses during the excavation process, based on the empirical methods and numerical method for the Nam Phoun hydropower station project in Laos. In this paper, RQD proposed by [3] and GSI were used to classify the rock masses and the finite element method was used for modeling the behavior of rock masses code called phase^2. v.8.0 developed by Rocscience [4]. The project is under a feasibility study, and the diversion tunnel will be constructed on the right bank slope. To accomplish the task, field investigation, measurements, and laboratory test were performed on the different samples collected from the dam site to determine the engineering properties of rock masses.

2. Geology of the study area

The dam site in this study is located at Nam Phoun River near the Nam Phoun Village in Paklay District of Xayaburi Province in Northwestern Laos, approximatively 2.6km upstream of the convergence points of the tributary Nam Gnam River. The latitude and longitude coordinates of the dam project location are 18°27'48.23"N and 101°28'04.45"E (Figure 1).

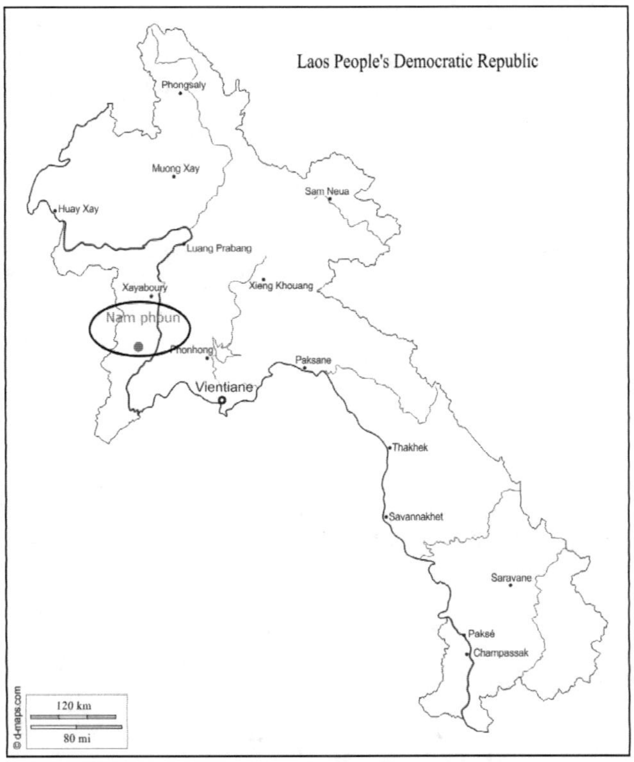

Figure 1. Location of the study area [5]

Diversion tunnel should be constructed at the right bank and is a total of 635.7 m long; inlet floor elevation is 243m and outlet floor elevation 242m; the cross-section is portal shaped, 6.0 m wide and approximatively 7.0 high. Field investigation performed in the study area showed that the geology of bedrock in the dam area is mostly comprised of tuff, tuffaceous sandstone and tuffaceous slate (CPz2) from Upper Devonian to Lower Carboniferous. The dam site area has the major presence of Quaternary overburden which is mostly composed of alluvial and diluvial-eluvial deposits. The upper part of overburden on both banks is composed of diluvial-eluvial gravelly clayey soil and lower part eluvial gravelly soil. The upper part is composed of deluvial-eluvial gravelly clay and lower part eluvial gravelly soil. The bedrock at diversion channel is lithologically composed of tuff, tuffaceous sandstone, and tuffaceous slate.

3. Rock mass classification

In the literature, There are various rock mass classification system developed in the field of rock, civil and mining engineering by many researchers such as Rock mass Quality Designation (RQD), Rock Mass Rating (RMR), Geological Strength Index (GSI), Tunneling quality index (Q-system) proposed by [2-6-7-8] and others, as New Australian Tunneling Method (NATM), Rock Structure Rating (RSR) for assessment and classification of rock mass; this provides some necessary input data for performing the numerical modeling of the rock mass. In this study RQD, and GSI were used because of its flexibility in terms of input parameters.

Rock Quality Designation (RQD) is defined as the ratio of the sum of the length of pieces of core longer than 10cm to a total length of the core and expressed by the following equation:

$$\mathrm{RQD} = \frac{\sum \text{lenght of core pieces} > 10\text{cm}}{\text{total length of core}} \times 100\% \quad (1)$$

Table 1 illustrate the classification of rock mass according RQD. The Geological Strength Index (GSI) is estimated using qualitative and quantitative factors depending on the petrographic nature of the rock and its degree of invoicing.

Table 1. Rock mass quality classification based on the RQD (Deere et al., 1967)

RQD value	Rock mass quality	Distribution	Lithologies	Rock Mass
90-100	Excellent	-	-	-
75-90	Good	65%	Tuff, tuffaceous sandstone, Tuffaceous slate	RM-1
50-75	Fair	25%	Tuffaceous phyllite Tuffaceous slate	RM-2
25-50	Poor	10%	Tuffaceous slate, tuffaceous sandstone, tuff	RM-3
<20	Very poor	-	-	-

Thus, this system was established to estimate the resistance of the rock mass, by reducing the resistance of the intact rock for the different geological conditions. For this system, the maximum resistance of a rock depends on the properties of the intact blocks and also depends on the freedom of the blocks to slide and rotate under a range of conditions of imposed stresses; figure 2 illustrates the chart of GSI [9].

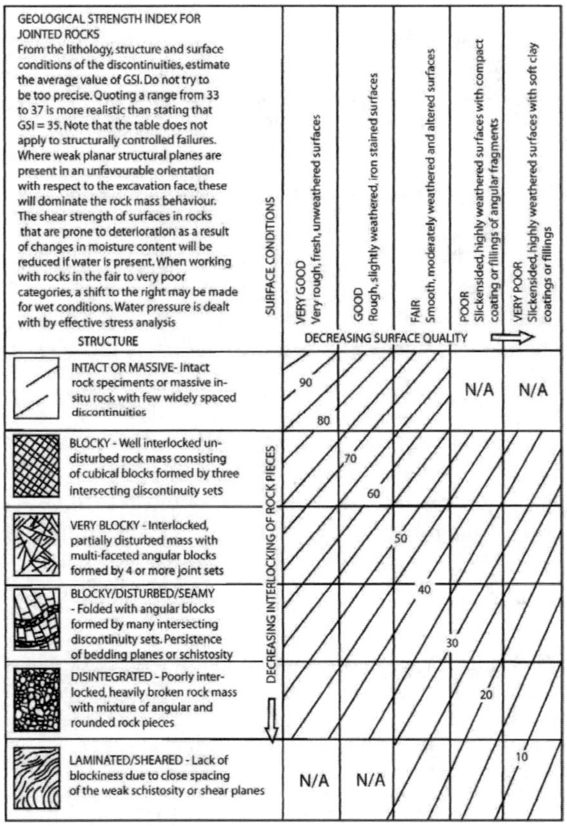

Figure 2. Chart of Geological Strength Index [9]

In rock engineering, in situ stress is determined by direct and indirect methods. Indirect methods, in situ stress, are determined by the methods like flat jack, overcoming and under coring, and hydraulic fracturing. These methods are costly time-consuming, the procedures used in the determination of these stresses are difficult, and the result may be questionable [10-11-12]. In the indirect method, the stresses (vertical and horizontal) are determined through empirical models developed in rock engineering by the following equations:

$$\sigma_v = \gamma . H \quad (2)$$

$$\sigma_h = \left(\frac{v}{1-v}\right)\sigma_v + \frac{\beta ErmG}{1-v}(H+100), \quad (3)$$

Where the unit weight of rock mass and H is the overburden, ν is Poisson's ratio, β is the coefficient of thermal expansion which the value for rocks is 8 x 10^{-6} / °C, Erm is the young's modulus of intact rock in MPa, G(°C/m) is the thermal gradient of rock (However, the simplified relationship of equation (2) for the determination of the horizontal stress is:

$$\sigma_h = \left(\frac{v}{1-v}\right)\sigma_v \quad (4)$$

In this study, the vertical and horizontal stresses were determined using equations (2) and (4) for each type of rock mass; the result is illustrated in the table 2.

4.1 Numerical modeling

Numerical modeling has been developed and applied in rock engineering because of its ability to better simulate actual slopes failure mechanisms [13]. Great progress has been realized in the last three decades in rock mass analysis. This allows us to classify numerical methods into the continuum and discontinuous [14-15]. In the continuum modeling, plastic softening and damage are the most techniques for capturing failure and localization [16-17]. The continuum method includes the Finite element Method (FEM), Finite Difference Method (FDM), and the Boundary Element Method (BEM). The discontinuous methods include Distinct Element Methods (DEM) and the Discontinuous Deformation Analysis (DDA). All these methods provide an approximate solution to problems. However, the finite element method (FEM) is one of the most used in the field of landslide and slope stability analysis; it provides more advantages such as:

a. the stress-strain relation of soil or rock is considered and thus more accurate mechanical behavior can be computed, such as non-linear deformation and the influence of water and earthquakes;
b. No assumptions are applied in advance related to the interslice forces and their directions, or the shape or location of the slip surface. The critical slip surface is determined automatically, and the slope fails naturally;
c. Complex slope geometries can be addressed, and parametric studies can be conducted [18-19].

In this study, to predict the behavior of surrounding rock masses around the diversion tunnel during the excavation process, a two-dimensional hybrid finite element model code Phase^2 [4] is used. The modeling concerns each type of rock mass identified around the tunnel site project.

Table 2: Parameter for numerical modeling Input parameters

Rock mass	Unit weight (MPa)	Modul of elasticity (MPa)	Poisson's ratio (ν)	Generalized Hoek -Brown parameters					Vertical stress (MPa)	Horizontal stress (MPa)
				m_b	s	a	m_i	*GSI (%)*		
RM-1	27,45	10000	0.28	2.60	0.0067	0.5040	13	55	0.21	0.816
RM-2	22.5	40000	0.34	0.821	0.0012	0.514	7	40	0.2	0.10
RM-3	19.61	40000	0.40	0.402	0.00013	0.5437	7	20	0.1568	0.104

5. Results and Discussion

Rock masses modeling was carried out for the construction of the diversion tunnel of the Nam Phoun hydropower station project site in this paper. The purpose of this study was to perform the modeling of surrounding rock masses for the stability of the diversion tunnel in the aim to predict the behavior of rock masses during the excavation process. Field investigation, measurements, and laboratory tests were conducted in the study area according to IRSM methods [20]. RQD and GSI as an empirical method were used.

According to RQD classification, the surrounding rock mass around the diversion tunnel site is the following distribution: 65% fair, 25% poor and 10 % very poor (Table 1). The GSI was used as an input parameter for numerical modeling and the result is illustrated in table 2.

The Finite element method was performed to model the rock masses using a commercial finite element code Phase^2 developed by Rocscience. The input parameters for the numerical modeling include uniaxial compressive strength, stresses (vertical and horizontal), cohesion, deformation modulus, Poisson's ratio, modulus elastic of intact rock, unit weight (Table 2). Each category of rock mass was modeling according to elastoplastic behavior using Generalized Hoek-Brown criterion and the tunnel modeling is 2D considering plane strain problem. It's important to notify that, the numerical modeling was selected for a sequential excavation, and the field stress using in the model was gravity loading for all rocks mass.

5.1. Numerical modeling for RM-1
The result of the Finite element modeling of the Rock mass-1 is illustrated in figure 3. Before excavation, the virgin sigma 1 value was 0.21 MPa. After excavation, the maximum sigma 1 value is 0.16 MPa at the crown, 0.96 MPa at sidewalls and the maximum value is 1.6 MPa at the corner sides (Figure 3a). The value of sigma 3 was 0.081 MPa before excavation; However, after excavation, the value of sigma 3 values is respectively 0.01 MPa at the crown, the base, and the sidewalls. After the comparison of the stress's virgin sigma 1 and sigma 3 before and after excavation, this is shown the increasing virgin sigma 1 of 1.39 MPa and the decreasing of virgin sigma 3 of 0.071 MPa. It's important to notify that for this section, the maximum stress concentration is developed at the corner sides of the tunnel (Figure 3b).

The maximum total displacement which constitutes the maximum deformation is 7.65e-004 m, situated at the base of the tunnel after excavation model (Figure 3c); Although the displacements are not so important around the tunnel the thickness of plastic zone is important at the base (yields zone is 100%) as shown in the Figure 3d. So, it's important for safety during excavations operations for this tunnel to prevent an analysis concerning the design of the support system.

5.2. Numerical modeling for RM-2
In this section, the result of finite element modeling of surrounding rock mass in the tunnel is illustrated in figure 4. Before excavation, the virgin stress sigma1 was 0.2 MPa. After excavation, the virgin stress sigma 1 is respectively 1.80 MPa at the crown, 0.30 MPa at the sidewalls, and 0.60 MPa at the base of the tunnel (Figure 3a). The virgin stress sigma 3 before excavation was 0,1 MPa; However, after excavation, the virgin stress sigma 3 is 0.23 MPa at the crown, 0.04 MPa respectively at the sidewalls and the base of the tunnel (Figure 3b). The difference between the maximum virgin stress sigma 1 before and after the excavation is 1.6 MPa, and for virgin sigma 3 is 0.03 MPa. These results showed the increasing of stresses sigma 1 and sigma 3 after excavation. Although, for this section, the maximum concentrations for stresses are at the crown and the base of the tunnel.

The maximum deformation is 1.80e-004 m, concentrated at the sidewalls and the crown (Figure 3c). This is can explain the maximum plastic zones at the crown (yields zone 80%) and the base of the tunnel (yields zone 100%) according to the figure 3d; But, for more safety, it will be necessary to establish the support accord to the quality of rock mass.

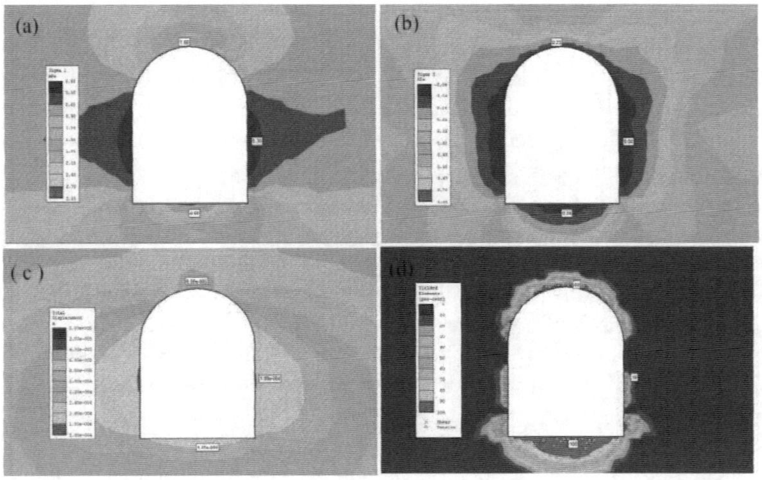

Figure 3. RM-2; (a) Sigma 1 contours; (b) Sigma 3; (c) Maximum displacement; (d) yielded elements.

The finite element modeling of rock mass 3 is represented in figure 4. The analysis of the virgin stress sigma 1 and sigma 3 before excavation was respectively 0.15 MPa and 0.10 MPa. After excavation, the virgin stress sigma 1 is 0.36 MPa at the crown of the tunnel, and 0.12 MPa at the sidewalls (Figure 4a). The virgin stress sigma 3 is 0.4 MPa at the crown, the sidewalls, and the base of the tunnel (Figure 4b). According to the results, there is an increase of stresses, respectively 0.21 MPa for sigma 1 and 0.3 MPa for sigma 3 after excavation around the tunnel.

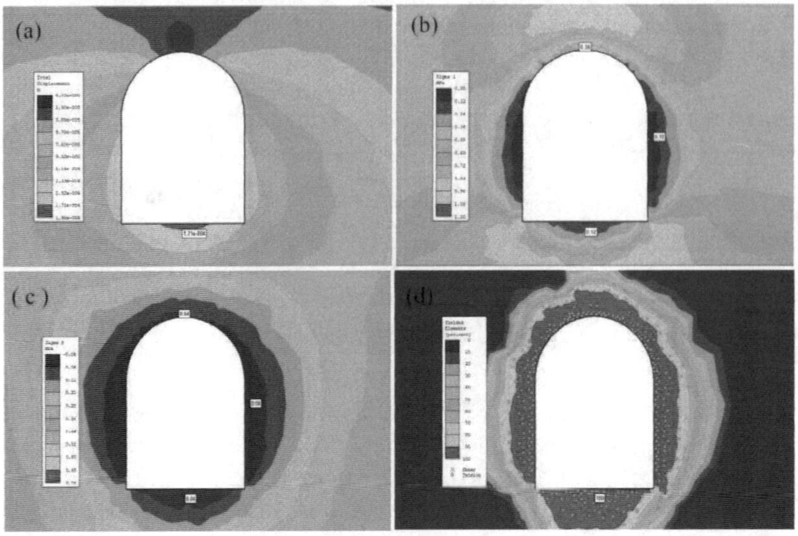

Figure 4. RM-3; (a) Sigma 1 contours; (b) Sigma 3; (c) Maximum displacement; (d) yielded elements.

According to figure 4c, the maximum displacement is 1.71e-004 m in this section. The maximum vectors deformation is focused around the tunnel; This is can be explained by the strength thickness of the yield's zones (at 100%) around the tunnel (Figure 4d). According to the results of yields elements analysis, the support design will be necessary for the rock mass during the construction processes.

6. Conclusion

In this paper, the empirical and numerical methods were used to modeling the behavior of rock masses in the aim to predict the stability of the tunnel during the excavation operations. Stresses redistribution and deformations around tunnel were analyzed before and after excavation for each type of rock masses. Finite element analysis code Phase^2 developed by Rocscience was used for modeling of surrounding rocks around the tunnel, the generalized Hoek-Brown criterion was used for the models. Based on the empirical and numerical modeling analyses performed for the construction of the diversion tunnel, the following conclusions can be drawn: a) For the RM-1, the stresses analysis showed the increase of the virgin sigma 1 of 1.39 MPa and the reduced of the virgin sigma 3 of 0.071 MPa after excavation in the modeling; the total displacement was observed as 7.65e-004 m where the yield zone was observed as 100% more focused at the base of the tunnel model. b) For the RM - 2, the stresses analysis performed showed the increase of the virgin sigma 1 of 1.6 MPa and sigma 3 of 0.03 MPa after excavation, the maximum deformation was observed as 1.80e-004 m, at the sidewalls and the crown of the tunnel model where the thickness of yielded elements were respectively 80% and 100%. c) For the RM- 3, the stress analysis performed showed the increased value of the virgin sigma 1 of 0.21 MPa and 0.3 MPa for sigma 3 after excavation around the tunnel model. The maximum deformation was observed as 1.71e-004 m and the large thickness yielded zones as 100% around the tunnel; and for all Rock masses in this case of study, it will be necessary to perform the analysis concerning the support design before and during the construction of the diversion tunnel for more safety. The findings study allowed a clearer understanding of the importance to assess a predictive analysis of slope stability during the feasibility phase of a project by engineers to have an idea of instabilities and its significant in preventing the impact on the cost of the project and the protection of human lives.

References

[1] Andriani G F and Praise M. 2017. "Applying rock mass classification to carbonates rocks for engineering purposes with a new approach using the rock engineering system,'' Journal of Rock Mechanics and Geotechnical Engineering, vol. 9, no. 2, pp. 364-369,

[2] Bassari H, Ozsan A, and karakus M. 2005. "Analysis of support requirements for a shallow diversion Tunnel at Guledar dam site Turkey,'' Engineering Geology, vol. 8, no 2, pp. 131-145,

[3] Deere, D., Hendron, A., Patton, f., and Cording, E. (1967). Design of surface and near surface construction in rock: failure and breakage of rock. 8th U.S Symposium on rock Mechanics. New York: Society of Mining Engineers, American institute of Mining, Metallurgical, and Petroleum Engineers.

[4] Rocscsiences. 2014. A 2D finit element program for calculating stress and estimating support around the underground excavations. Geomechanics software and research, Rocscience inc., Toronto, Ontario, Ca, nada.

[5] Mboussa D J W, Sun S., Zhang Y., Elenga H. I. 2019. Rock slope stability analysis with a kinematical method for the Nam Phoun dam site project, Laos: IOP Conf. Series: Earth and Environmental Science 362, 012137.

[6] Bieniawski Z. T., 1989. "Classification of Rock Mass for Engineering: The RMR System and Future Trends", in Rock Testing and site Characterization, Pergamon, Oxford, UK.

[7] Hoek, E. and Brown, E.T. 1997. Practical estimates of rock mass strength. Int. J. Rock Mech., 34,1165-1186. Min.Sci.

[8] Barton, N., Lien, R., and Lunde, J. 1974. Engineering Classification of rock for design of tunnel support. International Journal of Rock Mechanics and Mining Sciences, 6(4), 189-236.

[9] Marinos, P., and Hoek, E. (2000). GSI: a geologically friendly tool for rock mass strength estimation. In ISRM international Symposium. International Society for Rock Mechanics.

[10] Gurocak Z, Solanki P. and Zaman M.M. 'Empirical and numerical analyses of support requirements for a diversion tunnel at the Boztep dam site eastern urkey," Engineering Geology, vol, 91.no. 2-4, pp. 194-2007.

[11] Kabbazi A, and Gaffoori M, "Estimation of rock mass deformation modulus using a rock mass classification system," Geomechanical and Geoengineering, vol, 8, no. 1, pp. 46-52, 2013.

[12] Ur Rehman Z., Mohammad, S. Hussain, and Tahir M. 2017. "Numerical modeling for the engineering analysis of rock mass behavior due to sequential enlargement of Lowari Tunnel Chitral Khyber Pakhtunkhwa, Pakistan,'' International Journal of Geotechnical Engineering, pp. 1-7,

[13] Jiang M., Jiang T., Crosta G.B, Shi Z. Chen H., Zhang N. 2015. Modeling failure of jointed rock slope with two main joints set using a novel DEM bond contact model, Eng. Geol. 193 79-96.

[14] Jing, L., and Hudson, J. A. 2002. "Numerical methods in rock mechanics. Int. J. Rock Mech. Min. Sci., 39, 409-427.

[15] Jing, L. 2003. "A review of technics, advances and outstanding issue in numerical modelling for rock mechanics and rock engineering. Int. J. Rock Mech. Min. Sci., 40, 283-353.

[16] De Borst, R., and Muhlhaus, H.B. 1992. "Gradient-dependent plasticity: Formulation and algorithmic aspects." Int. J. Numer. Methods Eng., 35, 521-539.

[17] De Borst, R. 2002. "Fracture quasi-brittle materials: A review of continuum damage-based approaches." Eng. Fract. Mech., 69, 95-112.

[18] Crosta G B, Imposimato S, Rodeman D G. 2003. Numerical modelling of larges landslides stability and runout, Nat. Hazards earth syst. Sci. 3. 523-538.

[19] Tschuchnigg F., Schweiger H.F., Sloan S W. 2015. Slope stability analysis by means of finite element analysis and finite element strength reduction technics. Part I: Numerical studying

considering non-associated plasticity, comput. Geotech. 70 169-177.
[20] International Society for rock Mechanic IRSM, 1981. Rock Characterization, testing and monitoring. In: Brown, E.T. (Ed.), IRSM Suggested Method. Pergamon Press.

Acknowledgments
The research is financially supported by the Fundamental Research Funds for the Central Universities (No: 2109B42214). The authors thank the China Council Scholarship (CSC) for the support the and colleagues for a good collaboration.

The application of vapour equilibrium technique to control suction to study the shrinkage and water retention of compacted Claystone-Bentonite mixtures

Y F Arifin[1,2], M Arsyad[1], M Afdi[1], H Muslim[1]

[1]Civil Engineering Study Program, University of Lambung Mangkurat, Banjarbaru, Indonesia
[2]Wetland Based Material Research Center, University of Lambung Mangkurat, Banjarbaru, Indonesia

y.arifin@ulm.ac.id

Abstract. Understanding soil shrinkage and retaining water is essential for learning more about the possibility of cracking of liner. Moreover, the factors that influence it are also important to know to improve the material tested as hazardous waste liners. The vapour equilibrium technique is widely used to control suction of compacted soils experiencing drying-wetting phenomena. It is considered to be inexpensive, simple, and has the ability to adequately control the suction applied to soil samples. This paper, therefore, describes its application in studying the shrinkage and water retention in compacted claystone-bentonite mixtures. This involved using five saturated salt solutions including potassium sulphate (K_2SO_4), potassium chloride (KCl), sodium chloride (NaCl), potassium carbonate (K_2CO_3), and magnesium chloride ($MgCl_2.6H_2O$). The sample was allowed to be in equilibrium with the relative humidity salt solution and a calliper was used to measure the dimensions every day up to when this was achieved. The results showed the bentonite in the mixture affects the amount of shrinkage and water retention while the sample's initial moisture content was also found to be very influential on the magnitude of the primary and residual shrinkage. Moreover, the sample's ability to hold water was almost the same without differentiating the initial water content at a total suction of more than 41084.91 kPa.

1. Introduction

Claystone is a material usually avoided and discarded due to its unfavourable behaviour and the existence of its layers considered to be a problem in construction work have been reported in some publications [1–5]. The use of this material has, however, been widely recommended, especially as a barrier to hazardous waste and this requires mixing it with bentonite to improve its hydro-mechanical properties [6–9]. Zang and Kröhn [9] reported the superiority of the claystone-bentonite mixture's hydro-mechanical properties compared to compacted pure bentonite and bentonite-sand mixture which included higher density, stiffness, and stability, and very low permeability as well as its ability to release gas at lower pressure [9]. The characteristic absorption of clay minerals in the claystone/clay shale [10] also adds to its advantage compared to the use of sand to mix the clay liners. It is even more beneficial when the claystone is obtained at the project site due to the consideration for the use of local material by several researchers in recent times [11–13].

The two mixtures are made from clay and one of the characteristics considered is their high shrinkage which was divided by Cornelis et al. [14] into four parts which are structural, normal, residual, and zero.

Meanwhile, Mishra et al. [15] also classified shrinkage in the compacted clay and bentonite-sand mixtures into initial, primary, and residual stages. It is important to note that both structural and initial shrinkages are influenced by the bimodal pore distribution in the form of macro and micropores which were reported by Arifin [16] to be strongly influenced by the initial moisture content of the samples. Generally, the initial and residual shrinkages are not too significant and this means there is a need to study the primary shrinkage in claystone and its mixture with bentonite.

Mishra et al. [15] conducted used a drying process to obtain a shrinkage curve and this involved allowing the sample to be air-dried for two months. It is, however, difficult to implement this method in areas with high relative humidity (*RH*) such as in Indonesia due to the possible absorption of water from the air by the sample when the *RH* rises. Therefore, a better method is to control the *RH* around the sample using the Vapor Equilibrium Technique (VET) to ensure the sample is balanced at a lower *RH*. This approach has also been widely applied to test water retention behaviour or water characteristic curves of soils, especially clay apart from its application in shrinkage behaviour [16–18]. It has also been applied in testing the hydro-mechanical behaviour of unsaturated compacted clay to control the suction applied to the sample [7], [16], [19–21]. VET, however, generally requires using a saturated salt solution with some researchers observed to have used molal salt solutions such as sodium chloride [16], [18]. The sample also needs to be in equilibrium with the relative humidity above the solution [7], [16–20], [22] and this means temperature needs to be controlled due to its strong influence on the *RH*. The method is also usually combined with the axis translation technique at low suction [16], [21] and the changes in the temperature gradient affect the *RH* with further causes a variation in the total suction being applied [16], [23], [24]. This paper, therefore, discusses the shrinkage behaviour and water retention of compacted claystone-bentonite mixtures using VET.

2. Methods and Procedures

2.1. Materials Used

The materials used in this study include claystone and bentonite with the claystone obtained from the Banjarbaru city area while the bentonite is available in the market. The Engineering Properties of the two materials are summarized in Table 1 and the claystone observed to have a liquid limit (LL) of 50.76% and a plasticity index (PI) of 20.95% while and these are twice as much as those obtained from Belencito formation in eastern Andes mountain, Columbia, as reported by Espitia et al. [1].

2.2. Sample preparation

The Standard Proctor compaction test (ASTM D698-07) was conducted to determine the relationship between water content and claystone density used as a primary material and the optimum moisture content and maximum density were found to be 15% and 16kN/cm^3, respectively. All the samples were prepared using these values and those with lower and higher water contents at 10% and 15% were used to study the effect of water content.

Claystone was mixed with 5%, 10%, 15%, and 20% bentonite on dry weight basis while water was added to reach the pre-selected level of 10%, 15%, and 20% and value. Meanwhile, the water contained in each sample was also calculated before mixing. The mixed sample was left in equilibrium for one day and statically compacted to reach a density of 16 kN/m^3 and the dimension used was, however, thinner than usual as observed with the 10 mm height and 64.2 mm diameter.

There are two possible conditions in the clay liner or cover layer while on the field and the first involves the experience of an increase in water content due to rain in the sample which means the initial process is wetting. It is, however, possible to also experience immediate drying due to the exposure to air, thereby, leading to shrinkage. The sample was simulated in this study and this led to the second condition with the vapour equilibrium technique (VET) method.

The total suction of the initial sample was determined before being tested by measuring the *RH* using a sensor corrected for accuracy using the chilled-mirror hygrometer technique which has been confirmed to be the most reliable tool to measure *RH* [16], [23–25]. Meanwhile, the total suction sample was calculated using Equation 1 while the initial conditions of the claystone and bentonite mixture are presented in Table 2.

Table 1. Engineering properties of materials used

Properties		Claystone	Bentonite
Specific gravity		2.60	2.71
Water content	%	2.75	14.17
Grain size distribution:			
Gravel (> 2 mm)	%	0.02	0.00
Coarse sand (0.6-2.0 mm)	%	0.08	0.00
Medium sand (0.2-0.6 mm)	%	0.10	0.00
Fine sand (0.05-0.2 mm)	%	4.30	1.39
Silt and Clay (0.002-0.05)	%	43.94	8.33
Clay (<0.002mm)	%	51.55	90.28
Atterberg Limits:			
Liquid limit	%	50.76	351.71
Plastic limit	%	20.95	44.68
Shrinkage limit	%	9.74	41.89
Plasticity Index	%	29.81	307.03

$$s_t = -\frac{\rho_w RT}{M_w} ln\left(\frac{RH}{100}\right) \quad (1)$$

where s is total suction, ρ_w is the volumetric weight of water, R is a universal gas constant which is 8.31432 J/mol.K, T is the temperature in Kelvin, and M_w is molecule mass of water vapour which is 18.016 g/mol.

2.3. Vapour Equilibrium Technique

The five saturated salt solutions used in this study include potassium sulfate (K_2SO_4), potassium chloride (KCl), sodium chloride (NaCl), potassium carbonate (K_2CO_3), and magnesium chloride ($MgCl_2.6H_2O$). The *RH* or total suction of each solution calculated using Equation 1 is summarized in Table 3. The solution was later placed in an airtight glass equipped with rubber seals and clamp lids as shown in Figure 1.

Table 2. Initial conditions of claystone-bentonite mixtures

Sample code	Bentonite content (%)	Dry density (kN/m³)	Water content (%)	Total suction (kPa)
100CS-1	0	16	10	2663.56
100CS-2	0	16	15	2237.56
100CS-3	0	16	20	1816.35
95CS-5B-1	5	16	10	4089.92
95CS-5B-2	5	16	15	2664.36
95CS-5B-3	5	16	20	2104.87
90CS-10B-1	10	16	10	5231.79
90CS-10B-2	10	16	15	4233.12
90CS-10B-3	10	16	20	3236.72

Table 3. *RH* and total suction of saturated salt solution

Solution	K_2SO_4	KCl	$NaCl$	K_2CO_3	$MgCl_2.6H_2O$

RH (%)	95.2	81.5	74.3	46	34.2
T (°C)	27.7	28.1	27.8	28.2	28.0
Total suction (kPa)	6801.21	28318.56	41084.91	107528.77	148484.33

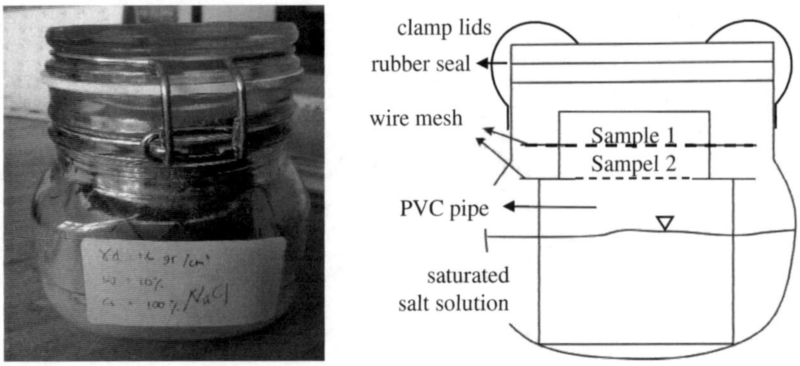

Figure 1. (a) Photo of an airtight glass used, and (b) sketch of components in the glass

The VET was generally conducted using a desiccator [17], [22] and, apart from the larger volume, the samples placed in the desiccator influenced each other and this slowed down the time to reach equilibrium [16].

The compacted samples were placed into a 615 cm^3 glass containing a saturated salt solution, as shown in Figure 1, and allowed to stay until they attain equilibrium. The weight as well as the dimensions including the height and diameter of the samples were measured every day using Vernier callipers through the direct measurement method [26]. They were placed in the oven for 24 hours to obtain the dry weight and two samples were tested to represent the conditions presented in Table 1. Moreover, smaller glasses were used to ensure each is filled with different solutions and due to its ability to allow concurrent testing of identical samples to reduce the time needed for the test.

3. Results and Discussion

3.1. Time to reach equilibrium

The samples were considered to have reached equilibrium condition when no change was recorded in their weight and dimensions. Figure 2 shows a typical change in the weight, diameter, and height of the samples 95CS-5B in (a) and 90CS-10B in (b) for a glass containing a saturated K_2SO_4 solution with a total suction of 6848.64 kPa. This means the weight decreases by time followed by the diameter and height. Meanwhile, the weight at 95CS-5B-3 and 90CS-10B-3 with an initial water content of 20% was constant on the 11th and 12th day while the value was also observed to be constant on the 9th day for 95CS-5B-1 and 90CS-10B-1 with an initial water content of 10%. Furthermore, the diameter and height measured for the 95CS-5B-1 and 90CS-10B samples -1 were relatively constant at the 9th and 10th days while it was 11th and 12th day for 95CS-5B-3 and 90CS-10B-3. These results showed the equilibrium time was affected by the initial water content such that a longer time is required to achieve equilibrium at a higher value for the initial moisture content.

Figure 3 shows the change in the weight by time at higher total suction of 39055.36 kPa for 95CS-5B in (a) and 90CS-10B in (b). The weights of the 95CS-5B-1, 95CS-5B-2, and 95CS-5B-3 were observed to be constant at the 7th, 8th, and 9th day while those for 90CS-10B-1 to 90CS-10B-3 were on the 8th, 9th, and 10th day. This means the time to reach equilibrium was relatively faster at high suction than lower suction.

Figure 3 also shows the sample's weight under equilibrium conditions was almost the same without differentiating the initial moisture content. This result is typical for samples which allowed equilibrium at a total suction of more than 41084.91 kPa.

3.2. Shrinkage behaviour of compacted claystone-bentonite mixtures

The soil shrinkage behaviour is usually depicted in the relationship between void ratio and water content [14–16, 26] and this is shown for claystone and bentonite mixture in Figure 4. The data is consistently displayed using different shape markers and colours as observed in the triangles, squares, and circles applied for the initial moisture contents at 10%, 15%, and 20% respectively. Moreover, green, blue and red colours were used to distinguish the percentage of bentonite at 100% claystone, 5% bentonite, and 10% bentonite respectively. The sample shrinkage curve is shown in the figure to only have two distinct stages which are primary and residual with no initial shrinkage obtained for all the samples tested. Generally, a sample with three stages usually has a high initial moisture content or become saturated in the swelling process as reported by Mirsha et al. [15]. A two-stage shrinkage sample was, however, found for the bentonite-sand mixture at optimum moisture content and maximum density conditions.

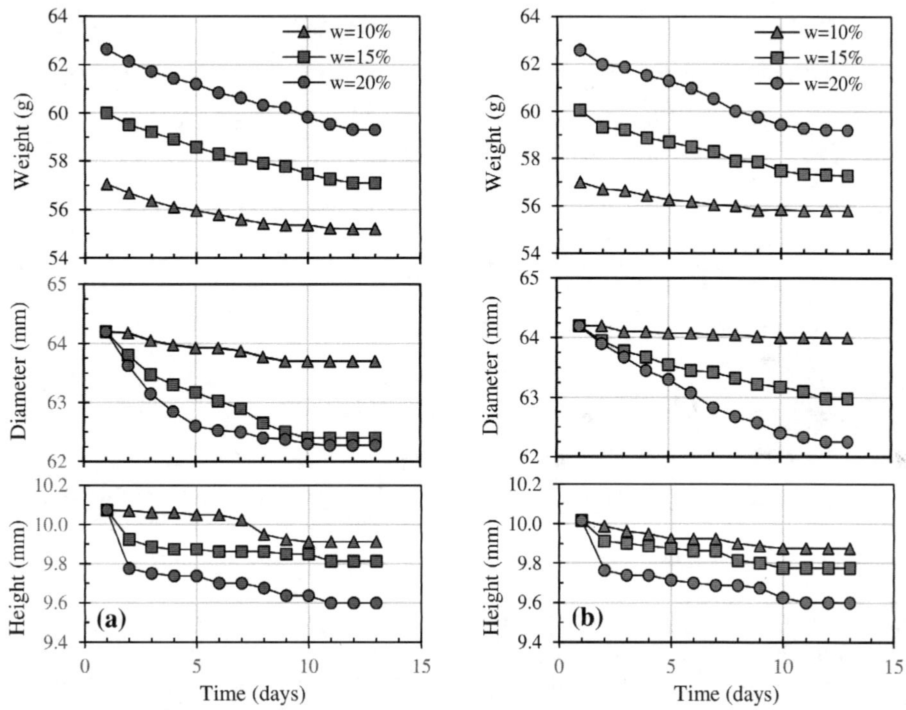

(a) (b)

Figure 2. Typical change of weight, diameter, and height of samples at total suction of 6848.64 kPa at (a) 95CS-5B, and (b) 90CS-10B

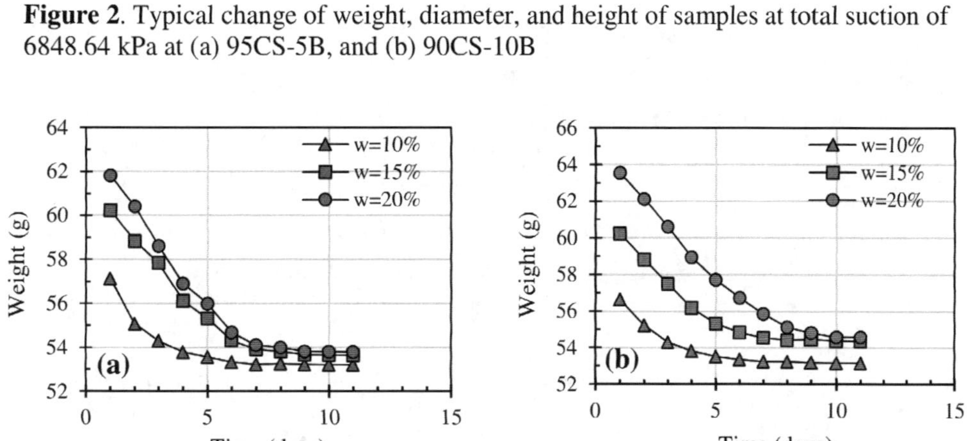

Figure 3. Weight by time curves at total suction of 41084.91kPa at (a) 95CS-5B, and (b) 90CS-10B

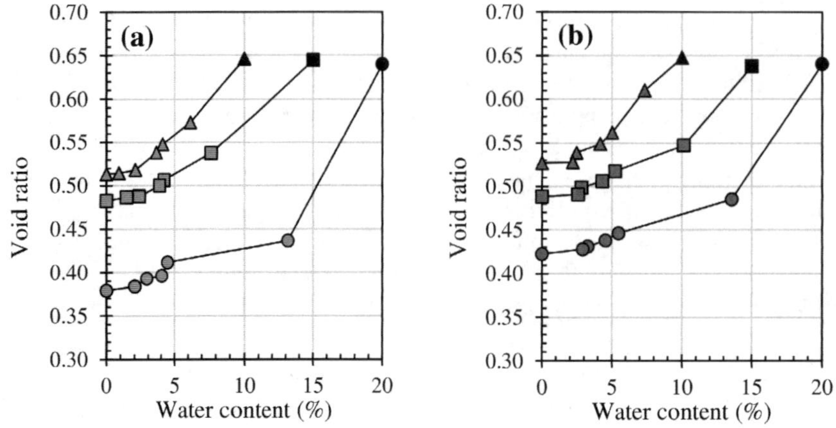

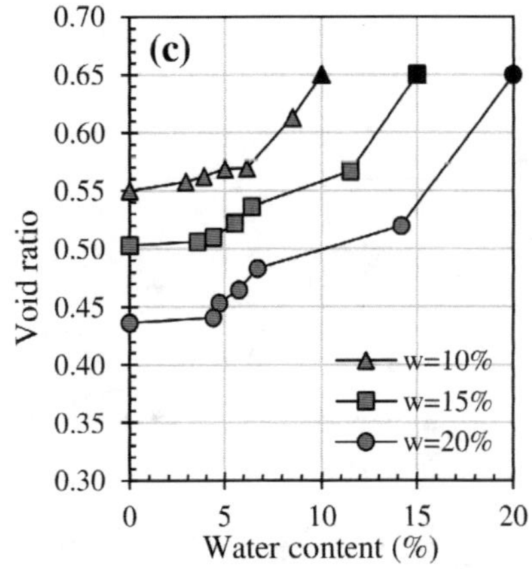

Figure 4. Water content-void ratio relationship of samples due to drying (a) 100CS (green), (b) 95CS-5B (blue), and (c) 90CS-10B (red)

The water content at which the primary shrinkage ended or the residual shrinkage started (w_r), as well as the magnitude of each, are summarized in Table 4. Moreover, the w_r, *PS, and RS* are used to represent the water content at residual shrinkage, primary shrinkage, and residual shrinkage, respectively. The information on Table 4 shows samples 100CS-1, 2, and 3 with an initial water content of 10%, 15%, and 20% have a w_r of 2,068%, 4,196%, and 4.47% respectively. The content at residual shrinkage was observed to have increased with the initial value and this was also the same for the claystone samples with bentonite such as 95CS-5B and 90CS-10B.

Table 4 also indicates the influence of the initial moisture content through the *PS* and *RS*. The 100CS-1, 2, and 3 experienced primary shrinkage of 19.79%, 21.37%, and 35.65%, and residual shrinkage of 0.9%, 4.74%, and 7.99%, respectively. At the same density, the *PS* and *RS* were recorded to have increased with the initial water content and the same trend was also obtained for claystone-bentonite mixtures.

Table 4. Shrinkage properties of claystone-bentonite mixtures

	100CS			95CS-5B			90CS-10B		
	1	2	3	1	2	3	1	2	3
w_r (%)	2.068	4.196	4.478	2.827	4.313	4.542	6.160	6.436	6.725
PS (%)	19.79	21.37	35.65	16.79	20.69	32.92	12.40	17.49	25.67
RS (%)	0.90	4.74	7.99	2.19	3.40	3.79	3.39	6.18	9.70

A comparison was made on the samples to determine the effect of bentonite at the same initial moisture content. The w_r of 100CS-1, 95CS-5B-1, and 90CS-10B-1 at initial content of 10% were 2,068%, 2,827%, and 6.16 while the primary shrinkage was 19.79%, 16.79%, and 12.40% and the residual shrinkage was 0.9%, 2.19%, and 3.39% respectively. These data showed the w_r and *PS* increased while *RS* decreased with an increase in the bentonite percentage. The samples with 10% and 20% initial water content also showed the change in void ratio as shown in Figures 5(a) and (b). The figures also show the effect of bentonite content on the shrinkage behaviour of the samples.

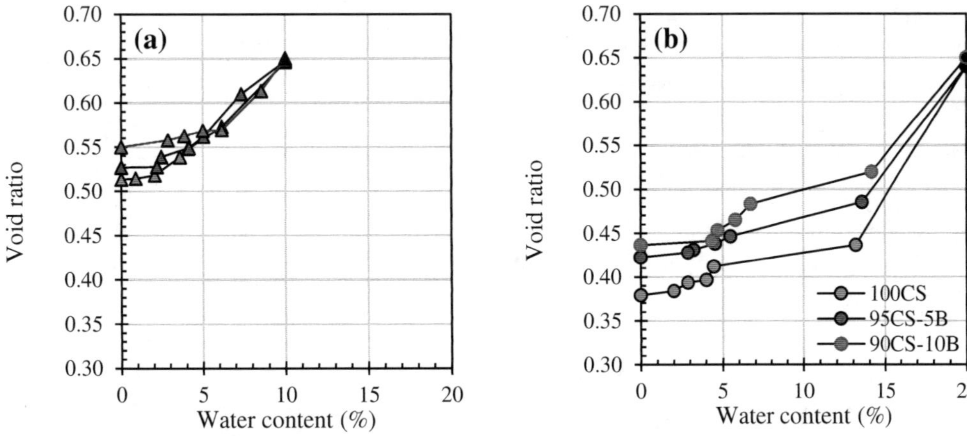

Figure 5. Water content-void ratio relationship of samples due to drying (a) w= 10%, and (b) w=20%

3.3. Water retention of compacted claystone and bentonite mixtures

The water retention curves of claystone and bentonite mixtures are indicated in the relationship between water content and total suction as shown in Figure 6. In consistence with the previous figures, the 100CS, 95CS-5B, and 90CS-10B are displayed in green, blue, and red respectively while the initial condition is in black. Moreover, those with an initial moisture content of 10%, 15%, and 20% are presented in triangle, square, and circle respectively. Figure 6(a), therefore, shows the water content of 100CS-1, 2, and 3 was initially 20%, 15%, and 10% and later decreased as the suction was increasing. Meanwhile, the amount of water content retained by the sample is almost the same at the total suction of 41084.91 kPa and the same trend is observed in other samples as shown in Figures 6(b) and (c). Arifin [16] found a similar result for bentonite and bentonite-sand mixtures. Furthermore, dry samples from saturated conditions have almost the same water content at suction higher than the air entry value (AEV) even though they are compacted at different water levels.

The difference in water retention curve at low total suction is due to the pore size distribution from the samples compacted at lower and higher moisture content than the required optimum. At the lower value, the macropores became dominant while it was the micropores at high values [16]. Moreover, water does not only fill the intra-particles or micropore but also the inter-particles or macropores at low total suction and also affected by the fabric. However, at the high total suction of 41084.91 kPa recorded in this study, the water was in the intraparticle or micropore and this reduced the influence of the fabric due to the adsorption of the water with hydration force by the clay surface.

Samples with the same conditions are plotted in a graph as shown in Figure 7 to determine the effect of the bentonite with those at 10% presented in Figure 7(a) and 20% in (b). The curve of the sample containing 5% bentonite is observed to be above the curve for the sample without bentonite or 100CS while the highest curve was recorded with 10% content. Moreover, the water content was found to have increased with the bentonite content at the same total suction and similar behaviour is present in Figure 7(b) showing samples with higher bentonite content have the ability to absorb more water.

IOP Conf. Series: Materials Science and Engineering 1212 (2022) 012029 doi:10.1088/1757-899X/1212/1/012029

4. Conclusions

The results of study on the shrinkage properties and water retention of compacted claystone and bentonite mixtures have been presented. The sample shrinkage curve has only two distinct stages which are the primary and residual shrinkage. Water content at residual shrinkage was observed to have increased with the initial water content. The amount of primary and residual shrinkages also increased with the sample's initial moisture content at the same density. The Primary shrinkage also increased with the bentonite content while the residual shrinkage was observed to have decreased. The water retention curve was strongly influenced by the initial moisture content at a total suction lower than 41084.91 kPa while the ability to hold water was almost the same at higher values. The results shows that the addition of bentonite to the claystone material increases the ability of the sample to hold water and a higher bentonite content holds more water content at the same total suction.

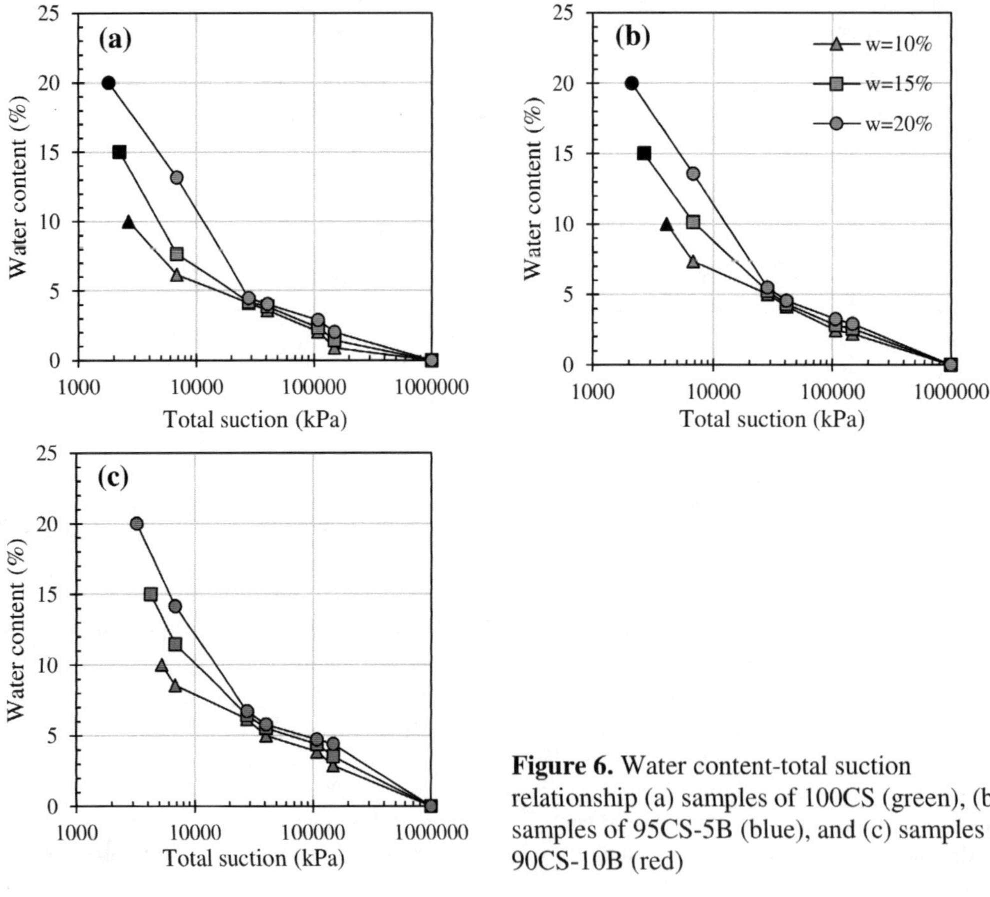

Figure 6. Water content-total suction relationship (a) samples of 100CS (green), (b) samples of 95CS-5B (blue), and (c) samples of 90CS-10B (red)

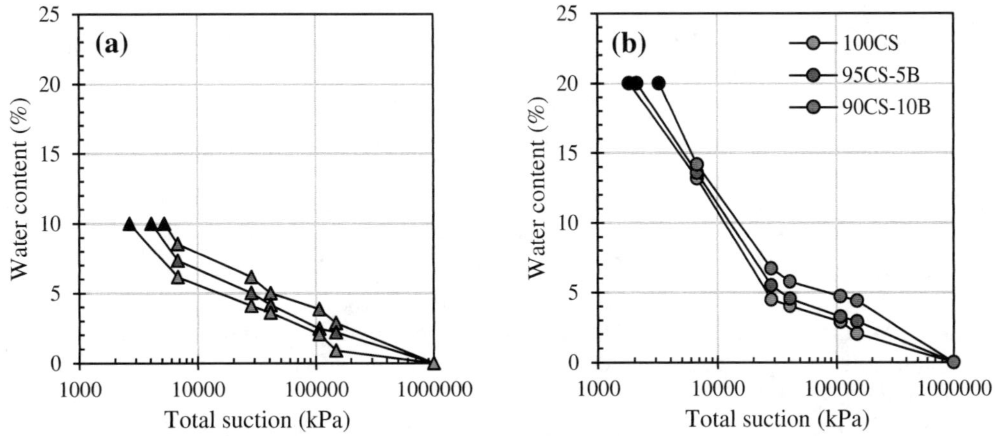

Figure 7. Bentonite effect on water retention behaviour of samples with the initial water content of (a) 10% and (b) 20%.

Acknowledgment
The authors are grateful for the financial support from the University of Lambung Mangkurat through "Dosen Wajib Meneliti" Program.

References

[1] Espitia J M, Caicedo B and Vallejo L 2019 Comparison of the Uniaxial Compressive Strength of the Belencito Claystone Under Stress Control and Suction Control Paths *Rock Mech. Rock Eng.* **52** 19–34 doi: 10.1007/s00603-018-1588-9

[2] Wan M, Delage P, Tang A M and Talandier J 2014 The water retention properties of the Callovo-Oxfordian claystone *Unsaturated Soils Res. Appl. - Proc. 6th Int. Conf. Unsaturated Soils, UNSAT 2014* **2** 1011–16

[3] Liu Z and Shao J 2016 Moisture effects on damage and failure of bure claystone under compression *Geotech. Lett.* **6** 182–6

[4] Özbek A and Gül M 2015 The geotechnical evaluation of sandstone–claystone alternations based on geological strength index *Arab. J. Geosci.* **8** 5257–68 doi: 10.1007/s12517-014-1541-5

[5] Sadisun I A, Shimada H, Ichinose M and Matsui K 2005 Study on the physical disintegration characteristics of Subang claystone subjected to a modified slaking index test *Geotech. Geol. Eng.* **23** 199–218 doi: 10.1007/s10706-003-6112-6

[6] Zhang C-L 2017 *Sealing Performance of Fractured Claystone and Clay-Based Materials* Gesellschaft für Anlagen- und Reaktor- sicherheit (GRS) gGmbH

[7] Wang Q, Tang A M, Cui Y J, Delage P and Gatmiri B 2017 Experimental study on the swelling behaviour of bentonite/claystone mixture *Eng. Geol.* **124** 59–66 doi: 10.1016/j.enggeo. 2011.10.003.

[8] Zhang C L 2014 Characterization of excavated claystone and claystone-bentonite mixtures as backfill/seal material *Geol. Soc. Spec. Publ.* **400** 327–37 doi: 10.1144/SP400.28

[9] Zhang C L and Kröhn K P 2019 Sealing behaviour of crushed claystone–bentonite mixtures *Geomech. Energy Environ.* **17** 90–105 doi: 10.1016/j.gete.2018.09.004

[10] Gu Y, Ding W, Yin S, Yin M and Xiao Z 2018 Adsorption characteristics of clay minerals in shale *Pet. Sci. Technol.* **36** 108–114 doi: 10.1080/10916466.2017.1405031

[11] Arifin Y F and Sambelum 2019 Bentonite Enhanced Soil as an Alternative Landfill Liner in Rikut Jawu, South Barito *IOP Conf. Ser. Earth Environ. Sci.* 239 doi: 10.1088/1755-1315/239/1/012003.

[12] Arsyad M, Mochtar I B, Mochtar N E and Arifin Y F 2020 Road embankment full-scale investigation on soft soil with geotextile stabilization *Int. J. GEOMATE.* **19** 145–152 doi: 10.21660/2020.71.04022.

[13] Correia S, Caldas R and Oluremi J R 2020 Feasibility of using CDW fine fraction and bentonite mixtures as alternative landfill barrier material *J. Mater. Cycles Waste Manag.* doi: 10.1007/s10163-020-01075-6.

[14] Cornelis W M, Corluy J, Medina H, Díaz J, Hartmann R, Van Meirvenne M and Ruiz M E 2006 Measuring and modelling the soil shrinkage characteristic curve *Geoderma.* **137** 179–191 doi: 10.1016/j.geoderma.2006.08.022.

[15] Mishra A K, Dhawan S and Rao S M 2008 Analysis of swelling and shrinkage behavior of compacted clays *Geotech. Geol. Eng.* **26** 289–298 doi: 10.1007/s10706-007-9165-0

[16] Arifin Y F 2008 *Thermo-hydro-mechanical Behavior of Compacted Bentonite Sand Mixtures: An Experimental Study* Bauhaus Universitaet Weimar, Germany

[17] Tang A-M and Cui Y-J, 2005 Controlling suction by vapour equilibrium technique at different temperatures, application to the determination of the water retention properties of MX80 clay *Canadian Geotechnical Journal* **42** 287-96

[18] Wu D, Li J and Zhang C Total Suction Characteristics of Reticulate Red Clay under Isopiestic Humidity Control *EDGE* **23** 613–628.

[19] Pintado X, Autio J and Punkkinen O 2010 Vapour Equilibrium Technique in Oedometer Tests: Measure, Control, and Numerical Simulation *Clay in Natural Engineered Barriers for Radioactive Waste Confinement 4th International Meeting* 307–308

[20] Pintado X, Lloret A and Romero E 2009 Assessment of the use of the vapour equilibrium technique in controlled-suction tests *Can. Geotech. J.* **46** 411–423 doi: 10.1139/T08-130

[21] Agus S S, Arifin Y F, Tripathy S and Schanz T 2013 Swelling pressure-suction relationship of heavily compacted bentonite-sand mixtures *Acta Geotech.* **8** doi: 10.1007/s11440-012-0189-0

[22] Blatz J A, Cui Y J and Oldecop L 2008 Vapour equilibrium and osmotic technique for suction control *Geotech. Geol. Eng.* **26** 661–673 doi: 10.1007/s10706-008-9196-1

[23] Agus S S and Schanz T 2007 Errors in Total Suction Measurements *Exp. Unsaturated Soil Mech.* 59–70 doi: 10.1007/3-540-69873-6_6

[24] Agus S S and Schanz T 2007 Comparison of Four Methods for Measuring Total Suction *Vadose Zo. J.* **4** 1087–1095 doi: 10.2136/vzj2004.0133

[25] Rahardjo H, Nong X F, Lee D T T, Leong E C and Fong Y K 2018 Expedited soil-water characteristic curve tests using combined centrifuge and chilled mirror techniques *Geotech. Test. J.* **41** 207–217 doi: 10.1520/GTJ20160275

[26] M. Wijaya, E. C. Leong, and H. Rahardjo, "Effect of shrinkage on air-entry value of soils," *Soils Found.*, vol. 55, no. 1, pp. 166–180, 2015, doi: 10.1016/j.sandf.2014.12.013

Development of Concrete Damage Classification in Beam-Column Joint based on Ultrasonic Pulse Velocity

A S Dermawan [1], S M Dewi [2], Wisnumurti [2], A Wibowo [2]
[1] Civil Engineering Department, Achmad Yani University Banjarmasin, Banjarmasin, Indonesia
[2] Civil Engineering Department, Brawijaya University, Malang, Indonesia

Corresponding author email: as_dermawan@yahoo.com

Abstract. The concreted conditions assessment of the systems is an essential aspect of security assessment programs. In situ measurements of Ultrasonic Pulse Velocity (UPV) may be indicative of the level of damage in the original concrete. UPV influenced by the specific characteristics of the mixture. In situ UPV measurements can be indicative of the level of damage in the original concrete. The research purpose is the damage classification, UPV test interpretation (strength, density, elasticity modulus, Concrete Quality Designation (CQD)), and determines the level of structural damage visually so that more accurate inspection results. The research result showed that the plastic hinge was more damaged than other parts of the beam-column joints. The UPV test obtained density 0.84-1.03 g/cm^3, CQD 10% -20%, static elastic modulus 7.68-8.39 Gpa according to [3],[4] including very poor and visually is included in category IV spalling off of covering concrete (crack width > 2mm). The use of UPV as supporting assessment for classification, repair, and maintenance of structures. If density, CQD, and elastic modulus of defining very poor classification, the structure that needs immediate repair. The use of UPV is faster, without damaging parts of the structure, and also induces damage to the core specimens as a result of the coring process, making it faster and more economical.

Keywords: concrete_damage, UPV, NDT, structure_assessment

1. Introduction

Many types of damage occur in concrete, including the expansion of thermal, alkali-silica reaction (ASR), plastic shrinkage, settlement, and electrochemical corrosion. The methods of testing are needed to find damage in concrete structures due to determine the internal integrity of these structures. A nondestructive test (NDT) methods for damage detection are available for assessing concrete strength in-situ. UPV methods were used to estimate the in-situ strength of concrete, density, and modulus of elasticity. The velocity of the wave propagation in the UPV method to identify the material quality and obtaining relatively conclusive results. The research method for improving test results was recorded by the ACI 228.1R-03[1] and ACI 318-14[2]. Recent studies have also indicated that UPV tests provide accurate results for estimating compressive strength and are influenced by different concrete properties[3],[4],[5]. The study uses the UPV approach, which is mostly accurate, efficient and scalable, which allow for an in-depth examination of material homogeneity. It is possible to assess concrete uniformity, to regulate its consistency, to track corrosion, to verify the existence of internal crack and voids, to allow comparisons with reference specimens, and to measure the capacity for compressive strength.

2. Nondestructive test (UPV) methods

The UPV approach has been known to provide a reliable means of measuring properties for some time and provide a unique potential for simple, precise, fast, safe, low cost and non-invasive quality control of buildings and other concrete destroyed by earthquake, fatigue or other catastrophic scenarios [6]. The UPV relis on the stiffness of the substances and its elastic properties. UPV is often used to calculate the overall strength of the material and to determine its elastic properties [7]. One of the uses of UPV is the information of the Stiffness (Elasticity Modulus) [8],[9],[10],[11].

3. Damage classification of structural members

Classification of damage undertaken for the identification and measurement of damage to columns and beams. The load capacity of the vertical load and the lateral resistance of the column adjacent to the affected beam is reduced. Harm evaluation shall be conducted until the harm is observed in the projector. The definition of the injury type and the summary of the injury can be found in Table 1.

Table 1. Damage Class Definition of R.C. Columns and Beams [13]

Damage Class	Description of Damage
I	- Visible narrow cracks on a concrete surface (Crack width is less than 0.2 mm)
II	- Visible clear cracks on a concrete surface (Crack width is about 0.2 -1.0 mm)
III	- The local crush of concrete cover - Remarkable wide cracks (Crack width is about 1.0 - 2.0 mm)
IV	- The remarkable crush of concrete with exposed reinforcing bars - Spalling off of concrete cover (Crack width is more than 2.0 mm)
V	- Buckling of reinforcing bars - Cracks in core concrete - Visible vertical and lateral deformation in columns and/or walls - Visible settlement and/or leaning of the building

4. Materials and Experimental Program

The fourth specimen of reinforced concrete beam-column joints that weakness after hitting the ultimate load was examined for this analysis to assess the relative conditions of concrete based on measured pulse path length in the beam-column joints can be seen in **Table 2**.

Table 2. Description of beam-column connection specimens [12]

Description	Beam	Column
Dimensions (mm)	150 x 200 x 1000	200 x 200 x 750
f'c (MPa)	21	21
Longitudinal reinforcement	4 Ø 13	4 Ø 13
Fy (MPa)	580	580
Stirrup	Ø 8 – 100	Ø 10 – 100
	Ø 8 – 50	Ø 10 – 50
Fy (MPa)	440	440

The beam-column connection specimen is divided into three zones, namely the beam-column connection area, beams along the X direction of the plastic hinge, and beams along the Y direction of plastic hinge. The classification of beam-column connection specimens, namely Monolithic (SK-A0), and Non-monolithic with notches and support-plate (A1-B1). The damage specimen is examined visually with UPV to achieve the degree of damage in the main beam-column joint, the x-direction beam, and the y-direction in the plastic hinge area show in Fig. 1.

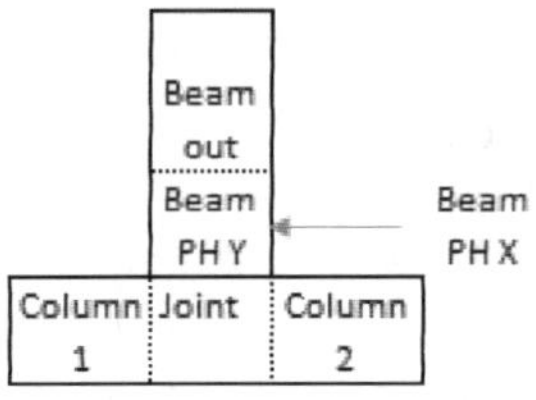

Fig. 1. Test location on reinforced concrete beam-column joints with UPV

5. Results

5.1 Visual Inspection

Results of visual inspection specimens of reinforced concrete beam-column joints damaged:

- SK.A0 Specimen

The SK.A0 specimen is a beam-column joint specimen that is made monolithically and has damaged after experiencing a peak load of 1.686 kg. A 5 cm x 5 cm grid specimen is made to facilitate visual observation. Visually damaged specimens have cracked on grid 2 with crack width of 1-2 mm, grid 4 with crack width <2 mm, and grid 6 with crack width < 2 mm. On the beam and column joints, cracks occur from the left to right side view with crack width> 2 mm. On the right side, a concrete cover was destroyed, and reinforcement was visible, the description seen in Fig. 2.

Front view Back view Right view Left view

Fig. 2. The Damaged Specimen of SK.A0.L1

- A1.B1 Specimen

The A1.B1 specimen is a specimen beam-column joint that made non-monolithically with notch and pedestal plate strengthening and has damaged after having experienced a peak load of 1.662 kg. A 5 cm x 5 cm grid specimen is made to facilitate visual observation. Visually damaged specimens have cracked on grid 3 with crack width > 2 mm and grid 7 with crack width < 2 mm. In the beam and column junction, cracks occur from the left to right side view with crack width > 2 mm and cracks on the column surface, and on the right side, a concrete cover destroys on grid 1 to grid 2, On the right side looks a shattered concrete cover (Fig. 3).

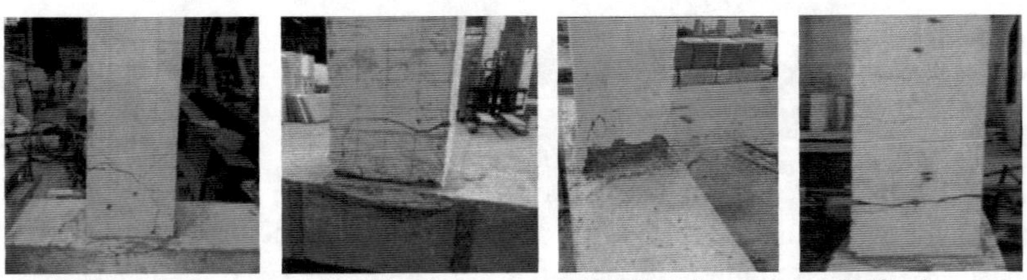

Front view Back view Right view Left view

Fig. 3. The Damaged Specimen of A1.B1.L3

5.2 NDT Inspection Using UPV

The UPV wave propagation test results for each section of the beam-column joints seen in **Table 3**.

Table 3. Test Result of UPV wave propagation test

Name of Specimens	Location	Rec. Num	Column 200 x 200 mm			Beams 150 x 200 mm		
			transit time (μs)	width (mm)	P-Wave (m/s)	transit time (μs)	width (mm)	Wave (m/s)
SK.A0	Joints	1	45.8	200	4258			
		2	35.8	200	5447			
	Beam PH	3				74.8	150	2607
	Y-Direct	4				120.8	150	1614
		5				62.8	150	3105

Name of Specimens	Location	Rec. Num	Column 200 x 200 mm			Beams 150 x 200 mm		
			transit time (μs)	width (mm)	P-Wave (m/s)	transit time (μs)	width (mm)	Wave (m/s)
	Beam PH	6				120.8	200	2607
	X-Direct	7				133.8	200	1614
A1.B1	Joints	1	49,8	200	5221			
		2	47,8	200	5439			
	Beam PH	3				155.8	150	1252
	Y-Direct	4				67.8	150	2876
		5				38.8	150	5026
	Beam PH X-Direct	6				116.8	200	2226

5.3. Interpretation of results

To classify the level of damage, the authors propose a classification of concrete damage with the use of UPV and damage classification:

SK.A0. Beam-column joints SK.A0.L1 in the x-direction have an average P wave UPV of 2.05 km/s (very Poor) and an average P wave in the y-direction of 2.44 km/s (Poor) in the plastic hinge region. The concrete beam in the x-direction has a density (ρ) 1 gr/cm^3, CQD 18.25%, concrete strength of 2.6 MPa and static elastic modulus (Es) of 8.3 Gpa and dynamic elastic modulus (Ep) of 8.68 Gpa while in the y-direction density (ρ) 1.26 gr/cm^3, CQD 31.4%, concrete strength 3.91 Mpa and static elastic modulus (Es) 9.4 Gpa and dynamic elastic modulus (Ep) 12.35 Gpa. The Nakano classification is included in category 4 Spalling off of covering concrete (crack Width > 2 mm) then the cracked specimen is examined in detail by tapping the concrete blocks in x and y directions with the hammer on grid 1, and 2 and experience spalled concrete loose or unsound concrete so has to be removed, can be seen in **Fig.4**.

Fig. 4. The specimen after knocked chek with the hammer

A1.B1. The results of the UPV wave in the x-direction, the average P wave is 2.23 km/s (Poor), and the average P wave in the y-direction is 3.05 km/s (Questionable) in the beam of the plastic joint region. The concrete beam in the x-direction has a density (ρ) 1.12 gr/cm^3, CQD 24.2%, concrete strength 3.13 MPa, Es (8.7 GPa) and Ep (10.26 GPa) while in the y-direction density (ρ) 1.67 gr/cm^3, CQD 51.71%, concrete strength 7.32 MPa, Es (12.2 Gpa) and Ep (19.28 Gpa). The Nakano classification is included in category 4 Spalling off of covering concrete (crack width > 2 mm), cracked specimens are examined in detail by tapping concrete blocks x and y in direction with a hammer as high as 15cm and experiencing spalled concrete loose or unsound concrete so has to be removed, can be seen in **Fig. 5.**

Fig. 5. The specimen after knocked chek with the hammer

The relationship between UPV wave test with ρ (density), Sn (Strength of Concrete), Stiffness of concrete with Ec (Static Modulus of elasticity) and Ep (Dynamic Modulus of elasticity), CQD (Concrete Quality Designation) with damages classification for visual assessment of post Earthquake Damage Evaluation shows on **Table 4.**

Table 4. Relation Damage Class with Interpretation of UPV wave Test

Range UPV Km/s	Conc. Quality	Sn Mpa	Ec GPa	Ep GPa	ρ g/cm^3	CQD	Damage Class
>4.5	Excellent	>32.57	>27.96	>41.92	>2.63	>100	
3.6 - 4.5	Good	12.89 - 32.57	16.07 - 27.96	26.83 – 41.92	2.03 - 2.63	70 - 100	I
3.0 - 3.6	Questionable	6.95 - 12.81	11.89 – 16.07	18.63 – 26.83	1.63 – 2.03	50 - 700	II
2.1 - 3.0	Poor	2.75 - 6.95	8.39 – 11.89	9.13 – 18.63	1.03 – 1.63	20 - 50	III
1.8 - 2.1	Very Poor	2.02 - 2.75	7.68 – 8.39	6.71 – 9.13	0.84 – 1.03	10 - 20	IV
<1.8	Abnormal	<2.02	<7.68	<6.71	<0.84	<10	

6. Discussion

Many case studies on concrete structures indicate that the UPV analysis is suited to assess the degree of concrete damage. The findings, presented mainly in terms of UPV variations, help spot anomalies within a structure. When the building structure is damaged due to earthquake/overload, one of the quick ways to inspect the building is to classify each part of the structure visually. The use of UPV is faster, without damaging parts of the structure, and also induces damage to the core specimens as a result of the coring process, making it faster and more economical. The UPV test obtained density 0.84-1.03 g/cm^3, CQD 10%-20%, static elastic modulus 7.68-8.39 Gpa including very poor and visually is included in category IV spalling off of covering concrete (crack width > 2 mm)[3]. The use of UPV as supporting assessment for classification, repair, and maintenance of structures. If density, CQD, and elastic modulus of defining very poor classification, the structure that needs immediate repair.

7. Conclusion

The analysis of UPV and visual (rapid vulnerability assessment) to assess the structural condition of the beam-column joint). The study showed that the plastic hinge was more damaged than other parts of the beam-column joint and estimated the concrete quality. The use of UPV, along with the visual observations proposed here, provides additional information so that the lack of visual observations can complement the use of UPV for the inspection of concrete parts. The use of UPV wave interpretation (density, CQD, Strength & elastic modulus). Higher accuracy for estimating the concrete quality with higher accuracy. The use of UPV, along with the visual observations suggested here, offers additional details so that lack of visual observations can complement the use of UPV for the inspection of concrete parts.

Acknowledgements

The research was conducted at the Brawijaya University for a laboratory project with the BPPDN grant program and the research grant program of The Ministry of Research, Technology, and Higher Education of the Republic of Indonesia.

References

[1] ACI Committee 228, "In-Place Methods to Estimate Concrete Strength Reported", American Concrete Institute. p. 44, 2003.
[2] ACI Committee 318-14, "Building Code Requirements for Structural Concrete (CAI 318–08) and Commentary", American Concrete Institute, Farmington Hills, MI,. 2014.
[3] F. Saint-Pierre, and A. Philibert, "Concrete Quality Designation based on Ultrasonic Pulse Velocity", *Cons. And Building Material*, 125. 1022-1027, 2016.

[4] DM. McCann, and MC Forde, "Review of NDT methods in the assessment of concrete and masonry structures", *NDT & E Int.* 34 (2) 71–84. 2001.
[5] J. Zhang, and X. Gao, "The detection and analysis of reservoir dam crack depth and grouting quality analysis", *Appl. Mech. Mater.* 204–208. 3329–3332, 2012.
[6] E. Marfisi, CJ. Burgoyne , MHG. Amin and LD. Hall, "The use of MRI to observe the structure of concrete", *Magazine of Concrete Research.* Vol. 57. No. 2, pp. 101–109, 2005.
[7] C. Vipulanandan, and V. Garas, "Electrical resistivity, pulse velocity, and compressive properties of carbon fibre reinforced cement mortar". *Journal of Material in Civil Engineering*, pp. 93-101, 2008.
[8] ACI International, BRE, Concrete Society, and ICRI, "Diagnosis of Deterioration in Concrete Structures Concrete Repair Manual", 2nd Edn, 1. 988-1027. 2003.
[9] AK. Chandrappa, and KP. Biligiri. "Influence of mix parameters on pore properties and modulus of pervious concrete: an application of ultrasonic pulse velocity". *Mater.Struct.*, 49, (12). 5255–5271. 2016.
[10] D. Barazanchy, W. Roth, and V. Giurgiutiu, "A nondestructive material characterization framework for retrieving a stiffness matrix using bulk waves". *Composite Structures.* 185. 27-37. 2018.
[11] S. Shahbazpanahi, F. Hejazi, M. Paknahad, A. Rahimipour, and MR. Nassimi. "Modelling Crack Propagation in RC Beam-Column Joints". *Teh. Vjesn.* 25 (4). 1183–1189. 2018.
[12] NCE. Yuliati, SM. Dewi, A. Wibowo, and Wisnumurti., "Experimental Investigation of crack on a Non-Monolithic Exterior Beam-Column Joint Of Reinforced Concrete Under Static Load", *Intern.J.of Civ.Eng and Tech.* 9. 2929-2937. 2018.
[13] Y.Nakano, M.Maeda, H.Kuramoto, and M.Murakami, "Guideline for post-earthquake damage evaluation and rehabilitation of rc buildings in Japan", World Conference on Earthquake Engineering Vancouver, B.C., Canada August 1-6, 2004

.

IOP Conf. Series: Materials Science and Engineering 1212 (2022) 012031 doi:10.1088/1757-899X/1212/1/012031

Climate change impact to dam Operation, case study of Darma Dam, West Java

A A Kuntoro[1,5], D T Ramadhani[2], A M S Idris[3], M Farid[4], M B Adityawan[4], Widyaningtias[1] and G O N Annisa[1],

[1] Centre for Water Resources Development, Institut Teknologi Bandung, Indonesia
[2] Independent Monitoring Team (IME), Dam Operational and. Improvement Safety Project (DOISP), Indonesia
[3] Directorate of Irrigation and Water Resources, Ministry of National Development Planning (BAPPENAS), Indonesia
[4] Centre for for Marine and Coastal Development, Institut Teknologi Bandung, Indonesia

[5]corresponding author: arnoak@ftsl.itb.ac.id

Abstract. Darma Dam is located in the upstream of Cisanggarung River, Kuningan Regency, West Java Province. Darma Dam construction dated from about 1922. Indonesian government continued the construction and began operating in 1962. With effective storage of about 40 million m^3, Darma Dam provides water for about 22 thousand irrigation areas and bulk water for several cities and regencies. Several problems encountered in Darma Dam operation and water management are 1) increasing water demand from domestic and industrial sectors, and 2) high inflow variation during the dry and wet season, resulting in a large amount of water spill from the dam spillway. This paper addressed the impact of climate change on the inflow variation of Darma Dam in the dry and wet seasons. Further analysis shows average water spills from the spillway during the wet season may increase from about 12 million m^3/year in the present condition to about 20 million m^3/year in 2020-2050, while the average water volume during the dry season may reduce from 22.5 million m^3 in the present condition to about 20.7 million m^3 in 2020-2050. This study suggests that dam operation need adjustment in the future as part of adaptation to climate change.

1. Introduction

West Java Province is one of the primary rice producers in Indonesia. In 2018, this province produced 9.5 million unhusked rice, nearly 17% of Indonesia's total rice production. However, rice production facing challenges from rapid urban development, high population growth, increasing domestic and industrial water demand, which compensated by more water extraction from surface water and groundwater [1], [2], [3]. Climate change may also increase the complexity of water resources problems. As intense extreme rainfall and prolonged drought tend to appear more frequent, the conflict between water users such as agriculture, domestic, and industry may intensify in the future [4]. Darma Dam is one of the dams in West Java with the primary purpose of irrigation water supply. Besides agriculture, this dam is also used for fish farming (using floating net cages), a raw water source of Kuningan Regency Water Supply Company, and a tourist destination. The decline in the service functionality of Darma Dam will give the highest economic losses to the agriculture and fisheries sectors [5]. Dam inflow may be profoundly affected by climate change as the rainy season tend to be wetter and dry season tends to be dryer. With the decreasing effective volume due to sedimentation, the probability of water spills from the dam's spillway during the rainy season and water level drop during the dry season may increase in the future. Therefore, it is necessary to update the dam operation to adapt to climate change, including the Darma Dam operation.

2. Materials and Method

2.1. Location

The maximum water surface of Darma Dam is about 0.425 km^2, with effective storage in the about 40 million m^3. Darma Dam is the primary water source for about 22,000 hectares of the irrigation area in the downstream of the dam. The location of the Darma Dam is shown in Figure 1.

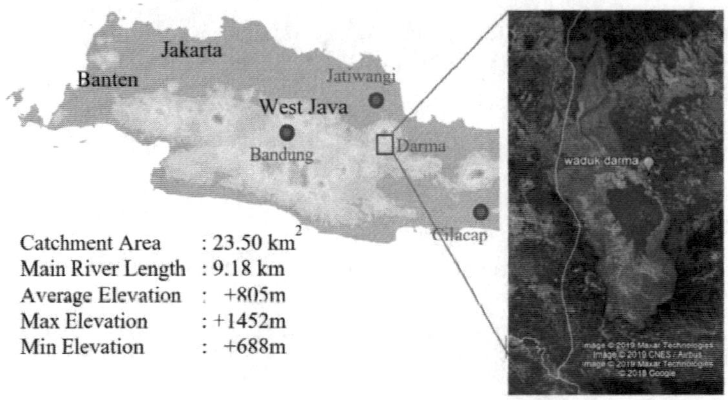

Figure 1. Location of Darma Dam

The average annual rainfall in Darma Dam Rainfall station from 2009-2018 is about 2680mm/year. The average monthly rainfall in the rainy season is about 364mm/month, while the average monthly rainfall in the dry season is about 83mm/month. Based on data from Dam Operation and Improvement Safety Project (DOISP), in 2013, about 54% of the catchment area of Darma Dam was rice fields. Other land-use types, such as production forest, settlement, dry farmland, and mixed plantation, covered about 38% of the catchment area, with nearly a similar proportion. Waterbody covered about 8% of the total Darma Dam catchment area.

2.2. Climate Projection

The World Climate Research Programme's (WCRP) Working Group on Coupled Modelling (WGCM) started to promote the fifth phase of the Coupled Model Intercomparison Project (CMIP5) in 2008. CMIP5 provides a multi-model context for assessing climate change feedbacks, predictability, and range of responses among the climate models [6]. This study compares 15 CMIP5 model outputs with the average rainfall of 21 stations in West Java Province for the 1980-2009 rainfall data series. CMIP5 datasets are available on the KNMI Climate Explorer website [7] (https://climexp.knmi.nl). This study uses only RCP8.5 for rainfall future prediction scenario.

Two main methods for downscaling climate data are dynamical and statistical. While dynamic downscaling uses a physical-based climate model to get higher resolution output, statistical downscaling uses a statistical relationship between climate variables. Reference [8] use Multivariate Adapted Constructed Analogs downscaling method in a case study of Western US, which gives better results than direct interpolation method due to its ability to capture relative humidity and winds more accurately. The study shows that the statistical downscaling method performs well to capture variations in meteorological parameters. However, data reconstruction based on statistical correlation requires a fairly long time series of observation station data (e.g.,> 10 years). In this study, data samples from more than 20 rainfall stations from 1980-2009 (30 years) covering the Upper Citarum Basin in West Java were used as basis for downscaling calibration.

Reference [9] use four statistical downscaling methods to downscale the native-scale output of NCEP/NCAR Reanalysis into a 12km grid. This study uses wet day fraction, extreme events, and weather patterns as comparison criteria for the downscaled data. Reference [10] use three statistical downscaling methods to downscale daily precipitation over the Yellow River Region, comparing spatial dependence, wet and dry spell, and inter-annual variability. This study suggests that more than one downscaling method may give a better interpretation of climate change projection better. Reference [11] uses fifteen different RCP models and is downscaled using different statistical downscaling methods to study the projection of extreme river flows in 11 catchments in Europe. References [9], [10], and [11] use several methods to downscale a global model output, i.e., NCEP/NCAR reanalysis data. Reference [9] and [10] use the downscaled results to evaluate climate change scenarios based on IPCC AR4 (2007). This study takes a different approach, where one downscaling method is used to evaluate several GCM outputs based on IPCC AR5 climate change scenarios (2014).

Reference [12] uses HBV and XAJ rainfall-runoff model to simulate discharge in Upper Hanjiang Basin, China, using statistically downscaled GCM data as input. This study suggests using multiple GCM outputs and downscaling methods to study climate change's impact on runoff. HBV and XAJ models are rarely used in Indonesia. Therefore, this study uses the The National Rural Electric Cooperatives Association (NRECA) model [13] as one of the rainfall-runoff models widely used in Indonesia. Model calibration uses inflow-outflow data of Darma Dam. The bias-corrected data from the selected CMIP5 model for 2030 to 2050 are used as input of the NRECA Rainfall-Runoff Model for future river discharge projection. No further land cover change is assumed from 2015 to 2050.

3. Result and Discussion

3.1. *Bias Correction of CMIP5 Rainfall*

CMIP5 provides climate change scenario simulations from several models. This study uses average rainfall data from 1980-2009 as a reference for bias correcting CMIP5 rainfall data output. The steps for preparing the bias correction are as follows: 1) Create a frequency duration curve (FDC) of historical data; 2) Create a CMIP5 and FDC graph in the same data period; 3) Create a correction equation to produce CMIP5's FDC close to historical data's FDC.

Goodness of fit assessment of CMIP5's FDC and historical data's FDC is performed using NSE and Pearson Correlation Coefficient. Comparison between monthly historical rainfall data and GCM output during the period of 1980-2009 is used to select which GCM output is best to be used as input for future discharge projection, which is shown in Table 1.

Table 1 shows that the Multi-Model Mean of 32 CMIP5 models computed in KNMI gives the best statistical fitness in NSE (=0.206) and r (=0.600) compared to other CMIP5 output. Based on this result, Multi-Model Mean CMIP5 data is used to simulate future discharge using NRECA Rainfall-Runoff Model.

Data obtained from CMIP5 is monthly rainfall data. However, the rainfall-runoff simulation is carried out using daily data. In this study, the transformation of monthly data into daily data is carried out through the following steps: 1) split historical rainfall data from 1980-2009 into the criteria of wet and dry months 2) plot the daily data in point 1 into duration curve; 3) approximate the duration curve in point 2 using the gamma two-variable function; 4) approximate the function of the number of wet days in a month; 5) Generate daily rainfall using a Monte-Carlo Simulation using monthly rainfall input, the average number of rainy days and gamma functions at point (3); 6) Adjust sum of daily rainfall in a month to be equal with monthly rainfall data.

3.2. *Rainfall-Runoff Model Calibration and Future Discharge Simulation*

Daily river discharge data in Darma Dam in 2015 is used for NRECA Rainfall-Runoff Model calibration. A comparison between river discharge data and the calibrated model result is shown in Figure 3. Calibration by using daily data gives NSE=0.70 and r=0.85. Based on the above calibration,

IOP Conf. Series: Materials Science and Engineering 1212 (2022) 012031 doi:10.1088/1757-899X/1212/1/012031

the model then used for simulation of river discharge in 2030-2050. This simulation is conducted by assuming no further land cover change from 2015 data to 2050.

Table 1. Bias Correction Function of CMIP5 Model for Case Study of West Java

Model	Bias correction function	NSE	r
Multi Model Mean	$R=-2.9835P^3+3.6875P^2-1.9716P+1.1802$	0.206	0.600
ACESS1-0	$R= 1.2703P + 0.9798; P\leq 0.8$; $R= -0.1006P + 0.0971; P>0.8$	-0.300	0.363
ACESS1-3	$R=5.0503P^3 -7.8013P^2 + 1.7908P + 1.0325$	-0.648	0.155
MIRROC-ESM	$R=-0.1173P^2 - 0.9808P + 1.0703$	-1.012	-0.03
BCC-CSM1-1	$R= 3.3307P^2 +0.3694P + 0.8675; P\leq 0.6$ $R= -5.7172P^2 + 4.18P + 1.852; P>0.6$	-0.168	0.425
BNU-ESM	$R=-0.4762P^2 - 0.154P + 0.9437$	-0.075	0.460
CESM-BGC	$R=-1.7193P^3 + 1.7169P^2- 0.7929P + 0.9223$	0.009	0.504
CMCC-CM	$R=-1.673P^2 + 0.319P + 0.6213; P\leq 0.7$ $R=0.027; P>0.7$	0.064	0.540
FGOALS-g2	$R=-0.9816P^2 + 0.5058P + 0.9404; P\leq 0.95$ $R= -8.3356P + 8.4038; P>0.95$	-0.192	0.415
GFDL-CM3	$R=-0.0483x + 0.7986; P\leq 0.9$ $R= -2550P^3+7169P^2-6718.7P+2100; P>0.9$	-0.018	0.506
GFDL-ESM2g	$R=0.607P$	-0.055	0.454
GFDL-ESM2M	$R=1.84P^3-3.111P^2 +0.673P + 0.608; P\leq 0.7$ $R=0.000154e^{12.512P}$	-0.032	0.482
GISS-E2-Hp3	$R=-0.178P^3 -0.4569P^2 + 0.1584P + 0.5118$	-0.535	0.226
GISS-E2-RP3	$R=-3.961P^3 + 1.320P^2 + 1.620P + 1.0315$	-0.188	0.418
HadGEM2-A0`	$R=-2.635P^3 -0.688P^2+0.071P+1.11; P\leq 0.65$ $R=-10.3P^3 +27.88P^2-25.164P+7.58; P>0.65$	-0.215	0.405
HadGEM2-CC	$R=4.192P^3-6.391P^2+1.191P+0.912; P\leq 0.7$ $R= 0.0004707P^{-13.322}$	-0.127	0.446

R = monthly rainfall (mm), P = probability of occurrence

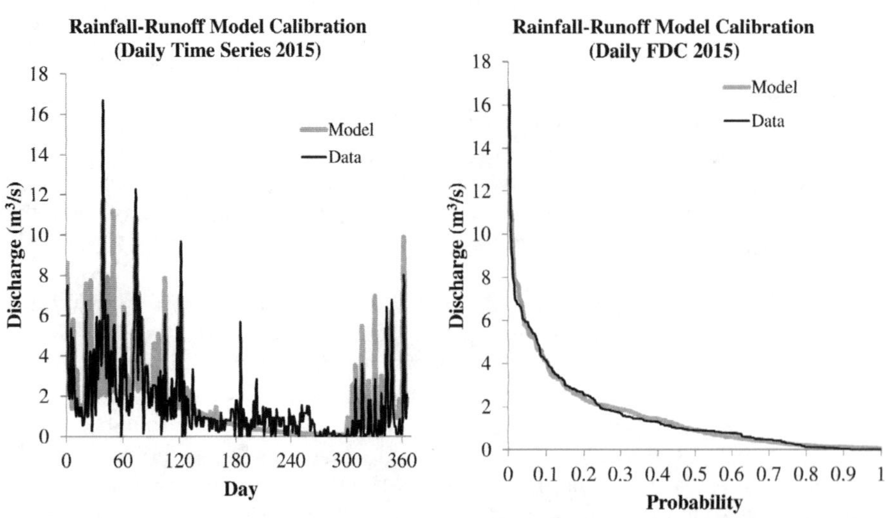

Figure 2. Rainfall-Runoff Model Calibration

Figure 3 shows the trend of low flow decrease within the observation period 2009-2018 and the projection 2030-2050. Compared to data series 2009-2018, Q50% decreases from 1.34m^3/s to 1.11m^3/s. Meanwhile, Q80% decreases from 0.33m^3/s to 0.05m^3/s in 2030-2050. The simulation shows the trend of longer drought and wetter rainy season in the future compared to what had happened in the past.

3.3. Dam Operation

The average water level of Darma Dam, irrigation supply, and spillway discharge is shown in Figure 4. Figure4 (left) shows that the average water level in Darma Dam is at a high level from April to June. Figure 4 (right) shows the discharge from the tunnel for irrigation supply and from the spillway. The average volume of water spills from the spillway in a year is about 12.2 million m^3. During this period the average volume of water flows from the tunnel outlet to supply irrigation demand and volume of water from spillway were nearly equal. The water level started to drop from July to October, which corresponds to high water demand from irrigation during the third planting season. This figure shows that there is room for further water optimization, for example, if the water, which usually spills, can be used for other water users.

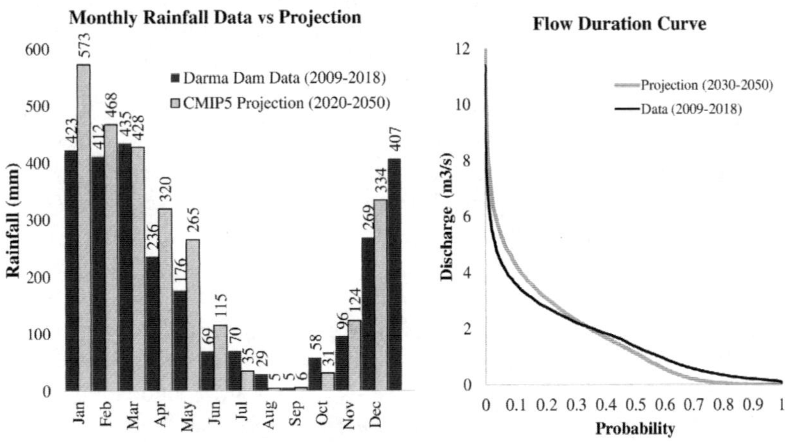

Figure 3. Future rainfall and discharge projection

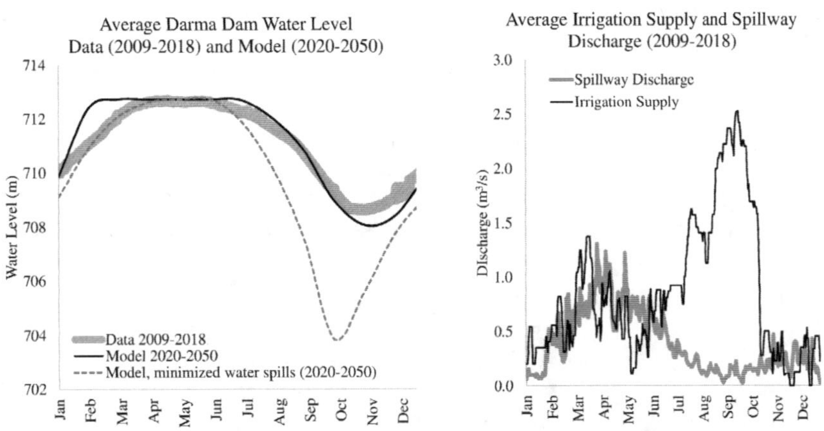

Figure 4. Future rainfall and discharge projection

Figure 4 (left) also shows a comparison between the average Darma Dam water level from 2009-2018, and simulated water level based on future discharge projection from 2020-2050. With the assumption of constant irrigation and domestic water demand, the probability of water spills and deficit may increase in the future. With the current dam operation rule, the average volume of water spills from the spillway may increase by 20 million m^3 in a year. Meanwhile, water elevation in the dry season (Sept-Nov) may decrease, lower than the level in the current condition. Water demand will continue to increase in the future. Based on the above simulation to minimize water spills from the spillway, Darma Dam can supply more water to the domestic sector up to 0.74 m^3/s. However, the water level in the dry season may drop 5m below the average level in the current operation rule. This simulation shows the importance of dam operation as one of the adaptation measures to climate change.

4. Conclusion

High economic development and population growth may lead to an increase in water demand. In this study, river discharge projection in the future is simulated by using NRECA Rainfall-Runoff Model, with data input of the bias corrected model output of the multi-model mean of CMIP5 for the RCP8.5 scenario. Simulation period of 2020-2050 shows that the average water spills from the spillway during the wet season may increase from about 12 million m^3/year in the present condition to about 20 million m^3/year in 2020-2050, while the average water volume during the dry season may reduce from 22.5 million m^3 in the present condition to about 20.7 million m^3 in 2020-2050. This study suggests that the probability of a long drought period in the Darma Dam may increase in the future. If land cover change is considered, the simulation result might give even lower discharge. There is still an opportunity to mitigate the impact of climate change by optimizing the present dam operation rule, for example, by preventing water spill out from the spillway and by managing water demand in the dry season.

References

[1] Nurcahyo H, Soekarno I, Hadihardaja I K, Rosyidie A, *Hydrologic Alteration in Watershed Using Flow Duration Curve, Case Study Upper Citarum Watershed, Indonesia* 2016 *International Proceedings of Chemical, Biological and Environment Engineering*, V94 24

[2] Julian M M, Nishio M, Poerbandono P, Ward P J, *Simulation of River Discharges in Major Watersheds or Northwestern Java from 1901 to 2006* 2011 *International Journal of Technology* **1** pp 37-46

[3] Agaton M, Setiawan Y, Effendi H, *Land use/land cover change detection in an urban watershed: a case study of upper Citarum Watershed, West Java Province, Indonesia* 2015, *LISAT-FSEM* Procedia Environmental Sciences **33** pp 654 – 660.

[4] Yoshida K, Azechi I, Hariya R, Tanaka K, Noda K, Oki K, Hongo C, Honma K, Maki M, and Shirakawa H, *Future Water Availability in the Asian Monsoon Region: A Case Study in Indonesia* 2013 *Journal of Developments in Sustainable Agriculture* **8** pp. 25-31.

[5] Pratama D S, Syaukat, Y, and Ekayani M *Estimasi Nilai Ekonomi dan Eksternalitas Negatif Pemanfaatan Waduk Darma* 2017 *Risalah Kebijakan Pertanian dan Lingkungan* Vol4 1 pp 13-27

[6] Taylor E T, Stouffer R J, and Meehl, G A, *An Overview of CMIP5 and the Experiment Design* 2012 *American Meteorological Society* pp 485-498.

[7] KNMI Climate Explorer https://climexp.knmi.nl.

[8] Abatzoglou J T, and Brown T J, *A comparison of statistical downscaling methods suited for wildfire applications* 2012 *Int. J. Climatol* **32** pp 772–780.

[9] Gutmann E, Pruitt T, Clark M P, Brekke L, Arnold J R, Raff D A, and Rasmussen R M, *An intercomparison of statistical downscaling methods used for water resource assessments in the United States* 2014 *Water Resour. Res* **50** pp.7167–7186

[10] Hu Y, Shreedhar M, and Uhlenbrook S *Downscaling daily precipitation over the Yellow River source region in China: a comparison of three statistical downscaling methods* 2013 *Theor Appl Climatol* **112** pp 447–460.

[11] Hundecha Y, Sunyer M A, Lawrence D, Madsen H, Willems P, Bürger G, Kiraučiūnienė J, Loukas A, Martinkova M, Osuch M, et al., *Inter-comparison of statistical downscaling methods for projection of extreme flow indices across Europe* 2016 *Journal of Hydrology* **541** pp 1273–1286

[12] Chen H, Xu, C, and Guo S *Comparison and evaluation of multiple GCMs, statistical downscaling and hydrological models in the study of climate change impacts on runoff* 2012 *Journal of Hydrology* **434–435** pp 36–45

[13] Crawford N H and Thurin S M 1981 *Hydrologic estimates for small hydroelectric projects* Small Decentralized Hydropower Program, National Rural Electric Cooperative Association

Acknowledgements
The authors would like to acknowledge the support given by the Ministry of Research and Technology of the Republic of Indonesia through 2020 Applied Research, DOISP and BAPPENAS.

The use of olive waste for development sustainable rigid pavement concrete material

M A Dahim[1], M Abuaddous[2], H Al-Mattarneh[3], A E Alluqmani[4] and R Ismail[5]

[1] Faculty of Engineering, King Khalid University, Abha, Saudi Arabia
[2] Civil Engineering Department, Yarmouk University, Irbid, Jordan
[3] Civil Engineering Department, Yarmouk University, Irbid, Jordan
[4] Department of Civil Engineering, Islamic University, Madinah, Saudi Arabia
[5] Civil Engineering Department, Jadara University, Irbid, Jordan

Hashem Al-Mattarneh the corresponding author : drhashem2010@yahoo.com ; hashem.mattarnch@yu.cdu.jo

Abstract. Recycle and reuse of agriculture and industrial wastes becomes a big chalenge in different parts of the world. The success in the waste recycle could lead to conserve the environment, reduce the use of cement, and improve health environment. This paper presents the potential use of fly ash from olive oil waste in Jordan to improve concrete material which could be used as a sustainable material for rigid pavement and building construction material. Olive oil ash was collected from olive oil mill and replace cement in producing concrete material. The range of cement replacement was 0% to 12.5% with increment 2.5%. The results indicate that olive oil reduces the workability of concrete material. The reduction of the slump of concrete increases with increasing olive ash content. Strength and durability of concrete improved and increased with increasing olive ash content in concrete up to 7.5 percent then the strength reduced. The results in this study show that the use of 7.5% was the optimum replacement of cement. This percent could produce concrete with higher strength and higher durability in comparison with the control concrete mix. Olive waste ash enhances both strength and durability because it reduces the effective water-cement ratio in concrete mix and filling the pore and void structure in concrete material. The benefits of this study could reduce the cost of concrete and recycle waste material and enhance concrete properties.

1. Introduction

Jordan is one of Mediterranean countries such as Spain, Italy, France Greece, Turkey, Tunisia, Syria, Lebanon. These countries have a vast olive tree plantation used to produce olive oil. The high demand for olive oil will continue to increase in the future leads to an increase in olive tree plantation. Jordan has large number of olive mills to produce olive oil. These mills produce a large amount of olive waste materials. Most of these wastes used in winter to produce heat and the by-product is olive oil waste fly ash. This ash is disposed in sanitary landfills. This may cause several environmental problems and pollute the ground water. Recycle these wastes may reduce these environmental problems and may also contribute to sustainable development of other construction materials such as concrete.

The concrete material is the most construction material used in the world. Concrete material mainly used in rigid pavement, road construction and as a building material. This material suffers from week strength and very porous structure make it deteriorate due to inclusion of liquids and chemical containing chloride. Several researchers investigate the use of additive to enhance its strength and durability. Fly ash used by many researchers to investigate its effect in improvement to the performance of concrete. If strength and durability of concrete enhanced using fly ash this could save a large amount of money in repair cost and rehabilitation of rigid payment and building constructed from concrete.

Fly ash is a by-product waste material produced from combustion of waste or combustion of materials to produce energy. Fly ash could be produced from coal combustion [1], municipal solid waste, burning of rice husk [2] and wheat straw ash [3] burning heavy oil [4]. Fly ash waste used to enhanced rigid pavement concrete material and building cement concrete material [5-7]. Fly ash waste used to enhance concrete durability [8]. In addition, fly ash waste used to enhanced asphalt flexible pavement material [9-18]. Limited studies were conducted to evaluate the effect of olive oil waste fly ash on the concrete properties. Olive oil waste ash used to improve properties of cement mortars [5,19], enhance alkali-Silica reaction of concrete material [20], and improve the performance of concrete exposed to elevated temperatures [21]. In addition, fly ash from industrial wastes was used to enhance properties of several types of concrete such as self-compacting concrete [22-24], light weight concrete [25-28] and geopolymer concrete [29-30].

To date, olive oil waste fly ash is dumped in sanitary land fill. This toxic fly ash is disposed of in landfills may leading to environmental problems. This study conducted to investigate the use of olive oil fly ash in rigid pavement concrete material on concrete performance such as workability, strength, and durability.

2. Materials and Methods

Olive oil waste was collected from Qumaym olive oil mill in Irbid, Jordan. The olive waste was burned in oven for two hours to ensure all wastes were converted to olive oil waste ash (OWA). The location of the olive wastes and the waste are shown in Figure 1. The OWA specific gravity was 2.05 and the diameter of the ash ranges from 1 to 10 μm. The chemical composition of the OWA is given in Table 1. Ordinary Portland cement was used to produced rigid pavement concrete material grade 30. The specific gravity of cement was 3.15. The chemical composition of cement is given in Table 1. Limestone aggregate with maximum aggregate size equal to 14 mm was used in this concrete. A control concrete mix was prepared with water cement ratio 0.5. Concrete control mix proportions were 360 kg/m^3 cement, 180 kg/m^3 water, 722 kg/m^3 fine aggregate and 1098 kg/m^3 coarse aggregate. The gradation of the aggregate is given in Figure 2. This particle size distribution of the aggregate meets the specification of ASTM 33- 93.

(a) (b)

Figure 1. Source of olive oil waste; (a) Jordan map shows the location of olive waste in Irbid; (b) Olive waste before combustion.

Table 1. Chemical composition of cement and olive oil waste ash (OWA).

Chemical Composition	Percentage in OWA (%)	Percentage in Cement (%)
SiO_2	21.0	20.5
Fe_2O_3	2.5	4.3
Al_2O_3	4.5	5.4
CaO	30.1	65.5
K_2O	30.0	-
Na_2O	0.4	-
SO_3	0.0	3.0
MgO	5.4	2.16
P_2O_5	6.0	-
LOI	2.81	3.0

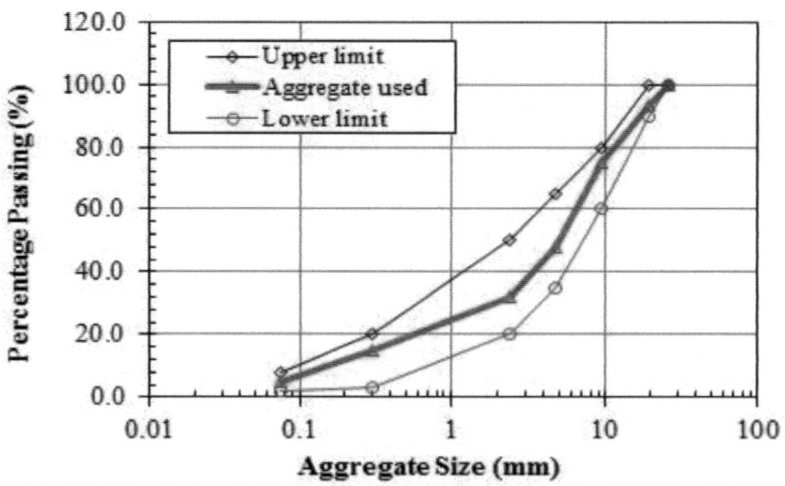

Figure 2. Gradation of limestone aggregate used in this study.

To evaluate the effect of adding OWA, five concrete mixes were prepared by partially replacement of cement in the control mix. The replacement or OWA content range from 2.5 kg to 12.5 kg with increment 2.5 kg.

Six concrete mix proportioning were produced with different percentages of OWA using ACI method [31]. For each mix cube specimens were prepared in the laboratory according to the standard [32]. Slump of the fresh concrete for all mixes were measured according to appropriate standard [33]. The concrete specimens were cured in water until the day of testing [34]. At 28 days of curing the concrete cubes tested for compressive strength and water absorption according standard test [35-36].

3. Results and discussion

To evaluate the effect of OWA on fresh properties and hardened properties of concrete material, slump, compressive strength and absorption of concrete mixes were tested and analyzed. The results of these properties are presented in the following subsections.

3.1. Effect of OWA on slump

The slump test was performed for all mixes with OWA content from 0% to 12.5%. samples of slump test are given in Figure 3. The results of slump of all mixes are presented in Figure 4. The results indicate that slump and workability of fresh concrete decreases with increasing OWA content. This may attribute to the smaller size of OWA and larger surface area. In addition, the specific gravity of OWA is less than specific gravity of cement. Replacement of cement by weight will lead to larger volume OWA which will consume more water from the mix. This will significantly reduce the workability as the slump reduced. Larger OWA content will cause a larger reduction in slump. The relationship between slump and OWA content could be modeled as a linear relationship and took the form:

$$Slump = \beta_0 + \beta_1 x \quad (1)$$

Where β_o and β_1 are model parameters. X is the OWA content. The model parameters and correlation coefficients are given in Figure 4. The square correlation coefficient was 0.9844. This indicate a good fit of the results using linear regression model.

Figure 3. Slump test of concrete containing OWA, (a) OWA 0%, (b) OWA 7.5%, (c) OWA 12.5%.

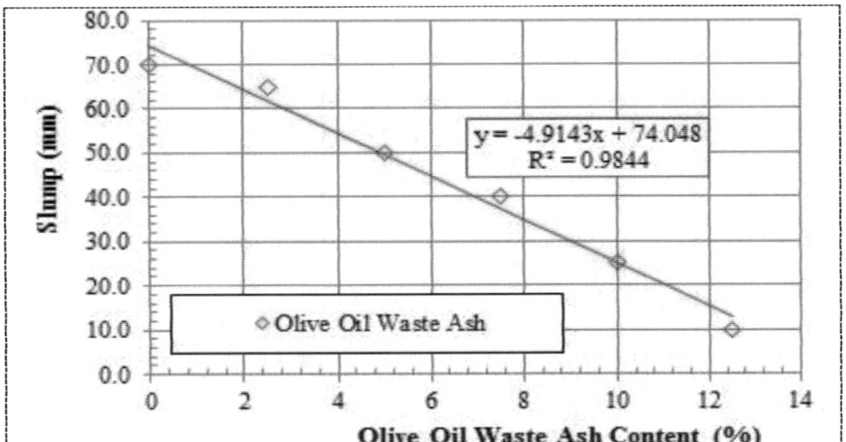

Figure 4. Relationship between slump value and OWA content.

3.2. Effect of OWA on compressive strength

The compressive strength at different time of curing for all concrete mixes was evaluated. The results of compressive strength versus curing time at all OWA content are given in Figure 5. The results indicate that compressive strength increases with increasing OWA content up to 7.5 percent.

Compressive strength decreases when the OWA content above 7.5 percent. These trends were observed over all curing time. The result show that compressive strength reaches higher value at OWA content 7.5 percent. The improvement in compressive strength because OWA has specific gravity less than cement therefore, the volume of OWA is larger than the same weight of cement. In addition, OWA has a smaller particle size and these reasons could reduce the effective water cement ratio of concrete containing OWA. Water cement ratio (w/c) is the most factors affecting strength of concrete and OWA reduce w/c leads to increase compressive strength according to Abraham laws.

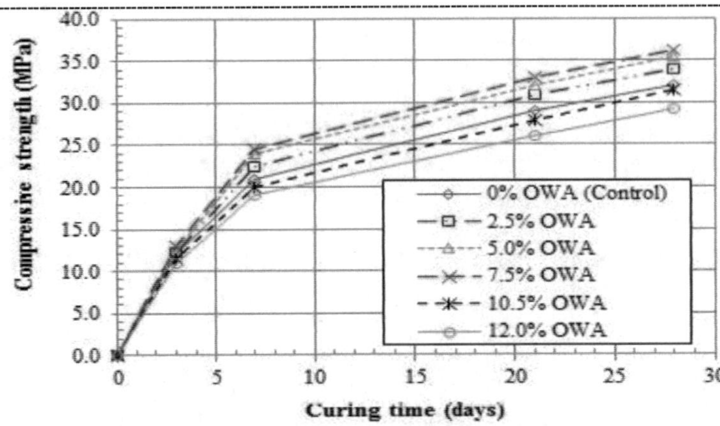

Figure 5. Compressive strength of concrete containing OWA at different curing time.

To better visualize the effect of OWA content on concrete compressive strength, the relationship between compressive strength of concrete and the OWA content was established using regression analysis. The results are shown in Figure 6. The best model to fit the result was quadratic formula given in equation 2.

$$Compressive\ Strength = \beta_0 + \beta_1 x + \beta_2 x^2 \quad (2)$$

Where β_o, β_1, and β_2 are model parameters equal to 31.807, 1.3957, and -0.1309 respectively. X is the OWA content. The square correlation coefficients R^2=0.8937.

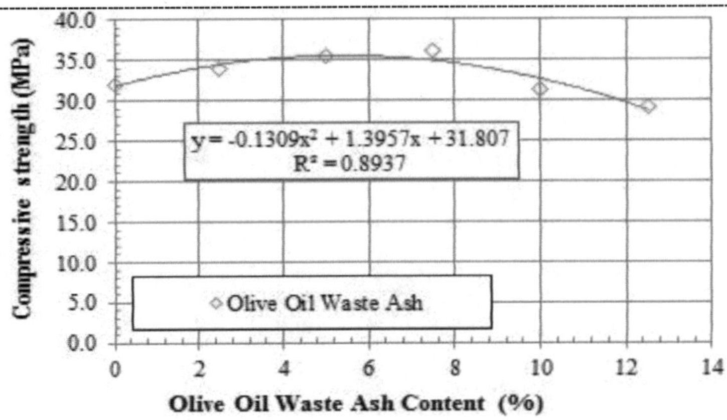

Figure 6. Relationship between concrete compressive strength at 28 days and OWA content.

3.3. *Effect of OWA on durability*

The durability of concrete was measured using concrete absorption. The absorption of concrete incorporated OWA is presented in Figure 7. The results indicate that absorption of concrete decreases with increasing OWA content until 7.5% then the trend reverse. This is because the fine particle of OWA could fill the voids and porous structure of concrete specially the interfacial zone. Also, OWA reduces effective w/c ratio which reduce the pore structure of concrete. This lead to enhance the durability of concrete.

To better visualize the effect of OWA content on concrete absorption, the relationship between absorption of concrete and the OWA content was established using regression analysis. The best model to fit the result was quadratic formula given in equation 3.

$$Absorption = \beta_0 + \beta_1 x + \beta_2 x^2 \quad (3)$$

Where β_o, β_1, and β_2 are the model parameters equal to 1.5432, -0.0691 and 0.0059, respectively. X is the OWA content. The square correlation coefficients R^2=0.8261. It is clear from the result that 7.5% of OWA is the optimum replacement content to enhance concrete durability.

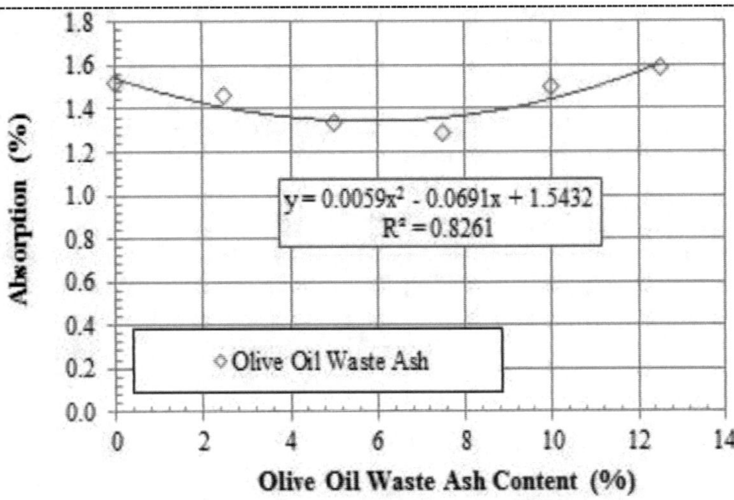

Figure 7. Relationship between concrete absorption at 28 days and OWA content.

4. Conclusions

The result in this study indicate the feasibility of using and recycling olive oil waste ash in concrete industry. Replacement of cement in concrete material used for rigid pavement and building construction could improve strength of concrete and enhance concrete durability. The results also indicate that 7.5 percent of olive waste ash is the optimum replacement of cement in concrete to achieve the best strength and durability. Olive waste ash enhance both strength and durability because it reduces the effective water cement ratio in concrete mix and filling the pore and void structure in concrete material. Olive oil waste ash content decrease the workability of concrete this problem could be solved by addition of superplasticizer to achieve the same concrete slump.

Funding statement
This research received no external funding.

Conflicts of interest
The authors declare that there is no conflict of interest regarding the publication of this paper.

References

[1] Leiva C, Vilches L F, Vale J, Fernández-Pereira C 2005 Influence of the type of ash on the fire resistance characteristics of ash-enriched mortars *Fuel* **84**, 1433–1439.

[2] Khana M N, Jamil M, Karimc M, Zain M 2014 Strength and durability of mortar and concrete containing rice husk ash: a review *World Appl. Sci. J.* **32** (5), 752–765.

[3] Al-Akhras N and Abu-Alfoul B, 2002 Effect of wheat straw ash on mechanical properties of autoclaved mortar, *Cem. Concr. Res.* **32** (6) 859–869.

[4] Al-Osta M A, Baig M G, Al-Malack M H and Al-Amoudi O S B 2016 Study of heavy fuel oil fly ash for use in concrete blocks and asphalt concrete mixes *Advances in Concrete Construction* **4** (2) 123-143. http://dx.doi.org/10.12989/acc.2016.4.2.123

[5] Alkheder S, Obaidat Y T and Taamneh M 2016 Effect of olive waste (Husk) on behavior of cement paste *Case Studies in Construction Materials* **5,** 19–25.

[6] Hardjito D and Rangan B V 2005 Development and Properties of Low-Calcium Fly Ash-Based Geopolymer Concrete *Research Report GC1*, Faculty of Engineering, Curtin University of Technology, Perth.

[7] Chindaprasirt P, Chareerat T and Sirivivatnanon V 2007, Worka-bility and strength of coarse high calcium fly ash geopolymer *Cement and Concrete Composocite* **29** (3), 224–229.

[8] Swaroop A H L, Venkateswararao K and Kondandaramarao P 2013 Durability studies on concrete with Fly ash and Ggbs, *Int. J. Eng. Res. Appl. (IJERA)* **3** (4), 285–289 (ISSN: 2248-9622).

[9] Parvez M A, Al-Abdul Wahhab H I, Shawabkeh R A and Hussein I A 2014 Asphalt modification using acid treated waste oil fly ash *Construction and Building Materials* **70** , 201–209.

[10] Woszuk A, Bandura L and Franus W 2019 Fly ash as low cost and environmentally friendly filler and its effect on the properties of mix asphalt *Journal of Cleaner Production* 235, **20**, 493-502

[11] Hoy M, Horpibulsuk S, and Arulrajah A 2016 Strength development of Recycled Asphalt Pavement – Fly ash geopolymer as a road construction material *Construction and Building Materials* **117** (1), 209-219

[12] Pasandín A R, Pérez I, Ramírez A, and Cano M M 2016 Moisture damage resistance of hot-mix asphalt made with paper industry wastes as filler *Journal of Cleaner Production* **112** (1), 20 853-862

[13] Laskar A I and Bhattacharjee R 2013 Effect of plasticizer and su-perplasticizer on rheology of fly-ash-based geopolyrner concrete *ACI Material Journal* **110** (5), 513–518.

[14] Xiao F, Shivaprasad P and Amirkhanian S 2012 Low-volume road wma mixtures: moisture susceptibility of mixtures containing coal ash and roofing shingle with moist aggregate. *Journal Material Civil Engineering* **24** (1), 48–56

[15] Zimmer F V 1970 Fly ash as bituminous filler. Washington (DC): United States Department of Interior, Bureau of Mines

[16] Jeffry S N A, Ramadhansyah P J, Abdul Hassan N, Yaacob H, and Mohd Satar M K I 2018 Mechanical performance of asphalt mixture containing nano-charcoal coconut shell ash *Construction and Building Materials* **173**, 40–48.

[17] Dahim M 2018 Enhancement the Performance of Asphalt Pavement Using Fly Ash Wastes in Saudi Arabia *Intern.ational Journal of Engineering & Technology* **7** (3.32), 50-53

[18] Dahim, M. 2018, The use of heavy oil fly ash as a filler in asphalt pavement mixture, Urban Transitions 2018 25-27 November, Sitges, Barcelona, Spain.

[19] Al-Akhras N M and Abdulwahid M Y 2010 Utilization of olive waste ash in mortar mixes, *Struct. Concr. J.* **11** (4) 221–228 (Thomas Telford, UK).

[20] Al-Akhras N M 2012 Performance of olive waste ash concrete exposed to alkali-Silica reaction, *Struct. Concr. J.* **13** (4) 221–226 (Wiley Group, USA).

[21] Al-Akhras N M, Al-Akhras K M and Attom M F 2009 Performance of olive waste ash concrete

exposed to elevated temperatures *Fire Saf. J.* **44** 370–375.
[22] Al-Qadi A, Mustapha N K, Al-Mattarneh H, Al-Kadi Q 2009 Central composite design models for workability and strength of self-compacting concrete *Journal of Engineering and Applied Science* **4** (3), 177-183.
[23] Narendra H, Muthu K U, Al-Mattarneh H M, Naidu N V R, Sowmya B S 2008 Optimization of self compacting concrete mixes using Taguchi method *International Conference on Construction and Building Technology*, Kuala Lumpur, Malaysia **1**, 191-204.
[24] Mohammed B S, Nuruddin M F, Aswin M, Mahamood N, Al-Mattarneh H 2016 Structural behavior of reinforced self-compacted engineered cementitious composite beams *Advances in Materials Science and Engineering* **2016**, 1-12. http://dx.doi.org/10.1155/2016/5615124
[25] Hassan A H, Al-Mattarneh H, Abdullahi M, Abu Hassan M, Mohammed B, Mustafa K 2008 Compressive Strength and Absorption of Concrete Composed of Natural and Palm Oil Clinker (POC) Aggregates *International Conference on Construction and Building Technology*, Kuala Lumpur, Malaysia **1**, 93-100.
[26] Abdullahi M, Al-Mattarneh H M A, Mohammed B S 2009 A MATLAB program for diagnosis and adjustment of mix proportions of structural lightweight concrete European *Journal of Scientific Research* **31** (1), 106-123.
[27] Abdullahi M, Al-Mattarneh H M A, Mohammed B S 2009 Statistical modeling of lightweight concrete mixtures *European journal of scientific research* **31** (1), 124-131.
[28] Alhasanat M B, Al-Qadi A N, Haddad H, Al-Mattarneh H 2016 Effect of Aggregate Size on the Engineering Properties of Palm Oil Clinker Concrete *GSTF Journal of Engineering Technology* (JET) **3** (4), 72-80.
[29] Nuruddin M, Malkawi A, Fauzi A, Mohammed B, Al-Mattarneh H Effects of alkaline solution on the microstructure of HCFA geopolymers, in: Zawawi N A (Ed.), Engineering Challenges for Sustainable Future: *Proceedings of the 3rd International Conference on Civil, Offshore and Environmental Engineering (ICCOEE)*, CRC Press, Taylor & Francis Group, London, 2016, pp. 501-505.
[30] Malkawi A B, Nuruddin M F, Fauzi A, Al-Mattarneh H, Mohammed B S Effect of plasticizers and water on properties of HCFA geopolymers *Key Engineering Materials* **733**, 76-79.
[31] ACI Committee 211.1-1991 Standard Practice for Selecting Proportions for Normal, Heavyweight and Mass Concrete. Detroit, American Concrete Institute.
[32] BS 1881: Part 108: 1983. Method for making test cubes from fresh concrete. British Standards Institution, Her Majesty Stationery Office, London.
[33] BS 1881: Part 102: 1983. Method for determination of slump. British Standards Institution, Her Majesty Stationery Office, London.
[34] BS 1881: Part 111: 1983. Method of normal curing of test specimen. British Standards Institution, Her Majesty Stationery Office, London.
[35] BS 1881: Part 116: 1983. Method for determination of compressive strength of concrete cubes. British Standards Institution, Her Majesty Stationery Office, London.
[36] BS 1881: Part 122: 1983 Method for determination of water absorption. British Standards Institution, Her Majesty Stationery Office, London.

Acknowledgements
The authors thank several universities for helping and allowing the researchers to use the lab facilities including King Khalid University, Alzaytoonah University, Islamic University in Madinah and Yarmouk University.

A new slip surface in noncohesive slopes

J A Repadi [1], F Bari[1], F A Ismail[2], Andriani[2], and A Hakam[2]

[1]Doctoral Program, Dept. Of Civil Eng., Faculty Of Engineering, Andalas University, Indonesia.

[2]Lecturer of Dept. Of Civil Eng., Faculty Of Engineering, Andalas University, Indonesia.

Email : julitaandrinirepadi@gmail.com

Abstract. Determining the slope slip surface is difficult, prominent and uncertain. Slip-surface are utilized to estimate slope safety. This study aimed at investigating the slip surface caused by statistical loads by using experimental testing methods. The experimental method was conducted by providing a statistical load on the sand slope The load was used until the slope collapsed. Sand slopes were modeled in a glass box measuring 110 cm x 40 cm x 10 cm. The sand material used was sand that passed through sieve no.10 and was held by sieve no. 200. At a distance of 2.5 cm horizontally and at a distance of 5 cm vertically, colored marks were used to see the pattern of grain movement. Observations were conducted until the slope collapsed. The collapse pattern revealed in the experimental method is non-circular.

Keywords :noncohesive, eksperimental, statis load, slip-surface.

1. Introduction

Every construction on or below the slope often involves slope stability analysis [1]. The analysis of slope stability is an unsolved problem in engineering mechanics [2]. In slope stability analysis, determining the depth, shape and slope of the slip surface is the most fundamental and important parameter[3]. It is very difficult to know the slip surface and is still a constraint for practicing engineering because the slip surface location is closed. Slip-surface determination remains an obstacle to this day. The slip surface is used to determine the slope safety factor. The safety factor of a slope will decrease if the slope force balance is disrupted [4]. This instability can occur due to external factors and internal factors that cause a decrease in the value of the safety factor [5]. There are 2 methods used to investigate slip surfaces, namely using experimental methods and numerical methods. Many researchers prefer to conduct the research numerically than experimentally because experimental research required expensive cost and also time consuming [5].

Researchers who have conducted numerical research were [6][7] [8] 9] [10] [11][12]][13]. [11] The landslide fields in the fellenius and bishop methods are assumed to be circular. The following is the slice surface that has been carried out by fellenius and bishop. The researchers assumed that the slice surface was rotational. Analytic methods of fellenius and bishop determine safety factors for the stability of slopes on homogeneous and layered soils using slope division methods with soil segments. Where the center point of the surface slip is determined by trial and error then describing the slip field in a trial and error manner [14].

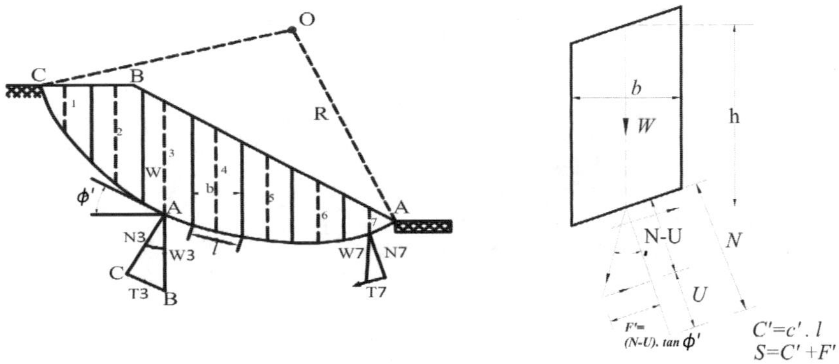

Figure 1. Slip circle for fellenius method.

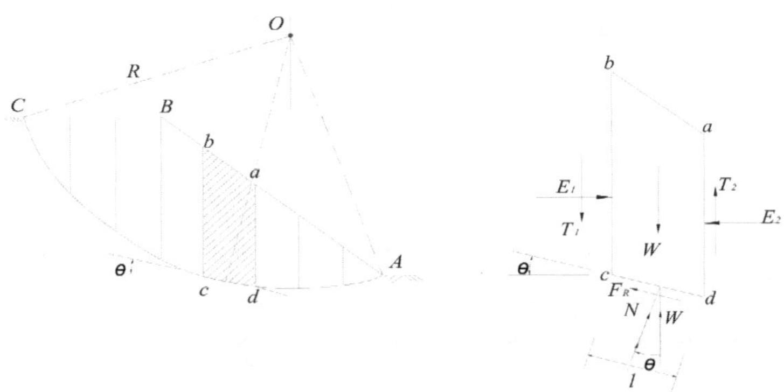

Figure 2. Slip circle for bishop method.

Researchers are constantly trying to locate surface slips. Using geophysical exploration methods, monitoring, mechanical search, logging, and numerical simulation methods obtained results of 5 possible surface slips, and still require a comprehensive method to determine the slip surface [3]. Finding a noncircular slice model with numerical methods [15]. The calculation of safety factors compared to the morgenstrean and spencer methods in which the new slip surface model is obtained has met all static equilibrium conditions, minimum safety factors, but slightly ignores the deformation.

This research was present to look for slip surface on noncohesive soils with experimental methods. Experiments are conducted carefully and carefully so that the results obtained are accurate.

2. Research Methodology

2.1 Research preparation process

Place the model slope on the floor and prepare the camera in front of and beside the model. In this study, the research preparation process was done by placing a glass box measuring 110 cm x 40 cm x 10 cm on a table, then on the front side and a camera was placed on the side to see the movement of the grain material [16]. Its are shown in **Figure 3.**

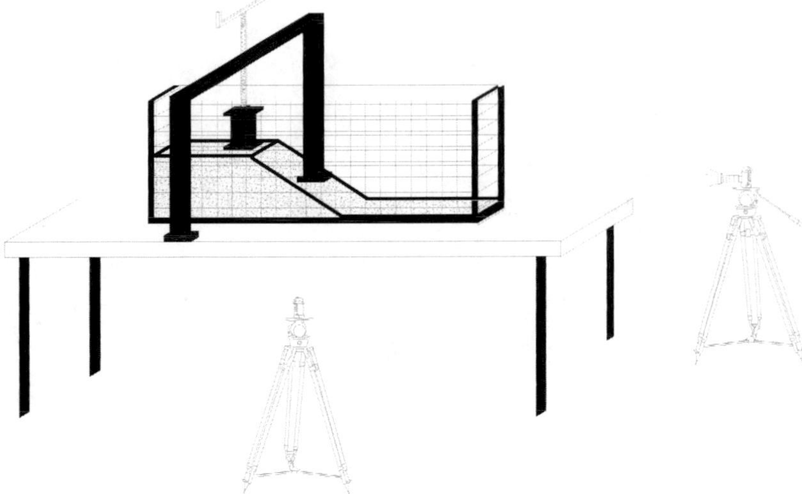

Figure 3. Table position, glass box, suppressor and camera.

2.2 Equipment used

The equipment used in the study were as follows:

1. Glass box measuring 110 cm x 40 cm x 10 cm with a glass thickness of 10 mm and used as a container for making slopes

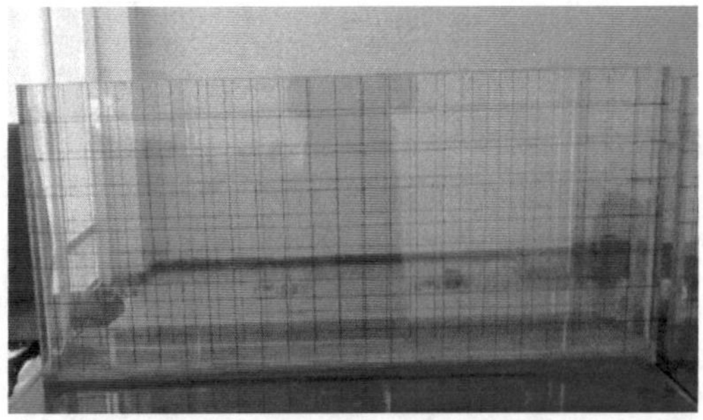

Figure 4. Glass box measuring 110 cm x 40 cm x 10 cm.

2. Pieces of iron used as static load

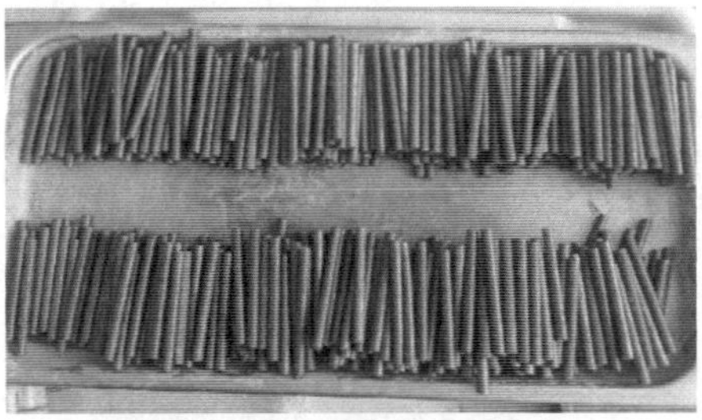

Figure 5. iron used as static load.

3. [17]Colored marks were utilized to see the movement of the grains when the load was used

Figure 6. Colored marks.

4. Loading cells were attached to the tool to measure the size of load that the presser put on the slope

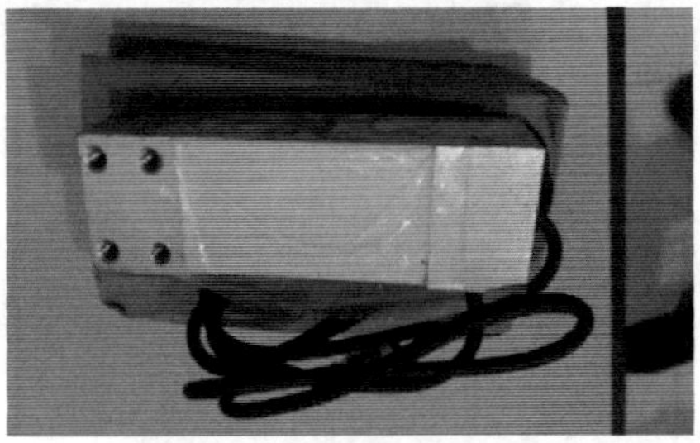

Figure 7. Loading cell.

IOP Conf. Series: Materials Science and Engineering 1212 (2022) 012033 doi:10.1088/1757-899X/1212/1/012033

5. Pressing tool (made by researchers)
   A pressing tool was used to provide additional loads on the slope so that the slope will collapse.

Figure 8. Alat penekan

6. Shieve Shaker was used to separate material based on grain size

Figure 9. Sieve no.10.

Figure 10. Sieve no.200.

Figure 11. Shieve Shaker.

2.3 Preparing material samples

Firstly, the sand was washed to remove sludge and other organic materials. The material used was fine sand, which passed through sieve no.10 and was held by sieve no.200. The sand was filtered using a shieve shaker.

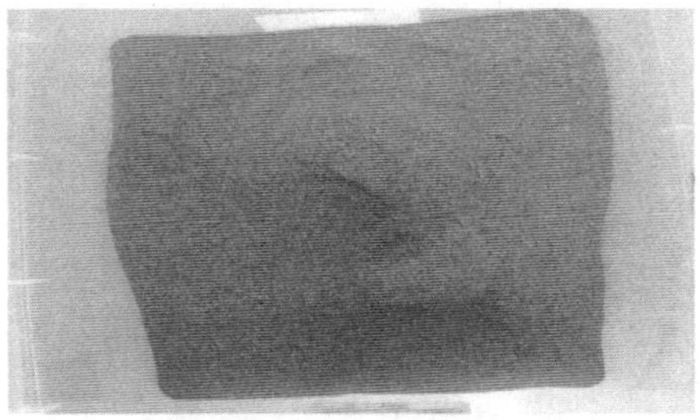

Figure 12. Sand escape sieve 10 restrained sieve 200.

2.4 Conducting soil parameter testing.

Soil property values total weight (γ) cohesion (c), and inner shear angle ($\emptyset$). It are shown in **Table 1.**

Table 1 . Soil property values

Type of Parameter	Value	Unit
Total unit Weight (γ)	1,45	Gram/cm^3
The Cohesion of Soil (c)	0	Gram/cm^3
Inner shear angel ( )	29,31	°

2.5 Load-giving experiments

Sand was poured into the glass box with the angle of the slope equal to the angle of shear in the slope. Color marks were given each 2.5 cm horizontally and 5 cm vertically. At the edge of the slope. a board was placed as a barrier so that the load acting on the slope was only vertically.

The loading experiment was conducted by arranging the pieces of iron, then placing the loading cell on top of the iron. The loading cell was pressed using a pressing tool. The loading cell produced the results of the load strength given by the pressing tool. Observations were made on the movement of the grains on the slope. The arrangement of the tools can be seen in **Figure 13** :

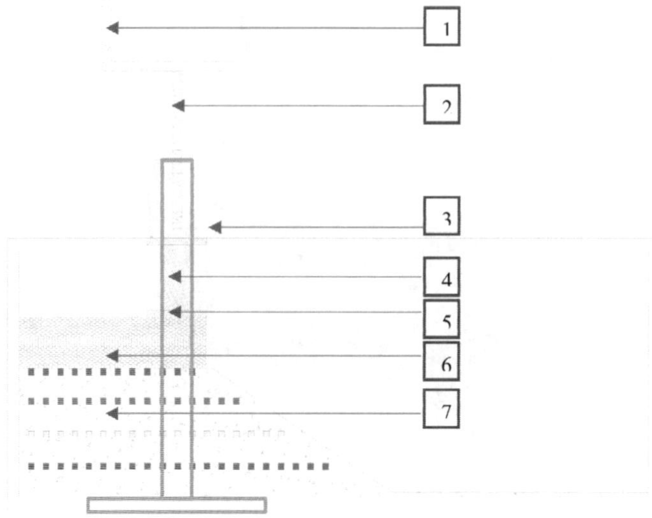

Figure 13. Model slope when slope is mounted suppressor

Note:

1	=	Crank
2	=	Bolt
3	=	Yellow barier
4	=	*Loading Cell*
5	=	Iron plate (load)
6	=	Iron (load)
7	=	Slope sand

3. Discussion

The thick pink lines indicate the collapse field that occurred as a result of the experiment, the pink lines indicate the collapse fields according to Fellenius and Bishop. The researchers overestimate the value of the slope safety factor [16]. It can be seen that the slice area on the slopes of the fellenius and bishop methods is larger than the experimental results (Figure 14). The collapse form that occurred in the experiment was noncircular, which is different from the Fellenius collapse form, bishop assumed the slice surface was rotational (Figure 15-16).

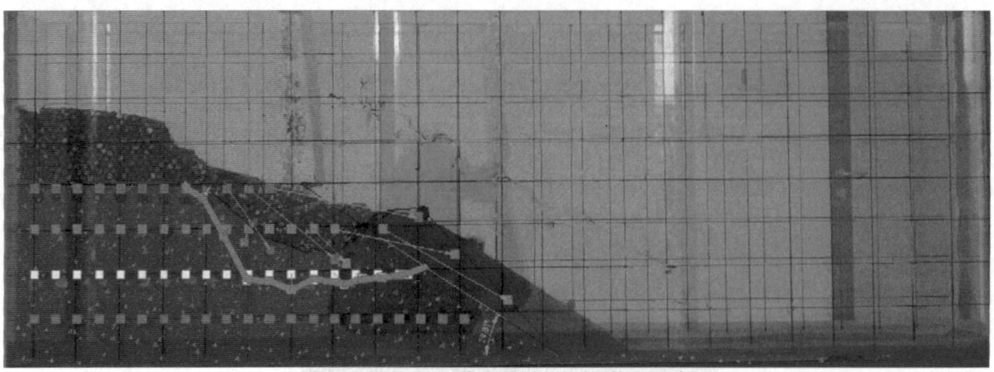

Figure 14. Experimental result slip-surface

IOP Conf. Series: Materials Science and Engineering 1212 (2022) 012033 doi:10.1088/1757-899X/1212/1/012033

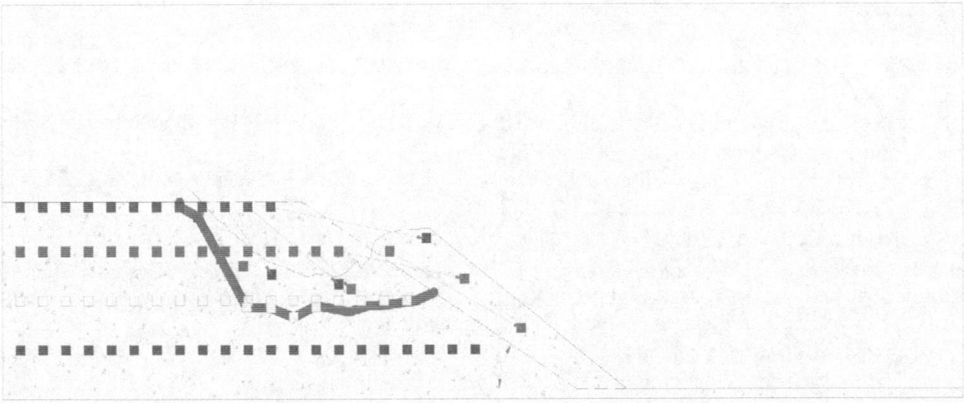

Figure 15. Slip surface experimental result

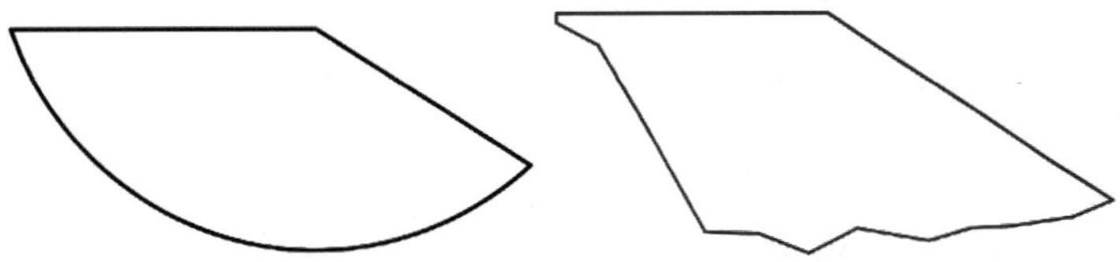

(a) (b)

Figure 16. (a) Slip surface results of fellenius and bishop, (b) Experimental slip surface

4. Conclusion

This laboratory research shows:

The slip surface found in this experiment the was non-circular, which is different from Fellenius' theory and the bishop is circular. It is suggested that future research studies calculate the safety factor with the new surface and compare it using numerical analysis

5. Acknowledgment

Thank you to Andalas University for the support provided and appreciation given, so that this experiment can run smoothly

References

[1] Liong G.T and Dave J.G.H 2012 Limit Equilibrium Vs Finite Element (Jakarta: HATTI –PIT-XVI) p

[2] Xue H et al 2018 Unified Overload Method of Slope Stability Analysis Based on Potensial Sliding Direction (China: KSCE) pp 3254-3262

[3] Xu G et al 2019 Determination Methods of A Tuff Landslide Slip-Surface (Bali: Proceedings of International Conference) pp K1-1 – K1-8

[4] Hakam A 2010 Stabilitas Lereng dan Dinding Penahan Tanah (Padang: Universitas Andalas Press) p

[5] Cen X and Wang H 2019 Slope Failure of Noncohesive Media Modelled with the Combined Finite-Discrete Element Method (China: applied sciences) pp 1-13

[6] Fellenius W (1936) Calculation of the stability of earth dams. In:Proceedings of the second congress of large dams, Wash- ington DC, vol 4, pp 445–463

[7] Bishop, A.W., 1955. The Use of the Slip Circle in the Stability Analysis of Slopes. *Geotechnique*, Vol. 5(1), pp. 7-17.

[8] Morgenstern N R and Price V E 1965 The Analisys of the Stability of General Slip Surfaces (London: The Institution of Civil Engineers) pp 79-93

[9] Spencer E 1967 A Method of Analisys of the Stability of Embankments Assuming Parallel Inter-Slice Forces (Inggris: Manchester University Institut of Science and Technology) pp 11-26

[10] Sarma S K 1973 Stability Analysis of Embankments and Slopes (London: Imperial College of Science and Technology) pp 423-433

[11] M Jafri et al 2020 Slope Stability Analysis in Ulubelu Lampung Using Computational Analysis Program (Lampung: Civense) pp 051-059

[12] Raj D and Yogendra S.2016. Effect of Building Loads on the Stability of Hill Slopes. ASCE.pp 638-647

[13] Janbu, N. 1973. *Slope Stability Computations. Embankment Dam Engineering - Casagrande Volume*, R.C. Hirschfeld and S.J. Poulos, eds., John Wiley and Sons, New York, pp 47-86.

[14] Paulus. 2015.Manual Kestabilan Lereng dan Penangan Longsor.Bandung.Edisi 3

[15] Liu T et al 2012 A New Closure to Slice Model for Slope Stability Analysis (China: Scientific Research) pp 394-399

[16] Tang L et al 2019 What is the Role of tensile cracks in cohesive slopes (China: Science Direct) pp 314-324

[17] Mazni, Deni Irda, *An Alternative Model Ofretaining Wallson Sandy Area To Prevent Landslides*, (jurnal ICEEDM, 2019)

IOP Conf. Series: Materials Science and Engineering 1212 (2022) 012034 doi:10.1088/1757-899X/1212/1/012034

The flexural strength of glue laminated timber beams based on deflection and strain rate with four point bending loading system

S Maricar[1], K Sulendra[2], H Listiawaty[3], and H O Baide

Department of Civil Engineering, Tadulako University, Palu, Central Sulawesi

Email: maricarshyama@gmail.com[1], ketutsulendra5@gmail.com[2], hlistiawaty@yahoo.com[3]

Abstract. The development of utilization of low quality wood as construction material is needed to reduce the exploitation of natural forests. However, low quality wood species have disadvantages in terms of mechanical properties. The mechanical properties of Sengon wood are relatively low, so it does not qualify as a structural element. Therefore, the system *glulam* can be applied to overcome this problem. The system *glulam* can produce relatively light structural elements with adequate performance. This system has been extensively developed, even at the stage of applying external reinforcement, to improve the performance of structural laminated beams. On that basis, this study aims to determine the flexural strength of laminated beams of Sengon wood as a low quality wood species.In order to achieve this goal, the laminated beam was tested using method *four point bending test* method. Tests were carried out on long span laminated beams (L = 2750 mm) to observe flexural strength. There are five (5) laminated blocks tested, namely (BLS-1, BLS-2, BLS-3, BLS-4 and BLS-5). Each group has dimensions of 55 mm in width and 155 mm in height. Each specimen consists of six layers of wood boards with a density Falcata 0.3 g / cm^3 The thickness of each layer was 26 mm and bonded with resin *urea formaldehyde* cold setting. Double-sided adhesive laying of 350 gr / m^2 at a compressive force of 2 MPa.The analysis result shows that the load-deflection relationship between BS-L consists of linear and nonlinear phases. The load performance characteristics of the two types of laminated beams are expressed as the ratio of the proportional limit load to the maximum load. The ratio value is expressed in the form $P_{e\,BL\text{-}S} = 0.7P_{max\,BL\text{-}S}$ and $M_{e\,BL\text{-}S} = 0.7M_{max\,BL\text{-}S}$. This form is similar to previous studies with a P_e to P_{max} ratio of 0.80.9. In this case, the average flexural strength of the laminated beam is 17 MPa with a maximum strain of 0.004.

Key words: ***Laminated Beam, Low Quality Wood, Flexural Strength***

1. Introduction

It is known that low quality wood has a relatively low weight. This of course will be very beneficial, especially to anticipate earthquake loads. However, the problem is that the inadequate performance of low quality wood is a major limitation of its use as structural beams. For this reason, lamination technology is an alternative that can be done to optimize the performance of wooden beams. The objective of this research is to determine the bearing capacity of beams made of low quality wood, so the problem of how laminates are treated to the flexural strength of low-quality beams and how is the difference in flexural strength of laminated beams based on deflection and strain data. Based on the background and problems that have been described, the specific objectives of this study are: Determine the change in the flexural strength of low-quality wood beams due to lamination treatment and determine the difference in the flexural strength of laminated wood beams based on deflection and strain data so that there are alternative raw materials for structural beams with different weights. small. and additional information for the development of construction material technology.

As a natural material, low quality wood has physical and mechanical properties with relatively low intensity. For example, Sengon wood, has a reference flexural strength value of ± 6000 MPa with a parallel tensile strength of 12,000 MPa [2]. Low quality wood is generally in the form of fast growing tree species, with a diameter of ± 60 cm. A log ductility analysis can generally be based on deflection (D) or curvature (f). Deflection is obtained through a series of vertical deflection measurements while curvature can be based on the relationship between strain and flexural stress.

To increase efficiency and reduce the effect of wood defects that are common in low-quality wood species, a lamination system either without reinforcement or with external reinforcement was developed. However, the lamination process itself has a reinforcing effect on beams, especially those designed with tensile bending collapse. There are three types of reinforcement that are being developed, namely: (1) flexural reinforcement, (2) shear reinforcement, (3) combined bending and shear strengthening with GFRP on the underside of the beam causes ductile collapse with an increase in MOR 11%, MOE 5% *Flexural strengthening* [15]. The laminated beam system can reduce the negative effect of wood defects [14]. The bending capacity of laminated blocks with a combination of lamina types will increase along with the increase in MOE in the outer layer [18]. With the strengthening of 0.82% of the steel bar area, there was an increase in stiffness of 25.9%, ultimate load of 48.1% and ductility of 43.8% [13]. Strengthening laminated beams with GFRP increases the flexural strength. is very significant and causes a shift in the location of the crack from the shear area [15]. Beams with FRPs reinforced behave more ductile and experience significantly increased stiffness. In addition, there is also an increase in the ultimate moment of capacity and beam compressive strength [16]. Reinforcement will reduce the need for high-strength wood in the outer layer of the laminate beam (may use low grade). The volume of wood can be reduced and will reduce product variability [11]. The increase in flexural strength with fiber glass reinforcement on the lower fibers of the laminated beam is 9.52% - 12.53%. Specifically, for the application of 1.2% FRP with a thickness of 0.7 mm combined with fl axfies, the upright tensile capacity of laminated wood beams increased by 74% and increased in stiffness by 41%. In this case the collapse that occurred was semiductile[6].

2. Research Methods

2.1 Research Framework and Data Collection Methods

The research was carried out by using experimental methods in the laboratory. Experiments were carried out through the bending test of a full-scale laminated beam structure model. Experiments are carried out in accordance with the stages as presented in the research framework. The data analysis includes:

1. Interpretation of tensile and compressive test results parallel in the context of their effect on the flexural strength of solid wood.
2. Interpretation of full-scale beam test results by comparing the performance of the beam from the reading of the strain rate and deflection rate and formulating conclusions based on the results of both

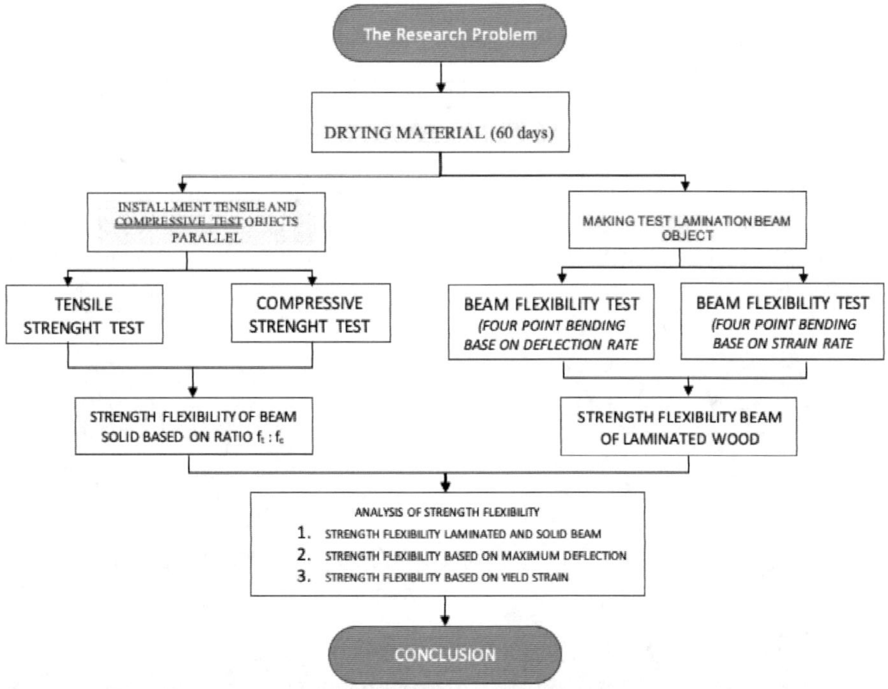

Figure 1. Research flow chart

2.2. Research Materials

The main material used is Sengon wood planks with a width of 6 cm, a thickness of 3 cm and a length of 300 cm. The sheets of the board are first dried for about 3 months. This is done in order to achieve a balanced moisture content of 10-12%. After the sheets are in equilibrium or air dry, then the tensile, compressive and laminated beams are made. The adhesive used for the manufacture of laminated blocks is cold setting urea formaldehyde resin. One beam requires 120 grams of resin powder and 60 grams of water. The raw material for Sengon wood and the adhesive used is presented in Figure 2.

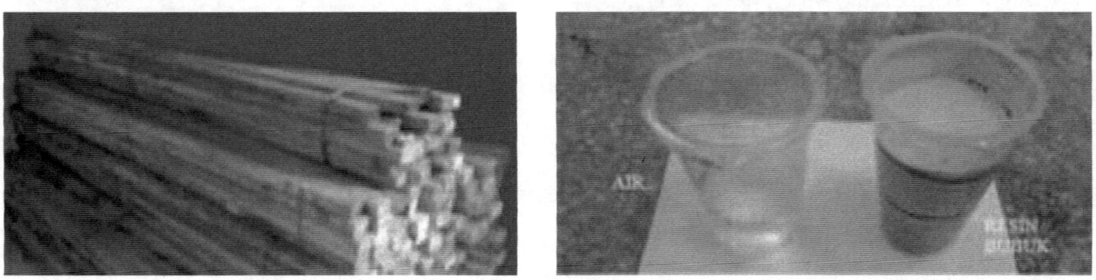

(a) (b)

Figure 2. (a) Sengon Wood Planks, (b) The Adhesive

2.3 Making Test Objects

Tensile specimens, compressive tests and laminated beams are made in accordance with ASTM D-143 requirements. 02. As for the number of each specimen as presented in Table 1 and figure 3.

Table 1. Types of Testing and Dimensions of Test Objects

No	Types of Testing	Total Sample (n)	Test Results
1	Strong Pull History	3	Tensile Strength and Stress-Strain Curves
2	Strong Press Along	3	Parallel Strength and Stress-Strain Curves
3	Block Bible Test with The Four Point Bending System	3	Flexural Strength, Load-Strain Curves and Deflection-Load Curves

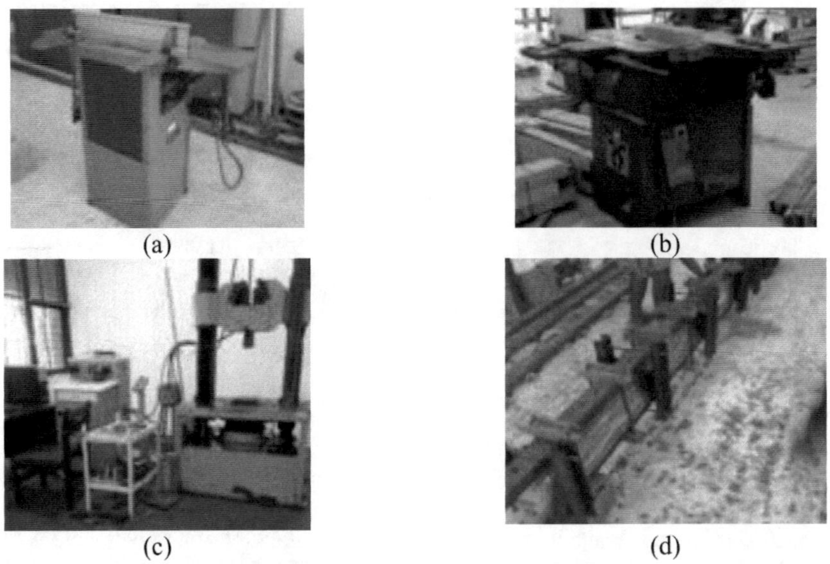

Figure 3. Tool for Making Test Objects. (a) Planer Machines for Leveling and Grading (b) Cutting Machine (Cyrcular Saw) (c) UTM for Laminated Block Pressing (d) Sett Clamp Tool

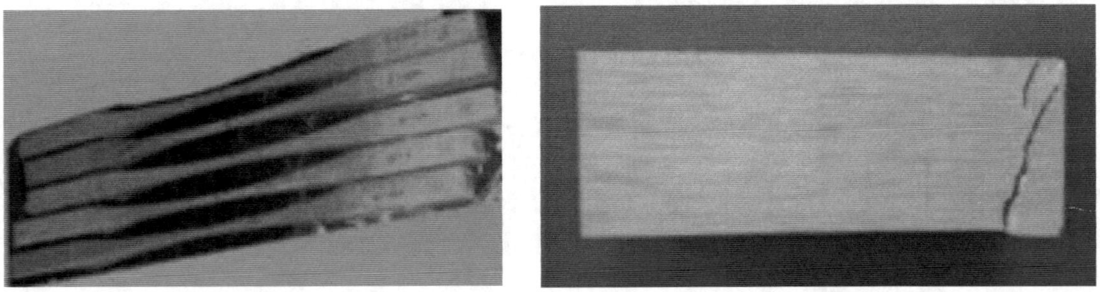

(a) (b)

Figure 4. Shape of tensile and compressive specimens according to ASTM. D143-03 (a) Fiber Parallel Tensile Test Object (b) Test specimens are parallel to the fibers.

The manufacturing of laminated blocks is carried out in stages: 1) Leveling of the laminate; 2) Gluing; 3) Pressing and 4) finishing. In this case, the laminate beam is formed from six sheets of laminate board which are glued together to form a complete structural beam, with dimensions of 55 mm in width, 155 mm in height and 2750 mm in span. Because there must be an oversteck at both ends of the pedestal, the total length of the beam made is 3000 mm. Schematically the whole process of making laminated blocks is presented in Figure 5.

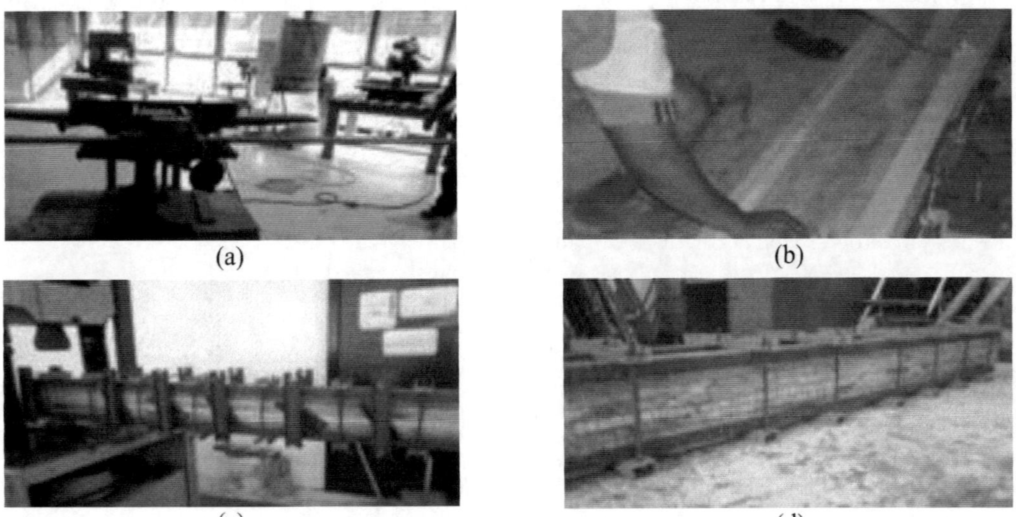

(a) (b) (c) (d)

(e)

(f)

Figure 5. The process of making laminated blocks. (a) Leveling Lamina (b) Adhesives (250 gr / block) (c) Deployment of 2 MPa per 25 cm beam length (d) Pressing (The blocks are put in place clamp is flat for 24 hours) (e) Finishing (f) End Result

Lamination leveling is carried out using a planer machine to clean the surface of the bonding plane, as well as to achieve the target net thickness of 26 mm. In addition, this leveling or cleaning process will remove the extract content that is on the surface of the lamina material. At this stage, alignment is only carried out on the two sides of the wide surface of the lamina. This process is carried out 24 hours before the gluing process, so as not to give the wood material the opportunity to extract the extract. In this case, the extract content in the bond area will reduce the quality of the adhesive.

After all the laminations are ready, then gluing is carried out with an adhesive material in the form of a mixture of powdered resin and water in a 2: 1 ratio. The adhesive resurfacing is done manually on both sides of each lamina with the duration of laying each block (six layers of lamina) a maximum of 15 minutes. This is to avoid exceeding the setting time of the adhesive. After the adhesive laying process, the clamping device is then installed for further pressing and pressing. The duration of this pressing is 24 hours in order to achieve the optimal level of adhesion layer maturity between each lamina material.

After 24 hours, the clamp is opened and the finishing process is continued. At this stage, the leveling process is carried out again for each beam to achieve a width dimension of 55 mm. The leveling process is carried out with a planer machine on only two sides of the beam height, because the beam height achieved during the pressing process is in accordance with the target, namely 155 mm.

2.4. Implementation of Testing

2.4.1. Tensile and Test Parallel

The tensile and parallel compressive tests are carried out according to the procedure specified by ASTM. D143-02 as presented in Figure 6. In the implementation of the tensile test, the main tool is used in the form of a strain meter to read the length increase that occurs in the test object due to loading.

Tensile axial loading was applied using the Universal Testing Machine. The parallel compression test is carried out using a compression test machine as a loading application tool. Meanwhile, to read the deformation used LVDT -150 mm and data logger as data reading instruments. The loading application in the parallel compression test is carried out until there is no additional loading value.

(a) (b)

Figure 6. (a) Parallel Tensile Test (b) Parallel Compression Test

2.4.2 Test of Laminated Beams with the Four Point Bending System

The setting up of a full-scale laminated beam bending test is presented in Figure 6. Implementation of tensile and compressive tests parallel to Sengon wood. In this case the loading application is implemented using load cell and hydraulic jack. The loading is carried out slowly based on a deflection rate of 1 mm per minute. The loading application is applied until the beam collapse occurs.

For deflection and strain reading, the LVDT and the TDS-650 data logger were used. In this case, the average beam collapse occurs after a loading application for 25 minutes. In this test, three LVDTs were used which were placed at three points in the middle area of the beam span.

Figure 7. Schematic of beam testing with four point bending test

3. Results of Testing and Discussion

3.1 Tensile and Compressive Strength of Sengon Wood

Based on Table 2. It is known that the parallel tensile strength of Sengon wood fibers in solid form is 34.3 MPa and the maximum parallel tensile strength is 0.0046. There is no yield stress and the curve tendency is linear as illustrated at figure 8. In this context, the collapse of Sengon wood is brittle and sudden.

Table 2. Result of Parallel Tensile Test for Sengon Wood

No	Test Object Name	P (kN)	$f_{t\ max}$ (MPa)	ε_{max}	Trend Curva
1.	TR1	15,6	35.3	0.00436	Linear
2.	TR2	15,4	33.6	0.00464	Linear
3.	TR3	15,5	34	0.00484	Linear
	Average		34.3	0.00461	-

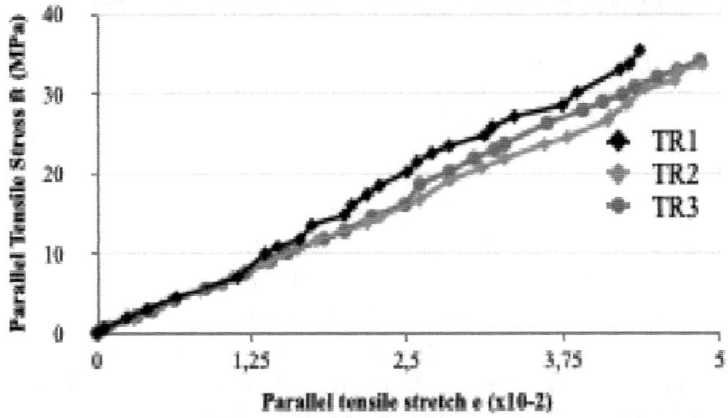

Figure 8. Parallel Tensile Stress Curve of Sengon Wood

Based on Table 3. It is known that the parallel compressive strength of Sengon wood fibers in solid form is 24.97 MPa and the maximum parallel compressive strain is 0.055. There is an average yield stress of 15.53 MPa at 0.0098 strain, and the trend of the curve is linear-nonlinear as seen at Figure 9. In this context, the collapse of Sengon wood is ductile and gradual.

Table 3. Result of Parallel Compressive Test for Sengon Wood

No	Object Name	Condition	P (kN)	f (MPa)	ε	Trend Curve
1	Sengon 1	Yield	36,9	14,76	0,0085	Linear - non Linear
		Ultimate	63	25,2	0,0141	
2	Sengon 2	Yield	40,9	16,36	0,0113	Linear - non Linear
		Ultimate	64	25,84	0,0191	
3	Sengon 3	Yield	38,7	15,48	0,0096	Linear - non Linear
		Ultimate	59,7	23,88	0,0098	

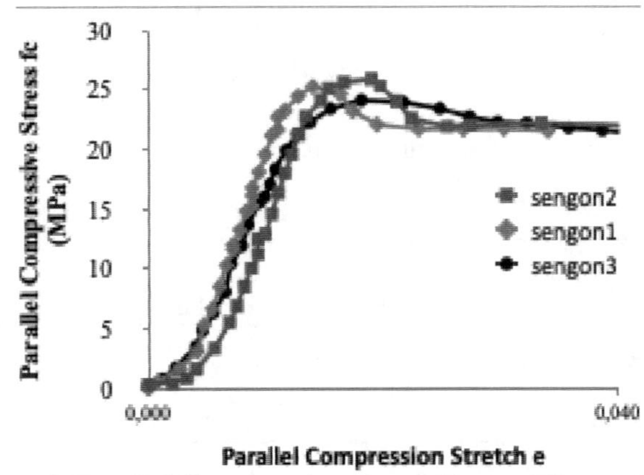

Figure 9. Parallel Compressive Stress Curve of Sengon Wood

3.2. Flexural Strength of Sengon Laminated Beams based on Deflection Rate

Next is the test for the bending results of laminated beams with a four point bending loading system to determine the flexural strength of the beam, stress, strain and deflection and beam collapse patterns.

Based on Table 4. It is known that the flexural strength of Sengon wood laminated beams is 21.33 MPa and the maximum deflection is 47.3 mm. There is an average yield stress of 17.31 MPa at 27.61 deflection, and the trend of the curve is linear-nonlinear as Figure 10. Deflection Load Curve of Sengon Laminated Beams. In this context, block collapse is ductile and gradual.

Table 4. Laminated Beam Flexure Test Results with the Four Point Bending System

No.	Name Test Object	$f_{b\,yield}$ (MPa)	$f_{b\,max}$ (MPa)	δ_{yield}	δ_{max}	Eb (MPa)
1.	BLS-1	18	20	24.98	32.32	7,534
2.	BLS-2	17.53	24.32	26.09	43.68	6,913
3.	BLS-3	20	23.19	28.3	35.76	6,711
4.	BLS-4	20	30.34	27.68	70.67	7,582
5.	BLS-5	21	28.8	30.99	54.22	7,000
	Average	17.31	21.33	27.61	47.33	7,148

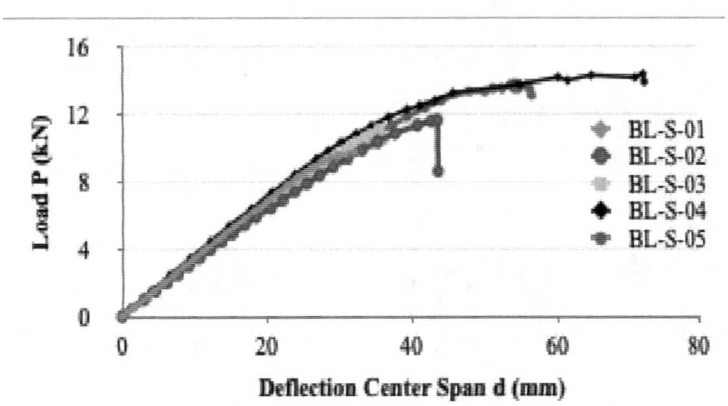

Figure 10. Deflection Load Curve of Sengon Laminated Beams

3.3 Flexural Strength of Sengon Laminated Beams based on the Strain Rate

Table 5 shows for the BL-S-2 specimen that the largest compressive stress is on the Lam-2 fiber at 21.81 MPa and the largest tensile stress is at the lowest fiber at 43 MPa. and the trend of the curve is linear - non-linear as illustrated in Figure 11. The failure pattern can be seen in Figure 12

Table 5. Flexural Strength of Laminated Beams with a Four Point Bending System

BEAM LAYER	$f_{b, collapsed}$ MPa	$f_{b, elastic}$ MPa	$f_{b, max}$ MPa
above	-16.73	-20.29	-16.80
lam-1	-21.59	-16.63	-21.60
lam-2	-21.81	-12.76	-21.81
lam-3	-12.05	-5.00	-11.98
lam-5	18.36	12.74	18.39
under	43.00	27.00	43.00

The sign (-) indicates the compressive stress

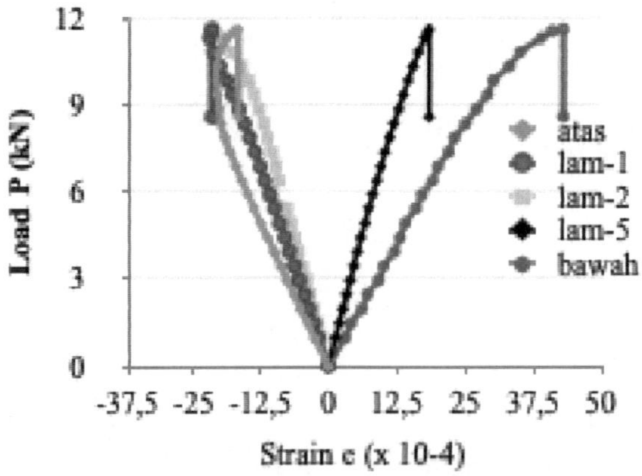

Figure 11. Strain Curve of Sengon Woods Laminated Beam BL-S-2

(a)

(b)

Figure 12. Beam Failure Pattern with four point bending system

4. Conclusion

From the test results it can be concluded thathload relationship - BS-L deflection consists of linear and nonlinear phases. The load performance characteristics of the two types of laminated beams are expressed as the ratio of the proportional limit load to the maximum load. The ratio value is expressed in the form Pe BL-S = 0.7Pmax BL-S and Me BL-S = 0.7Mmax BL-S. This form is similar to previous studies with a Pe to Pmax ratio of 0.8-0.9. In this case, the average flexural strength of the resulting laminated beam is 17 MPa with a maximum strain of 0.004.

5. Suggestion

For the manufacture of composite beams it is advisable to increase the variety of wood types to obtain significant strength and more efficient materials.

Further analysis is needed regarding the manufacture of composite blocks using other waste that is more environmentally friendly at low cost and significant strength.

Reference

[1] Ahmad, Y. 2010. Bending Behavior of Timber Beams Strengthened Using Fiber Reinforced Polymer Bars and Plates. Phd. Thesis. Kuala Lumpur: Universiti Teknologi Malaysia.

[2] Awaludin, A. 2011. Research on the Physical Properties and Mechanics of Wood Glugu and Sengon in Merapi Area to Accelerate the Economic Recovery of the Merapi Community after the 2010 Merapi Eruption. Yogyakarta.

[3] Blass, HJ, P. Aune, BS Choo, R. Gorlacher, DR, Griffiths., And G. Steck, 1995. Timber Engineering Step I. Centrum Hout, The Nederland.

[4] Bourreau D., Y. Aimene, J. Beauchêne and B. Thibaut, 2013. Feasibility of Glued Laminated Timber Beams With Tropical Hardwoods. European Journal of Wood and Wood Products. Vol. 71, No. 5.p: 653-662.

[5] Breyer, DE, KJ Fridley, and KE Cobeen, 1998, Design of Wood Structures ASD. Mcgraw-Hill Inc. New York.

[6] Bustos, C., Mohammad, M., Hernandez, RE and Beauregard, R. 2003. Effects of Curing Time and End Pressure on The Tensile Strength of Finger-Joined Black Spruce Lumber. Forest Products Journal, Vol. 53, No. 11.p: 85-89.

[7] Čizmar, D, D. Damjanović, K. Pavković, and V. Rajčić, 2014. Ductility Analysis of Laminated Timber Beams of Small Section Height, Građevinar Vol. 66, No.5, p: 395-406. Doi: 10.14256 / Jce.874.2013

[8] Falk, RH, & Colling, F., 1995. Laminating Effects in Glued-Laminated Timber Beams. Journal of Structural Engineering. 121 (December): 1857–1863.

[9] Fathi, L. 2014. Structural and Mechanical Properties of The Wood From Coconut Palms, Oil Palms and Date Palms. PhD Thesis. Fakultät Für Mathematik, Informatik Und Naturwissenschaften, Universität Hamburg.

[10] Forest Products Laboratory. 2010. Wood Handbook Wood as An Engineering Material. Robert J. Ross, (Ed). Madison, WI: Department of Agriculture, Forest Service, Forest Products Laboratory.

[11] Issa, CA & Kmeid, Z. 2005. Advanced Wood Engineering: Glulam Beams. Construction and Building Materials., 19 (2005): 99-106.

[12] Krisnawati, H., Eveliina, V., Maarit, K., Markku, K. 2011. Paraserianthes Falcataria (L.) Nielsen: Ecology, Silviculture and Productivity. Bogor: Center for International Forestry Research.

[13] Luca VD, and C. Marano, 2012. Prestressed Glulam Timbers Reinforced with Steel Bars. Journal Construction and Building Materials. Vol. 30, p: 206–217

[14] Nadir Y & Nagarajan P. 2014. The Bahavior of Horizontally Glued Laminated Beams Using Rubber Wood. Journal Construction and Building Materials. 55: 398 – 405.

[15] Osmannezhad, S., Faezipour, M. & Ebrahimi, G., 2014. Effects of GFRP on bending strength of glulam made of poplar (Populus deltoids) and beech (Fagus orientalis). Construction and Building Materials, 51, pp. 34–39.

[16] Raftery, GM, & Harte, AM 2011. Low-Grade Glued Laminated Timber Reinforced with FRP Plate. Composites: Part B.42 (4): 724-735.

[17] Tomasi, R., Maria, AP, & Maurizio, P. 2010. Ductile Design of Glued-Laminated Timber Beams. Practice Periodical on Structural Design and Construction. 14 (3): 113-122.

[18] Yang, TH, Song-Yung, W., Cheng-Jung, L., & Ming-Jer, T. 2008. Evaluation of The Mechanical Properties of Douglas-Fir and Japanese Cedar Lumber and Its Structural Glulam by Nondestructive Techniques. Construction and Building Materials. 22 (2008):

Shear strength behaviour of liquefiable sand of petobo on treated by agarose under direct shear test

A Beddu, H Setiawan, Nuraida, M R Anugrah

Civil Engineering Department, Tadulako University, Indonesia

arifin.b@untad.ac.id, hendra3909@gmail.com, mohreskyanugrah1@gmail.com

Abstract Liquefaction process is associated with the loss of the shear strength of the saturated loose sands caused by strong earthquakes. Due to mitigitation of liquefaction hazard, an appropriate mitigation of liquefaction using environmentally friendly methods is critical and becoming increasingly important and unavoidable. The laboratory investigation was carried out to study the shear strength behaviour of liquefiable sand of Petobo treated by agarose on different concentration 1%,3% 5%. A series of direct shear test were conducted under three level of vertical stress 10 kPa, 20 kPa, and 30 kPa on the specimen. It was found that the optimum content of agarose which can be considered is at 1% -3%, using stress ratio ($\tau/\sigma v$) analysis shows that stress ratio decreases with increasing the vertical stress on the same agar content. The implication this result that the application of this method must consider variation of material source and characteristic, and the suitable level of vertical stresses.

Keywords: Liquefaction, Agarose; Shear strength, Petobo Sand, Direct Shear.

1. Introduction

Petobo sub-district is one of the areas that was affected by liquefaction on due to the 7.5 Mw Palu Central Sulawesi Earthquake in 2018. Lateral spreading, surface fracture and flow sliding occur massively and cover a large area. Although, the mechanisms and effects of liquefaction have been studied by a number of researchers [1][2] but because the coverage area affected is very broad, massive research in various reviews to study further this phenomenon includes behavior of shear strength characteristic of soil sand which can be useful for the purposes of numerical analysis, empirical methods, experimental studies for the purpose of mitigation and remediation.

The mechanism that occurs during the liquefaction process is associated with the loss of the shear strength of the saturated loose sands as a result of the loss of effective stress as a consequence of increased pore water stress due to seismic force propagation [3][4]. Currently, the most common methods propose to improve the mitigation and remediation of liquefaction hazards are vibration, grouting and injection methods, chemical stabilization using lime, cement and fly ash, geosynthetic stabilization [5][6]. However, due to environmental considerations some of these methods are not recommended, therefore the need to find sustainable technology is becoming increasingly important and unavoidable. In this regards, several alternative environmentally friendly methods that have been researched, the use of biopolymers such as agarose seems quite promising and effective [7].

Recent previous studies related to the use of agarose as a polymer to improve sandy soil strength have done, include by [8] using the unconfined compression test, using 1%-4% agarose concentration, have certain the strength and deformation characteristics behavior in stiffness and ductility. Likewise other reseacher [9]the results of their study as passive stabilization using coloidal silica, while [10][7] which specifically focused their research on the dynamic loading aspect using the triaxial test concluded that increasing the behavior of shear strength in agorase content 0.5%, 1% and 2%. However, from these research, the type and source of the test materials used were slightly limited, so to see its behavior, it still requires various sources of sand material and the type of study, includes material from the liquefaction site and using other types testing such as direct shear tests.

Therefore, in this study, experiment were conducted to explore the behavior of agarose using a different sand source from previous studies where in this case the sand was taken from a location that affected liquefaction. Sand were taken from Petobo, testing using a direct shear test with three levels of vertical stress, respectively are 10 kPa, 20 kPa and 30 kPa. Using agarose content of 1%, 3% and 5%, and untreated was 0% agarose content. The effectiveness of using agarose as a stabilizing agent through variations in the sand source being tested and certain types of conditions such as different loading and testing conditions, so that agarose could be used as option for liquefaction mitigation according to the conditions and properties of the material being tested.

2. Method experiment

2.1 Material Used

2.1.1 Sand

Sandy soil presented in this research were taken from Petobo, located in Palu city, Central Sulawesi at the depth of 1.5m-2.0m. The physical properties are presented in Table 1, with Cu 4.67 and Cc 0.87 was categorized as poorly graded sand soils. Minimum and maximum void ratio figures seem appropriate value that has a very large range of 0.58 to 0.98, so that the soil can be in a very loose and can be very dense conditions. The plot range of grain size materials in Figure 1 shows the compares of the susceptible curves according to Japanese liquefaction susceptibility criteria for port and harbour facilities [11]

Tabel.1 The physical properties of sandy soil

Property	Value
Effective size (D_{10})	0.15
Effective size (D_{30})	0.30
Effective size (D_{60})	0.70
Uniformity coefficient (Cu)	4,67
Curvature coefficient (Cc)	0,87
Initial void ratio (e_o)	0.78
Minimum void ratio (e_{min})	0.58
Maximum void ratio (e_{max})	0.98
Specific Gravity (Gs)	2,66
Plasticity index, PI (%)	NP

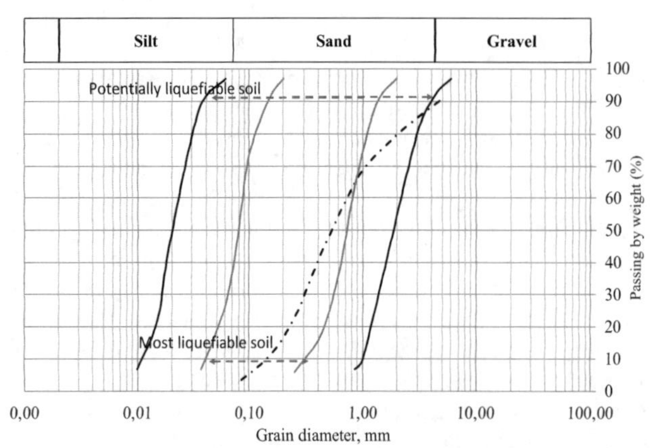

Figure. 1 Over lay gradings boundaries of liquefaction susceptible soil according to according to Japanese liquefaction susceptibility criteria for port and harbour facilities[11], where 68% portion grain size distribution of soil was lay in most liquefiable soil range border and 32% portion was in potentially liquefiable border.

2.1.2 Agarose

The commercial agarose was used in this study in powder form produced by PT Agar Swallow, without preservatives and food hardener. Agarose powder products begin to dissolve at 85°C and solidify at 32°C - 40°C where the viscosity are constant at this temperature. A preliminary test was carried out to obtain the viscosity of agarose solution in a mixture of 1%, 3% and 5%.

2.2 Sample Preparation

A quantity of natural sandy soil was sieved to separately sand passed sieve number #8(2,36mm) and retained sieve number #100 (0,15mm) to obtain some clean sand samples with zero fines content. The clean sand soil specimens are prepared using the wet mixing method. The required dry weight of clean sand to meet of the specimens mixing weight then is recalculated as function of the relative density and saturated water content.

The content of agarose water mixture solution were carried out for all of the test specimens by weight which ranged at 1%, 3% and 5% portion. The mixture of agar and water which has previously heated at temperatures over 85°C, was poured into a cup that has been filled with a number of dry sand, then stirred softly until evenly mixed well.

The mixture was filled into specimen molds, 63 mm in diameter by 35 mm height, with the weight adjusted to produce wet densities of 15,0 kN/m^3 to 15,5 kN/m^3, relative densities of 36%–47%. The same initial wet densities were maintained during test preparation. Trial test was undertaken to meet a sample preparation method that was simultaneously easy and reproducible. All sample has been cured for one day before being tested.

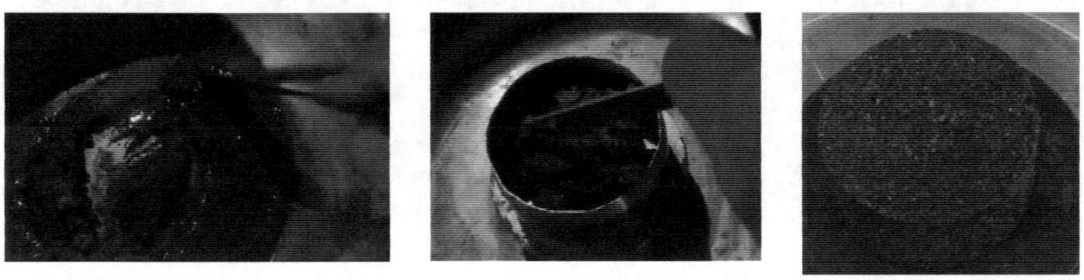

a. *b.* *c.*

Figure 2. The test specimen preparation process; (a). mixing agarose solution and dry sand; (b). pouring the mixture into a direct shear test mold; (c). specimens that have been left for one day and prepared are put in a direct shear test box

2.3 Test procedure

The testing apparatus used in the current study is direct shear test of soils. All tests were conducted with the ASTM D-3080 procedures. The shear loading is controlled using automated system at rate of 1.0 mm/minute. The specimen were tested under three different vertical normal stress values of 15 kPa, 20 kPa and 30 kPa.

3. Results and discussions

3.1 Effect of agar content on shear strength parameter

The direct test results at three levels vertical stress for all specimen with specified agar concentration is tabulated in Table 2. The calculation of shear strength parameters, cohesion (c) and angle of internal friction (ϕ) refers to the Mohr Coulomb formula. Meanwhile, peak shear strength is obtained from the shear stress vs shear displacement relationship curve in the direct shear test. It was found that 1% agar content increased soil cohesion by 11.84% from 8.86 kPa to 9.91 kPa and for 3% agar content the increase was 9.81% from 8.86 kPa to 9.67 kPa while 5% agar content decreased the cohesion value by 8.24% from 8.86 kPa to 8.13 kPa. These results as shown in Table 2, indicate that the volume of agarose mixed in the soil grains that fill the pores forms a proportional condition that changes the cohesion and friction values between soil particles, high concentrations cause weak bonds and friction between grains.

Table 2. Strength parameter of soil sampel tested by direct shear test

Agar Concentration %	Strength Parameter		Peak Shear Strength at Vertical Stress		
	c (kPa)	ϕ (°)	10 kPa	20 kPa	30 kPa
0	8.86	35.35	16.33	22.19	29.99
1	9.91	35.83	17.80	23.89	32.67
3	9.67	34.86	17.31	21.21	30.72
5	8.13	32.33	14.63	19.50	26.82

3.2 *Effect of agar content on Peak Shear Strength behaviour*

The mixtures of specimen samples were tested under three different normal vertical stress values of 10 kPa, 20 kPa and 30 kPa in the direct shear test as shown in Table 2. The maximum shear strength was 32.67 kPa and the minimum value was 14.63 kPa which is lower than minimum value of untreated sample. The peak of shear strength on the different level of vertical stress values and variation of agarosa content are plotted Figure 2. The shear test result found that the optimum value of shear strength was reached at 1% for all concentration of agarosa. The curves at 10 kPa and 20 kPa vertical stresses change slightly at each given agar content, which differs to the 30 kPa vertical stresses. For 5% agarose content, the shear strength fell below the untreated soil at all levels of applied vertical stress. At 5% level of agarose content, the achieved shear stress was lower than untreated soil at all vertical stress levels, this is due to the volume of agaraso is no longer effective in forming cohesion and friction bonds between grains, from this result the optimum content of agarosa between 1% to 5%.

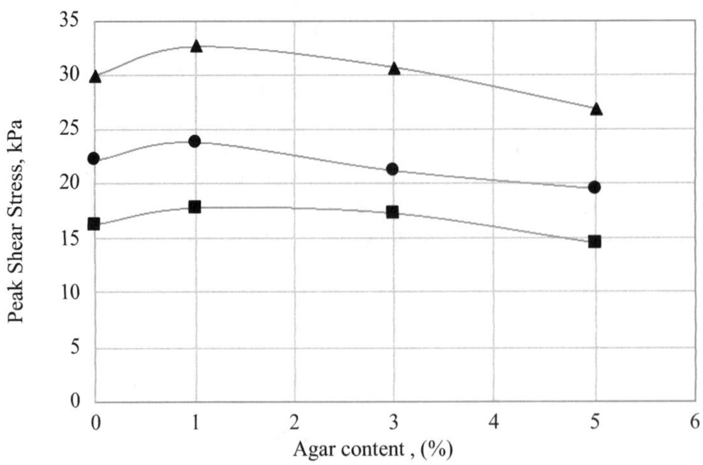

Figure.3. Peak shear strength of specimen mixtured agar content at three levels vertical stress respectively are 10 kPa (■), 20 kPa (•).and 30 kPa (▲); for untreated sampel 0%, agar content 1%; 3%, 5%.

3.3 *Effect of agar content on vertical normal stress*

According to the coulomb formula, the shear strength of the soil is a combination of the cohesion value, the shear angle and the vertical stress given. As a result of changing the value of the shear angle and soil cohesion, it gives a different value of shear strength. The analysis was carried out using stress ratio (τ/σv) in term normalized shear stress to each vertical stress level applied for each speciments test. It was found that peak shear stress depends on the level of vertical stress applied, where the value of shear stress rato decreases with increasing the vertical stress, as can be seen from Figure 4-6. These results show that the increase in vertical stress increases the shear strength due to the confining stress of the shear ring, and the greatest is at 1% level, and decreases drastically at 5% level which is lower than the shear stress of untreated soil. This indicates that the high vertical stress causes the adhesions and bonds formed between the grains and the agarose gel to break. The implication is that the application of this method must consider vertical stresses, or in other words its application is more suitable in conditions of low vertical stress, this result relevant to previous research by [10].

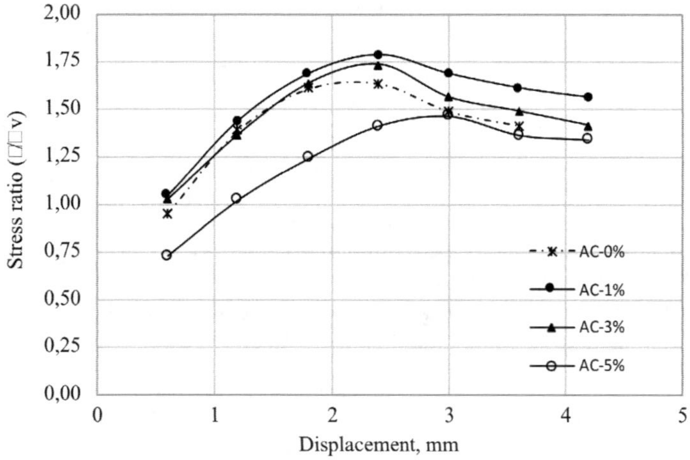

Figure 4. Stress ratio (τ/σv) versus shear displacment at vertical stress σv 10 kPa

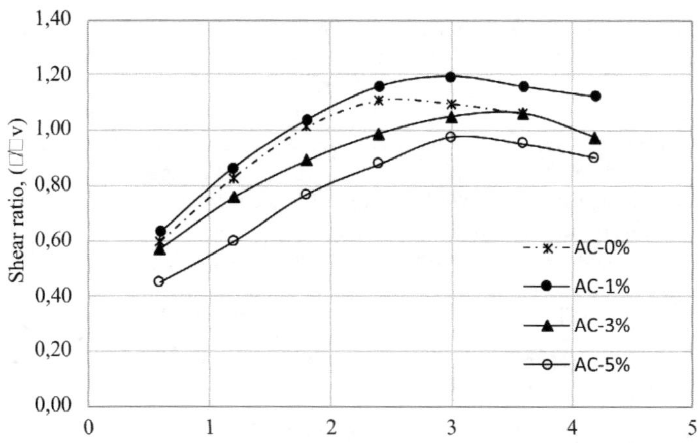

Figure 5. Stress ratio (τ/σv) versus shear displacment at vertical stress σv 20 kPa

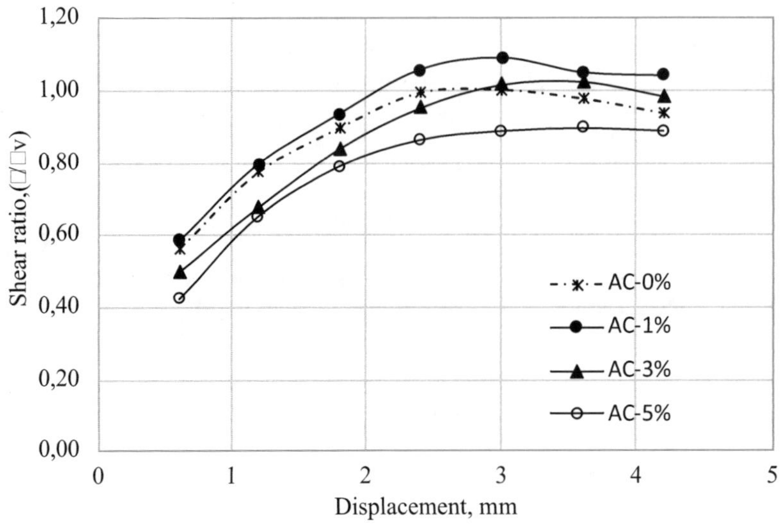

Figure 6. Stress ratio (τ/σv) versus shear displacment at vertical stress σv 30 kPa

4 Summary and Conclusions

This study investigated the shear stress behavior of a mixture of sand and agarose using a direct shear test under 10kPa, 20 kPa and 30 kPa vertical stress. Mixing solution of agarose content tested at 1%;3% and 5% for 24 hours curing using sand which has previously been examined for criteria of most liquefiable soil, taken from the Petobo village, previously exposed to liquefaction due to the 7.5Mw earthquake in 2018 so that it represents a liquifiable soil. The effectiveness so far has been observed using the ratio of normal stress to shear stress according to the agarose content. The significant conclusions drawn from this study include the following. The optimum content of agarose which can be considered is at 1% -3%. Whereas at 1% agarosa content, the cohesion value decreases but the shear angle does not change significantly, but for 3% the cohesion and sediment of soil shear changes significantly. So that the choice of 1% level can be selected if the target of modification is to change the cohesion value without changing the friction angle, while 3% changes both the cohesion value and the friction angle. Peak shear strength shows the highest value at 1% content of agarose, for all vertical stress levels, while for 5% peak shear strength is lower than untreated soil at all levels of vertical stress, this is due to the high volume of agaraso is no longer effective in forming cohesion and friction bonds between grains. The high vertical stress causes the adhesions and bonds formed between the grains and the agarose gel to break. The implication is that the application of this method must consider vertical stresses, or in other words its application is more suitable in conditions of low vertical stress. This conclusion is drawn based on a limited number of tests and methods, therefore it can be developed using more test samples, with variations in moisture content and mixed viscosity and saturation, the test method on dynamic loading with the measurement of pore water pressure and long term of treatment must be considered in the next develop research. The results of this study may be different or same as previous research, therefore it is important to conduct more research using different sand sources to assess the significance of agarose as a stabilization material for mitigating of liquefaction hazard.

References

[1] R. F. Hidayat, T. Kiyota, N. Tada, J. Hayakawa, and H. Nawir, "Reconnaissance on liquefaction-induced flow failure caused by the 2018 Mw 7.5 sulawesi earthquake, palu, Indonesia," *J. Eng. Technol. Sci.*, vol. 52, no. 1, pp. 51–65, 2020.

[2] I. M. Watkinson and R. Hall, "Impact of communal irrigation on the 2018 Palu earthquake-triggered landslides," *Nat. Geosci.*, 2018.

[3] M. W. Greenfield, S. S. Sideras, and S. L. Kramer, "Identification of time of liquefaction triggering," *J. Soil Dyn. Earthq. Eng.*, vol. xx, no. October, pp. 1–15, 2019.

[4] X. Bao, Z. Jin, H. Cui, X. Chen, and X. Xie, "Soil liquefaction mitigation in geotechnical engineering: An overview of recently developed methods," *Soil Dyn. Earthq. Eng.*, vol. 120, no. June 2018, pp. 273–291, 2019.

[5] R. W. Boulanger, M. Khosravi, A. Khosravi, and D. W. Wilson, "Remediation of liquefaction effects for an embankment using soil-cement walls: Centrifuge and numerical modeling," *Soil Dyn. Earthq. Eng.*, vol. 114, no. December 2017, pp. 38–50, 2018.

[6] Y. Huang and Z. Wen, "Recent developments of soil improvement methods for seismic liquefaction mitigation," *Nat. Hazards*, vol. 76, no. 3, pp. 1927–1938, 2015.

[7] S. Smitha and Rangaswamy, "Effect of Biopolymer Treatment on Pore Pressure Response and Dynamic Properties of Silty Sand," *J. Mater. Civ. Eng.*, vol. 32, no. 8, pp. 1–15, 2020.

[8] H. R. Khatami and B. C. O'Kelly, "Improving mechanical properties of sand using biopolymers," *J. Geotech. Geoenvironmental Eng.*, vol. 139, no. 8, pp. 1402–1406, 2013.

[9] G. I. Agapoulaki and A. G. Papadimitriou, "Rheological properties of colloidal silica grout for passive stabilization against liquefaction," *J. Mater. Civ. Eng.*, vol. 30, no. 10, pp. 1–11, 2018.

[10] S. Smitha, K. Rangaswamy, and D. S. Keerthi, "Triaxial test behaviour of silty sands treated with agar biopolymer," *Int. J. Geotech. Eng.*, pp. 1–12, 2019.

[11] R. A. Green and K. Ziotopoulou, "Overview of screening criteria for liquefaction triggering susceptibility," *Proc. Ninth Pacific Conf. Earthq. Eng.*, no. November, pp. 1–8, 2015.

Acknowledgment

The authors thank the Faculty of Engineering, Tadulako University, for funding this research under grant No.646.z/UN28.2/PL/2020

The properties of preplaced aggregate concrete technology contain the industrial waste-material and the various shapes and sizes of coarse aggregate

N Chairunnisa*, H Ruzhanah, Hairida and L.S Daniel

Engineering Faculty, Lambung Mangkurat University, Indonesia

*Corresponding author: nursiah.chairunnisa@ulm.ac.id

Abstract. The success of preplaced aggregate concrete technology depends on two main factors which are potential grout and coarse aggregate. This research was conducted experimentally to determine the effect of using two different fly ash sources as an alternative for the partial replacement of cement and several size and shapes of coarse aggregate on the compressive and tensile strength of PAC specimens. This involved the use of seven concrete mixes with a low water-cement ratio of 0.4 and cement to sand ratio of 1:0.75 to produce standard cylinder specimens of concrete containing rounded and crush aggregate. Moreover, fly ash was added at a dosage of 5% and 10% of cement weight while three shapes and sizes of a rounded and crushed aggregate at 20 mm, 30 mm, and a mixture of the two were also applied. The results showed the compressive strength of specimens with different sizes or a mix of rounded aggregate in PAC exhibited a similar performance with 30 mm of crushed coarse aggregate. Furthermore, the specimen with a higher content of calcium fly ash demonstrated a more rapid strength at an early age of seven days than those with lower content. Therefore, the partial replacement of cement with industrial waste material in the form of fly ash in preplaced aggregate concrete has the ability to save up to 10% of cement and also produce certain environmental benefits.

1. Introduction

Preplaced Aggregate Concrete is a simple concept of concrete technology consisting of two-stage casting. The first involves placing coarse aggregate into formworks while the second is injecting the potential grout into the void [1]. This method provides several benefits compared to the conventional concrete due to its flowability, absence of internal vibration, lower energy consumption, and cement weight [2-4]. However, one of the essential factors affecting PAC Technology is the quality of coarse aggregate and grout especially due to the influence of several factors such as the type of sand, cement-sand ratio composition, water-cement ratio, and cement for conventional grout [5,6].

The success of PAC depends on two main factors which are grout workability and coarse aggregate size void, especially due to the expectation of potential grouts to fulfill all coarse aggregate voids. According to [7], the recommended size of coarse aggregates in PAC was 38 mm. Previous research proposed using coarse aggregates of sizes smaller than 38 mm, with the aggregates being gap-graded and of multiple sizes [8]. Single-graded aggregates of different sizes were also applied in another study [9]. Meanwhile, the effects of using fly ash as a partial replacement of cement and sand in a 0.33 low water binder ratio grout and coarse aggregate gradation for PAC mixtures have been studied [10]. Fly ash, as a pozzolanic material, was also applied to improve the fresh properties of low water binder grouts due to its abilities [2,11,12]. Its abundance as an industrial waste has led to its common use as concrete material.

This research was, therefore, conducted to determine the effect of size and shape of coarse aggregates, and fly ash as partial replacement of cement on the mechanical properties of PAC through the use of a series of laboratory experiments on grout workability, cylinder compressive strength, and splitting tensile strength.

2. Materials and Method

2.1. Coarse and fine aggregates

Coarse aggregate is one of the important factors in achieving a successful preplaced aggregate concrete. Therefore, the coarse aggregates used in this study were in two shapes including rounded and crushed with three different sizes which were 20 mm, 30 mm, and a mixture of the two sizes which were produced locally in South Kalimantan as shown in Figure 1. Meanwhile, the fine aggregates used include natural river sand with zone 3 gradation from Barito river, South Kalimantan Province.

(a) Rounded aggregates (b) crushed aggregates

Figure 1. Coarse Aggregate: rounded and crushed stones.

2.2. *Industrial waste- fly ash*

Fly ash was locally sourced from two different locations in Kalimantan and added at 5% and 10% by weight of cement to determine their use as pozzolanic materials in a preplaced aggregate concrete. The FA-1 was obtained from PLTU Asam Asam while FA-2 was from Tamiyang Layang as indicated in Figure 2.

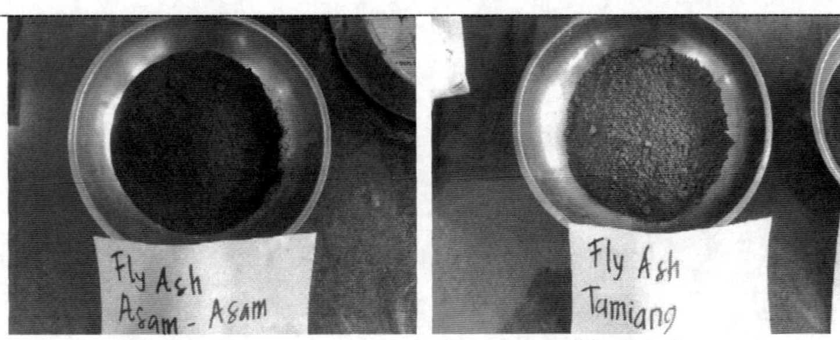

(a) FA-1 (b) FA-2

Figure 2. Locally-sourced Fly Ash

2.3. Chemical Admixture

Viscocrete 1003 type of superplasticizer imported by SIKA was used for all the mixes. This liquid chemical admixture was added at 0.7% by weight of cement [5].

3. Mix design and preparation of specimens

3.1. Mix design of specimens

A total of seven different mix proportions were designed to investigate the mechanical properties of preplaced aggregate concrete due to the effect of coarse aggregate size, shape, and fly ash. Moreover, 0.4 water-cement ratio, 1:0.75 sand cement ratio, and 0.7% weight cement of superplasticizer in accordance with previous experiments. The mix designs of PAC specimens are, therefore, summarized in Table 1.

Table 1. Mix design of PAC Specimens.

Type of mix	weight per cubic meter m^3							Superplasticizer
	cement	sand	crush aggregate	rounded aggregate	FA-1	FA-2	water	% of cement weight
R-20	479.28	359.46	1455	-	-		191.71	0.7
R-30	455.31	359.46	1455	-	-		191.71	0.7
R20+30	409.78	359.46	1455	-	-		191.71	0.7
CFA-0%	455.31	359.46	1455	-	-		191.71	0.7
CFA1-5%	390.60	292.95	-	1641	23.96		156.24	0.7
CFA1-10%	390.60	292.95	-	1641	47.93		156.24	0.7
CFA2-5%	390.60	292.95	-	1641	-	23.96	156.24	0.7

3.2. Preparation of Specimen

The PAC specimens used were prepared using two stages with the first involved placing of coarse aggregates in 150 mm x 300 mm cylinder molds while the second was injecting the grout into the molds to fill the voids as shown in Figures 3 and 4.

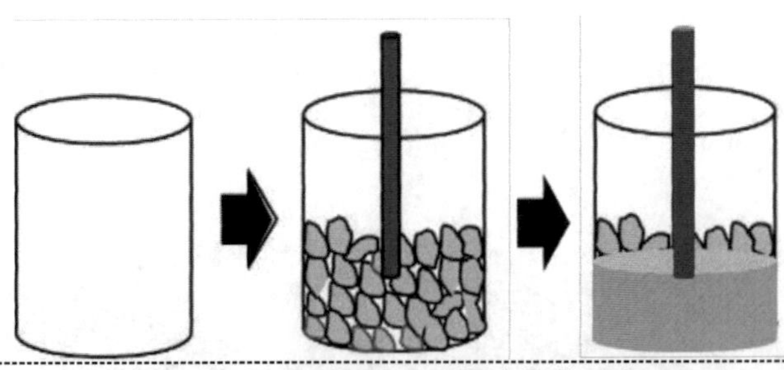

Figure 3. Preparation of PAC Specimens

Figure 4. PAC Specimens in the laboratory

4. Results and Discussion

4.1. Fresh properties of the grouts

The physical properties and chemical of different types of fly ash are presented in Table 2 and FA-1 was found to have a higher specific gravity compared to FA-2. It was also discovered to have more calcium oxide content which is classified between Class F and C according to the ASTM C618 [13]. Meanwhile, the instrument used to determine the flow time is shown in Figure 5.

Table 2. Physical and chemical properties of fly ash (Laboratory data)

Parameter	Unit weight (%)		Parameter	FA-2	
	FA-1	FA-2		unit	value
SiO^2	39.08	29.94	K_2O	0.77	0.84
Al_2O_3	11.96	15.44	MnO_2	0.34	0.22
Fe_2O_3	22.22	15.84	TiO_2	0.65	0.81
CaO	14.33	23.68	P_2O_5	0.16	0.51
MgO	7.26	9.93	SO_3	1.89	1.14
Na_2O	0.80	1.11			

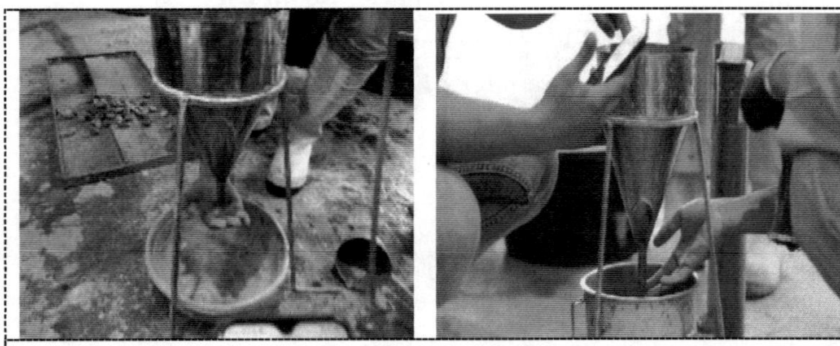

Figure 5. Flow cone test.

The fresh properties of Preplaced Aggregate Grout mixtures used in determining the grout consistency are indicated in Table 3 with the flow times observed to be within the range of 26.12 to 33.38 seconds. Meanwhile, ASTM C939 [14] requires an efflux time of 8 seconds for pure water and 8-35 seconds for grout of PAC of flow time to drain 1725 ml using a flow cone test. This means the time reduced due to the partial replacement of cement with fly ash in the mixtures and all the specimens were found to have fulfilled the criteria of flowability from the test.

Table 3. Fresh Properties of Grout

Type of Grout	Superplasticizer (%)	Flow time (second)	Grout compressive strength	Type of coarse aggregate
R-20	0.7	32.64	-	rounded
R-30	0.7	27.77	-	rounded
R20+30	0.7	27.84	-	rounded
CFA-0%	0.7	26.12	34.73	crush
CFA1-5%	0.7	33.38	32.83	crush
CFA1-10%	0.7	29.62	29.42	crush
CFA2-5%	0.7	29.23	20.80	crush

4.2. Hardened concrete of PAC

The compressive strength of all specimens is shown in Figures 6 and 7 as well as Table 4 and they were all observed to have increased with curing days. Moreover, C-FA1-5%, C-FA1-10%, and C-FA2-5% were discovered to have lower values on the seventh day compared to the C-FA1-0% which is the control specimen without any fly ash content. Meanwhile, the C-FA1-5% specimen had a similar result with control at a later curing age of 28 days.

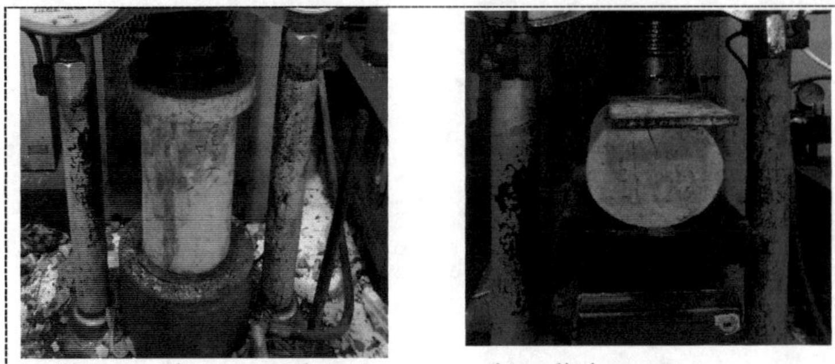

(a) Compressive strength test (b) splitting test

Figure 6. detailed of the test set up

Table 4. Compressive and splitting tensile strength of specimen PAC.

Specimens (ID)	Compressive strength on days			Tensile strength
	7	14	28	28
	MPa	MPa	MPa	MPa
R-20	12.08	12.83	18.12	1.910
R-30	11.42	12.93	18.12	2.087
R20+30	14.91	17.74	19.82	1.839
CFA-0%	14.15	16.04	20.38	2.310
CFA1-5%	8.68	12.64	20.00	1.930
CFA1-10%	8.12	11.32	17.74	1.840
CFA2-5%	10	8.00	11	1.840

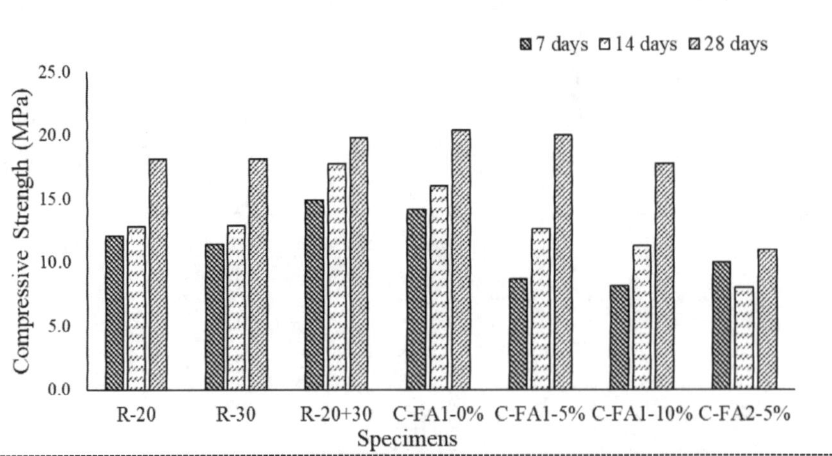

Figure 7. Compressive strength of the specimens

Figures 8(a) and (b) showed the influence of the shape of coarse aggregate on PAC's compressive strength and no significant difference was found with the increase in rounded aggregate sizes from 20 mm to 30 mm. Meanwhile, the round aggregate specimen with mixed sizes or R-20+30 was found to have increased the compressive strength by ± 10%. Figure 8(b) further shows the uniformly rounded aggregates tend to exhibit lower performance compared to the crushing aggregate [9] while round mixed size (R-20+30) had similar values with C-FA1-0%. This was associated with the better mechanical interlocking between the different sizes which produced an efficient stress transfer through the aggregates [9, 15,16].

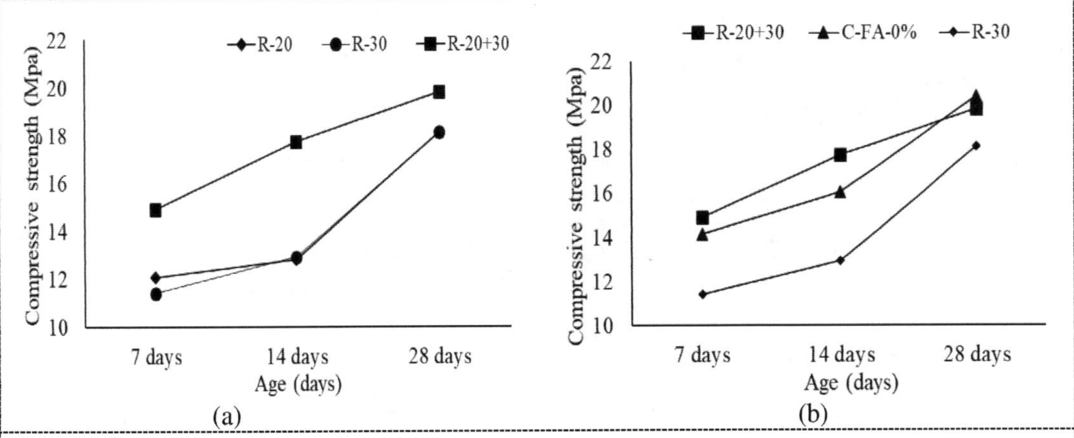

Figure 8. Compressive strength of the specimens

Figure 9(a) shows the increment in the percentage of fly ash in a mixture from 5% to 10% reduced the compressive strength of the PAC. This is similar to the findings of previous research which showed an increase in the partial replacement of OPC by FA decreased the compressive strength measured at the same date [12].

Figure 9(b) shows CFA2-5% had a lower performance compared to the CFA1-5% in terms of compressive strength. This is due to the different chemical properties of fly ash presented in Table 2 where the total percentage of CaO for FA-2 was observed to be more than 20% and classified as a cementitious material while FA-1 varied between 10 and 20% and categorized as a cementitious and pozzolanic material [14]. Therefore, CFA2-5% has a more rapid strength at the early ages of seven days than concrete with lower CaO such as CFA1-5% and this is in line with the findings of previous experiments [17]. Table 4 also shows the tensile strength of the specimens had a similar trend with compressive strength and analogous to the values obtained for the normal concrete. Moreover, the splitting tensile strength was found to be reducing as the fly ash content in the mixture increased.
The splitting tensile strength is normally estimated using the compressive strength but it also several equations such as ACI 318-05 [18], ACI 363 [19], Eurocode 2 [20], Neville [8], Sofi et al. [21], Lee and Lee [22] formulated using Table 5 with *ft* used to represent the splitting tensile strength (MPa) while *fc* is the average compressive strength (MPa).

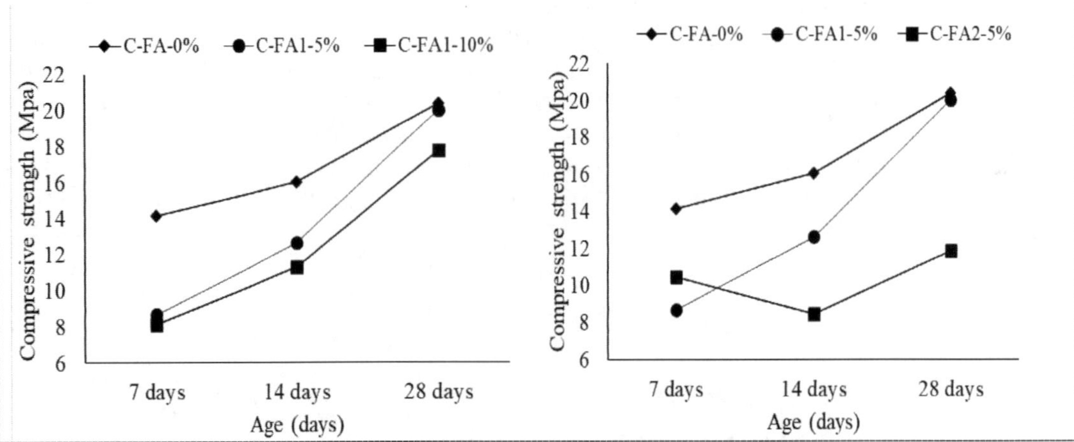

Figure 9. Compressive Strength of specimens

Table 5. Prediction models for splitting tensile strength

code	Equation (SI Unit)	Equations
ACI 318	$f_t = 0.56\sqrt{f_c}$	Equation 1
ACI 363	$f_t = 0.59\sqrt{f_c}$	Equation 2
Eurocode	$f_t = 0.33 f_c^{\frac{2}{3}}$	Equation 3
Sofi et al.	$f_t = 0.48\sqrt{f_c}$	Equation 4
Lee and Lee	$f_t = 0.45\sqrt{f_c}$	Equation 5
Neville	$f_t = 0.23 f_c^{0.67}$	Equation 6

Figure 10 shows the splitting tensile strength value obtained in this research is lower than all prediction models with ACI 318, ACI 363, and Eurocode [18-20] significantly overestimated. Meanwhile, the tensile strength for all the PAC specimens was observed to be slightly closer to the value for both Equations 5 and 6. It is, however, important to note that the proposed prediction models are strongly influenced by several factors such as the chemical and physical properties of raw materials.

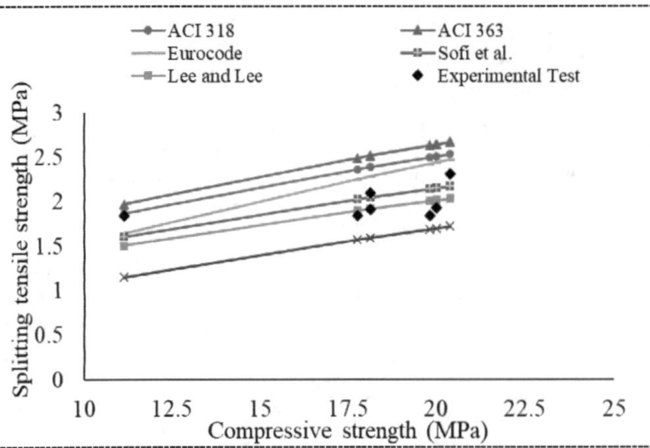

Figure 10. Comparison of the experimental and predictions model of splitting tensile strength

5. Conclusion

The relationship between the compressive and splitting tensile strengths of preplaced aggregate concrete produced using fly ash as a partial replacement of cement and coarse aggregate was determined in this study. Specimens with rounded shape aggregates at different sizes exhibited a similar performance with a 30 mm crushed aggregate. Moreover, specimens with higher calcium fly ash had more rapid strength at an early age of seven days while the tensile strength for all samples was slightly closer to the value for prediction models and previous experiments. Therefore, the partial replacement of cement with industrial waste material in the form of fly ash in preplaced aggregate concrete has the ability to save up to 10% of cement and also produce certain environmental benefits.

References

[1] ACI Committee 304.1 2005 *Guide for the use of preplaced aggregate concrete for structural and concrete applications* (Farmington Hills, Michigan, USA)

[2] Mirza J, Mirza M S, Roy, V, Saleh K. 2002, Basic rheological and mechanical properties of high-volume fly ash grouts. *Constr. Build. Mater.* **16** 353–363.

[3] Najjar M F, Soliman A M and Nehdi M L 2014 Critical overview of two-stage concrete: Properties and applications. *Constr. Build. Mater.* **62** 47–58

[4] Jing Lv, Tianhua Zhou and Kunlun Li 2020. Development and investigation of a new low-cement consumption concrete—Preplaced Aggregate concrete *Sustainability* **12** 1080

[5] Chairunnisa N and Fardheny A F 2019 The study of flowability and the compressive strength of grout/mortar proportions for preplaced concrete aggregate (PAC) *MATEC web of conference* **280** 04010

[6] Satyarno I, Solehudin A P, Meyarto D, Hadiyatmoko D, Muhammad P, and Afnan R 2014. *Procedia Eng.* **95**, 356–365

[7] S. Champion and L. Davis. 1958 Grouted concrete construction," *Reinf. Concrete Rev.*

[8] Neville A M 1995 *Properties of concrete* 4th Ed London Pearson Education Limited 443 (846) London

[9] Kunal K Das and Siu S. E. Lam 2019 Effect of Coarse Aggregate Size and Grouting Process on Properties of preplaced Aggregate Concrete *Proc. Of. the 4 World Congr on Civil, Structural, and Environmental Engineering (CSEE'19)*

[10] Coo M and Pheeraphan T 2015 Effect of sand, fly ash, and coarse aggregate gradation on preplaced aggregate concrete studied through factorial design *Constr.Build. Mater* **93** 812-821

[11] Ganaw A 2012 *Rheology of grout for preplaced aggregate concrete* PhD Thesis (UK, University of Bradford)

[12] Najjar M F 2016 *Innovating two-stage concrete with improved rheological, mechanical and durability properties* PhD Thesis (The University of Western Ontario)

[13] ASTM C618-12a 2012 *Standard Specification for Coal Fly Ash and Raw or Calcined Natural Pozzolan for Use in Concrete* ASTM International West Conshohocken PA

[14] ASTM C939 / C939M-16a 2016 *Standard Test Method for Flow of Grout for Preplaced-Aggregate Concrete (Flow Cone Method)* ASTM International, West Conshohocken PA

[15] Abdelgader H S 1999 How to design concrete produced by a two-stage concreting method, *Cem Concr Res*, **29** 3 331-337.

[16] Malley J O and Abdelgader H S 2010 Investigation into viability of using two-stage (pre-placed aggregate) concrete in Irish setting, *Front Struct Civ Eng in China* **4** 1 127-132.

[17] Burden D 2006 *The durability of concrete containing high levels of fly ash.* MSc Thesis (Illinois: Portland Cement Association University of New Brunswick)

[18] ACI Committee 2011 *Building Code Requirements for Structural Concrete and Commentary.* ACI Farmington Hills Mich pp 111.

[19] ACI 363 R-92, 1999. *State-of-the-Art Report on High-Strength Concrete* American Concrete Institute, Farmington Hills, Mich

[20] Eurocode 2 2004 *Design of Concrete Structures-Part 1–2: General Rules-Structural fire Design* (London: European Standard).
[21] Sofi M, Van Deventer J S J, Mendis P A, Lukey G C 2007 Engineering properties of inorganic polymer concretes (IPCs), *Cem. Concr. Res.* **37 (2)** 251–257
[22] Lee N K and Lee H K 2013 Setting and mechanical properties of alkali-activated fly ash/ slag concrete manufactured at room temperature, *Constr. Build. Mater.* **47** 1201–1209.

IOP Conf. Series: Materials Science and Engineering 1212 (2022) 012037 doi:10.1088/1757-899X/1212/1/012037

Determination of earthquake prone zones at university of tadulako based on dominant periods and peak ground acceleration (PGA)

I U Meidji[1*], S Mulyati[2], N R Janat[1], H Jayadi[1], Asrafil[1]

[1]Geological Engineering, Civil Engineering Faculty, University of Tadulako
[2]Architect Engineering, Civil Engineering Faculty, University of Tadulako

[*]corresponding e-mail : ichauntari@untad.ac.id

Abstract. University of Tadulako is the largest State University in Palu City. When an earthquake with a magnitude of 7.4 Mw on September 28, 2018, occurred, many buildings were damaged and even collapsed at that time, even causing casualties due to the rubble. Research on the Local Site Effect is essential for the assessment of seismic hazard. In this study, the local site effect was analyzed using the HVSR method based on microtremor data. The predominant Period (To) ranges between 1.709 s to 3.816 s, indicates that this area consists of alluvium and has a very thick sediment layer. Another parameter calculated in this paper is the peak ground acceleration (PGA) with values from 0.914 g to 0.924 g. This value is the first indicated soil damage level due to ground motions. The results of this study can be used as a consideration in the development of regional spatial planning and building structures based on earthquake analysis.

1. Introduction

The earthquake that occurred on 28 September 2018 caused various damage in various places in Palu City and its surroundings, including one of the state university in Palu City, University of Tadulako. The greatest strength of the earthquake at that time was reaching 7.4 mw, shaking with an epicenter distance about 70 km, causing damage to buildings on this university. The earthquake caused secondary disasters such as liquefaction, tsunami, landslides and collapsed buildings [7][20]. This following disaster caused humans to be injured even caused death. The collapsed building incident also occurred on a University of Tadulako, there were casualties due to the collapse of buildings. However, it turns out that not all university buildings collapsed during the earthquake, some buildings still survived well, some were cracked, badly damaged and even collapsed. This assumes that the level of damage to buildings is not fully influenced by the magnitude, epicenter distance and earthquake duration but also the characteristics of the soil at the local site [12]. Certain soil characteristics can increase the damage rates during the earthquakes [10].

One of the soil characteristic parameters that can be analyzed related to earthquake shocks and its impact on buildings is the dominant period. The method used to obtain the dominant period value in this study was by analyzing the HVSR curve from microtremor data. Similar studies that have been conducted before are research by Konno et al [11], Nakamura et al [15], Parolai, at al [17], Marjiyono [16], Kanai [2] and many more studies using dominant period to see the soil characteristics. According to [10], the HVSR method introduced by Nakamura is considered to be very economical and effective for studying the dynamic characteristics of the surface soil that cause the local site effect during an earthquake.

Another analysis to determine the impact of an earthquake on a local site that can be done is peak ground acceleration (PGA), the value of PGA is an important factor in determining amount of the impact of earthquake damage [22].The PGA value can be determined through a ground acceleration measuring device or an accelerograph, but this method requires the absence of a lot of tools, time and costs [9], another way is by using an empirical approach that can provide PGA values according to the points needed [6] as presented in this paper. In this study, the empirical formula used was introduced by Fukushima and Tanaka [5] to determine the risk of vulnerability at the location based on the magnitude and hypocenter distance of the earthquake on 28 September 2020.

According to Cipta, et al. [4] in Abdullah [1] a fault in Palu City can provide earthquake risk to the X level scale Modified Mercalli Intensity (MMI) by the United State Geology Society (USGS) scale. Therefore, by analyzing the two parameters, both the dominant period and the PGA can be used as an evaluation material for the disaster-based development plan of University of Tadulako to reduce the risk if an earthquake occurs.

2. Method

2.1 Regional Geology

Based on a regional geological map by Sukamto, et al., [19] at 1: 25,000 scale, the regional stratigraphy of Palu City is divided into 4 units, which are the Tinombo Formation, Intrusive Rock Complex, Molasa Celebes, and Alluvium and Coastal Deposits. University of Tadulako itself is in 2 formations, namely the alluvium and molasa formation. The alluvium formation in the west and further to the east is a molass formation, the lithology of the two formations is sandstone and gravel. Alluvium and coastal sediment estimated to be in Holocene age consisting of gravel, sand, mud, and coral limestone, each formed in river, delta and shallow sea environments, which are the youngest sediments in this area [19].

Tinombo formation is the oldest unit in the study area. This formation is exposed in the west and east bunds. This formation also overlaps the metamorphic rock complex, the oldest rock in the Central Sulawesi region, inconsistently. Inside there are rubble originating from metamorphic rocks. These deposits consist of shale, conglomerates, sandstones, chert, and volcanic rocks, which are deposited in the marine environment. Near the intrusion there are slates and scraped rock and near the contact there are filites and quartzites. The western part of the bund contains the most sandstones and chert. Volcanic rock rubble is generally found in sandstones [19].

2.2 Horizontal perVertical Spectral Rasio (HVSR)

Collecting data in this study using the microtremor method that produces data signals, this signals then can be analyzed using the HVSR (horizontal per vertical ratio) method. This method is one of the easiest and cheapest methods to determine subsurface characteristics without damaging the surface. Technically, this data relies on a spectrum of ambient (micro seismic) vibrations that are spread over the earth's surface in the horizontal (North-South and East-West) and vertical (Up-Down) direction. The microtremor method was introduced by Kanai and Tanaka [8] to determine the local geological characteristics associated with the subsoil structure which is now developing and is used to identify the dominant period value at the site point. The HVSR technique itself was developed by Nakamura [14], he found that the resonant frequency increases linearly with the H / V ratio from the microtremor until a frequency peak is obtained. The H / V value is considered as the amplification of the ground motion at the measurement point.

As previously explained, the HVSR method analysis produces resonant frequency and ground motion amplification values. These two values can be used to analyze the geological character of the local area related to the soil response that can damage buildings due to earthquakes. This value is obtained from the peak of the H / V ratio as in equation 1.

$$(H/V)_m = \frac{(\{S_{NS}\}^2+\{S_{EW}\}^2)^{1/2}}{S_{UD}} \quad (1)$$

where (H / V) m is the spectrum of the H / V ratio, S_{UD} is the spectrum of natural vibrations of vertical movement, S_{NS} and S_{EW} are two horizontal movements [14]

A simulation test was conducted by Lachet and Brad [13] which showed that the peak frequency value (dominant frequency) changes with variations in geological conditions. In this study, the value of the dominant period to be analyzed is obtained by equation 2.

$$T_0 = \left(\frac{1}{f_0}\right) \quad (2)$$

Where T_0 is the dominant period of the soil and f_0 is the dominant frequency of the soil. The dominant period is the value of time when a microtremor wave travels from the surface through the sedimentary layer until it meets the bedrock or experiences reflection against its reflected plane [21]. Based on the value of this dominant period, the soil can be classified by Kanai classification as showed in table 1 [2].

Table 1. Soil Classification based on the Soil Dominant Period [2]

Soil Classification	Dominant Periods	Kanai Classification	Explanation
Type I	0.05 – 0.1	Tertiary rocks	The layer is not thick and consists of hard rocks (< 5m)
Type II	0.1 – 0.25	Pebbly sand	The layer's thicknesses thicker than that of type I (5-10)
Type III	0.25 – 0.4	Alluvial rocks and buff formation	The layer's thickness is 10-30m
Type IV	> 0.4	Alluvial rocks formed from sedimentation	The layer's thickness is very thick.

In addition to the dominant period, this research also calculates the value of peak ground acceleration (PGA) which is the response of the maximum ground motion acceleration to earthquakes that have occurred in the sediment layer in that place. PGA is an important parameter to determine the impact of earthquake shocks [3]. PGA is expressed in g (gravity acceleration) in cm/s^2 or gal, where 1gal is equal to 0.01 m / s^2. The calculation of PGA in this study uses the Fukushima and Tanaka [5] as shown in equation 3.

$$\text{Log}\ \alpha = 1.3 + 0.41\ M_w - \text{Log}\ (R \,.\, 0.32 \,.\, 10^{\,0.41M_w}) - 0.0034\ R \quad (3)$$

Where: α is the peak ground acceleration value of the observation point in cm/s^2, M is the Moment Magnitude and R is the Hypocenter distance.

3. Results and Discussion

The dominant period value is obtained from the HVSR spectrum in Figure 1 through microtremor data collection. Figure 2 shows how the window selection is in accordance with the instructions from the Site Effects Assessment Using Ambient Excitations [18] in order to obtain a reliable HVSR curve that represents the subsoil conditions of the study area, this curve is obtained through equation 1. In this study, 14 points microtremor data were taken with a spacing of less than 250 meters. Figures 3a and 3b represent the results of the HVSR spectrum analysis of the study area. Points 6, 7, 8, 9, 10 and 12 have a curved shape like Figure 3a, which is a curve without a clear peak. While the other 8 points have a clear peak curve shape as shown in Figure 3b.

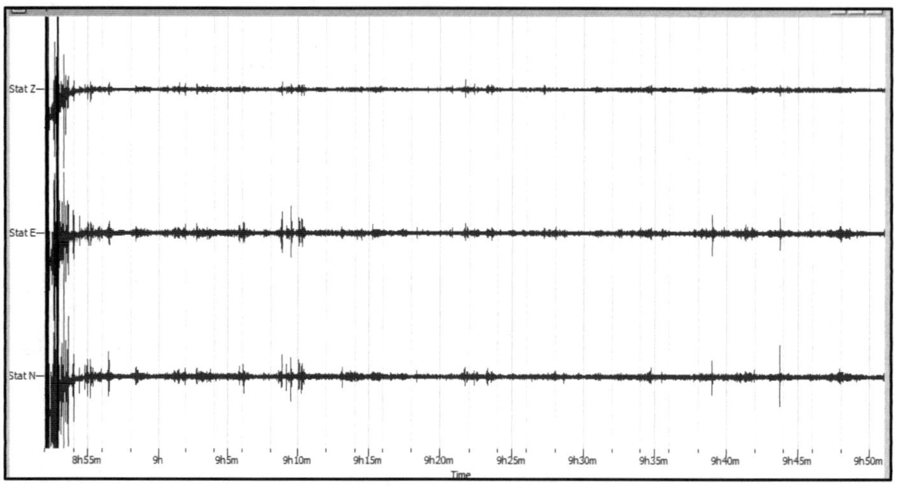

Figure 1. 3 component data signals from microtremor measurement

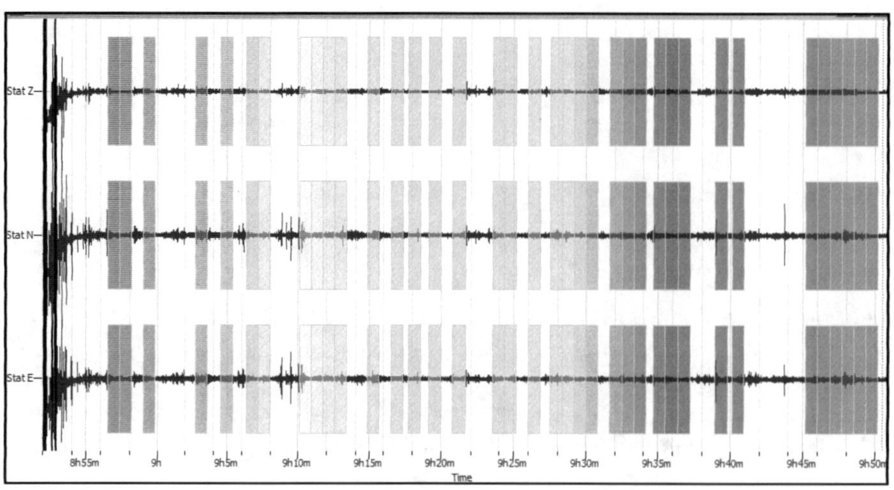

Figure 2. Window Selection

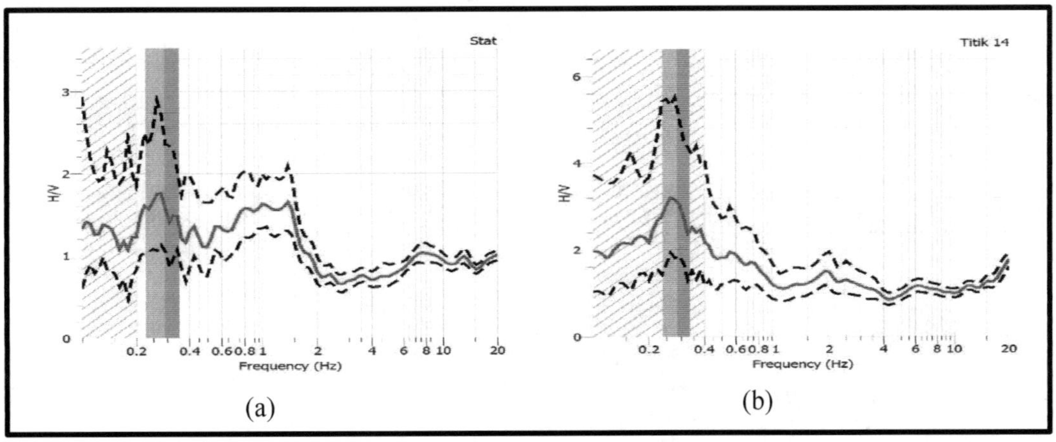

Figure 3. HVSR Curve (a) without clear peak and (b) clear peak

The information obtained from the HVSR curve is the value of the resonant frequency or the dominant frequency (f_0), then the dominant period value is calculated using equation 2. The calculation results show that the value of the dominant period varies from 1.709 s to the highest dominant period is 3.816 s. This indicates that the entire study area has a type 4 soil classification consisting of alluvial rocks formed from sedimentation, in accordance with the geological conditions of the research area based on[19] which states that this location consists of weathered alluvial and molasses. In addition, the value of the dominant period which is more than 0.4 indicates that the thickness of the sediment layer is very deep. Conditions like this are relatively dangerous for the buildings that stand on them.

Figure 4 is a map of the period dominant distribution values which provides an overview of areas that have the potential to cause more serious damage to buildings. The north and east sides of this area have a greater potential for building damage if an earthquake occurs compared to other locations. Building construction is recommended in areas with blue to green area on the map that have a lower dominant period value. Construction of buildings in yellow to red areas on the map should be avoided or strengthening the structure of buildings in these areas because of the higher risk location.

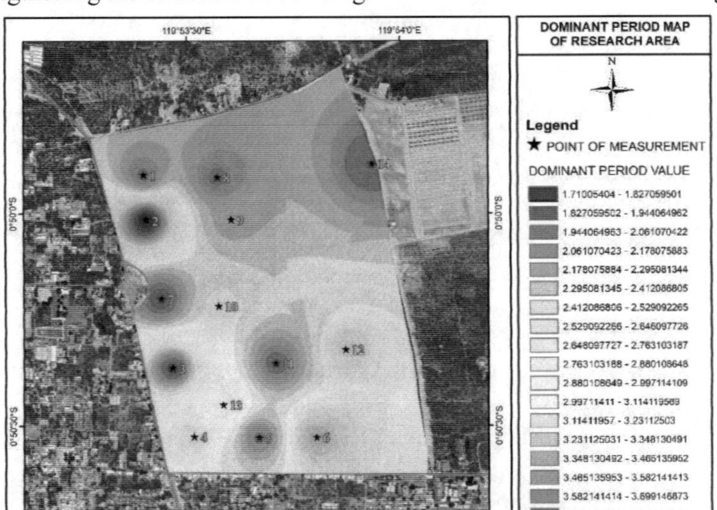

Figure 4. Map of dominant periods distribution around the University of Tadulako

The PGA map in Figure 5 is the response of the ground to the earthquake that occurred on 28 September 2020 with a magnitude of 7.4 MW. The calculation of PGA value for Fukushima and Tanaka equation only depends on the hypocenter position and the earthquake magnitude. So that the value distribution pattern corresponds to the hypocenter distance of the earthquake at that time. This can be seen from the large PGA value in the north and smaller towards the south because it is farther from the epicenter. The PGA value in each location was not so significant, the difference was from 0.914 g to 0.934 g. The location which have high dominant period and high PGA value causes greater damage to buildings at the local site.

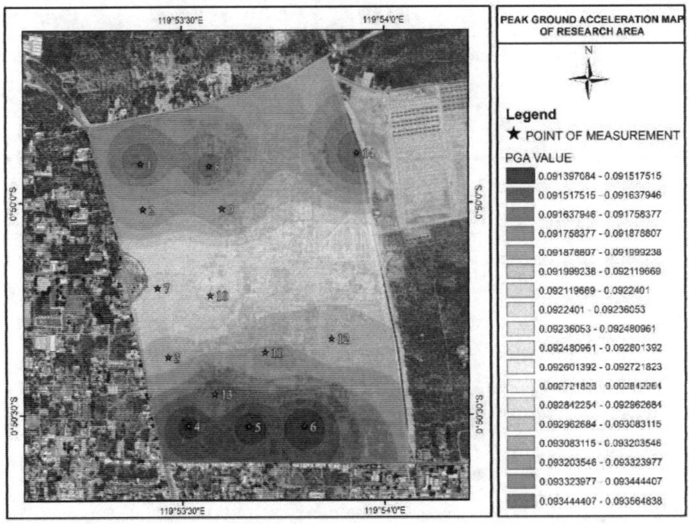

Figure 5. Map of Peak Ground Acceleration (PGA) distributions around the University of Tadulako

4. Conclusion

This study presents a brief result of the risk of soil vulnerability at University of Tadulako. Based on the HVSR analysis, the study area has a relatively high dominant period value varying from 1.7 s to 3.8 s and is associated with soil type 4 it means Alluvial rocks formed from sedimentation and The layer's thickness is very thick. Based on this, it can be said that the thickness of the sediment in the north and east of the study location is greater than the other locations. The PGA values did not vary much with the values from 0.914 g to 0.924 g. Earthquakes have the potential to occur in the future, evaluation of spatial plans and building structures at University of Tadulako is needed to reduce the risk of damage to existing building construction.

References

[1] Abdullah A. I, Abdullah. 2019. *A field survey for the rupture continuity of Palu-Koro fault after Donggala earthquake on September 28th, 2018*. Journal of Physics: Conference Series. doi:10.1088/1742-6596/1434/1/012009

[2] Kanai, K., dan Tanaka, T. (1961). On Microtremors VIII. *Bulletin of the Earthquake Research Institute* 39, 97–114.

[3] Campbell, Kenneth W, Bozorgnia, Yousef. 2006. *Campbell-Bozorgnia NGA Empirical Ground Motion Model for theAverage Horizontal Component of PGA, PGV, PGD and SA at Selected SpectralPeriods Ranging from 0.01–10.0 Seconds(Version 1.1)*. Pacific Earthquake Engineering Research Center, College of Engineering, University of California, Berkeley.

[4] Cipta A, Robiana R, Griffin J D, Horspool N, Hidayati S and Cummins P. 2016. *A probabilistic seismic hazard assessment for Sulawesi, Indonesia*. Geological Society of London Special Publications, 441, https://doi.org/10.1144/SP441.6

[5] Fukushima, Y. and Tanaka, T. .1990. *A New Attenuation Relation for Peak Horizontal Acceleration of Strong Earthquake Ground Motion in Japan*. Bulletin of the Seismological Society of America, 80(4), 757 - 783

[6] K. S. Vipin, P. Anbazhagan, and T. G. Sitharam. 2009. Estimation of peak ground acceleration and spectral acceleration for South India with local site effects: probabilistic approach. *Nat. Hazards Earth Syst. Sci*., 9, 865–878

[7] Abdullah A I, Abdullah, Musa M D T, Uno I, Jayadi H and Sahabudin 2019 Resistivity model of hallow subsurface to find the path of geothermal manifestation in Bora Village of Sigi Regency of Central Sulawesi *J. Phys.: Conf. Ser.* **1242** 012048

[8] Kanai K, Tanaka T. 1961. *On microtremors VIII*. Bulletin of the Earthquake Research Institute 1961; 39:97e114.

[9] Sunaryo. 2017. Study of Peak Ground Acceleration (PGA) by Means of Microzonation Data: Case Study on Batubesi Dam of Nuha, East Luwu, South Sulawesi, Indonesia. *AIP Conference Proceedings* 1908, 030013 (2017); https://doi.org/10.1063/1.5012713

[10] Kivan zorlu. 2018. *Determination of predominant site period of loose terrestrial Units (caliche) with microtremor measurements upper cretaceous-tertiary geology/stratigraphy of pertek and its vicinity (tunceli, turkey)*. Bull. Min. Res. Exp. 156: 205-220

[11] Konno, K. and Ohmachi, T. 1998. *Ground-motion characteristics estimated from spectral ratio between horizontal and vertical components of microtremor*. Bulletin of the Seismological Society of America 88(1), 228–241

[12] Kramer, S.L., 1996. *Geotechnical Earthquake Engineering*, Prentice -Hall, Inc., h.1 - 2.

[13] Lachet, C., dan Brad, P.Y., 1994. *Numerical and Theoretical Investigations on The Possibilities and Limitations of Nakamura's Technique*. J. Phys. Earth, 42, 377-397. Research Institute 1989;30(1):25e33

[14] Nakamura Y. 1989. *A method for dynamic characteristics estimation of subsurface using microtremor on the ground surface*. Railway Technical Research Institute/Tetsudo Gijutsu Kenkyujo Volume 30.

[15] Nakamura, Y. 2000. *Clear identification of fundamental idea of Nakamura's technique and its applications*. Proceedings of the 12th World Conference on Earthquake Engineering. Auckland, New Zealand.

[16] J. W. Lane, J. R., E. A. White., G. V. Steele and J. C. Cannia, "Estimation of bedrock depth using the horizontal-to-vertical (H/V) ambient-noise seismic method," in Symposium on the Application of Geophysics to Engineering and Environmental Problems (Colorado, Environmental and Engineering Geophysical Society, Denver, Colorado, 2008).

[17] Parolai, S., and Galiana-Merino, J.J. (2006). *Effect of transient seismic noise on estimates of H/V spectral ratios*. Bull. Seism. Soc. Am., 96, 228-236.

[18] SESAME Project. 2004. *Guidelines for the implementation of the H/V spectral ratio technique on ambient vibrations measurements, processing, and interpretation*. http://sesame-fp5.obs.ujf-grenoble.fr/Papers/HV_User_Guidelines.pdf

[19] Sukamto, R., Sumadirdja, H., Suptandar, T., Hardjoprawiro, S., Sudana, S., 1994. *Peta Geologi Lembar Palu, Sulawesi, Skala 1: 250.000*. Pusat Penelitian dan Pengembangan Geologi, Bandung

[20] M. Choudhury, S. Verma, and P. Saha, "Effects of earthquake on the surrounding environment: An overview," *Proceedings of International Conference on Recent Advances in Mechanics and Materials (ICRAMM-2016)*, vol. RR03, p. 8, 2016.

[21] Seo, K., Haile, M., Moriuchi, M., Mori, K., Sueoka, H., Mualchin, L., Sheng, L. H. and Enomoto, T., 1994. *Microtremor measurement in Los Angeles area –A Preliminary Study of Seismic Microzonation*. Technical Report, Tokyo Institute of Technology.
[22] Yih-Min Wu, Tzay-Chyn Shin, and Chien-Hsin Chang. 2001. *Near Real-Time Mapping of Peak Ground Acceleration and Peak Ground Velocity Following a Strong Earthquake*. Bulletin of the Seismological Society of America, 91, 5, pp. 1218–1228

Acknowledgment
Central Sulawesi Province Meteorology and Geophysics Agency (BMKG) for their cooperation in data collection and discussion.

Potential of tidal power plants with least square method on Toaya Beach

H Azikin, Nurhidayat, E Affandy, S S Syahruddin

Department of Civil Engineering, Tadulako University, Middle Sulawesi, Indonesia

Email : h_azikin@yahoo.com, hidayatsaja@gmail.com, sarahsahnazsyahruddin@gmail.com

Abstract. The use of energy, especially electrical power, is needed by the wider community—lots of alternative energy from nature, especially in Indonesia that can be utilized to produce electricity. One of the newest alternative examples is the energy produced by tides. Tidal energy is a type of renewable energy that is relatively more predictable in number. One method that can process the waves is the Least Square Method. Where the results of data processing using this method can be known based on tidal observations for 15 days on Toaya Beach located in Donggala Regency, Central Sulawesi, the types of tidal types that occur based on the results of Formzahl's count is 0.357 that the tidal types that occur are varied types leaning to double daily, with a Root Mean Square Error (RMSE) of 0.028, which means that the difference between predictive results and observational data is accurate. Then also obtained the value of the design water level elevation that is the highest high water level (HHWL) is 2.7 meters, and the lowest low water level (LLWL) is -0.3 meters with a height difference is 3.0 meters. So with the high height difference, the total energy generated by two times a day with a sea area of 9 km2 which is capable of producing energy movement from seawater can be calculated the potential energy generated from the waves of Toaya Beach, which is 5,53 MW.

Keywords: Tidal, Least Square method, formzahl, elevation, energy potential.

1. Introduction

Indonesia is a country that is developing towards changes towards the better for advancing the nation. However, in achieving these change goals, Indonesia is faced with various problems. One of the issues that occur is that along with the development of human civilization, Indonesia's population is increasing so that the need for energy is also growing. So far, most energy needs come from non-renewable energy, such as coal, oil and natural gas, which are usually used as fuel for electricity generation.

Therefore, the existence of ideas that can generate electrical energy is vital. In this case, the issue of tidal power generation will be discussed both from the generator, raw materials to facilitate the generation process and how the generator works so that it can generate sufficient electrical energy.

The use of tides for electricity generation is the right step to minimize the use of fossil fuels as fuel for conventional power plants. One method that can process the waves of seawater is the Least Square Method. This method is one of the ways that can be used to process tidal data with high data frequency accuracy. This method is often used after the invention of modern numerical calculators. This is because this method requires a matrix calculation process with large matrix dimensions (Gumelar, 2016).

Toaya is one of the villages in Sindue Subdistrict, Donggala Regency, which throughout the town is located on the coast of the open sea so that the authors chose this village as a reference that the beach will research Tidal Power Plants (TPP).

IOP Conf. Series: Materials Science and Engineering 1212 (2022) 012038 doi:10.1088/1757-899X/1212/1/012038

The purpose of this study is to determine the type of tides that occur at Toaya Beach based on the calculation of Formzah numbers using the Least Square method, to obtain the water level elevation of the HHWL and LLWL plans on Toaya Beach, and to determine the potential generated by tidal power plants. Sea on Toaya Beach.

Based on the description above, the formulation of the problem of this study is how the type of tides that occur at Toaya Beach based on the calculation of Formzahl numbers using the Least Square Method, what is the elevation of the HHWL and LLWL design water levels on Toaya Beach, and whether there is potential for tidal power plants. Low tide at Toaya Beach.

2. Overview Of The Research Location

The research location is planned to be located in Toaya Village, Sindue District, Donggala Regency, Central Sulawesi Province, approximately 40 km from Palu City. With an area of 23,540 hectares, which consists of 5 hamlets. Coordinates 00 ° 36'7,1 / 119 ° 47'28,8 ', and the typology is coastal / fisherman. The height of the land above the sea level is 100 asl.

Figure 1 Research Sites at Toaya Beach (Google Earth, 2020)

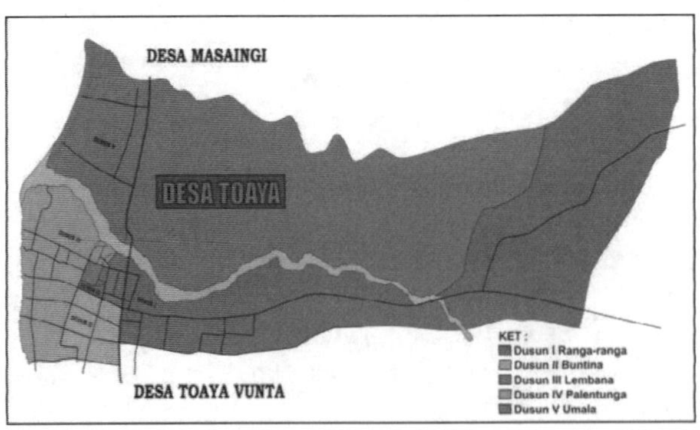

Figure 2. Map of Toaya Village (Village Profile, 2016)

3. Literature Review

3.1 Tidal

Ocean tides are a phenomenon of the periodic up and down movement of sea levels caused by a combination of forces and attractive forces from astronomical objects, especially by the sun, earth and moon. The influence of other celestial bodies can be ignored because of their greater distance or smaller size (Dronkers, 1964).

IOP Conf. Series: Materials Science and Engineering 1212 (2022) 012038 doi:10.1088/1757-899X/1212/1/012038

3.2 Tidal Curve

Tidal conditions can be displayed visually in the form of tidal curves. The curve describes the sea level at a particular time. The x-axis shows the time, while the y-axis shows the sea level. Tide height is the vertical distance measured from the top of the highest water to the lowest water position. The tidal period is the time it takes from the highest (or lowest) water level to the next highest (or lowest) water level. The period when Fenomena sea level moves up is called the Tide period, while the period when the sea level moves down is called the low tide period.

According to Nontji (1987), there are four types of ebb and flow based on their period and regularity. Within a month, the daily variation of the tidal range changes systematically with the lunar cycle. The tidal range also depends on the shape of the waters and the configuration of the ocean floor. Here below the tides of seawater in Indonesia are divided into four types, namely: Regular Double Daily, Mixed Semi-Diurnal Tides, Mixed Semi-Diurnal Tides, Regular Single Daily Tides.

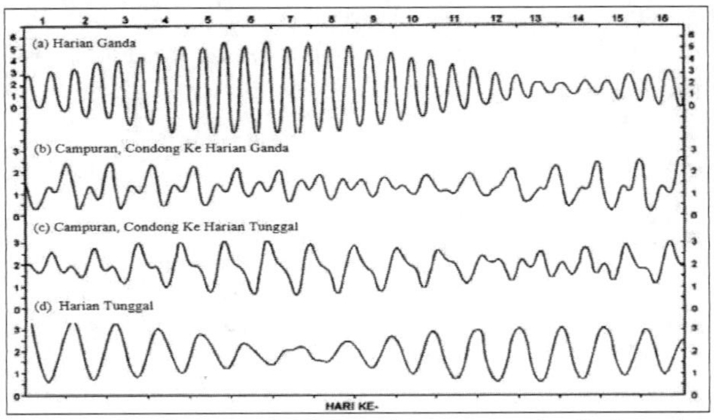

Figure 3. Types of Tides that occur in Indonesia (Triatmodjo, 1999)

Apart from looking at the tide data plotted in the form of a tidal type graph, it can also be determined based on the Formzahl (F) number which is expressed in the form of the following equation:

$$F = \frac{AK_1+AO_1}{AM_2+AS_2} \tag{3.1}$$

Where:

F : *Formzahl* or tidal constant

AK_1 : The amplitude of the single daily mean tidal wave chicks are affected by the declination of the moon and the sun

AO1 : The amplitude of the mean available daily tidal wave children as affected by the declination of the sun

AM2 : The amplitude of the sub-daily mean double tidal wave affected by the moon

AS2 : Amplitude of the mean daily double tidal wave child affected by the sun The tidal type classification can be seen in **Table 1.**

Table 1. Tidal Type Classification

Form Value	Type of Tide	Phenomenon
$F \leq 0,25$	Semi Diurnal Tides	2x pairs a day with the same relative height
$0,25 < F \leq 1,5$	Mixed Mainly Semidiurnal Tides	2x pairs a day with different heights and different intervals

IOP Conf. Series: Materials Science and Engineering 1212 (2022) 012038 doi:10.1088/1757-899X/1212/1/012038

$1,50 < F \leq 3,0$	Mixed Mainly Diurnal Tides	1x or 2x pairs a day at different intervals
$F > 3,0$	Diurnal Tides	1x pair a day, when Spring can occur 2x pairs a day

(Source: Triatmojo, 2010)

3.3. Tidal Power Plant

Tidal Power Plants Seawater is a power plant that utilizes tidal energy in seawater to make energy in other forms, especially electricity, where tides can drive turbines to run a generator. Tidal power generation for seawater is renewable energy. There are two types of generator types, namely Tidal Dam (Barrage Tidal System) and Offshore Turbines (Onshore Turbines). The Sea Water Tidal Power Generation System is very suitable for use in Indonesia, especially in Central Sulawesi, because Indonesia is known for its ocean waves.

3.4. The advantages of TPP

1. Once built, tidal energy can be obtained free of charge.
2. Does not produce greenhouse gases or other waste.
3. Does not require fuel
4. Low operating costs.
5. The tides are predictable.
6. Offshore turbines have low installation costs and do not cause a massive environmental impact

3.5. Disadvantages of TPP

1. Depends on the characteristics of the Tides.
2. Need a precise location.
3. The tools must be sturdy so they can withstand any weather conditions.
4. A dam that covers a river mouth is costly to build and covers a vast area so that it changes the environmental ecosystem both upstream and downstream for many kilometres.
5. Can only supply energy for approximately 10 hours per day, when the waves move in or out.

3.6. Tidal Power Plant (TPP) Type Onshore Turbines

This tidal power plant uses an offshore turbine method which is more like an underwater version of a wind power plant. The shape of the tidal turbine is as varied as the wind turbine.

This system does not require a dam but is directly installed in the open seas; the thrust is generated from the movement of the kinetic energy of ocean currents because the density of water is higher than the wind, the onshore turbine can produce more incredible power with the same size as the wind turbine where the tides can also be utilized so that they can drive turbines and generators. This explanation can be seen in figure 4

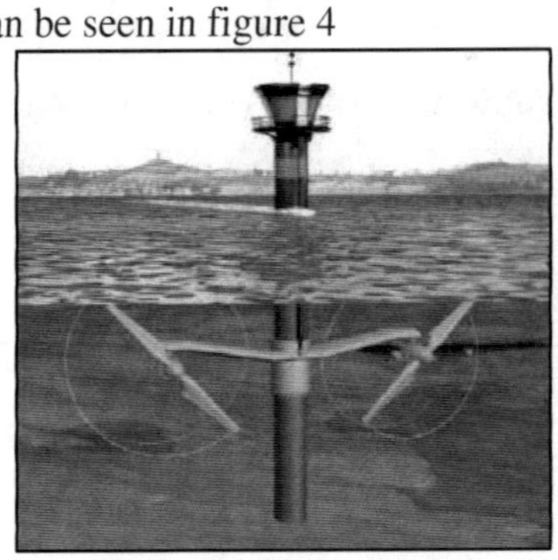

Figure 4. Tidal Turbine in the Sea

Where to calculate the amount of energy produced can be calculated using the following equation:

$$E = \frac{1}{2}\, x\, A\, \mathrm{x}\, \rho\, x\, \mathrm{g}\, x\, H^2 \qquad (3.2)$$

Where:
E = Energy (Joule)
A = Sea area (m^2)
ρ = Sea water density (kg/m^3)
g = Gravity (m/s^2)
H = Head (m)

For the sea area (A) which is estimated to be able to push and produce energy movement in the sea area, it is assumed that the surface of the tidal energy generator is 9 km² (Theocsbury, 2020).

3.7. Tidal Harmonic Components

The tides are periodic so they can be predicted. To predict tides, the amplitudes and faces of each component of the tidal generator are needed. The main features of waves consist of the daily central part, daily tides and long period tides. However, due to the interaction with the shape of the beach morphology and the superposition between the primary tidal waves, new tidal components are formed. These main components, also called astronomical frequencies, are used in the creation of interactive programming for the decomposition of tidal components. In the tidal analysis using the least-squares method, the amplitude and phase components of the tidal harmonic equation are calculated based on observational data on the position of the water level within a certain period, using ingredients with known frequencies. The number of tide components that can be decomposed depends on the length of the data, the longer the observation data, the more tidal details that can be produced. For example, the number and pairs of components resulting from observation with a length of 1 month will differ from an observation length of 1 year. The tidal parts can be seen in Table 3.2 as follows.

3.8. Least Square Method

The principle of tidal analysis with the least-squares method is to minimize the difference in the composite signal and the signal size. The equation for the least-squares process can be seen in the following equation:

$$h(t) + v(t_n) = hm + \sum_{i=1}^{k} A_1 \cos(\omega_1 t - g_i) \qquad (3.3)$$

Where:
h(t) : water level a function of time
Ai : the amplitude of the ith component
ω_1 : the angular velocity of the ith component
gi : the ith component phase
Hm : mean water level
t : time / time interval
k : the number of components
$v(t_n)$: residue

3.9. Error Rate

Root Mean Square Error (RMSE) is the average value of the squares of the error; it can also express the size of the error generated by a forecast model. A low RMSE value indicates that the variation in the amount generated by a forecast model is close to the variation in its observation value, meaning that the closer to zero the RMSE cost, the more accurate the elevation prediction results are. RMSE is the simulation result error rate. The calculation of the error rate aims to determine how much difference the simulation results from data with the measurement data. The analysis of the RMSE value is calculated using the following equation:

$$RMSE = \sqrt{\frac{\sum_{i=1}^{n}(Yi-\hat{Y}i)^2}{n}} \tag{3.4}$$

Information:
Yi : initial data (observational data)
Ŷi : final data (predictive data)
N : the amount of data

3.10. Water Level Elevation

Given the sea level elevation is always changing every time, it is necessary to set a height based on tidal data, which can be used as a guide in planning a port. Some of these elevations are; High Water Level, (HWL), Low Water Level (LWL), Mean High Water Level (MHWL), Mean Low Water Level (MLWL), Mean Sea Level (MSL), Highest High Water Level (HHWL), Lowest Low Water Level (LLWL), Higher High water level (HHWL), dan Lower Low Water Level (LLWL).
Sea level elevation can be calculated from the value of the tide components obtained from the calculation of the tide analysis. The following are some equations for determining the design water level (Dalpan et al., 2015).

3.11. Bathymetry

A bathymetric map is a map of the depth of the sea which is expressed in several depths or depth contours measured against a vertical datum. Bathymetry (from the Greek meaning "depth" and "measure") is the study of the centres underwater and the three-dimensional study of the floor of an ocean or lake. A bathymetric map generally displays base or direct relief with contour lines called depth contours (isobaths) and may have additional information in the form of surface navigation information.

In general, the seabed relief is not very varied compared to land relief. This is due to weak erosion and sedimentation.

4. Results And Discussion

4.1. Tidal Data Analysis

The data to be analyzed is tidal data from observations in the field which take place every hour for 15 days, from March 14, 2020, to March 28, 2020.

From the data observation process, table compilers can be made to facilitate tidal data analysis. Then from this table, we can obtain a graph of the tidal observations as follows in figure 5:

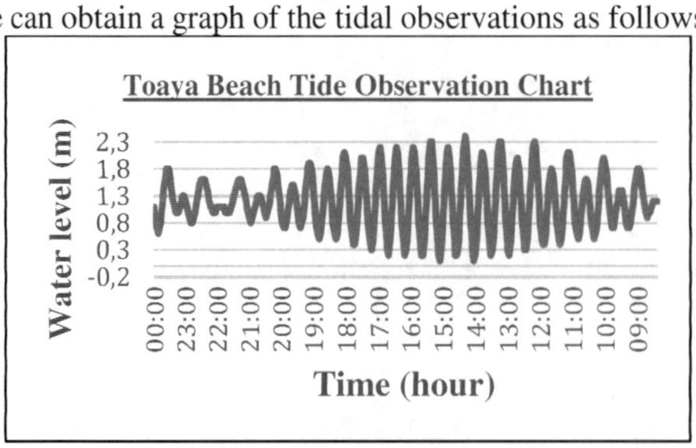

Figure 5. Tide Observation Graph

4.2. Results of the Least Method of Tidal Data Analysis

Based on the results of tidal data analysis at Toaya Beach with the Least Square method, the tidal constituent value of 2 harmonic constants is obtained, namely, amplitude and phase difference. The following is the final result of harmonic constants between Amplitude (A) and Phase Difference (g°) using the Least Square method in Table 2.

Table 2. Final Result Harmonic Constants Least Square Method

No.	Tidal constituents	Amplitude (A)	Different Phase (g)
		(m)	(°)
1	K1	0.199	174.394
2	O1	0.127	189.575
3	M2	0.545	348.719
4	S2	0.369	126.771
5	N2	0.072	66.343
6	M4	0.003	222.122
7	K2	0.091	94.080
8	P1	0.076	15.351
9	MS4	0.001	154.844

By experimenting with 9 constituents, the results are the Formzahl number, RMSE, water level elevation, as follows:
Formzahl number (F) : 0,357
RMSE : 0.028
HHWL : 2,7 m
MSL : 1,2 m
LLWL : -0,3 m

The results of the calculation using the Least Square method, can be obtained the type or type of tides based on the Formzahl number with comparisons **[F = $(AK_1) + (AO_1) / (AM_2) + (AS_2)$]**. Here's how to calculate it with a comparison of the F value:

F = [(AK1) + (AO1) / (AM2) + (AS2)]
= [(0,199) + (0,217) / (0,545) + (0,369)
= 0,357

Terms:
$F \leq 0.25$: Double daily pairs (semi diurnal)
$0.25 < F < 1.50$: Mixed tide prevailing semi diurnal
$1.50 < F < 3.00$: Mixed tide prevailing diurnal
$F \geq 3.00$: Single daily tide (diurnal tide)

Based on the requirements for the Formzahl number, the F value for the Least Square method is obtained, namely 0.357 ($0.25 < F < 1.50$), so the type or type of tide for the Toaya Beach area, Donggala Regency, Central Sulawesi is mixed tide prevailing semi diurnal). This means that in one day there are two tides and two ebbs with different heights and intervals.

4.3. Calculating the Potential Electrical Energy Generated

To calculate the electric power that can be generated based on the tidal height of Toaya Beach, the description of the calculation of the power / energy that can be generated is as follows, by comparing the energy produced by forecasting sea water waves.

Known :
ρ = 1025 kg/m^3
g = 9,81 m/s^2
htidal = HHWL – LWLL = 3,0 m
A = 9 km^2 (assumed that the surface area utilizes tidal energy)
Etidal value = ½ x A x ρ x g x h^2 = 39,8 x 10^{10} J

IOP Conf. Series: Materials Science and Engineering 1212 (2022) 012038 doi:10.1088/1757-899X/1212/1/012038

From the results of the previous discussion that on Toaya Beach there were 2 tides and 2 receding times a day, then the potential energy produced when the sea water receded was zero (0). So that the total electrical energy that can be generated in one day can be: 79,6 x 10^{10} J

From the values obtained in the above calculations, it can be calculated that the amount of electrical energy that can be produced per day in energy units is as follows:

E/day = 79,6 x 10^{10}J / (24 x 3600)
= 9,22 x10^{6} Watt
= 9,22 MW

It is assumed that the efficiency value of the generator is 60%., So the potential generated by the tidal effect of sea water on Toaya Beach with an estimated sea area can encourage and produce energy movement in the sea area covering an area of 3x3 km2 and a head as high as 3 meters which is roughly 5, 53 MW.

4.4. Determining the Position of TPP

Bathymetry maps can be interpreted as a measure of the depth of the sea, both regarding the size of the seafloor elevation which is a source of information and an overview of the seabed, and provides many clues about the structure of the sea (Nurjaya, 1991).

The following is a bathymetry map of Toaya beach which can be seen in Figure 6.

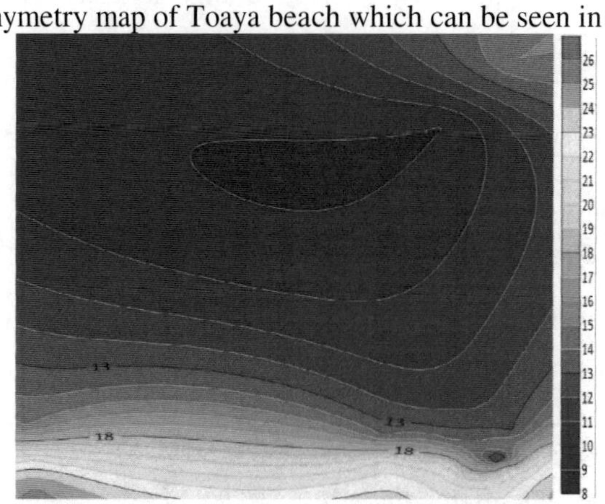

Figure 6. Bathymetry Map of Toaya Beach

The image above is the Toaya Beach Contour Map. The tool used when measuring is a depth meter; this is a tool to measure the depth of the sea, which is flashed from above the sea level to the bottom of the sea. The purpose of bathymetry measurement is to determine the condition of the seafloor depth which will then be used as the basis for determining the position of the power plant so that it can be known and also as a reference point for determining the height of MSL, HHWL and LLWL. In the data collection process, the depth of seawater is measured at an interval of 10 meters from the coast, and then the surfer software is used to make the results of the bathymetric map image.

5. Conclusion

Based on the tidal constituent harmonic constant value using the Least Square method, the type or type of tides at the research location at Toaya Beach with an observation time of 15 days, based on the calculation of Formzahl (F) numbers, the F value is 0.357 which indicates that the type The tides that occur are mixed wave prevailing semi-diurnal, where every day there are two tides and two ebbs a day but sometimes there are one tide and one ebb with different heights and times. With the error rate in this study calculated using RMSE, indicating that the RMSE value is 0.028, the difference between the predicted data and the observation data is accurate.

Based on the results of the planning water level elevation data, the MSL value = 1.2 m, HHWL = 2.7 m, LLWL = -0.3 m, so that the resulting head value reaches 3.0 m.

Toaya Beach is one of the beaches in Central Sulawesi which has the potential to be a place for a Tidal Power Plant, so based on the calculation of the potential energy generated by tides with a head value of 3 meters and a sea area which is thought to encourage and produces energy movement in the sea area covering an area of 9 km2, namely the total energy potential of around 5.53 MW.

References

[1] Dalpan, E., Pratomo, A., & Adil, I. 2015 Analisis Pasang Surut Di Dermaga Sungai Enam Kijang Kabupaten Bintan Provinsi Kepulauan Riau Abstrak
[2] Dronkers, J. J. 1964 Tidal computations in rivers and coastal waters
[3] Google Earth. 2020 Toaya Beach Area Map Donggala Regency
[4] Gumelar, J., Sasmito, B., & Amarrohman, F. J. 2016 Analisis Harmonik Dengan Menggunakan Teknik Kuadrat Terkecil Untuk Penentuan Komponen-komponen Pasut Di Wilayah Laut Selatan Pulau Jawa Dari Satelit Altimetri Topex/poseidon Dan Jason-1. Jurnal Geodesi Undip
[5] Hidayati, N 2017 Dinamika Pantai. UB Press. Malang
[6] Nontji, A.,1987 Laut Nusantara. Penerbit Djambatan, Jakarta Dinamika Pantai. UB Press. Malang.
[7] Nurjaya, I.W. 1991 Penuntun Praktikum Dasar – dasar Oseanografi. Bogor: Institut Pertanian Bogor
[8] Theocsbury 2020 Tidal Barrage. https://en.m.wikipedia.org/wiki/Tidal_barrage . Retrieved 22 June 2020
[9] Triatmodjo, B. 1999 Teknik Pantai. Yogyakarta:Beta Offset
[10] Triatmodjo, B. 2010 Hidrologi Terapan. Beta Offset, Yogyakarta
[11] Village Profile 2016 Profile of Toaya Village, Donggala Regency

Potential of tidal power plants on Tibo Beach with spektrum method

Setiyawan, A Rusdin, T Amaliah, Olphino

Department of Civil Engineering, Tadulako University

Email : setiyawanvip@yahoo.co.id, arus33@yahoo.com, toetiamaliah83@gmail.com, Boeadiolphino@gmail.com

Abstract. As time evolved, the demand for electrical energy also increased. As a result, renewable energy is needed to replace fossil fuels as fuel for electricity generation. Tidal water is one of the renewable alternative energy sources used as fuel for electricity generation. The tides will never run out and can also be predicted. One method that can predict data about waves is the Spectrum Method. Tide prediction data can be indicated by carrying out tidal height observations for 15 days at Tibo Beach, Donggala Regency, Central Sulawesi. Based on observations, the data obtained are the highest high water level (HHWL) is 2.4 meters, and the lowest low water level (LLWL) is 0.1 meters. Also, the type of tides on the Tibo coast is a type of mixed waves that tend to double daily, there are two tides with high water and two tides with low water with different water levels. Because the value of formzahl is 0.312. With a 2.4 meter high HHWL and a sea area of 35.4 km2, the electricity generated is 13.08 MW.

1. Introduction

Along with the development of human civilization, the need for energy is also increasing. So far, most of the energy needs come from non-renewable energy, such as coal, which is usually used as fuel for electricity generation.

However, this energy cannot always fulfill all human needs in the long term, given that energy reserves are getting depleted and the production process, which takes millions of years. Therefore, an in-depth study of the use of renewable alternative energy is needed to answer this challenge. Apart from being renewable, alternative energy sources that are environmentally friendly and at low cost are also needed. (Saputra,2019)

An innovation by utilizing alternative energy sources, which has not been used to overcome electricity problems, is developed through this research. One of the abundant and inexhaustible energy sources is energy in the oceans, including tidal power, which is commonly found in Indonesian waters. (Sangari,2014)

Tibo is one of the villages in Sindue District, which has relatively large sea waves. This is used as a reference that the beach, according to the author, is feasible in the research of Tidal Power Plants (PLTPS). For the village of Tibo itself, the technology to be developed is a tidal power plant using On Shore technology. This is because On Shore technology is very suitable for building areas that do not require extensive construction. The sea waves in Tibo Village are large enough to make it possible to produce electrical energy.

The purpose of this study was to determine the type of tides at Tibo Beach based on the calculation of the Formzahl number with the Spectrum method, to obtain the design water level elevation of the HHWL and LLWL at Tibo Beach, and to determine the potential generated by the tidal power plant on the coast of Tibo Beach.

Based on the description above, the problem formulation of this research is how the type of tides that occur at Tibo Beach based on the calculation of Formzahl numbers using the Spectrum Method, what is the water level elevation of the HHWL and LLWL plans on Tibo Beach, and whether there is potential for tidal power plants at Tibo Beach.

2. Overview Of The Research Location

The research location is planned to be located in Tibo Village, Sindue District, Donggala Regency, Central Sulawesi Province, approximately 55 km from Palu City, with coordinates 00029'12,59" S, 119049'1,97" E

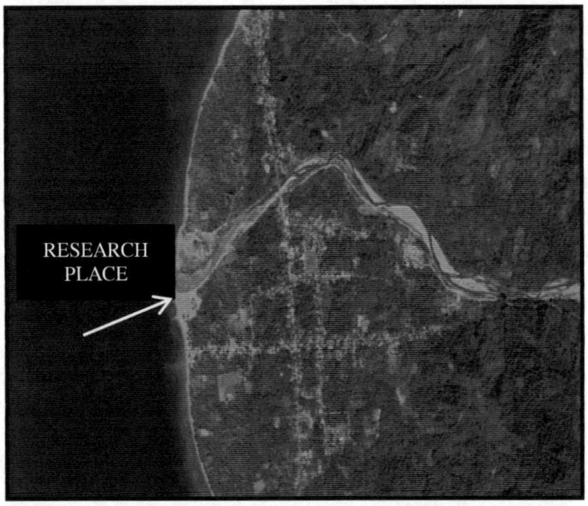

Figure 1 Research location on the beach of Tibo (*Google Earth, 2020*)

3. Literature Review

3.1 Tidal

Tides are fluctuations in sea level due to the attraction of objects in the sky, especially the sun and the moon, to seawater masses on earth. The moon's interest, which affects the tides, is 2.2 times greater than that of the sun (Triatmodjo, 2012).
This is what causes when one area is experiencing high tide; other regions will experience receding seawater.

3.2 Tidal Curve

Tidal conditions can be displayed visually in the form of tidal curves. The curve describes the sea level at a particular time. The x-axis shows the time, while the y-axis shows the sea level. Tide height is the vertical distance measured from the top of the highest water to the lowest water position. The tidal period is the time it takes from the highest (or lowest) water level to the next highest (or lowest) water level. When the sea level moves up is called the Tide period, while the sea level moves down, it is called the quiet tide period.

According to Nontji (1987), there are four types of ebb and flow based on their period and regularity. Within a month, the daily variation of the tidal range changes systematically with the lunar cycle. The tidal range also depends on the shape of the waters and the configuration of the ocean floor. Here below the tides of seawater in Indonesia are divided into four types, namely: Regular Double Daily Tides (Semi Diurnal Tide), Mixed Semi-Diurnal Tides, Mixed Semi-Diurnal Tides, Mixed Tides Leaning to a Single Daily (Mixed- Diurnal), Regular Single Daily Tides (Diurnal Tide)

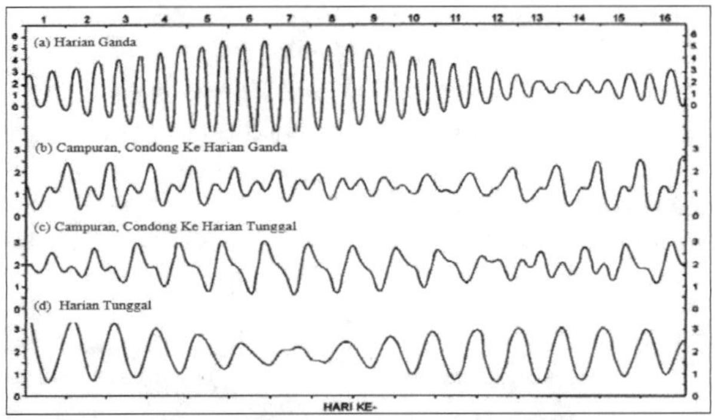

Figure 2 Types of Tides that occur in Indonesia (Triatmodjo, 1999)

Apart from looking at the tide data plotted in the form of a tidal type graph, it can also be determined based on the Formzahl (F) number, which is expressed in the form of the following equation:

$$F = \frac{AK_1+AO_1}{AM_2+AS_2} \tag{3.1}$$

Where:
F : *Formzahl* atau or tidal constant
AK_1 : The amplitude of the single daily mean tidal wave chicks are affected by the declination of the moon and the sun
AO_1: The amplitude of the mean single daily tidal wave children as affected by the declination of the sun
AM_2: The amplitude of the sub-daily mean double tidal wave affected by the moon
AS_2: Amplitude of the mean daily double tidal wave child affected by the sun
The tidal type classification can be seen in Table 1.

Table 1. Tidal Type Classification

Form Value	Type of Tide	Phenomenon
F ≤ 0,25	Semi Diurnal Tides	2x pairs a day with the same relative height
0,25 < F ≤ 1,5	Mixed Mainly Semidiurnal Tides	2x pairs a day with different heights and different intervals
1,50 < F ≤ 3,0	Mixed Mainly Diurnal Tides	1x or 2x pairs a day at different intervals
F > 3,0	Diurnal Tides	1x pair a day, when Spring can occur 2x pairs a day

(Source: Triatmojo, 2010)

3.3. Tidal Power Plant

Tidal Power Plants Seawater is a power plant that utilizes tidal energy in seawater to make energy in other forms, especially electricity, where tides can drive turbines to run a generator. Tidal power generation for seawater is renewable energy. There are two types of generator types: Tidal Dam (Barrage Tidal System) and Offshore Turbines (Onshore Turbines). The Sea Water Tidal Power

Generation System is very suitable for Indonesia, especially in Central Sulawesi, because Indonesia is known for its ocean waves.

3.4. The advantages of using PLTPS

1. Once built, tidal energy can be obtained free of charge.
2. Does not produce greenhouse gases or other waste.
3. Does not require fuel
4. Low operating costs.
5. The tides are predictable.
6. Offshore turbines have low installation costs and do not cause a large environmental impact

3.5. The lack of using PLTPS

1. Depends on the characteristics of the Tides.
2. Need a precise location.
3. The tools must be sturdy so they can withstand any weather conditions.
4. A dam that covers a river mouth is very expensive to build, and covers a very large area so that it changes the environmental ecosystem both upstream and downstream for many kilometers.
5. Can only supply energy for approximately 10 hours per day, when the waves move in or out.

3.6. Tidal Power Plant (PLTPS) Type Onshore Turbines

This tidal power plant uses an offshore turbine method, which is more like an underwater version of a wind power plant. The form of a tidal turbine is as varied as the wind turbine.
This system does not require a dam but is directly installed in the open seas; the thrust is generated from the movement of the kinetic energy of ocean currents; because the density of water is higher than the wind, the onshore turbine can produce tremendous power with the same size as the wind turbine where the tides can also be utilized so that they can drive turbines and generators.

Figure 3 Tidal Turbine in the Sea

Where to calculate the amount of energy produced can be calculated using the following equation:

$$E = \tfrac{1}{2} \times A \times \rho \times g \times H^2 \quad (3.2)$$

Where:
E = Energy (Joule)
A = Sea area (m^2)
ρ = Sea water density (kg/m^3)
g = Gravity (m/s^2)
H = Head (m)

For the sea area (A) which is estimated to be able to push and produce energy movement in the sea area, it is assumed that the surface of the tidal energy generator is 9 km² (Theocsbury, 2020).

3.7. Tidal Harmonic Components

predict tides, the amplitudes and faces of each component of the tidal generator are needed. The main features of waves consist of the daily middle part, daily tides, and long-period tides. However, due to the interaction with the beach morphology and the superposition between the main tidal wave components, new tidal features are formed. These principal components, also called astronomical frequencies, are used to create interactive programming for the decomposition of tidal components. In the tidal analysis using the least-squares method, the amplitude and phase parts of the tidal harmonic equation are calculated based on observational data on the water level position within a certain period, using components with known frequencies. The number of tide components that can be decomposed depends on the length of the data; the longer the observation data, the more tidal details can be produced. For example, the number and pairs of components resulting from observation with a length of 1 month will differ from an observation length of 1 year. The tidal parts can be seen in Table 2 as follows

Table 2. The frequency of 9 ocean waves

No	Type of Component	Frequency (deg / hr)	Period (hours)
1	K1	15,04	23,94
2	O1	13,94	25,82
3	P1	14,96	24,06
4	M2	28,98	12,42
5	S2	30	12
6	K2	30,08	11,97
7	N2	28,44	12,66
8	M4	57,97	6,21
9	MS4	58,98	6,1

Source: (Hidayati, 2017)

3.8. Least Square Method

In this study, the data processing method used the Spectral method. The Spectral Method is a transformation method which is presented as a Fourier Transform (Zakaria, 1998), as follows

$$P(f_m) = \frac{\Delta t}{2\sqrt{\pi}} \sum_{nN/2}^{nN/2} P(t_n) e^{\frac{-2.\pi.i}{M}} m.n \quad (3.3)$$

where :
P(tn) = Time series
P(fm) = frequency domain
tn = time series that shows the amount of data up to n
fm = Frequency domain.

3.9. Error Rate

Root Mean Square Error (RMSE) or error rate is the average value of the error; it can also express the size of the error generated by a forecast model. A low RMSE value indicates that the variation in the amount generated by a forecast model is close to its observation value variation. The closer to zero the RMSE value, the more accurate the elevation prediction results are. RMSE is the simulation result error rate.

The error rate calculation aims to determine how much difference the simulation results from data with the measurement data. The analysis of the RMSE value is calculated using the following equation:

$$\text{RMSE} = \sqrt{\frac{\sum_{i=1}^{n}(Yi-\hat{Y}i)^2}{n}} \tag{3.4}$$

Information:
Yi : initial data (observational data)
Ŷi : final data (predictive data)
N : amount of data

3.10. Water Level Elevation
The sea level elevation changes every time, it is necessary to have an elevation determined based on tidal data, which can be used as a guide in planning a port. Some of these elevations are; High Water Level, (HWL), Low Water level (LWL), Mean High Water Level (MHWL), Mean Low Water Level (MLWL), Mean Sea Level (MSL), Highest High Water Level, (HHWL), Lowest Low Water Level (LLWL**),** Higher high water level, and Lower low water level.

3.11. Bathymetry
A bathymetric map is a map of the sea's depth, which is expressed in several depths or depth contours measured against a vertical datum. Bathymetry (from the Greek meaning "depth" and "measure") is the study of the centers underwater and the three-dimensional analysis of the floor of an ocean or lake. A bathymetric map generally displays base or direct relief with contour lines called depth contours (isobaths) and may have additional information in the form of surface navigation information.

In general, the seabed relief is not very varied compared to land relief. This is due to weak erosion and sedimentation

4. Results And Discussion
4.1. Tidal Data Analysis
The data to be analyzed is tidal data from observations in the field that take place every hour for 15 days, from March 14, 2020, to March 28, 2020.
From the data observation process, table compilers can be made to facilitate tidal data analysis. Then from this table, we can obtain the tide observation graph as follows:

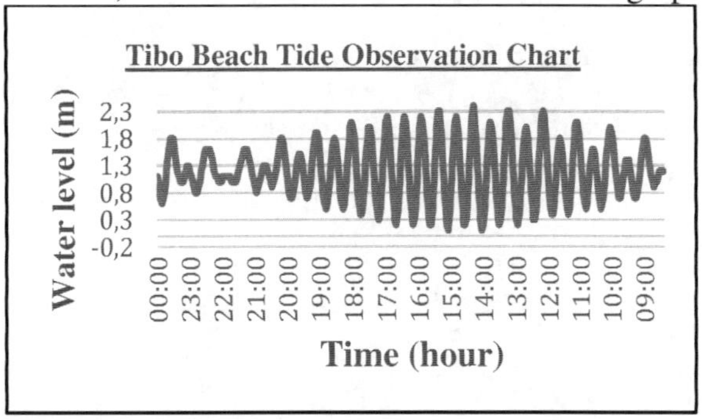

Figure 4. Tide Observation Chart

4.2. Results of the Least Method of Tidal Data Analysis
Based on the results of tidal data analysis at Tibo Beach with the spectrum method, the tidal constituent values of 8 harmonic constants are shown in Table 3 below.

Tabel 3. Result Harmonic Constants Spectrum Method

Tidal Component	Constant (A)
M_2	0.586
S_2	0.341
K_2	0.093
K_1	0.172
O_1	0.123
P_1	0.567
M_4	0.003
MS_4	0.0007

By experimenting with 9 constituents, the results of the Formzahl number, RMSE, water level elevation are as follows:

- *Formzahl* (F) : 0,312
- RMSE : 0.049
- HHWL : 2,4 m
- MSL : 1,2 m
- LLWL : 0,1 m

The results of calculations using this spectrum method can be obtained by the type or type of tide based on the Formzahl number with a comparison **[F = $(AK_1) + (AO_1) / (AM_2) + (AS_2)$]**. Here's how to calculate it with a comparison of the F value:

F = [(AK1) + (AO1) / (AM2) + (AS2)]
= [(0,172) + (0,123) / (0,586) + (0,341)
= 0,312

Requirement :

$F \leq 0{,}25$	: Semi Diurnal
$0{,}25 < F < 1{,}50$	: Mixed tide prevailing semi diurnal
$1{,}50 < F < 3{,}00$	: Mixed tide prevailing diurnal
$F \geq 3{,}00$	: Diurnal tide

Based on the requirements for the Formzahl number, the F value for the spectrum method is 0.312 (0.25 <F <1.50), so the type or type of tide for the Tibo Beach area, Donggala Regency, Central Sulawesi is mixed tide prevailing semi diurnal. This means that in one day there are two tides and two ebbs with different heights and intervals.

4.3. Calculating the Potential Electrical Energy Generated

To calculate the electric power that can be generated based on the tidal height of Tibo Beach, the description of the calculation of the power / energy that can be generated is as follows, by comparing the energy produced by forecasting sea water waves

Known :

ρ = 1025 kg/m^3
g = 9,81 m/s^2
Htide = HHWL – LWLL = 2,3 m
A = 35.4 km^2 (fetch value)

Amount of Etide = ½ x A x ρ x g x h^2 = 94 x 10^{10} J

From the results of the previous discussion that on Tibo Beach there were 2 tides and 2 receding times a day, then the potential energy generated when the sea water receded was 0. So that the total electrical energy that can be generated in one day can be 188 x 10^{10} J

E per day = 188 x 10^{10} J / (24 x 3600)

$= 21.79 \times 10^6$ Watt
$= 21.79$ MW

It is assumed that the efficiency value of the generator is 60%., So the potential generated by the tidal effect of sea water on Tibo Beach with an estimated sea area can push and produce energy movement in the sea area covering an area of 35.4 km2 and a head as high as 2.4 meters namely 13.08 MW.

4.4. Determining the Position of PLTPS

Bathymetry maps can be interpreted as a measure of the sea's depth, both regarding the seafloor elevation's size, which is a source of information and an overview of the seabed, and provides many clues about marine structure (Nurjaya, 1991).
The following is a bathymetric map of the Tibo beach which can be seen in Figure 5

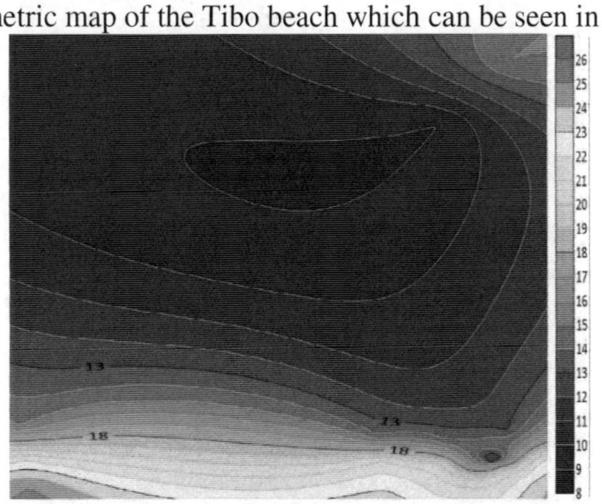

Figure 5. Bathymetry Map Tibo Beach

The image above is the Contour Map of Tibo Beach. The tool used when measuring is a depth meter; this is a tool to measure the sea's depth, centered from above sea level to below the seabed. The purpose of bathymetry measurement is to determine the seafloor depth's condition, which will then be used as the basis for determining the position of the power plant so that it can be known and also as a reference point for determining the height of MSL, HHWL, and LLWL. In the data collection process, the depth of seawater is measured at an interval of 10 meters from the coast, then the surfer software is used to create the bathymetric map image results.

5. Conclusion

Based on the calculation of Formzahl numbers, the Type of Tides at Tibo Beach is a type of tide. Mixed tide tends to double daily (composite wave prevailing semi-diurnal), where in one day, there are two times Tide a day with different heights and different intervals.
Based on the results of data processing using the Spectrum method, the HHWL value is 2.4 meters. And the LLWL value is 0.1 meter.
Tibo Beach is one of the beaches in Central Sulawesi, which can become a place for a Tidal Power Plant. Because based on the results of data processing, the power generated from the calculation of tidal power potential energy at the coast of Tibo is 13.08 MW. Also, Tibo Beach is one of the beaches in Central Sulawesi that has the potential and meets the requirements for constructing a tidal power plant.

References

[1] Arthur, R., S. 1950 Wave Forecasting And Hindcasting. Proceedings of First Conference on Coastal Engineering No 1 1950. Long Beach, California, October, 1950
[2] Google Earth 2020 Tibo Beach Area Map Donggala Regency

[3] Hidayati, N. 2017 Dinamika Pantai. UB Press. Malang
[4] Huda, A.N 2014 Membuat Fetch Menggunakan Arcgis. https://www.scribd.com /doc196392678/Membuat-Fetch-Menggunakan-Arcgis . Diakses pada 23 Juni 2020
[5] Ippen, A.T 1996 Estuary and Coastline Hydrodynamics, Mc. Graw Hill Book Company, Inc., New York
[6] Ilham. M 2020 Pengertian Pantai, Manfaat dan Ciri-ciri. https://materibelajar.co.id/pantai-adalah/, diakses pada tanggal 10 februari 2020.Triatmodjo, B. 1999
[7] Loebis, J 1992 Banjir Rencana Untuk Bangunan Air. Departemen Pekerjaan Umum. Jakarta
[8] Muchlis, M. & Permana, A.D (2003) Proyeksi Kebutuhan Listrik PLN Tahun 2003 S.D 2020, http://www.oocities.org/markal.bppt/publish/slistrk/slmuch.pdf, diakses pada tanggal 15 Februari 2020
[9] Nontji, A 1987 Laut Nusantara, Penerbit Djambatan, Jakarta.
[10] Nurjaya, I.W 1991 Penuntun Praktikum Dasar – dasar Oseanografi. Bogor: Institut Pertanian Bogor.
[11] Pariwono, J.I., 1989 Kondisi Pasang Surut Indonesia, Pusat Pengembangan dan Penelitian-Lembaga Ilmu Pengetahuan Indonesia (P3O-LIPI), Jakarta..
[12] Profil Desa 2019 Profil Desa Sikara", Kabupaten Donggala
[13] Saputra. A 2019 Studi Potensi Pembangkit Listrik Tenaga Gelombang Laut Tipe Pelampung di Pantai Lembasada Kabupaten Donggala. Universitas Tadulako, Palu..
[14] Sangari, F.J 2014 Perancangan Pembangkit Listrik Pasang Surut Air Laut, Universitas Negeri Manado. Manado.
[15] Siahaan, D.H 2009 Pengujian Sudu Rata Prototipe Turbin Air Terapung Pada Aliran Sungai. FT : Universitas Sumatra Utara Medan.
[16] Soewarno 1991 Hidrologi Pengukuran dan Pengolahan Data Aliran Sungai. PT Nova. Bandung
[17] Subarkah, I 1978 Hidrologi Untuk Perencanaan Bangunan Air. Idea Dharma, Bandung
[18] Theocsbury 2020Tidal Barrage. https://en.m.wikipedia.org/wiki/Tidal_barrage . Diakses pada tanggal 22 Juni 2020.
[19] Triatmodjo, B 1999 Teknik Pantai. Yogyakarta:Beta Offset.
[20] Triatmodjo, B 2010 Hidrologi Terapan. Beta Offset, Yogyakarta
[21] Triatmodjo, B 2012 Perencanaan Bangunan Pantai, Beta Offset, Yogyakarta.
[22] Zakaria, A 1998 Preliminary Study of Tidal Prediction Using Least Squares Method, Thesis (master), Bandung Institute of technology, Bandung, Indonesia.

Centralised and decentralised data communication scheme for voltage regulation in DC Microgrids

M A Setiawan[1], E Sulistyo

Politeknik Manufaktur Negeri Bangka Belitung, Sungailiat – Indonesia 33211

[1]Email : made.andik.s@gmail.com

Abstract. Voltage regulation is one of the main control issues in DC Microgrids (MGs). To achieve voltage regulation in MGs, exchange information between distributed generation units (DG) is inevitable. There are two types of data exchange proposed and discussed, centralised and decentralised data communication schemes. Many papers in the literature did not give attention to the type of data communication infrastructure that will have a significant impact on both centralised and decentralised schemes. This paper proposes centralised and decentralised data communication scheme and their impact on voltage regulation in DC MGs. The dynamic performance of a DC MG with loads fluctuations, operating with the proposed technique, is evaluated through simulation analyses, realized in MATLAB. Both the proposed centralised and decentralised methods are able to maintain the voltage within acceptable limit during loads fluctuations; however the centralised method is around five times faster than the decentralised method. The results show the superiority of the proposed method for the DC MGs operations during load demands fluctuations, loads varieties, communication delays and structures.

1. Introduction

A cluster of distributed generation units (DGs) with several loads and energy storage systems (ESSs) that are interconnected by a network of lines, and located at the vicinity of each other is defined as a Microgrid (MG) [1]. An MG can act as an independent power supply and distribution system (islanded mode) or operates in grid-connected mode whenever required [2]. For proper operation and control of DGs within an MG, each DG should be updated with the information about the MG operating mode [3], [4]. It is expected that in the near future all the operations in an MG will be fully automated. The automation system of the future MGs includes fetching data from the sensors, passing the data to the controllers and finally passing the control commands to the actuators. Therefore, MGs need a fast and accurate data communication system to transfer the measured data and command signals; and hence the use of reliable communication technology is essential [5].

Implementing communication technology into MGs calls for several considerations; this includes, compressing data to avoid long communication delay [6], [7], adopting hierarchical communication network to reduce the number of data traffics [8], and improving data exchanging format among DGs, ESSs, electric vehicles (EVs), loads and sensors [9]. At this stage, no attention has been given to these parameters in designing data communication for MG operations. The required communication technology in the MG application is required to have a capability for covering the scattered location of the DGs, ESSs, loads, handling numerous and massive number of sensors/meters, and spread in geographical area [10], [11]. However, establishing data communication infrastructure in the MG leads to a significant installation cost. Each communication technology (i.e. wire and wireless technology) has a specific coverage area, and hence for the MGs with units scattered over a wide geographical area, repeaters for improving data transmission power, quality and strengthening the signals are required but this will be on the account of cost [12]. While several papers presenting various techniques for DC MG voltage regulation and power sharing can be found in the literatures, no attention was given to the communication infrastructure effects on their methods [10, 11]. Communication infrastructure has a significant impact on communication delay, data transmission method, communication technology type, and the installation costs [15].

To perform a proper control for voltage regulation and power/current sharing; data exchange and communication among DGs, loads, and ESSs for updating and synchronizing the outputs via established communication infrastructure is essential. There are two types of data exchange and data communication schemes which are proposed and discussed in the literature, i.e. centralised and decentralised data communication schemes [12, 13]. Centralised scheme is easy to implement and maintain. However, it has limitation in scalability, requires complex communication networks and is more prone to communication failure [13]. On the other hand, decentralised scheme offers more robust to communication failure and flexibility, however it requires longer processing time, less precision and more complex algorithm [18]. In centralised scheme, all data required from DG units, ESSs and loads are transmitted to MG central controller. Contrary with centralised scheme, in decentralised scheme, data in the DG units and ESSs and loads are transmitted to others DG units and ESSs. The completed exchange data received by each DG units or ESSs are compared, processed and executed by each DG units or ESSs itself [12, 10]. Figure 1 shows an illustration for centralised and decentralised data communication in DC MG.

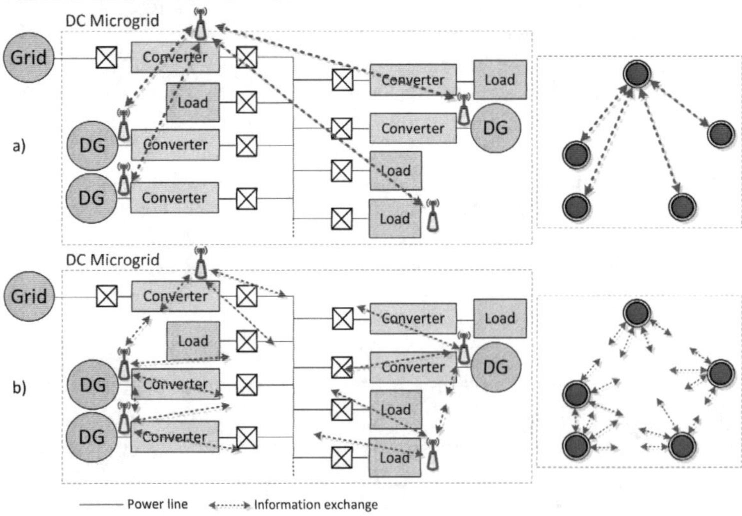

Figure 1. Illustration of data communication schemes in MG (a) centralised, (b) decentralised

A suitable data communication scheme can be determined by the established communication infrastructure which affects the communication delay that influences the power flow among DGs and in the interconnected lines [19]. A type of communication infrastructure might be suitable and has smaller communication delay for one type of data communication scheme which not be the same for another scheme. Most of the papers in the literature discussing voltage regulation in the MGs did not investigate the data communication delay effects on their proposed methods. These issues are the main gap of existing research that is addressed in this paper. The main contributions of this paper are:

- Proposing centralised and decentralised data communication scheme method for transmitting and exchanging data in DC MG regardless the type of communication technology, infrastructure and voltage regulation and power sharing method.
- Proposing a technique for voltage regulation in DC MG which can be employed for different data communication infrastructures and schemes.

2. Proposed Method

Consider the DC MG of Fig. 1(a) that consists of two DGs and one load [20]. In this figure, the voltage and current of the load can be expressed as:

$$V_{\mathrm{L}} = \sum_{i=1}^{N} \frac{V_{\mathrm{DG}i} \prod_{j=1,j\neq i}^{N} Z_{\mathrm{DG}j}}{\prod_{j=1}^{N} Z_{\mathrm{DG}j}} Z_{\mathrm{L}} \qquad (1)$$

$$I_{\mathrm{L}} = \sum I_{\mathrm{DG}k}, \forall k \in \{1 \cdots N\}$$

According to (1), following a change in the load demand, the DGs voltages $\sum V_{\mathrm{DGk}}$ can compensate for this change. However, due to the fact that it is impractical to measure Z_C, and hard to consider Z_L due to its dynamic changes, the iteration process of updating ΔV_{DGk} is required to be continuously adjusted until it compensates for the load voltage changes ΔV_L, which can be expressed as:

$$\Delta V_{\mathrm{L}} = \Delta V_{\mathrm{DGk}}, \forall \mathrm{k} \in \{1 \cdots \mathrm{N}\} \qquad (2)$$

ΔV_{DGk} needs to be updated discretely to prevent instability for the power electronics-based DC-DC converters due to continuous variations of their references [21]. This method requires a feedback signal from the voltage of the closest load to a DG to regulate the load voltage following any fluctuations in the load demand. However due to unknown line and load impedances, coordination and synchronization outputs of each DG is required for proper current load sharing among DGs. Updating without any coordination and synchronization among DGs will lead to un-proportional or over/under currents in the load generated by DGs as described in (1). The adjusted output current of each DG can be expressed as:

$$\Delta I_{\mathrm{DG}k} = \frac{\Delta(\sum I_{\mathrm{DG}k}, \forall k \in \{1 \cdots N\}), I_{\mathrm{DG}k}}{\sum I_{\mathrm{DG}k}, \forall k \in \{1 \cdots N\}} \qquad (3)$$

The method of (3) requires a MG central controller for collecting and computing all generated currents from DGs. Performing proper power/current sharing among DGs, therefore it also requires a weighting factor of the DG generated ratio (R_{DG}) in the range $0 \leq R_{\mathrm{DG}} \leq 1$. The proposed voltage regulation in (2) can be re-written as:

$$\Delta V_{\mathrm{DG}k} = \left(\sum R_{\mathrm{DG}}, \Delta I_{\mathrm{DG}k}\right) R_{\mathrm{DG}}\, \Delta V_{\mathrm{L}}, \forall k \in \{1 \cdots N\} \qquad (4)$$

R_{DG} of each DG is determined by the MG central controller based on capacity, or operational cost of the DGs. The renewable energy sources have an intermittent character and their energy output depends on environmental conditions. Hence, their generated power can be less than their capacity. The recent generated power by each renewable energy-based DG should be considered when determining a new value of R_{DG}.

The proposed voltage regulation for DC MG is illustrated in Fig. 1. It is to be noted that the distance between the DGs, sensors and MG central controller of the MG can be quite far, and thereby the utilizing a communication technology is essential and avoidable [17]. In this paper, the DC-DC converter of a DG is assumed as a semi-ideal converter which has the ability to produce a voltage output equal to its reference in a short interval. In Fig. 1, R_i and L_i represent the internal resistance and inductance of the converter's filter, while R_C and L_C represent the resistance and inductance of Z_C.

As discussed above, the data communication infrastructure and technology determines the type of the data communication scheme that can be applied and implemented on the MG. In this paper, data communication can be classified as a centralised data communication scheme, which is when the required data is transmitted from DGs, ESSs, or loads to the MG central controller [12, 13]. The MG central controller calculates and processes the received data, according to the implemented technique, before re-transmitting to the DGs, ESSs, and loads. In a decentralised data communication scheme, all the required data is transmitted and exchanged between the DGs and ESSs without any coordination from the MG central controller [10, 12].

According to (1), the load currents are supplied by the generated currents of all MG DGs. In a steady-state condition, the DGs generate the proportional current output according to their current ratio. However, voltage and current outputs generated proportionally require continuous data updating and proper synchronisation among DGs. Un-synchronised/randomly updating voltage and current outputs lead to an over/un-proportionally supplied current on the load side. Therefore, to achieve a proportional/designated voltage and current on the load side, periodic synchronisation among DGs for updating outputs is essential and performed via the proposed centralised and decentralised data communication schemes.

The required data to be exchanged and transmitted, using the proposed voltage method is presented in (1) to (4). In the proposed centralised data communication scheme, I_{DG} and V_{DG} are transmitted from the DGs to the MG central controller to determine R_{DG} and I_T/P_T. The data I_T/P_T is then calculated by the MG central controller and transmitted back to the DGs for re-calculating and updating of R^*_{DG}. To determine N, the data s_{DG} is transmitted to the MG central controller by the DGs. On the other hand, in the proposed decentralised data communication scheme, all the above required data are transmitted to the next closest DG. The updated DG outputs are performed only when the above data is received in full from the previous closest DG. The proposed data flow in a decentralised scheme is designed to prevent data collision; it also shortens queues in the communication line.

In summary, in the proposed centralised data communication scheme, DG outputs are updated simultaneously, while in the proposed decentralised data communication scheme, the DG outputs are updated sequentially. The data flow transmissions and flowcharts of the proposed centralised and decentralised MG data communication schemes are illustrated in Figure 2 and Figure 3 respectively in which P1, P2, P3, and P4 represent the communication device Points of DGs, ESSs, or loads in MG.

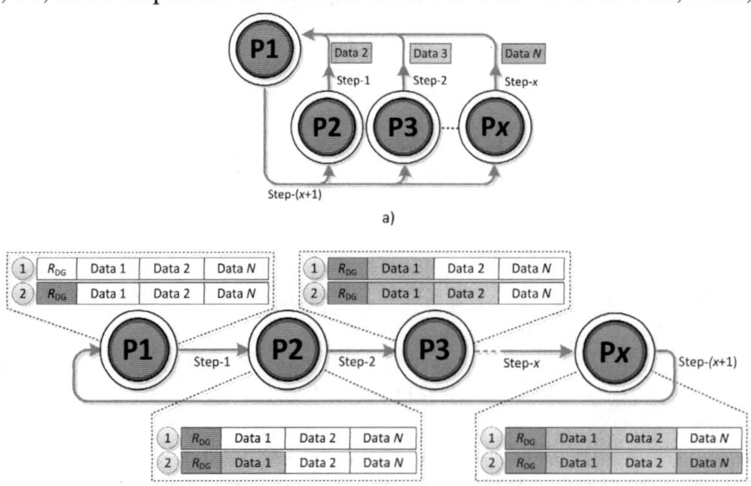

Figure 2. Data transmission flow of the proposed MG data communication scheme with: a) centralised scheme, b) decentralised scheme

In the proposed technique, all received data are saved in the DG controllers and in the MG central controller, and then used in the calculations until a new value is received. Upon failure in the communication system, the proposed control technique continues to function using the stored data. To prevent and eliminate communication failure problems, each communication technology has a protocol, such as the re-transmittal of the confirmation code to the transmitter [12]. Therefore, in this paper, we assume that all transmitted data are received in a good and acceptable condition by the receiver.

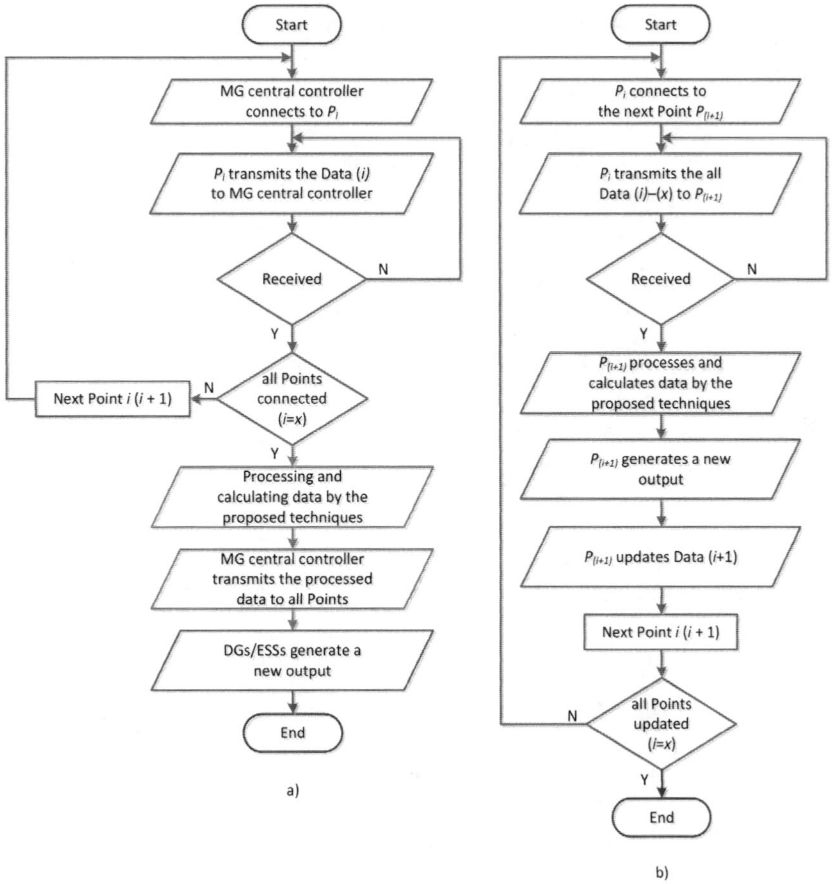

Figure 3. One-cycle processing flowchart of the proposed MG data communication scheme with: a) centralised scheme, b) decentralised scheme

3. Simulation Results

The impacts of data latencies on the dynamic operation of the DC MG can be evaluated using the DC MG network consisting of four DGs and six loads points shown in Figure 4. The technical parameters of the investigated DC MG under are summarized in Table 1.

The DC MG is initially assumed to be in a steady-state condition, and the voltages at the load points are within an acceptable limit. The employed voltage and power/current load sharing techniques are the proposed techniques discussed in (1) through (4), and ΔV_L is defined as $avg(V_{Li}^{max}, V_{Li}^{min}, \forall i \epsilon \{1 \cdots 6\})$.

Table 1. Considered technical parameters for data communication scheme simulation

Base parameters 220 V, 70 A, 34 kW.
Lines $R_{C1..10} = 0.1\,\Omega$, $R_{C2} = 0.2\,\Omega$, $L_{C1..10} = 10$ mH.
Loads capacity ≈ 5kW, $R_{L1..6} = 10\,\Omega$, $L_{L1..6} = 10$ mH, $C_{L1..6} = 1$ mF.
DG1 Capacity ≈ 6kW, Vmax = 250V, $L_i = 4$ mH, $C_i = 1.5$ mF, $R_i = 0.1\,\Omega$.
DG2 Capacity ≈ 11kW,
DG3 Capacity ≈ 11kW,
DG4 Capacity ≈ 6kW.

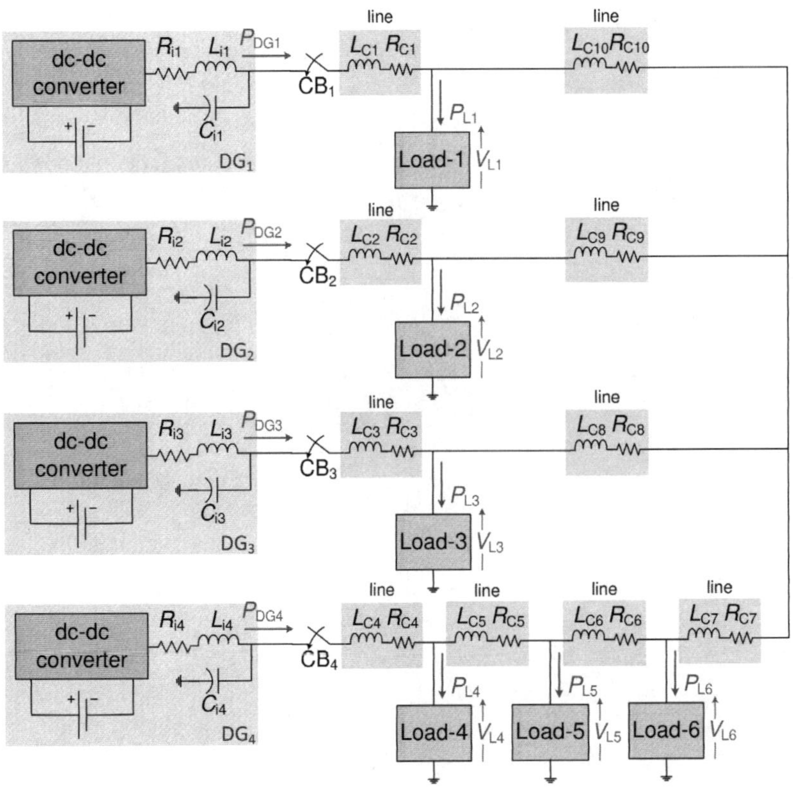

Figure 4. Investigated DC MG: power line connection diagram.

According to the proposed data flows in Fig. 2, the proposed sequential charts in Fig. 3, the communication delay for the proposed centralised and decentralised data communication scheme for the investigated DC MG network is used. However, to examine the performance of the proposed methods in this paper, the maximum delay is used instead of the minimum delay.

Fig. 3 shows that in the proposed centralised data communication scheme, the periods are implemented for just one-cycle to update the DG outputs. When the analysis is conducted on the proposed decentralised data communication scheme, the periods are implemented for the total number of DGs and load cycles. When updating each DG individually, the required periods differ between DGs, as illustrated in Figs. 3(a) and 3(b).

In this simulation, the DGs power ratios are proportionally defined based on their capacities, i.e. $\mathrm{R_{DG1}} = \mathrm{R_{DG4}} = \frac{6}{34}$, and $\mathrm{R_{DG2}} = \mathrm{R_{DG3}} = \frac{11}{34}$. The maximum loads demand is simulated at t = 3 s. The impacts of the proposed centralised and decentralised data communication scheme on the voltage regulation of the DC MG network of Fig. 4 is shown in Fig. 5 and Fig. 6.

Simulation results reveal that the proposed centralised and decentralised data communication scheme can maintain all load voltages within acceptable limits during fluctuations of load demands as shown in Fig. 5 and Fig. 6. Results also indicate that the DGs can precisely generate power at designated rate. The DG1, DG2, and DG3 generate their voltage outputs higher than DG4 to provide the required power for Load-4 and Load-5 which are closer to DG4. On the other hand, the power in DC MG can be achieved faster in the proposed centralised data communication scheme than the decentralised one. In the proposed centralised data communication scheme, the proportionally power generated by DGs are achieved in a time less than 10s of the simulation time while it takes 60s in the decentralised scheme to achieve the designated power as depicted in Figure 5 and Figure 6. It can be noticed that there are no circulating currents among DGs occurring during the simulation time even though there are dissimilarities of the outputs voltages and powers generated.

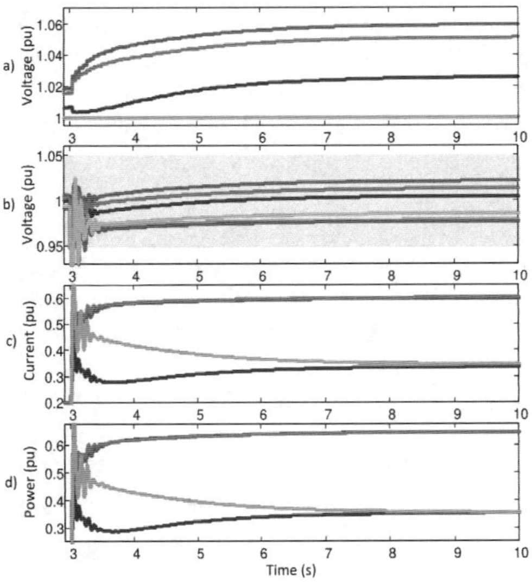

Figure 5. Voltage, current, and power profiles of the DC MG in the proposed centralised data communication simulation

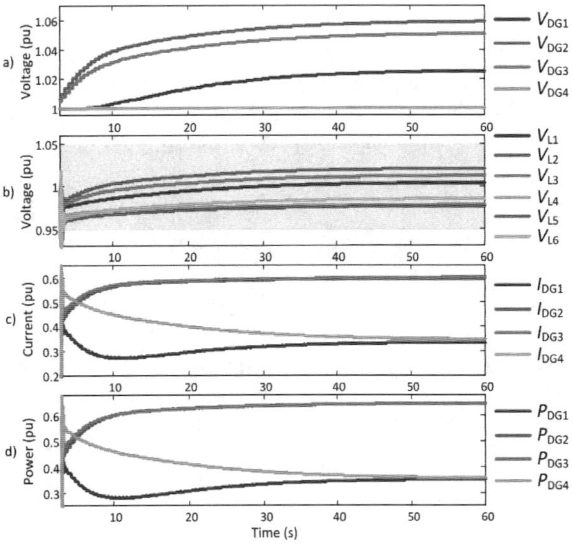

Figure 6. Voltage, current, and power profiles of the DC MG in the proposed decentralised data communication simulation

4. Conclusions

The impacts of the proposed communication infrastructure approaches in the DC MG operation are examined by implementing the proposed centralised and decentralised data communication scheme. In the proposed centralised scheme, the synchronization among DGs for updating their outputs is conducted by MG central controller by transmitting the required data to all DGs at the time. On the other hand in the proposed decentralised scheme, updating DG outputs are performed after receiving the required data from another DG. Simulation results indicate that the proposed centralised and decentralised data communication schemes can maintain all loads voltages within acceptable limits during fluctuations of load demands. Meantime the designated generated power by DGs can be achieved.

References

[1] E. Planas, J. Andreu, J. I. Gárate, I. Martínez De Alegría, and E. Ibarra, "AC and DC technology in microgrids: A review," *Renew. Sustain. Energy Rev.*, vol. 43, pp. 726–749, 2015.

[2] P. Zhang, H. Zhao, H. Cai, J. Shi, and X. He, "Power decoupling strategy based on 'virtual negative resistor'for inverters in low-voltage microgrids," *Iet Power Electron.*, vol. 9, no. 5, pp. 1037–1044, 2016.

[3] J. J. Jamian, H. A. Illias, K. Gia Ing, and H. Mokhlis, "Optimum distribution network operation considering distributed generation mode of operations and safety margin," *IET Renew. Power Gener.*, vol. 10, no. 8, pp. 1049–1058, 2016.

[4] M. A. Setiawan, A. Abu-Siada, and F. Shahnia, "Voltage regulation in DC microgrids with various circuit configurations," in *2016 IEEE 2nd Annual Southern Power Electronics Conference, SPEC 2016*, 2016.

[5] P. García, P. Arboleya, B. Mohamed, A. C. Vega, and M. C. Vega, "Implementation of a hybrid distributed / centralized real-time monitoring system for a DC / AC microgrid with energy storage capabilities," *IEEE Trans. Ind. Informatics*, vol. 3203, no. 1551, pp. 1–10, 2016.

[6] M. Saleh, Y. Esa, S. Member, A. Mohamed, and S. Member, "Impact of Communication Latency on the Bus Voltage of Centrally Controlled DC Microgrids During Islanding," *IEEE Trans. Sustain. Energy*, vol. PP, no. c, p. 1, 2018.
[7] N. C. F. Tse, J. Y. C. Chan, W. Lau, J. T. Y. Poon, and L. L. Lai, "Real-Time Power-Quality Monitoring With Hybrid Sinusoidal and Lifting Wavelet Compression Algorithm," *IEEE Trans. Power Deliv.*, vol. 27, no. 4, pp. 1718–1726, 2012.
[8] H. Y. Tung *et al.*, "The Generic Design of a High-Traffic Advanced Metering Infrastructure Using ZigBee," *IEEE Trans. Ind. Informatics*, vol. 10, no. 1, pp. 836–844, 2014.
[9] I. Hwang, D. Lee, and J. Baek, "Home network configuring scheme for all electric appliances using ZigBee-based integrated remote controller," *IEEE Trans. Consum. Electron.*, vol. 55, no. 3, pp. 1300–1307, Aug. 2009.
[10] M. A. Setiawan, F. Shahnia, R. P. S. Chandrasena, and A. Ghosh, "Data communication network and its delay effect on the dynamic operation of distributed generation units in a microgrid," in *Asia-Pacific Power and Energy Engineering Conference, APPEEC*, 2014.
[11] A. Abdali, R. Noroozian, and K. Mazlumi, "Electrical Power and Energy Systems Simultaneous control and protection schemes for DC multi microgrids," *Electr. Power Energy Syst.*, vol. 104, no. June 2017, pp. 230–245, 2019.
[12] M. A. Setiawan, F. Shahnia, A. Ghosh, and S. Rajakaruna, "Developing the ZigBee based data payload coding for data communication in microgrids," in *2014 Australasian Universities Power Engineering Conference, AUPEC 2014 - Proceedings*, 2014.
[13] V. Nasirian, S. Moayedi, A. Davoudi, and F. Lewis, "Distributed Cooperative Control of DC Microgrids," *IEEE Trans. Power Electron.*, vol. 30, no. 4, pp. 2288–2303, 2014.
[14] M. Sechilariu, B. C. Wang, and F. Locment, "Supervision control for optimal energy cost management in DC microgrid: Design and simulation," *Int. J. Electr. Power Energy Syst.*, vol. 58, pp. 140–149, 2014.
[15] A. Usman and S. H. Shami, "Evolution of Communication Technologies for Smart Grid applications," *Renew. Sustain. Energy Rev.*, vol. 19, pp. 191–199, Mar. 2013.
[16] C. N. Papadimitriou, E. I. Zountouridou, and N. D. Hatziargyriou, "Review of hierarchical control in DC microgrids," *Electr. Power Syst. Res.*, vol. 122, pp. 159–167, 2015.
[17] S. Moayedi and A. Davoudi, "Distributed Tertiary Control of DC Microgrid Clusters," *IEEE Trans. Power Electron.*, vol. 31, no. 2, pp. 1717–1733, 2016.
[18] J. Chi, P. Wang, J. Xiao, Y. Tang, and F. H. Choo, "Implementation of Hierarchical Control in DC microgrids," *IEEE Trans. Ind. Electron.*, vol. 61, no. 8, pp. 4032–4042, 2014.
[19] K. A. Alobeidli, M. H. Syed, M. S. El Moursi, and H. H. Zeineldin, "Novel Coordinated Voltage Control for Hybrid Micro-Grid With Islanding Capability," *IEEE Trans. Smart Grid*, vol. 6, no. 3, pp. 1116–1127, 2015.
[20] M. A. Setiawan, A. Abu-Siada, and F. Shahnia, "A New Technique for Power Sharing in DC Microgrids," in *IEEE Southern Power Electronics Conference*, 2016, no. 1, pp. 1–5.
[21] J. Wang, B. Bao, J. Xu, G. Zhou, and W. Hu, "Dynamical Effects of Equivalent Series Resistance of Output Capacitor in Constant On-Time Controlled Buck Converter," *IEEE Trans. Ind. Electron.*, vol. 60, no. 5, pp. 1759–1768, 2013.

Acknowledgments
This research is supported and funded by Penelitian Dasar scheme from Ministry of RistekBrin Indonesian Government, No: 131/SP2H/AMD/LT/DRPM/2020.

IOP Conf. Series: Materials Science and Engineering 1212 (2022) 012041 doi:10.1088/1757-899X/1212/1/012041

Module trainer of photovoltaic integrated low voltage grid (PV-LV grid) with a variety load

Y A Rahman, R Fauzi, M Sarjan, Sukri

Electrical Engineering Department, Tadulako University, Palu, Indonesia

Email : yuliasmi.rahman81@gmail.com

Abstract. This research is a simulation of the penetration of renewable energy at the primary source. The entire system is a compound system, especially on the load. Load types provide different voltage and current treatment. The simulation is carried out with a mini-grid device consisting of five buses with interface sensors to monitor voltage and current. Photovoltaic grid module (PV) -LV as a trainer, uses two power sources to supply the load to the voltage distribution network module. The loading scenario h two plans, where the difference lies in the type of inductive load (scenario A) and capacitive (scenario B). The types of load consist of incandescent lamps, fluorescent lamps, energy-saving lamps, drilling machines, and soldering tools. Variations in this load to meet predetermined loading scenarios. The PV measurement results show maximum power generation at noon and flowed to the load trainer via a 300W grid-tie inverter of 191.1W. The average load bus voltage increases for scenario A and scenario B by 1.045% and 1.36%, respectively. The result indicates that the integration of PV in low voltage systems positively impacts the voltage profile for both inductive and capacitive loads. The amount of voltage improvement depends on the value of the active power injection of the PV and the load supplied by the system.

1. Introduction

The utilization of the photovoltaic (PV) greatly extends its implementation. The PV energy was a clean energy source that is environmentally friendly and free of emissions is a strong reason to encourage the widespread use of PV. By 2040 the International Energy Agency (IEA) expects 40% of the installed capacity target of renewable energy in ASEAN countries to come from solar PV or 52 GW [1]. The implementation of PV is divided into three models are PV-on grid, PV-off grid, and PV hybrid. The use of PV is still dominated by off-grid PV, where the system works alternately between PV and the primary source. This scheme requires a storage system that has implications for additional cost components. Apart from saving on battery component costs and lowering bill costs [2], the PV-on grid also has a positive technical impact on the distribution system. The integration of PV on the grid will also reduce power losses and voltage drops [3-5]. The series of on-grid courses are connected to the distribution network by optimizing the solar panels' energy to produce as much electricity as possible. Besides that, the on-grid system also applies a load-sharing scheme in which the network and PV are integrated to complete consumers' electricity needs. Electrical equipment includes household, commercial, and industrial markets. The varying load will determine the value of the voltage and current parameters. Load variations from consumers directly affect the voltage's quality, which has implications for the power loss variable [6]. The types of inductive and capacitive loads scenario in this study provide different injection or absorption of additional reactive data. Research [7] has tested the effect of PV penetration on resident loads, which shows that loads with a fixed impedance or resistance only represent loads with a small voltage drop. This study investigates the effect of variations in the load modeled as residential, industrial, and commercial expenses, which is the scenario as RC load and RL load, which are expected to give different voltage drop variations. This is a representation of consumers who are served on the side of the low voltage distribution. Voltage and current monitoring are carried out using sensors connected to the microcontroller as a data logger. This research contributes to the observation of voltage and current under different load conditions by referring to real loading scenarios experienced by the distribution network. This trainer module is equipped with PV output data recording, main grid output, and external temperature conditions that affect PV [8,9]. This study traced the results in four different sessions, each introduction, material and method, results, and conclusion session.

2. Material and Method

The research conducted for this paper focuses on observing and collecting current. Before and after PV, voltage comparison data is integrated into the primary source on the low voltage (LV) grid trainer module. The steps of this study are illustrated in the flowchart in Figure 1.

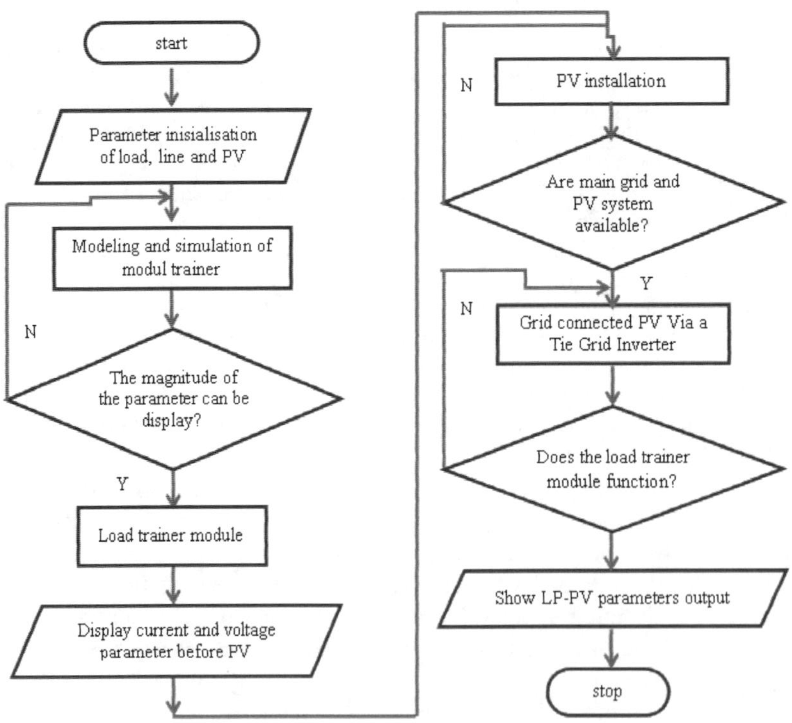

Figure 1. The procedure of PV-LV trainer module

2.1. PV-on grid

The solar panel used to generate electrical energy has a total capacity of 240 Wp, a monocrystalline type. This type is because it has actual power generation yield characteristics more than other technology [10,11]. The features and specifications of solar panels are presented in Table 1.

Table 1. Solar module FL-P-80W specifications

Peak power	80 Wp
Production tolerance	3%

Open circuit voltage (VOC)	21.60 A
Maximum power current (IMP)	4.45 A
Maximum power voltage (VMP)	18.00 V
Short circuit current (ISC)	4.74
Maximum system voltage	1000 VDC
Module dimension	1195x540x30 mm

The fabric tested all technical data in Table 1 were in standard condition. The PV panel output is connected directly through the Grid Tie Inverter (GTI) to the low voltage distribution network load without going through storage. This system works synchronously and automatically shares the load between PV and LV grid, as shown in the block diagram in Figure 2. The power meter is installed on bus four, which is the connection point of the PV. PV monitoring consists of current and voltage sensors used as a comparison for the analog measuring instrument installed on the module.

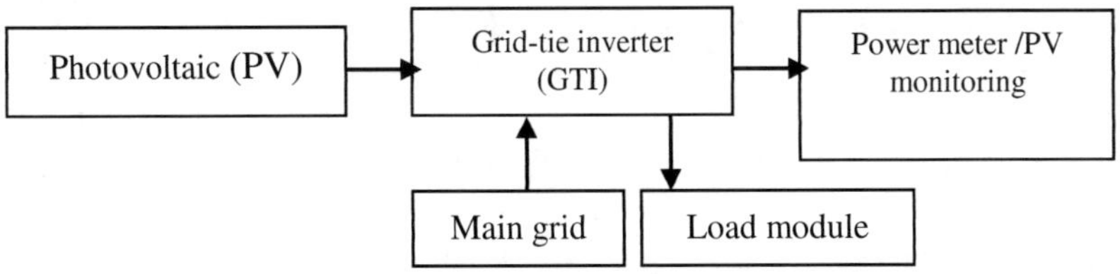

Figure 2. Diagram block of PV-LV trainer module

2.2. Grid-tie inverter

The GTI used in this study is a type of micro-inverter with soft switching considerations. Soft witching is an advantage because it requires a small power supply, increases inverter efficiency, and reduces heat dissipation. GTI uses a boost converter and an H-bridge inverter [12]. The GTI specifications are shown in Table 2. Boost converter functions to utilize diode components, capacitors, inductors, and MOSFETs.

Table 2. Technical parameter of GTI

Rated power	300 W
DC MAX input	360 W
DC input range	10.8-28 V
MPPT Voltage	15-23VDC
DC max. current	20A
AC max output	330W
AC output range	190-260 VAC
Peak efficiency	91%

The H-bridge inverter uses a MOSFET with a pulse width modulation (PWM) control system. The LC filter circuit is used to reduce harmonics from the sinusoidal waves generated from the inverter.

3. Result and discussion

The Solargis software carried out data retrieval using the trainer module was carried out with a temperature condition of 27.4 °C with the mapping of solar irradiation levels shown in Figure 3 concerning Solargis' record data [13].

Figure 3. Mapping solar data

The explanation of the daily solar radiation map is presented with the following index values:

- Direct normal irradiation (DNI) = 4.075 kWh/kWp per day
- Global horizontal irradiation (GHI) = 5.181 kWh/m^2 per day
- Diffuse horizontalirradiation (DIF) = 2.316 kWh/m^2 per day
- Global tilted irradiation at an optimum angle (GTIopta) =5.193 kWh/m^2 per day.
- Optimum tilt of PV modules (OPTA) = 4/0^0
- Air temperature (TEMP) =27.4 ^{0}C

The results of assembling the PV, GTI, and trainer modules are shown in Figure 4.

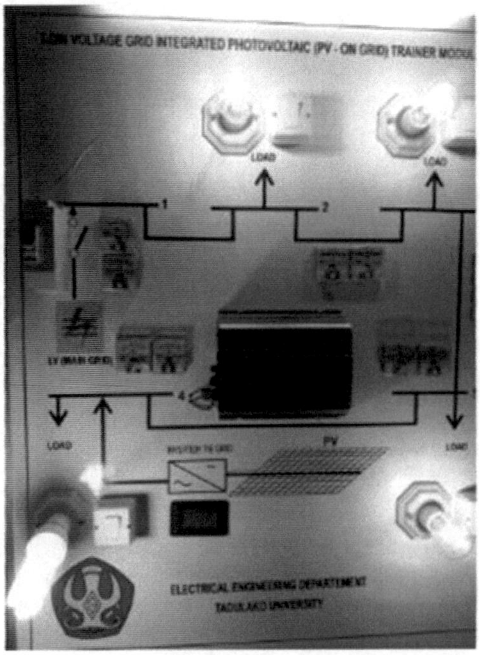

Figure 4. PV-LV trainer module

3.1. PV-on grid module trainer performance under load variation

The experiment was carried out using a combination of load models on the free, namely pure resistive, resistive-inductive, resistive-capacitive, and resistive-inductive-capacitive. The variations are carried out using two scenarios called scenarios A and B. In plan A, Bus 2, 3, 4 are given resistive, inductive loads with different inductive loads, while bus 5 uses a resistive load. In scenario B, further treatment is given to bus four by adding a capacitive load. The bus voltage recorded in the monitoring system for each scenario is shown in Figures 5a and 5b.

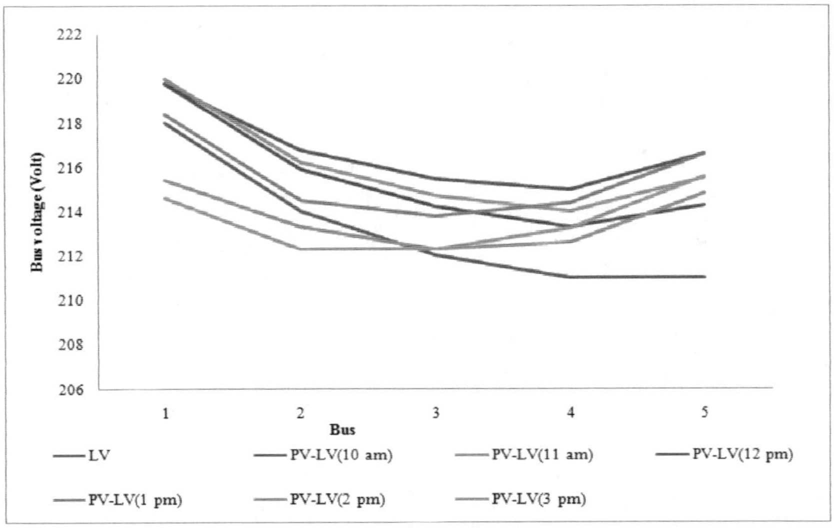

(a)

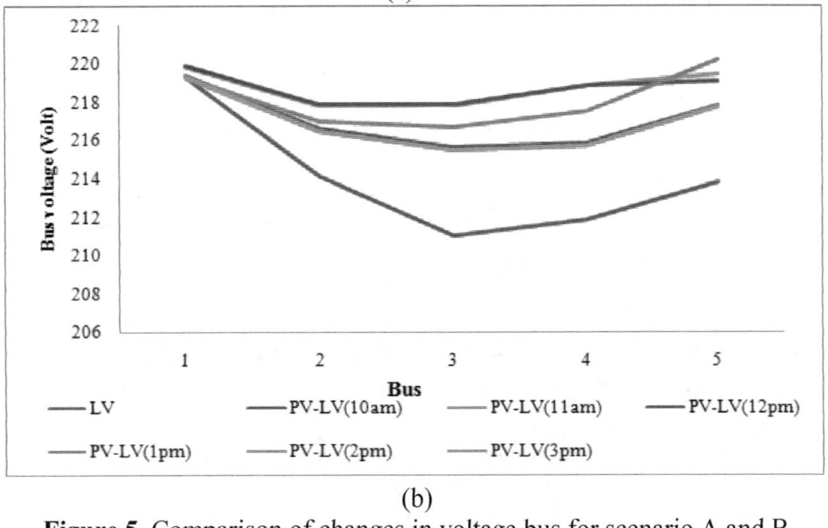

(b)

Figure 5. Comparison of changes in voltage bus for scenario A and B

The selection of the GTI connection as the output of PV synchronization and the primary grid is based on the enormous load served by the network and the furthest distance from the primary source. The researcher aimed at understanding the benefits of placing PV in an appropriate location [14]. Measurement results and data records show that the highest PV supply occurs from 11 am to 2 pm. The PV-LV supply through the GTI reaches 191.1 W, which can serve a total 360 W load.

3.2. Power quality monitoring

After the integration of PV into the LV system, changes in bus voltage parameters that affect the LV grid's power quality are observed. For scenario A, the most extensive voltage drop occurs on bus five by 4.09%. The integration of PV provides an improved voltage profile so that the voltage drop decreases to 1.55%. In scenario B, which contains a capacitive load, it affects the voltage drop that occurs. The nature of the capacitor gives the most significant reduction in a voltage drop of 2.68%. The integration of PV provides the most significant impact on improving the voltage profile to 2.045%. At the peak of the PV supply, the voltage drop on bus five reached 1.55% and 0.41% for each scenario. The improvement of the voltage profile is a positive impact that PV gives through the active power injection. This causes the change in the total active and reactive power in the power flow to decrease, which indirectly reduces the system's power loss. This will be true if the real power from injected PV is precisely [15]. Excessive injection of PV active power will harm the system.

4. Conclusion

Tests have been carried out using a low voltage distribution network integrated PV trainer module. Synchronization of the primary grid and PV is carried out by the tie grid inverter. The author collected data at seven different times. The most significant power is obtained at noon with the distribution of power from PV of 53.03% from the total load. The resistive-inductive load variable results in a more considerable voltage drop than the resistive-capacitive load variation. The integration of PV in miniature low-voltage systems has a positive impact on improving system voltage. The average voltage improvement obtained is 2.31% and 1.07% in each scenario.

References

[1] http://www.aseanenergy.org/ 2015 Status and challenges towards regional energy connectivity *ASEAN Centre for energy* Accesses: 25 March 2019.

[2] J Sreedevi, N Ashwin, M Naini R 2016 A study on grid connected PV system *Proc. national Power Systems Conference (NPSC)* 19-21 Dec.

[3] Yuli A R,Salama M, Yusran, Amil A I 2019 Selection of sensitive buses using the firefly algorithm for optimal multiple types of distributed generations allocation *International Journal of Advanced Computer Scienced and Applications* **10** p.316

[4] Priyanka C and M.Rizwan Voltage regulation mitigation techniques in distribution system with high PV penetration: A review 2018 *Renewable and sustainable energy reviews* **82** pp 3279-3287.

[5] Michael A C and Duncan SC Effects of distributed PV generation on California's distribution system, Part 1: Engineering simulations *Solar Energy* **128** pp. 126-138.

[6] Roberto Perillo Barbosa da Silva, Rodolfo Quadros, Fabricio Parra Santilio, Luiz Carlos Pereira da Silva, Bismarck Castillo Carvalho and Arnulfo Barroso de Vasconcellos 2017 Effects of electronic loads on electrical measurements power quality and billing *Journal of Control, Automation and Electrical Systems* **28**, pp. 654–663.

[7] Fetnando B. dos Reis , Kapil Duwadi , Robert Foutney ,Reinaldo Tonkoski ; Timothy M. Hansen ; Mohammad Asif Iqbal Khan , Sumit Paudyal 2019 Impact of Residential Load Models for Overvoltage Prevention Studies in PV-Rich LV Grids *2019 IEEE Milan PowerTech* p 1-6.

[8] A.R. Amelia, Y.M. Irwan, W.Z. Leow, M Irwanto, I. Safwati, M. Zhafarina Investigation of the Effect Temperature on Photovoltaic (PV) Panel Output Performance 2016 *International journal on advanced science and engineering information technology,* **6 (5)** pp 682-688.

[9] Wael Al-Kouz, Sameer Al-Dahidi, Bashar Hammad Mohammad Al-Abed 2019 Modeling Analysis Framework for Investigating the Impact of Dust and Temperature on PV Systems' Performance and Optimum Cleaning Frequency *Appl. Sci* **9(7)** 1397 pp. 1-22.

[10 Liang Jiang , Sai Cui , Peng Sun , Yanyan Wang , and Chao Yang Comparison of

Monocrystalline and Polycrystalline Solar Modules 2020 *IEEE 5th Information Technology and Mechatronics Engineering Conference (ITOEC)*.

[11] Wei Lin 2017 Selection analysis of polysilicon and monocrystalline silicon modules (in Chinese) *Electric Engineering China* **57–58**, pp. 74.

[12] Jose Luis D B and Arturo M A 2016 Simulation of a double-stage micro-inverter for grid-connected photovoltaic modules *13th International Conference on Electrical Engineering, Computing Science and Automatic Control (CCE)* Mexico.

[13] https://globalsolaratlas.info/map.

[14] Yuli A R, Salama M, Yusran, Amil A I 2018 An empirical metheuristic assesmet for solving of multi type distributed generation allocation problem *Proc. Of IEEE International Seminar of Research of Information Technology and Intelligent System* **19048390** p.699.

[15] Masoud Farhoodnea, Azah Mohamed, Hussain Shareef, and Hadi Zayandehroodi 2012 Power quality impact of grid-connected photovoltaic generation system in distribution networks *IEEE Student Conference on Research and Development (SCOReD)* p 1-6

Acknowledgements
Authors would like to express our highest appreciation to the Faculty of Engineering, the University of Tadulako, for research funding through DIPA number 646.r/UN28.2/PL/2020. The authors also expressed their gratitude to the Department of Energy and Mineral Resources of Central Sulawesi Province to monitor equipment testing facilities.

IOP Conf. Series: Materials Science and Engineering 1212 (2022) 012042 doi:10.1088/1757-899X/1212/1/012042

Smart water meter for automatic meter reading

A Amir, R Fauzi, Y Arifin

Electrical Engineering Department, Tadulako University, Palu, Central Sulawesi, Indonesia

Email : Yusnaini.arifin@gmail.com

Abstract. Clean water is one of the main sectors in smart city that need well management. One of the clean water management is utilization of water meters. The smart meter is more suitable applied for smart city. Recent Smart Water Meter allows water authorities to obtain water consumption data remotely. It also provides ability to collect and record the data in real time that can be utilised for multipurpose. However, in Indonesia, the water meters are used only to measure the total volume of clean water consumption for billing purpose only using mechanical water meter and requires labour intensive manual. Currently, many researches on smart meter design have been developed. However, the smart meter only measure and record the water consumption, without ability in which customer can determine the amount of water as needed. This paper describes design and development of smart water metering with Internet of Things. Flow meter is used as a sensor of water flowing through the pipe. The ability of the proposed smart meter is not only to measure and to record the volume water consumed, but also the customer can determine the water desired and required. The volume of water measured by the smart meter is compared with the manual measurement. The result shows that the water measured manually differs slightly from smart meter measurement using water flow sensor. The maximum difference, error, is 0.03 litres. The proposed smart meter has ability to close the main valve once the determined amount of water is reached.

1. Introduction

A smart city is a city area that has integrated information and communication technology in managing and monitoring the city effectively and eficiently[1]. The integration of technology in city management becomes possible with the existence of the internet of things. The management include transportation, energy, health, clean water, waste/disposal.

Among the physical infrastructure in smart cities, the clean water supply network is the most important because water is a very valuable urban resource. However, efficient changes in lifestyle and eating habits, as well as world population growth significantly increase the need for clean water worldwide [2]. The increase of clean water demand is more in urban areas. Thus, supply of clean water becomes challenging. This phenomenon is worsen by climate change. Climate change brings negative impacts not only on water availability quantity but also on water surface quality [3]. Water meter is one of the important devices in the supply water management [4].

Historically, the purpose of water meter installation is customer billing, and then water conservation [5]. The customer cannot gain the amount of water consumed at any time. Therefore, the consumers find it difficult to predict the water consumption at at any time. Information on the use of the amount of water consumed in real time, customers can access their patter water use. Pattern water use allows the consumer to limit their water usage when needed. This leads to water savings. Water meter may lead to supporting water conservation efforts with changes in water use behavior [2]. Controlling of tap water flow is one of the behaviors.

Most countries of the world have been installed fully mechanical water meter that measures total volume of water consumption [6]. In Indonesia, water meter used is the mechanical meters that require labor intensive reading, monthly reading, and tend to be inaccurate reading [7]. The inaccuracy in meter readings is exacerbated by the lack of access to the meter, such as a locked fence, that may adversely affect the customer.

Recently, smart meters have been widely used worldwide to overcome the disadvantage of the traditional meter. The smart meter allows the absence of labor and the reading is done remotely. Reading data can be stored in real time and utilized for various purposes, such as billing and water use patterns.

This paper focuses on the design and implementation of water meter based on Internet of Things (IoT). The proposed water meter applied IoT to send data from flow meter to the server and android of the customer.

The contribution of this project, generally, is twofold: Firstly, the proposed water meter is IoT base, as Automatic Meter Reading. Secondly, it proposed solenoid valve to close the water flow as the volume of water achieved the customer desire or requirement.

2. Related Work

2.1. Flow meter

Flow meter is a device that can detect and measure the fluid based on output of the flow sensor [8]. Flow meter, based on work principle, is classified for differential pressure, positive displacement, velocity, mass, and open channel.

The differential pressure flow meter, including orifice plate, measures the flow rate with the principle of pressure difference, called Bernoulli's principle, which says that there is a relationship between fluid pressure and fluid velocity. The relationship is inversely proportional. One of the implementation of orifice flow meter is in the geothermal industry to measure flow rate of single-phase liquid or steam [9].

Positive Displacement (PD) flow meter, including oscillating piston meters and notating disk meter, has the ability to measure volume directly. On the other hand, other flow meters can work with the principle of measuring the flow rate that will be converted to fluid volume. Using a PD flow meter, the output signal is directly related to the volume of fluid through the flow meter. Digital flow meter for fuel vendor has implemented of PD flow meter [10].

Velocity flow meter include electromagnetic and ultrasonic meter. Electromagnetic meter is based on Faraday's law that states variation in time of magnetic flux enclosed by the conductive loop generated electromotive magnetic force (EMF) or voltage[11]. The generated voltage is proportional to fluid velocity.On the other hand, ultrasonic technologies are used for volumetric flow measurement, including transmit time meters and Doppler effect meters [12].

2.2. Smart Meters

In more recent times, many researches have developed smart meter. Smart meter can be used for irrigation system, residential and industrial water consumption.

Measuring water need in field is essential for irrigation management system. Design and implementation low cost water flow meter is applied [8]. The design used turbine type flow meter, ATMega89S52 microcontroller, and pump. The pump supplies water to the crops based on the crop requirement. After the required water is fulfilled, the pump will automatically off. The prototype works based on the time when the required water is achieved. The design system allowed the absence of labor to monitor continuously. This leads to a reduction in costs and avoiding plants from damage. However, the limitation of design is the absence of Automatic Meter Reading (AMR). As a result, the water use cannot be monitored and read remotely in real-time.

In [6], a digital meter system has been developed enabling AMR. The meter system utilized turbine type flow meter, ATMega328 microcontroller, and GSM as interface. However, the consumers cannot set the desired or required water to the digital meter.

Enligtened by the design water flow that enable setting of water requirement [8] and the implemented AMR in smart meter [6], the design of low cost smart meter is proposed. The proposed smart meter enables the meter to close the valve automatically when the total water is achieved. Furthermore, to enable AMR, the proposed water meter used Internet of Things to inform the water flow in real time through a smart phone.

3. Proposed design and implementation

3.1. Block Diagram Proposed Water Meter

In the proposed smart meter, data processing is carried out by a microcontroller, Arduino. The data received by the microcontroller is based on the digital output from the water flow sensor. To enableAutomatic Meter Reading (AMR), the Wireless Transceiver is integrated with the proposed smart meter, Internet of Things. This allows the data received by Arduino from the water flow sensor to be sent wirelessly via an internet device and received by the customer on an android device connected to the internet. Once the desired or needed water achieved, the main valve will be closed automatically. The block diagram of the proposed smart meter and flowchart of the system are shown in figure 1 and 2, respectively.

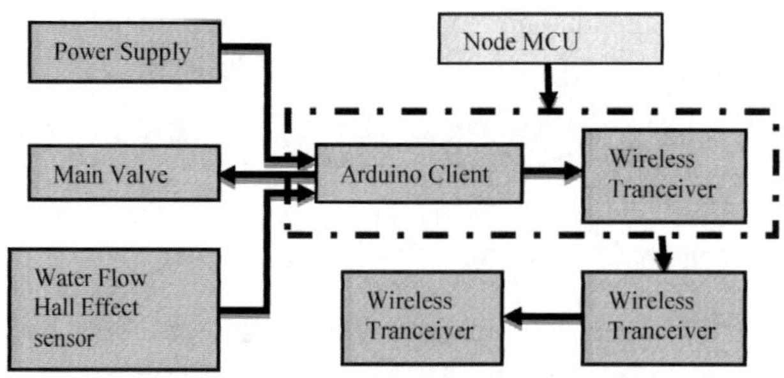

Figure 1. Block diagram of the proposed smart meter

AT89S52 microcontroller is used to monitor sensors. Here we have determined the flow rate with changes in water velocity. Speed depends on the pressure forcing the pipe to pass. Since the cross-sectional area of the pipe is known and remains constant, the average velocity is directly proportional to the flow rate. The relationship between the water flow rate and the average velocity of water flow in the pipe is shown in equation 1 [12].

$$Q = V * A \tag{1}$$

Where Q is the flow rate / total flow of water through the pipe, V is the average velocity of flow, and A is the cross-sectional area of the pipe. Other factors affecting water flow rate in the pipe are viscosity, density and friction of liquids in contact with the pipe.

3.2. Water Flow Sensor

A flow sensor is a device that detects and measures water flowing through pipes which is a necessary component of an efficient flow meter water system that is proposed. Mainly acts as a sensory organ for the brain in smart meter system, providing information for detecting, the amount of water consumed by the consumer. The basic working principle of the Flow meter is based on the output of the flow sensor. The flow sensor used in this proposed system is a sensor with hall effect working principles. As water flows through the rotor, the blades rotate. When the turbine rotates the magnetic field produced and corresponding to the pulse ac is generated which is then converted into digital output. The amount of credit generated per liter can be calculated by software programming. Thus the pulse produces an output frequency that is directly comparable to the volumetric flow rate/total flow rate through the meter. The spesification of flow meter sensor used, called Hall Effect sensor, canbe seen in the table 1 [13].

TABLE I. SPESIFICATION OF WATER FLOW SENSOR

Sensor Parameter	*range*
Sensor type	Hall effect sensor
Working voltage	5 to 18 V DC
Working flow rate	1 – 30 litres/minutes
Working temperature range	-25 to +80°C
Working humidity range	35% - 80%
Maximum water pressure	2.0 MPa

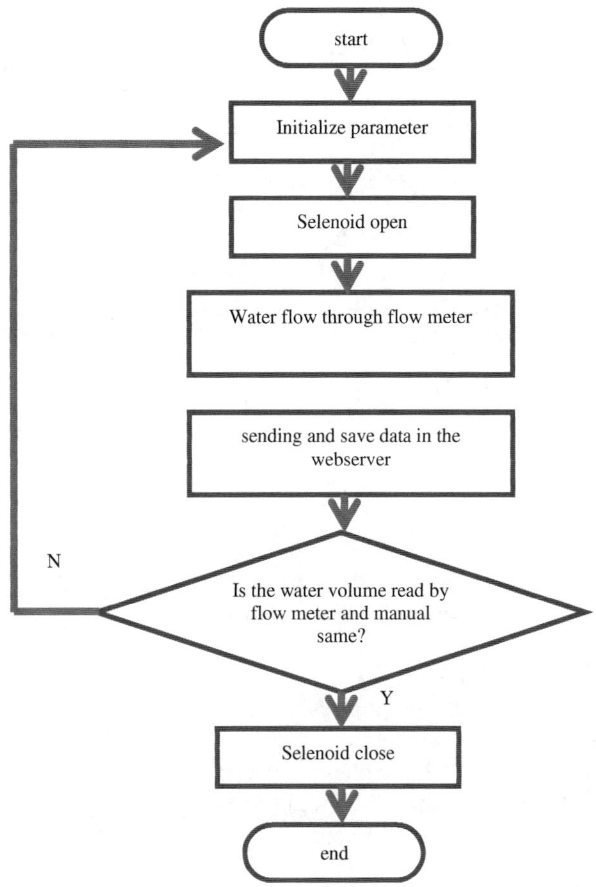

Figure 2. Flowchart of the proposed smart meter

4. Result and Discussion

A water tank and water pump are used to flow the water through the water flow. In this study, the amount of water that passes through the flow meter is compared with manual measurement with a measuring cup. The amount of water flowed is set from 10 to 50 litres in 5 litres increments. The result is shown in table 2.

Identify applicable funding agency here. If none, delete this text box.

Table 2. Comparison flow meter sensor and manual measurement

Measurements (litres)		**Error (Litres)**
Manual	Flow meter sensor	
10	10	0.0
20	19.99	0.01
30	29.99	0.01
40	39.98	0.02
50	49.97	0.03

The flow meter sensor measurement that has been observed slightly deviates from the manual measurement with maximum error 0.03 litres, shown in figure 3. This is caused by the transfer data delay.

This research in line with [14] . The designed system utilizes flow meter, hall efect sensor, in which the output will be an input to the microcontroller, PIC microcontroller. However, the research does not apply Internet of Things, so the data cannot be sent to microcontroller wirelessly. The data shows through the GSM. The result shows that there is no differences between LCD and GSM via readings. In addition, the maximum error of LCD and actual readings is 0.02 litres. Although the error in this study was smaller than the proposed study, this study was a GSM-based smart meter. Currently, internet-based information is developing very rapidly compared to information via GSM.

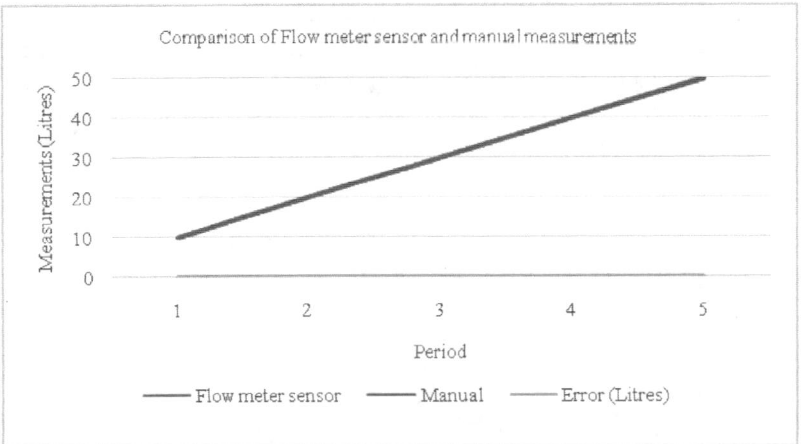

Figure 3. Flow meter sensor

The main valve will closed automatically when the certain amount of water is fulfilled. The amount of water is set at 10, 20, 30, 40, and 50 litres.

The smart meter is equipped with wireless transceiver integrated with microcontroller using node MCU. This module is compatible with smartphones to communicate with the microcontroller using universal asynchronous receiver transmitter (UART) protocol that is already available in the node MCU. The data will send to the web server and consumer can access via certain IP address.

5. Conclusion

The proposed smart water meter is designed using a sensor flow meter with the internet of things, which offers a cost-effective solution for water resource management in smart cities. The use of node MCU allows delivering real time information to the user's smart phone about the amount of water flowing through the sensor. The module of node MCU will send the data to the web server. The result shows that flow meter sensor operates accurately with maximum error 0.3 liters. The smart meter also allows the consumer to determine the water requirement. The main valve will close automatically when the requirement is achieved.

References

[1] J. Jin, J. Gubbi, S. Marusic, and M. Palaniswami, "An information framework for creating a smart city through internet of things," *IEEE Internet Things J.*, vol. 1, no. 2, pp. 112–121, 2014, doi: 10.1109/JIOT.2013.2296516.

[2] T. Randall and R. Koech, "Smart Water Metering Technology for Water Management in Urban Areas," *Water e-Journal*, vol. 4, no. 1, pp. 1–14, 2019, doi: 10.21139/wej.2019.001.

[3] I. Delpla, A. V. Jung, E. Baures, M. Clement, and O. Thomas, "Impacts of climate change on surface water quality in relation to drinking water production," *Environ. Int.*, vol. 35, no. 8, pp. 1225–1233, 2009, doi: 10.1016/j.envint.2009.07.001.

[4] X. J. Li and P. H. J. Chong, "Design and implementation of a self-powered smart water meter," *Sensors (Switzerland)*, vol. 19, no. 19, 2019, doi: 10.3390/s19194177.

[5] L. M. Irons, J. Boxall, V. Speight, B. Holden, and B. Tam, "Data driven analysis of customer flow meter data," *Procedia Eng.*, vol. 119, no. 1, pp. 834–843, 2015, doi: 10.1016/j.proeng.2015.08.947.

[6] K. Shrotriya, M. Jain, M. Mittal, L. Yadav, and N. Vijay, "Digital Water Meter Using Arduino," *IJEMR Int. J. Eng. Manag. Res.*, vol. 7, no. 2, pp. 276–279, 2017.

[7] Y. R. Putra, D. Triyanto, and Suhardi, "Rancang Bangun Perangkat Monitoring Dan Pengaturan Penggunaan Air Pdam (Perusahaan Daerah Air Minum) Berbasis Arduino Dengan Antarmuka Website," *J. Coding Sist. Komput. Untan ISSN 2338-493X*, vol. 05, no. 1, pp. 33–34, 2017, [Online]. Available: http://jurnal.untan.ac.id/index.php/jcskommipa/article/download/19172/16025.

[8] R. Sood, M. Kaur, and H. Lenka, "Design and Development of Automatic Water Flowmeter," *Int. J. Comput. Sci. Eng. Appl.*, vol. 3, no. 3, pp. 49–59, 2013, doi: 10.5121/ijcsea.2013.3306.

[9] M. H. Mubarok, S. J. Zarrouk, and J. E. Cater, "Two-phase flow measurement of geothermal fluid using orifice plate: Field testing and CFD validation," *Renew. Energy*, vol. 134, pp. 927–946, 2019, doi: 10.1016/j.renene.2018.11.081.

[10] P. Megantoro, D. A. Husnan, M. U. Sattar, A. Maseleno, and O. Tanane, "Validation Method for Digital Flow Meter for Fuel Vendors," *J. Robot. Control*, vol. 1, no. 2, pp. 44–48, 2020, doi: 10.18196/jrc.1210.

[11] C. V. Palau, G. V. Do Bomfim, B. M. De Azevedo, and I. B. Peralta, "Numerical study of upstream disturbances on the performance of electromagnetic and ultrasonic flowmeters," *Sci. Agric.*, vol. 77, no. 4, 2020, doi: 10.1590/1678-992x-2018-0208.

[12] G. Rajita and N. Mandal, "Review on transit time ultrasonic flowmeter," *2016 2nd Int. Conf. Control. Instrumentation, Energy Commun. CIEC 2016*, pp. 88–92, 2016, doi: 10.1109/CIEC.2016.7513740.

[13] WikiSeeed Studio, "YF-S201 Water Flow Sensor Datasheet," p. 1, 2013, [Online]. Available: http://www.seeedstudio.com/wiki/index.php?title=G1/2_Water_Flow_sensor&action.

[14] P. Mwangi, E. Mwangi, and P. M., "A Low Cost Water Meter System based on the Global System for Mobile Communications," *Int. J. Comput. Appl.*, vol. 142, no. 12, pp. 7–12, 2016, doi: 10.5120/ijca2016909945.

Acknowledgments

The authors would like to express their gratitude to Faculty Engineering of Tadulako University for funding the research. Our gratitude also goes to Fatur and Amran for their contribution in data collection.

IOP Conf. Series: Materials Science and Engineering 1212 (2022) 012043 doi:10.1088/1757-899X/1212/1/012043

Development and testing of a light dimming control using arduino uno

Waluyo, F Hadiatna, A Widura, P Setiana

Department of Electrical Engineering, Institut Teknologi Nasional Bandung (Itenas), Bandung, Indonesia.

E-mail: waluyo@itenas.ac.id

Abstract. Increasing population and human needs have an impact on increasing the need for electrical energy. One of them is for lighting needs. Therefore, it is necessary to save the lighting system so that energy consumption is minimum and the need for lighting is optimal, by controlling light dimming. This paper presents an implementation and testing of a dimming light control using an Arduino Uno microcontroller. The circuit used a 12-volt power supply, as a voltage source, to increase to 42 volts, to meet a lamp voltage, through a dc-dc converter. After obtaining the maximum voltage, a MOSFET cut off the voltage according to the desired light or performance level. The duty cycle was directly proportional to the output voltage, using a PWM coding to get the necessary light intensity. Some testing was conducted, including the measurement point shifting to the side. The testing results show that PMW and LDR decreased as the duty cycle increased. Nevertheless, both decreasing are different, the PWM decreased linearly with a gradient of -2.55 and the LDR decreased hyperbolically. While, the illuminance, current, and power rose as the duty cycle increased. The illuminance increased, tent to be saturated, as the power increased. However, the illuminance was reduced as the PWM and LDR increased. The illuminance decreased slightly as the measurement points shifted to the side.

1. Introduction

The need for humans and the population that is always increasing will have an impact on increasing the need for electrical energy. One of them is the electricity need for lighting. On other hand, the availability of electrical energy is limited. Based on this, the lighting system should be saved, where the provision of lighting is as optimal as possible, but the energy consumed is minimal. Electrical energy conservation, including lighting systems, is indispensable. Therefore, it is necessary for an efficient lighting system, and vital in homes, offices, industries, urban and rural areas [**1**].

Based on the procedure for design in buildings regulated in SNI 63-6575-2001, the lighting system can be differentiated into uniform, local, and combined lighting system. In the uniform or even lighting system, this system functions to provide an even and fairly constant level of light throughout the room. The even lighting can be accomplished by installing the armature simultaneously on the ceiling bar either directly or not. For a safe local lighting system, it is necessary to require a high level of light to provide more light than the ambient one. While the combined lighting systems are used for those who need high levels of light. By this, it can show the shapes and textures that require the light to come from a certain direction [**2**].

It can be carried out by measuring it directly using a lux meter measurement tool so that it can be known what the magnitude of the exposure level is in a given work. What is meant by the worksheet is an imaginary horizontal plane located 0.75 meters above the floor in the entire room [**2**].

There are two main light sources, natural and man-made. The latter has conventionally only two modes, on and off switches. There is not any intermediate-range, whether manually or automatically according to the surrounding condition. This case leads to electrical energy wastage and not effective and not efficient anymore [**3-5**] and may potentially dangerous [**6**].

In commercial and office sectors, major electrical energy consumption is caused by lighting [**7**]. The lighting control can be implemented in homes, smart cities, airports, agriculture, parking, malls, universities, industries, and so on [**4**]. Besides light, color has attracted mankind, especially white color [**8**]. The lighting is necessary also in a vehicle, including an automobile, and plays an important role to avoid accidents [**9-10**]. Street lights consume large electrical energy [**11**]. Therefore, a smart street light reduces energy consumption and maintains costs [**12**]. The lighting has an important role in outdoor and indoor environments [**13**], and huge energy is wasted due to lighting in rural and urban areas [**14**]. It creates also a comfortable atmosphere, such as warm and bright [**15**].

Based on these various reasons, it is important to conduct an automation system in lighting, for saving, comfort as well as for safety. The light automation system is a cost-effective and safe way to reduce power consumption [**16-17**], reducing unused electricity and lengthening light lifetime [**16**]. A dimming light is reducing wasting energy, overcome manual switching, primary cost, and maintenance decrease [**17**]. There are intermediate levels [**3, 6, 18**]. The system can be implemented in workstations, park street lighting, lights, agriculture field monitoring, home automation, etc [**3-4,16**].

A main controlling device for lamp dimming use a microcontroller [**16, 19**], and some other devices, such as light sensor, LDR (light dependent resistor) as an instant LED driver, DC source [**3, 9, 16, 19,**], and an infrared sensor [**4**]. Arduino microcontroller is very widely used in light dimming due to easily available in markets and relatively cheap. It was for the automation of the speed and light on automobiles [**10**], smart lighting [**1, 3, 13-14**], for touch dimming circuit [**6**], bright controlling [**5**], street lighting [**4, 11, 12**], temperature and color tuning [**8**], vehicle headlight automation [**9**], and light dimming [**15, 20**].

An Android-based spatial lighting arrangement system discussed efficient, practical, and appropriate lighting settings and it is complemented by lighting setting optimizations based on the standard SNI 03-6197-2000. It was also designed practically to be remotely controlled by using a smartphone [**21**].

For a recent lighting system, it is also necessary to light-emitting diode (LED) lamps. The LEDs consume low energy [**6, 17**], due to cool conditions, long lifetime, and energy-saving due to high illuminance [**16, 18**]. Some researches used LEDs, for smart light [**3**], touch dimming [**6**], brightness [**5**], and luminaire controlling [**7**], street, color, vehicle head, and rural lighting [**1, 4, 8-9, 11-14, 22**]. A High Power LED (HPL) is an LED that works several hundred mA to A. HPL is usually used for outdoor lighting or industrial lighting. The use of HPL aims to save costs, because HPL requires a relatively small current, but still has a larger lumen [**23**].

The light dimming controls usually use LDR, such as for smart light [**3**], brightness control [**5**], smart street light control [**4, 12**], and headlight of vehicles [**9**].

Pulse Width Modulation (PWM) is generally a means of reproducing the signal width expressed by a pulse in a period, to obtain a different average voltage. PWM can be applied to adjust the power or voltage to turn on the light. In the digital method, each PWM change is influenced by the resolution of the PWM itself. Suppose that the 8-bit digital PWM means that the PWM has a resolution of $2^8 = 256$, that is, this PWM output has 256 variations, the variation ranging from 0 - 255 which represents a duty cycle of 0-100% of the PWM output [**24**].

The addition of light dimming is usually a boost converter circuit. It increases DC voltage to a higher level. This circuit is required when the required voltage of electronic components is higher than the available power supply voltage [**25**].

There have been some researches regarding light dimming in various applications. Generally, they used Arduino Uno microcontroller, LDRs, and LEDs. Nevertheless, they are limited in testing. Based on the previous researches, it is necessary to conduct further research on light dimming design, implementation, and testing. This research involved an HPL, Arduino Uno microcontroller, MOSFET, DC-DC converter, and an LDR. Besides the design and implementation, the research also conducted testing on some parameters, such as PWM, duty cycle, LDR, illuminance, current, power, voltage, and side shifting.

2. Research Method

The light dimmer circuit uses a 12-volt power supply as a voltage source which then passes through the dc-dc converter to increase the voltage from 12 volts to 42 volts to meet the needs of the used lamps. After obtaining the maximum voltage at the maximum capacity at 100%, the IRF 540N MOSFET functions as a voltage cut off according to the desired light needs or performance level. In this process, the duty cycle will be directly proportional to the output voltage, in the process of setting this duty cycle, it is carried out in the basic software using coding to get the light intensity setting according to the need, as an example of a lamp can be turned on when it is a bright condition, a bit bright, dim, slightly dim and off. Figure 1 shows the block diagram of the research of the light dimming system. The main subsystems are DC supply, DC-DC converter, a lamp, a controller, and a MOSFET. Furthermore, the wiring circuit diagram is shown in Figure 2. In this case, the lamp is HPL 32 watts and an Arduino Uno microcontroller as the controller.

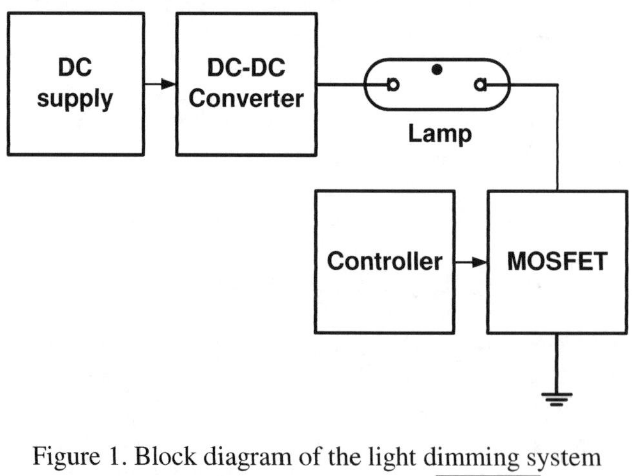

Figure 1. Block diagram of the light dimming system

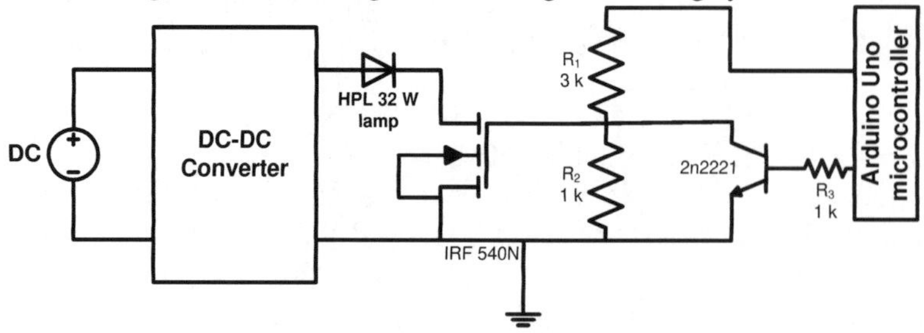

Figure 2. Light dimmer circuit

Figure 3 shows the lighting intensity measurement with a distance of 2.5 m, between the High-Performance Lamp (HPL), as the light source, and the light receiver wall. Moreover, on the wall, the measuring point for light acceptance was shifted up to 50 cm to the side. So that the variation of light received was obtained over that distance.

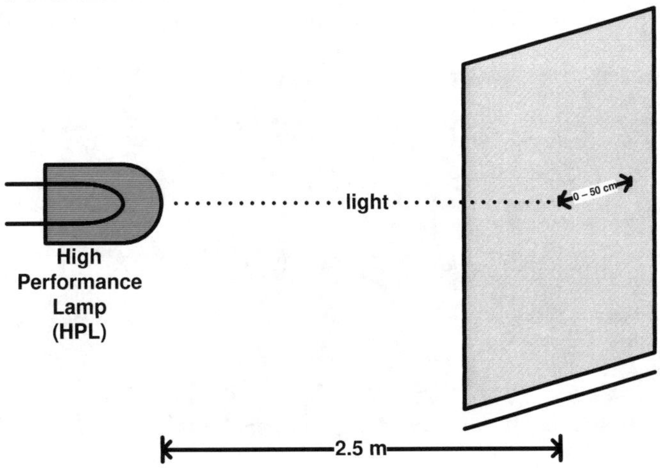

Figure 3. Lighting intensity measurement

A Light Dependent Resistor (LDR) and Lux meter or light meter were used to measure the lighting intensity or illuminance. The LDR value was measured by using a multimeter. Both LDR's and lux meter's values were noted while the lighting intensity was being reduced between 100% and 0% in a 10% step.

3. Research Results and Discussion

Figure 4 shows the dimming system circuit. The system consists of a power supply, for DC supply of power, DC-DC converter for the DC voltage regulation, MOSFET and voltage divider circuits for the light dimming for lowering the voltage to insert the microcontroller, an Arduino Uno for the main controller, a computer/laptop for coding of the microcontroller, and a High-Performance Lamp (HPL) for the lighting source.

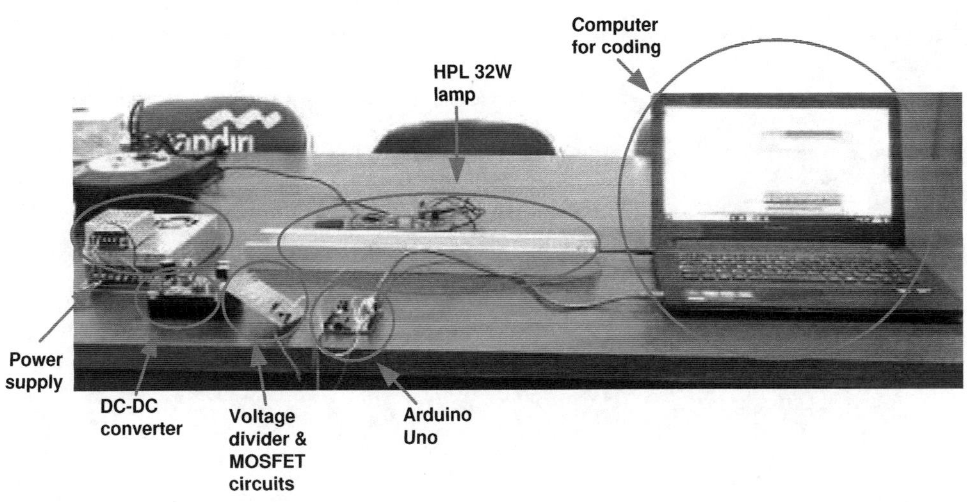

Figure 4. Picture light dimming system

Figure 5 shows the dimming system circuit when the dimming lighting of 80% (a) and 40% (b), which show a bit bright and a bit dim respectively. The system gives the lighting intensity was enough for the lighting. Thus, based on this testing, the system dimming process ran smoothly.

(a) 80%

(b) 40%

Figure 5. Picture light dimming system in 80% and 40% conditions

Figure 6 shows the PWM and LDR (Light Dependent Resistor) as the duty cycle function. Both parameters decreased as the duty cycle increased. Nevertheless, the PWM decreased linearly, while LDR decreased drastically, hyperbolically.

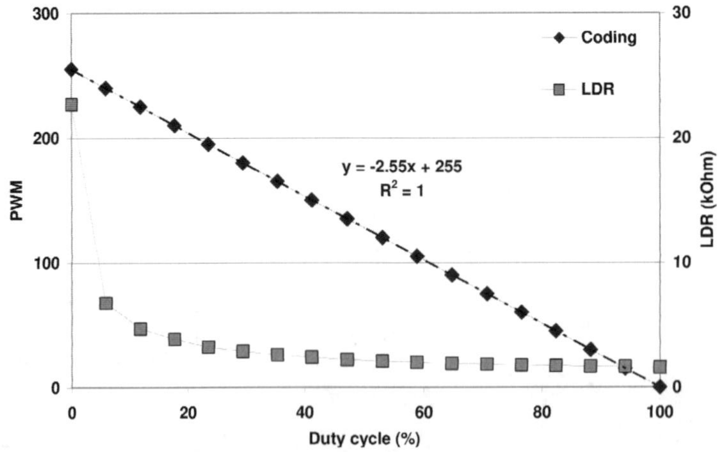

Figure 6. PWM and LDR function of duty cycle

Figure 7 (a) shows the illuminance and current as of the duty cycle function. Both illuminance and current parameters increased linearly, as the duty cycle increased. While 7(b) shows the illuminance as a function of current, as linear too. These characteristics are in line with that are revealed in [8].

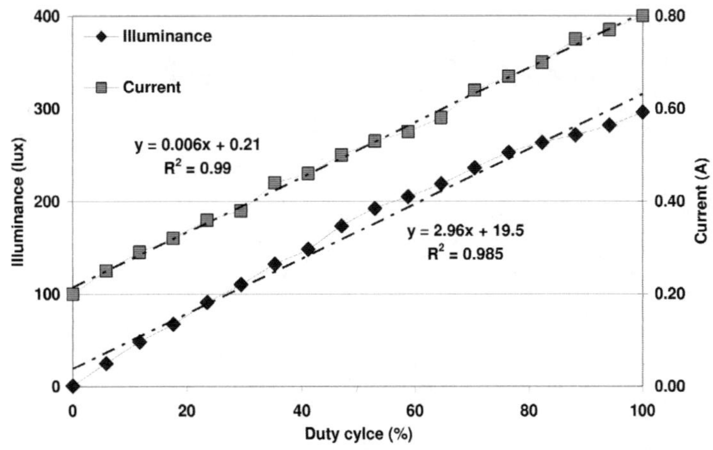

(a)

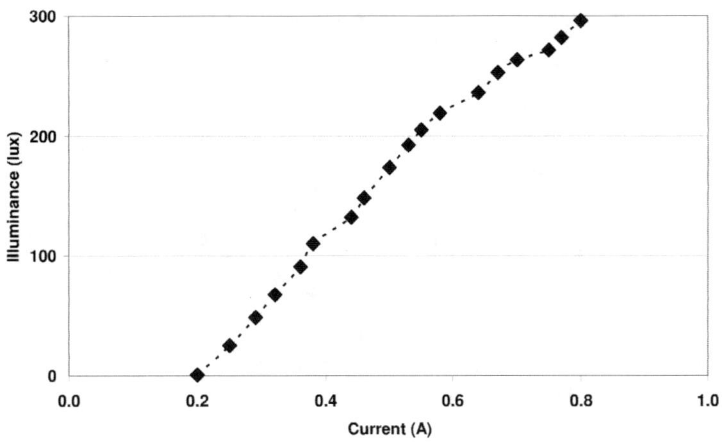

(b)

Figure 7. (a) Illuminance & current vs duty cycle, and (b) Illuminance vs current

Figure 8 shows the power and current as of the duty cycle function. Both power and current increased linearly as the duty cycle increased.

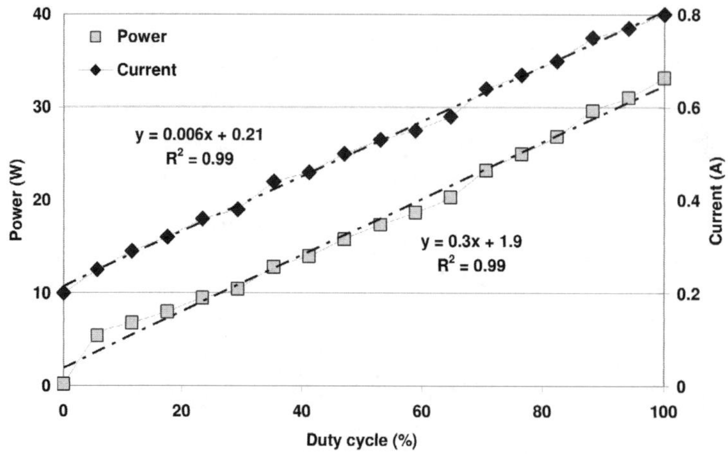

Figure 8. Power and current as duty cycle function

Figure 9 shows the illuminance as the function of power. The illuminance increased linearly, and to be saturated, as the power increased. Thus, the illuminance will tend to be constant as the power of the lamp continuously.

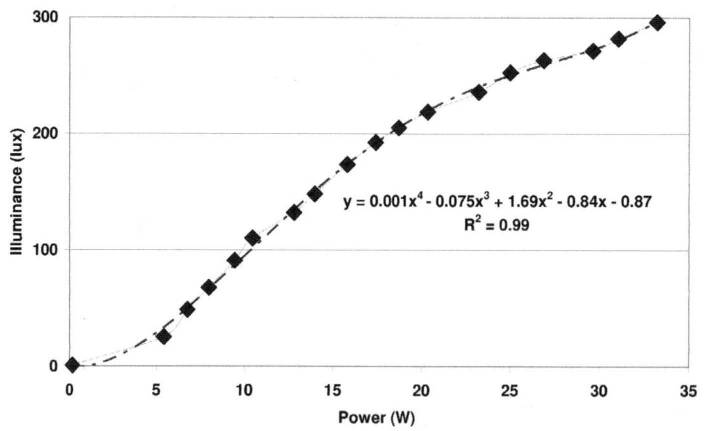

Figure 9. Illuminance function of power

Figure 10 shows the voltage and power as the current function. Both voltage and power increased linearly as the current increased. Nevertheless, both increasing parameters are different. For the voltage, it is necessary a threshold value of the current, for this case 0.25 ampere to get 21.7 volts of minimum voltage. This phenomenon, especially the voltage against the current, due to knee point existence, is similar to the chart that is revealed in [8]. Nevertheless, the direction of the knee pint is the opposite.

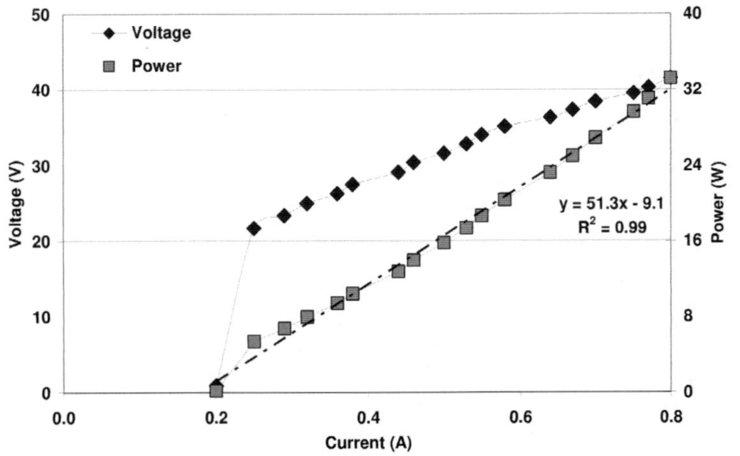

Figure 10. Voltage and power function of current

Figure 11 shows the illuminance as a function of the PWM coding. The illuminance decreases linearly as the PWM coding increases. The illuminance varies depending also on the distance to the perpendicular point of the board exposed to the light, as shown in Figure 13.

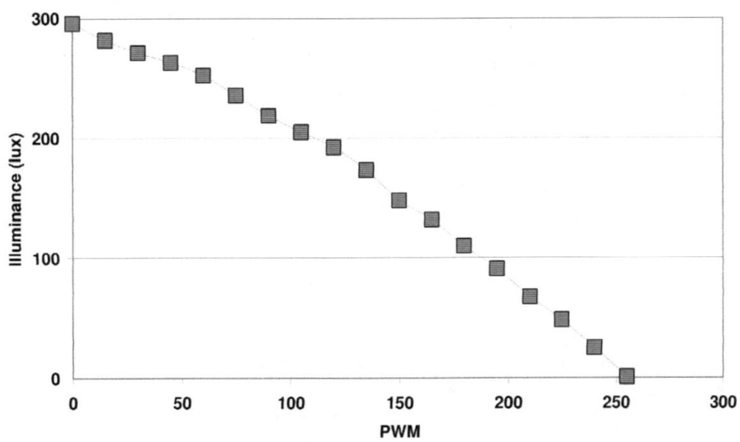

Figure 11. Illuminance function of coding

Figure 12 shows the LDR as the illuminance function. The LDR reduced drastically as the illuminance increased, followed by hyperbolic patterns on various distances from the center. This curve or chart is an improvement and more real than the previous concept, because this chart is obtained from the test results with the resistance measurements on the LDR on the wall illuminated by an HPL lamp [**9, 26**].

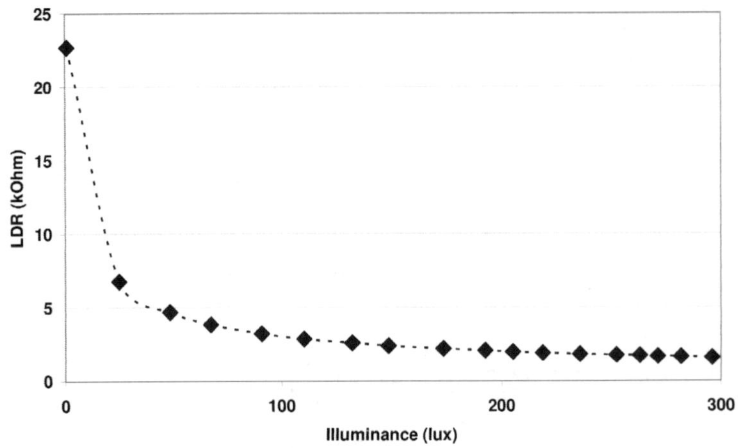

Figure 12. Illuminance function of LDR

Figure 13 shows the illuminances as a function of the distances from the center on various coding. The illuminance reduced slightly as the distance from the center increased.

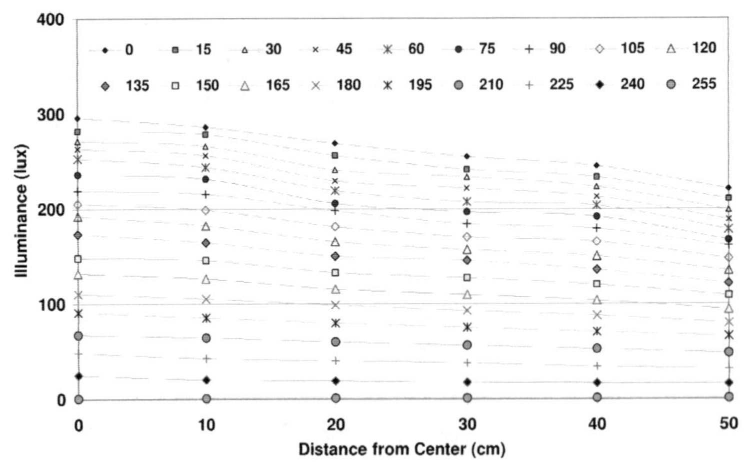

Figure 13. Illuminance function of distances from the center on various PWM

4. Conclusion

The microcontroller-based dimming circuit has been successfully implemented. The duty cycle increases, make the PWM coding and LDR decrease linearly and hyperbolically respectively. However, the duty cycle increases, so that the illuminance, current, and power increase too. The power and current increase linearly as the duty cycle increases. The illuminance increases as the power increase linearly. Nevertheless, increasing illuminance tends to be saturated. The voltage and power increase linearly as the current rises. Even though, the voltage chart has a knee point. The illuminance reduces fairly linearly, as then PWM coding increases. The LDR reduces drastically as illuminance increases. Finally, the illuminance decreases slightly as the distance of the measurement point from the center point to the side increases.

References

[1] N.S. Yoganathan, G. Lakshmi Priya, R. Monisha, A. Lavanya, S. Kowsalya, Density-Based Smart Lighting System Using IoT, International Journal of Advanced Research in Electrical, Electronics and Instrumentation Engineering, Vol. 7(2), Feb. 2018, 504-513.
[2] BSN, *Tata cara perancangan sistem pencahayaan buatan pada bangunan gedung* (The procedure for designing artificial lighting systems in buildings), BSN, 2001.
[3] D.K. Rath, Arduino Based: Smart Light Control System, International Journal of Engineering Research and General Science, Vol. 4(2), March-April, 2016, 784-790.
[4] Z. Mumtaz, S. Ullah, Z. Ilyas, N. Aslam, S. Iqbal, S. Liu, J.A. Meo, H.A. Madni, An Automation System for Controlling Streetlights and Monitoring Objects Using Arduino, Sensors 2018, 18, 1-14.
[5] S. Jeyavinotha, B.P. Deepika, R. Remya, N.V. Ajin, Design of Automation System to Control the Brightness of LED Using Arduino Uno and IoT, International Journal of Recent Trends in Engineering & Research (IJRTER), Conference on Recent Trends in Electrical and Electronics Engineering (RTEEE '19), Special Issue; March- 2019, 247-250.
[6] B.V. Kumar, G.M. Rao, J.Y. Kumar, S.H. Krishna, Touch Dimmer Switch Circuit, International Research Journal of Engineering and Technology (IRJET), Vol. 6(4), Apr 2019, 1971-1973.
[7] B.M.R. Rocha, Enhanced Networked Luminaire Controller for Sustainable Ambient Illumination, Thesis to obtain the Master of Science Degree, Electrical and Computer Engineering, Técnico Lisboa, Nov. 2015.
[8] S. Prathap, S. Sunny, A.S. Nair, Colour temperature tuning to improve efficacy of white light, International Conference on Emerging Trends in Engineering, Science and Technology (ICETEST-2015), Procedia Technology 24 (2016) 1186-1193.
[9] Lakshmi, K., Nevetha, R., Ilakkiya, S.N., Ganesan, R., Automatic Vehicle Headlight Management System to Prevent Accidents Due To Headlight Glare, International Journal of Innovative Technology and Exploring Engineering (IJITEE), Vol. 8(9), July 2019, 757-760.
[10] A.M. Jose, M. Philip, N. Ramakrishnan, V.R. Shalu Raj, B. Mathew, Implementation of Automatic Light and Speed Controllers for Automobiles, International Journal of Innovative Research in Computer and Communication Engineering, Vol. 5(3), March 2017, pp. 5773-5777.
[11] A. D'souza, O. Bhosale, M. Bhilare, S. Sawant, Arduino Based Solar Street Lighting, International Journal of Scientific & Engineering Research, Vol. 9(2), pp. 36-38.
[12] A.M. Sindhu, J. George, S. Roy, J. Chandra, Smart Streetlight Using IR Sensors, IOSR Journal of Mobile Computing & Application (IOSR-JMCA), Vol. 3(2), Mar.-Apr.2016, pp. 39-44.
[13] P.S. Raju, M. Mahalingam, R.A. Rajendran, Design, Implementation, and Power Analysis of Pervasive Adaptive Resourceful Smart Lighting and Alerting Devices in Developing Countries Supporting Incandescent and LED Light Bulbs, Sensors 2019, 19(9), 2032; pp.1-20.
[14] A. Vinutharani, C.K. Vanamala, Hybrid, Low Cost, Sensor-Based Smart Lighting for Rural Areas, International Journal of Advanced Research in Computer and Communication Engineering Vol. 5 (5), May 2016, pp. 260-263.
[15] S. Sakpal, P. Kulkarni, V. Chitre, S. Gharat, Review of LED DIMMER, 4th International Conference On Engineering Confluence & Inauguration of Lotfi Zadeh Center Excellence in Health Science And Technology (LZCODE) – EQUINOX 2018, IOSR Journal of Engineering (IOSRJEN), pp. 16-19.
[16] Z. Mumtaz, S. Ullah, Z. Ilyas, S. Liu, N. Aslam, J.A. Meo, H.A. Madni, Automatic Streetlights that Glow on Detecting Night and Object using Arduino, Computer Science, Other Computer Science, 28 Jun 2018.
[17] P.R. Prasad, V. Nani, M. Srinivas, M. Karthik, Automatic Street Light Dimmer Circuit Using Arduino, Pramana Research Journal, Vol. 9(4), 2019, 490-497.
[18] H.S. Walunj, R. Kawitkar, Km. A. Singh, Dimming Functionality of LED Bulb: A Working Model, International Research Journal of Engineering and Technology (IRJET), Vol. 5(5), May 2018, 3370-3371.
[19] Taufik, W. Xiong, J. Sato, Saidah, Ambient Light Adaptive LED Light Dimmer, Telkomika,

Vol.13 (2), June 2015, 281-290.
[20] A. Kumar, Fajr, T.N. Khajwal, N. Khera, Microcontroller Based Digital AC Dimming Techniques for Light Intensity Control, 2019 6th International Conference on Signal Processing and Integrated Networks (SPIN), 2019 6th International Conference on Signal Processing and Integrated Networks (SPIN), 7-8 March 2019, IEEE, Noida, India, pp. 452-455.
[21] M.D. Putro, F.D. Kambey, *Sistem Pengaturan Pencahayaan Ruangan Berbasis Android pada Rumah Pintar* (Smart Home Based Lighting Arrangement System), Jurnal Nasional Teknik Elektro, Vol. 5(3), Nov. 2016, 297-307.
[22] N. Yoshiura, Y. Fujii, N. Ohta,_Smart street light system looking like usual street lights based on sensor networks, 2013 13th International Symposium on Communications and Information Technologies (ISCIT), 4-6 Sept. 2013, IEEE, Surat Thani, Thailand.
[23] M. Walidain, I.D. Sara, M. Syukri, *Perancangan Sistem Penerangan LED Sebagai Sumber Cahaya Pada Pengujian Modul Surya* (Designing LED Lighting Systems as Light Sources in Solar Module Testing), KitEktro: Jurnal Online Teknik Elektro e-ISSN: 2252-7036, Vol.3(2), 2018: 46-52.
[24] G. Turesna, Zulkarnain, Hermawan, *Pengendali Intensitas Lampu Ruangan Berbasis Arduino Uno Menggunakan Metode Fuzzy Logic* (Room Light Intensity Controller Based on Arduino Uno Using Fuzzy Logic Method), Jurnal Otomasi Konmtrol dan Instrumentasi, ITB, Vol 7 (2), 2015, 73-88.
[25] *A. Mujadin, S. Rahmatia, Joule Thief Sebagai Boost Converter Daya LED Menggunakan Sel Volta Berbasis Air Laut* (Joulc Thicf As Boost Converter Power LED Using Seawater Based Volta Cell), Jurnal Al-Azhar Indonesia Seri Sains dan Teknologi, Vol. 4(2), Sept. 2017, 52-57.
[26] A.S. Jalan, A Survey on Automatic Street Lighting System on Indian Streets Using Arduino, International Journal of Innovative Research in Science, Engineering and Technology, Vol. 6(3), March 2017, pp. 4139-4144.

Acknowledgment

The authors are grateful to The Grant of the Higher Education Superior Applied Research, No. 387/B.05/LPPM-Itenas/IV/2020, for a part of the financially supporting research.

Evaluation the influence of distance-based K-means method for detecting moving vehicles

Y Sari[1], P B Prakoso[2], A R Baskara[1]

[1] Department of Technology Information, Universitas Lambung Mangkurat, Banjarmasin, Indonesia
[2] Department of Civil Engineering, Universitas Lambung Mangkurat, Banjarmasin, Indonesia

yuzlena@ulm.ac.id

Abstract : Detecting moving vehicles is one of important elements in the applications of Intelligent Transport System (ITS). Detecting moving vehicles is also part of the detection of moving objects. K-Means method has been successfully applied to unsupervised cluster pixels for the detection of moving objects. In general, K-Means is a heuristic algorithm that partitioned the data set into K clusters by minimizing the number of squared distances in each cluster. In this paper, the K-Means algorithm applies Euclidean distance, Manhattan distance, Canberra distance, Chebyshev distance and Braycurtis distance. The aim of this study is to compare and evaluate the implementation of these distances in the K-Means clustering algorithm. The comparison is done with the basis of K-Means assessed with various evaluation paramaters, namely MSE, PSNR, SSIM and PCQI. The results exhibit that the Manhattan distance delivers the best MSE, PSNR, SSIM and PCQI values compared to other distances. Whereas for data processing time exposes that the Braycurtis distance has more advantages.

Keywords: Detection of moving vehicles Distance-based method K-Means

1. Introduction

Detecting moving vehicles is one of important processes for the applications of Intelligent Transport System (ITS). Basically, detecting moving vehicles is part of the detection of moving objects. This task is an important underlying part of traffic monitoring and many other applications. The issue of computation in the detection of moving object has been studied, analyzed and the latest techniques have been proposed such as Adaptive Threshold Algorithm, Clustering Algorithms, etc. [1]–[6]. Clustering is a technique for categorizing data into groups. Distance metrics play a very important role in the grouping process. The more similarities between data in groups, the more likely certain data items belong to a particular group.

There are several algorithms available for clustering. One of them is K-Means clustering. In general, K-Means is a heuristic algorithm that partitioned the data set into K clusters by minimizing the number of squared distances in each cluster [7], [8]. The benefits of the K-Means algorithm are due to computational efficiency and ease of use. Distance metrics are assigned to find similar data objects that lead to the development of robust algorithms for data mining functionality such as classification and clustering [8]–[10]. This paper discusses the comparison results of solving two-dimensional data by applying the K-Means algorithm using different distance metrics; namely Manhattan distance, Canberra distance, Chebyshev distance and Braycurtis distance; as well as the basic K-Means algorithm applying Euclidian distance metrics. The results are displayed with the help of a histogram.

2. Design of the Proposed Method

First, the initial data in the form of video is converted into frames. Later a frame is extracted into images, which are then used as input for the K-Means method. K-Means is a clustering method, where the grouping process is based on different distances. After getting the cluster result in the form of a vector, it is converted into a matrix again and then the background is turned to white and the foreground is turned to black. The foreground is adjusted according to the color of the original image, while the background is left white. Next, the results are assessed using evaluation parameters such as MSE, PSNR, SSIM and PCQI. The flowchart of the proposed method is shown in Figure 1.

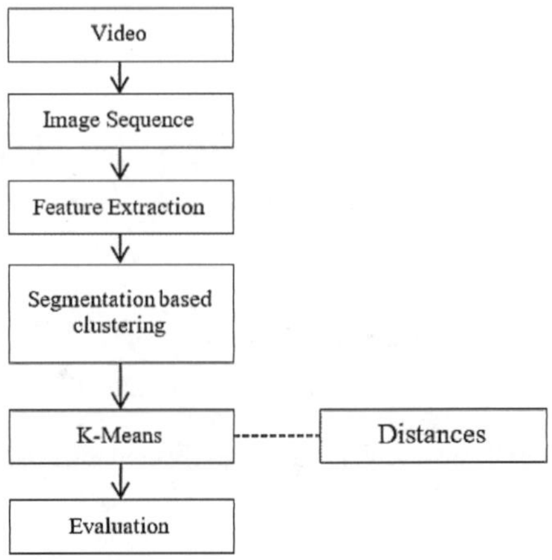

Figure 1. Flowchart of distance-based K-Means for object detection

2.1. Moving Vehicle Detection

The vehicle detection process is divided into several stages:

2.1.1. Data collection

Data comes from video data of vehicles on a highway. Data is retrieved using a camera, which is stored in video format. The reason for choosing test data taken in the form of video is to examine whether the application can run in real time.

2.1.2. Image Sequence

Video data is converted to sequence of images. The process of transforming the video into image sequences begins with inputting the video. Next is the video reading process, where the video will be read in each frame. After that, looping of all frames in the video is performed. The next operation is to label the frames. The labelled data used for analysis is 400 data frames. Data sample of one video frame is shown in Figure 2.

Figure 2. Example of the 200th image frame

2.1.3. Feature Extraction

The feature extraction technique converts a RGB matrix, which contains pixel into vector data. The image processing operation from the RGB matrix to vector data starts from storing RGB data. The feature extraction is done by extracting R, G and B into vectors and making 1 data for each frame.

2.2. Segmentation based Clustering

Image segmentation is considered as the basic operation that is important for meaningful analysis and interpretation of images obtained [11]–[13]. The segmentation process is applied to distinguish objects from the background and unwanted objects. This processing section is responsible for identifying moving vehicles from static objects or backgrounds. Images are partitioned into groups of pixels that are homogeneous with respect to several criteria. Different groups cannot intersect and adjacent groups must be heterogeneous. The bounding box is determined based on the ratio of the area of interest when conducting a clot analysis [14]–[19]. Figure 3 shows the results before and after segmentation.

(a) (b)

Figure 3. Segmentation result of original image (a) into segmented image (b)

2.3. K-Means

The K-Means clustering algorithm employs Euclidean distance [17], [18] to measure the similarity between objects. Besides using the Euclidean distance, other distance metrics can also be used and this method is called the median-K method. The median-K method tries to eliminate the weakness of the K-Means method because it is less sensitive to outliers. The stages of the K-Means algorithm are as follows:

- Let K be the number of clusters.
- Select K points to be taken as centroids for group K. These points can be chosen randomly. This criterion is valid unless the user has several detailed views in the data.
- Calculate the distance by using the Euclidian formula/other distance formula for each data point set from each selected centroid.
- Each object to the nearest cluster is based on the distance value calculated in the steps above.
- Calculate the centroid cluster again by calculating the average value of the attributes of all objects in eachcluster.
- Analyze the cluster membership whether it changes or does not change. If it does not change then this is the last step, if not then go back to the step of distance calculation.

2.4. Distance

In this study, there are several distance calculations that are applied and will be analyzed at the later stages of results. The distance calculation methods used are:

- Euclidean Distance

  Euclidean Distance is employed because it allows to determine the length of the shortest path between two points, so that it is able to measure the closeness between two features. The Euclidean Distance equation is shown in Equation 1 as follows:

$$ED = Sqrt\,\{(X_2 + X_1)^2 + (Y_2 - Y_1)^2\} \quad (1)$$

- Canberra Distance

  Canberra Distance is used because it is able to calculate the appropriate values in scattered data. Then the values are compared based on rank, so that they can consider the distance between two points and their relevance. The Canberra Distance formulation can be seen in Equation 2 as follows:

$$Canb\,(x,y) = \sum_{i=1}^{d} \frac{|x_i - y_i|}{|x_i| + |y_i|} \tag{2}$$

- Chebyshev Distance
  Chebyshev Distance is utilized because it can determine the biggest distance from the difference in each coordinate point. Chebyshev Distance formulation is depicted in Equation 3 as follows:

$$d_{chebyshev}\,(x, y) = max_i(\,|x_i - y_i|\,) \tag{3}$$

- Manhattan Distance
  Manhattan Distance is also known as square distance, where the distance of a city block is defined as a point segment to the coordinate axis. Manhattan distance formulation is shown in Equation 4 as follows:

$$d_M\,(x_1 - x_2) = \sum_{i=1}^{i=n} |x_1 - y_1| \tag{4}$$

- Braycurtis Distance
  Braycurtis Distance is used because it allows to measure inequality between two different locations, based on calculations in each location. Braycurtis Distance is exhibited as formulation in equation 5 as follows:

$$dBraycurtis = \frac{\sum |\,x_i - y_i\,|}{\sum (x_i + y_i)} \tag{5}$$

2.5. Evaluation

To compare and evaluate the distance calculation methods applied in K-Means, this study assigns several measurement parameters as follows:

- Mean Square Error (MSE)
  Mean square error (MSE) is a quality measurement that is calculated based on 2 images compared, so that the level of mismatch between the images can be seen [19], [20]. The smaller the MSE value that is generated, the more the image is compared. The MSE equation is described as follows:

$$MSE = \frac{1}{M*N} \sum_{i=1}^{M} \sum_{j=1}^{N} \big(x(i,j) - y(i,j)\big)^2 \tag{6}$$

- Peak Signal to Noise Ratio (PSNR)
  Peak signal to noise ratio (PSNR) is a quality measurement calculated based on 2 image objects compared to MSE [21]–[23]. Comparison between the maximum values of the signal is measured by the amount of noise that affects the signal. The results obtained from the PSNR are said to be more similar if the PSNR value is getting bigger. PSNR is usually measured in decibels (db) written in equation (7).

$$PSNR = 10 \log_{10} \left(\frac{MAX^2}{MSE}\right) \tag{7}$$

- Structural SIMilarity (SSIM)
  Structural similarity (SSIM) is used to calculate the similarity index between 2 image objects that are compared structurally [1], [24]. The value of the SSIM demonstrates that two images have structural similarities when approaching value of 1. The SSIM formula is based on a comparison of 3 measurements of luminance, contrast and structure. The SSIM formula can be summarized as in equation (8) provided that the values α, β and γ are 1.

$$SSIM(x,y) = \frac{\big((2\mu_x\mu_y + C_1)(2\sigma_{xy} + C_2)\big)}{(\mu_x^2 + \mu_y^2 + C_1)(\sigma_x^2 + \sigma_y^2 + C_2)} \tag{8}$$

- Patch-based Contrast Quality Index (PCQI)
  Patch-based contrast quality index (PCQI) has concept such as SSIM which is used to calculate similarity indexes between two image objects [13], [25], [26]. PCQI utilizes patch-based representations that allow each new input image to be divided into three components physically meaningful. The strength of the component is determined by the new coordinate system, but the evaluation component produces a value of more than 1 if there is an increase in image. The average of the PCQI formula is written in equation (9), and the PCQI

formula explains in equation (10) based on a comparison of three measurements of intensity, contrast and structure.

$$PCQI(x,y) = \frac{1}{M}\sum_{j=1}^{M} PCQI(x_j, y_j) \quad (9)$$

$$PCQI(x,y) = q_i(x,y) * q_c(x,y) * q_s(x,y) \quad (10)$$

3. Results and Analysis

The results of vehicle detection computation are done by calculating the distances, and the number of images used for the analysis is 400 data. Based on the results obtained by applying Euclidean Distance, Canberra Distance, Chebyshev Distance, Manhattan Distance, and Braycurtis Distance in K-Means, the comparison of the MSE, PSNR, SSIM and PCQI values is shown in Table 1 below:

Table 1. Result of Distance Evaluation

Euclidean	Canberra	Chebyshev	Manhattan	Braycurtis	Evaluation
1487.056704	2620.657	1502.033	1328.147	3159.799292	MSE
20.65784169	19.22424	20.74845	21.13607	18.83470784	PSNR
0.817551531	0.723066	0.823867	0.831358	0.715181703	SSIM
0.767853418	0.700277	0.776038	0.786532	0.699320642	PCQI

From the results above, the smallest MSE with a value of 1.328 is generated by Manhattan distance and Manhattan distance also produces the highest PSNR value of 21.14. The higher PSNR value implies a closer resemblance between the reconstruction result and the original image. Likewise, with the SSIM, the Manhattan distance produces a value close to 1, which is 0.83. The value of SSIM means that it has structural similarity if it is close to 1. For PCQI results, the Manhattan distance also has the best value compared to the other distances, namely 0.79, which is close to 1. Comparison of the processing time of different distances can be seen in Table 2 below:

Table 2. Processing Time Comparison

Distance	Euclidean	Canberra	Chebyshev	Manhattan	Braycurtis
Time(second)	2.889	23.339	0.381	0.348	0.300

Comparing the processing time, the distance of Braycurtis has the shortest time or fastest data processing, which is 0.3 second.

4. Conclusion

Evaluation of different distance approaches applied in the K-Means method demonstrates that the Manhattan distance produces the best values of MSE, PSNR, SSIM and PCQI compared to Euclidean distance, Canberra distance, Chebyshev distance and Braycurtis distance. Whereas for data processing time, Braycurtis distance has more advantages. It is suggested for further studies that comparisons of different distances applied in other cluster methods should be also conducted.

References

[1] M. I. Chacon-murguia, A. Guzman-pando, G. Ramirez-alonso, and J. A. Ramirez-quintana, "A novel instrument to compare dynamic object detection algorithms ☆," *Image Vis. Comput.*, vol. 88, pp. 19–28, 2019, doi: 10.1016/j.imavis.2019.04.006.

[2] G. Guido, A. Vitale, F. F. Saccomanno, V. Astarita, and V. Giofrè, "Vehicle Tracking System based on Videotaping Data," *Procedia - Soc. Behav. Sci.*, vol. 111, pp. 1123–1132, 2014, doi: 10.1016/j.sbspro.2014.01.147.

[3] C. Zhang, J. Ren, L. Fan, and C. Yu, "A Novel Method of Preceding Vehicle Detection and Vehicle Distance Measurement Based on Monocular Vision," *Rev. la Fac. Ing. U.C.V.,* vol. 32, pp. 92–98, 2017.

[4] L. Qin, H. Snoussi, and F. Abdallah, "Object tracking using adaptive covariance descriptor and clustering-based model updating for visual surveillance," *Sensors (Switzerland)*, vol. 14, no. 6, pp. 9380–9407, 2014,doi: 10.3390/s140609380.

[5] C. Tang *et al.*, "Real-time detection of moving objects in a video sequence by using data fusion algorithm," *Trans. ofthe Inst. ofMeasurement Control*, pp. 1–12, 2018, doi: 10.1177/0142331218773550.

[6] Z. Moutakki, I. M. Ouloul, K. Afdel, and A. Amghar, "Real-Time System Based On Feature Extraction For Vehicle Detection And Classification," vol. 19, no. 2, pp. 93–102, 2018, doi: 10.2478/ttj-2018-0008.

[7] M. Z. Rodriguez, C. H. Comin, D. Casanova, and O. M. Bruno, "Clustering Algorithms : A Comparative Approach Description of the clustering algorithms ' parameters," pp. 4–7.

[8] M. K. Rafsanjani, Z. A. Varzaneh, N. E. Chukanlo, M. K. Rafsanjani, Z. A. Varzaneh, and N. E. Chukanlo, "A survey of Hierarchical Clustering Algorithms," *J. Math. Comput. Sci.*, vol. 5, no. 3, pp. 229–240, 2012.

[9] E. A. Pambudi, P. N. Andono, and R. A. Pramunendar, "Image Segmentation Analysis Based On K-Means PSO By Using Three Distance Measures," vol. 9102, no. August, pp. 1821–1826, 2018, doi: 10.21917/ijivp.2018.0256.

[10] M. Algorithm and I. S. Nasir, "The Proposed Image Segmentation Method Based On Adaptive K- The Proposed Image Segmentation Method Based On Adaptive K-Means Algorithm," no. February, 2019.

[11] S. Gao, C. Zhang, and W. B. Chen, "An improvement of color image segmentation through projective clustering," *Proc. 2012 IEEE 13th Int. Conf. Inf. Reuse Integr. IRI 2012*, pp. 152–158, 2012, doi: 10.1109/IRI.2012.6303004.

[12] L. Christodoulou, T. Kasparis, and O. Marques, "Advanced statistical and adaptive threshold techniques for moving object detection and segmentation," *2011 17th Int. Conf. Digit. Signal Process. DSP*, pp. 1–6,2011, doi: 10.1109/ICDSP.2011.6004875.

[13] P. Spagnolo, T. D. Orazio, M. Leo, and A. Distante, "Moving object segmentation by background subtraction and temporal analysis," vol. 24, pp. 411–423, 2006, doi: 10.1016/j.imavis.2006.01.001.

[14] D. Realization, K. Futsuhara, Y. Zhang, J. Wang, and X. Yang, "Multiple Vehicle Detection and Segmentation in Malaysia Traffic Flow Multiple Vehicle Detection and Segmentation in Malaysia TrafficFlow," vol. vid, 2018, doi: 10.1088/1757-899X/318/1/012041.

[15] K. V. Abramov, P. V. Skribtsov, and P. A. Kazantsev, "Image Segmentation Method Selection for Vehicle Detection Using Unmanned Aerial Vehicle," vol. 9, no. 5, pp. 295–303, 2015, doi: 10.5539/mas.v9n5p295.

[16] L. Mou, S. Member, X. X. Zhu, and S. Member, "and Video Using a Multitask Learning Residual Fully Convolutional Network," *IEEE Trans. Geosci. Remote Sens.*, vol. PP, pp. 1–13, 2020, doi: 10.1109/TGRS.2018.2841808.

[17] R. Sindhu, R. Nandal, P. Dhamija, H. Sehrawat, and C. Science, "A Review On K-mean Algorithm And It ' S Different Distance," vol. 9, no. 2, pp. 1423–1430, 2017, doi: 10.21817/ijet/2017/v9i2/170902227.

[18] S. P. Singh and A. Yadav, "Study of K-Means and Enhanced K-Means Clustering Algorithm," *Int. J. Adv. Res. Comput. Sci.*, vol. 4, no. 10, 2013.

[19] "Methods of Evaluating Estimators Mean Square Error (MSE) of an Estimator," pp. 1–11.

[20] K. Kalirajan and M. Sudha, "Moving Object Detection for Video Surveillance," vol. 2015, 2015.

[21] O. F. Television, S. Programme, and M. Signals, "ITU-T Rec. J.340 (06/2010) Reference algorithm for computing peak signal to noise ratio of a processed video sequence with compensation for constant spatialshifts, constant temporal shift, and constant luminance gain and offset," 2010.

[22] T. Gwalior, "Performance Evaluation and Comparison of Different Noise , apply on PNGImage Format used in Deconvolution Wiener filter (FFT) Algorithm Kalpana Chaurasia and Mrs . Nidhi Sharma *," vol. 4, pp. 8–14, 2015, doi: 10.18052/www.scipress.com/ETET.4.8.

[23] W. Preedanan, T. Kondo, P. Bunnun, and I. Kumazawa, "A Comparative Study of Image Quality Assesment," in *2018 International Workshop on Advanced Image Technology (IWAIT)*, 2018, pp. 4–7.

[24] K. G. Larkin, "Structural Similarity Index SSIMplified :," no. 1, pp. 1–4, 2015.

[25] M. R. Maharaj and B. Naidoo, "An Analysis of Objective and Human Assessments in Contrast Enhancement," vol. 13, no. 22, pp. 15843–15859, 2018.

[26] A. Khan, S. Saad, A. Ali, A. Anwer, S. H. Adil, and F. Meriaudeau, "Subsea Pipeline Corrosion Estimation by Restoring and Enhancing Degraded Underwater Images," *IEEE Access*, vol. 6, pp. 40585–40601, 2018,doi: 10.1109/ACCESS.2018.2855725.

Design of Prototype Information System for Tracking & Tracing Fish Distribution Based on Mobile Agent

Rizka Ardiansyah[1], Yazdi Pusadan[2], Elimawaty Rombe[3], Rahmat Mubaraq[4], Suryadi Hadi[5], Aristo Indrajaya[6] and Iskandar[7]

[1,2] Department of Information Technology, Tadulako University, Indonesia
[3,4,5] Department of Management Tadulako University, Indonesia
[6] Department of Electrical Engineering Tadulako University, Indonesia
[7] Department of Mechanical Engineering Tadulako University, Indonesia

E-mail: rizka@untad.ac.id

Abstract. The National Fish Logistics System or often called SLIN is an Indonesian Ministry of Maritime Affairs and Fisheries program that aims to maintain the stability of the production and marketing systems and control the disparity in national fish prices. Central Sulawesi is the Province that becomes the main corridor of this program. The inefficient distribution monitoring process generally causes several problems in the field of fisheries distribution management that still often occur today by the regional Ministry of Maritime Affairs and Fisheries. It indicates that SLIN is not yet running optimally. This study purpose a prototype design of fish distribution tracking based on a mobile agent that can use to help consumers to track distribution channels and get information about the origin of the fish to be purchased. The data will further process for monitoring fish distribution in a real-time manner by the regional Ministry of Maritime Affairs and Fisheries. A proper monitoring mechanism will undoubtedly help the government in making policies and conducting supervision to make the SLIN implementation successful in Central Sulawesi. By the research, we found that the proposed method can gather data from every level fish distribution agent then processed the data to inform about distribution line and the origin of the fish for the consumer. The proposed solution framework could be implemented and nearly fit with current implementation criteria. The framework later can be a base framework for developing a more advanced information system for SLIN in Central Sulawesi Region.

Keyword: SLIN, tracking fish distribution, mobile agent, fish logistic system.

1. Introduction

The National Fish Logistics System (SLIN) is a system formed based on Fisheries Maritime Regulation No. 5 of 2014, which aims to maintain the stability of the production and marketing system and control the disparity in national fish prices. SLIN used as a management system for fishery product supply chains, including materials, production equipment. SLIN also used as information ranging from procurement, storage to distribution as a unity of a policy to increase capacity policies and stabilize fisheries production systems from upstream to downstream, control disparity prices, as well as to meet the needs of domestic consumers [1].

Central Sulawesi is the Province that becomes the main corridor of this program. The main problems in the fisheries sector that are still common today are: (1) the issue of inter-regional fish stock inequality, (2) monopolization of fish distribution by fish collectors, (3) scarcity of fish stocks in some regions. The problem shows that SLIN is not yet working optimally. The critical aspect of this problem is monitoring, which is not sufficient and efficient, because the methods applied have not been able to improve the efficiency and effectiveness of data distribution from distribution agents to the regional Ministry of Maritime Affairs and Fisheries. It is directly impacting the development of policies that are less accurate and strategic. Previous research conducted by Mubaraq [2] & Rombe [3] revealed a conceptual model of how information systems play an essential role in improving operational performance and distribution of fish logistics. According to them, the integration of information systems in logistics distribution operations as a traceability system can develop distribution schemes so that they can support various management policies and the satisfaction of buyers as consumers. The use of monitoring and tracking system is essential to reduce costs and smooth identification of bottlenecks and operation defects [4]. Most of the existent tracking techniques use RFID (Radio Frequency Identification), GPS (Global Positioning System), NFC (Near Field Communication), RTLS (Real-Time Locating System) [5].

Rombe [3] depict several requirements to implement a fish distribution trace & trackability system in the Province of Central Sulawesi, such as the system must be comfortable and inexpensive to implement and operate. The system must be able to adjust to how the distribution entity behaves. In other words, the system must be able to facilitate integration between the entities of the distributing entity.

2. Related Work

The implementation of the traceability system involves two main principles, namely tracing and tracking [6]. Tracing is the ability of the system to trace the origin of fish, while tracking is the ability to locate the fish after they are caught [7]. Traceability acts as a link to share relevant information between actors in the supply chain [8]. Currently has been several studies conducted related to food tracing & tracking information systems, including research conducted by Shamsuzzoha [9], Egharevba [10], M'hand [11], and Purwandoko [6]. There are several technologies commonly used in tracing and tracking systems such as GPS, GTIN, RFID, Barcodes, Q.R. Code, Cloud Server, and WSN [9-12][6]. GTIN, Barcode, Q.R. Code, and RFID are generally used as item identification media to track objects. The technology is capable of storing large amounts of data in small data formats. They offer labour savings, speed, and cost savings, among other benefits. Barcodes and Q.R. Codes are the most affordable and commonly used devices on a variety of electronic devices, specifically for the latest generation of smartphones [11]. Barcodes have limitations in terms of code readings that are longer and more complicated when the physical label is imperfect or damaged. Moreover, the Q.R. Code can store more data than a barcode [13][14]. Some system designs are suggested in several studies using GPS-based technology integrated into an autonomous IoT system based on Wireless Sensor Network [5][13][12]. The basis of such a system design is not following the implementation criteria mentioned by Rombe [3]. Using a self-contained system that works on a WSN network is too irrelevant to be realized as a Fish tracking & tracing system in the Province of Central Sulawesi because of being expensive, and the technology also requires a framework and policy readiness from the parties. Besides, the mechanism of fish distribution in the Central Sulawesi Province does not yet have a distribution warehouse, so the distribution channels are not fixed and randomly. The proposed design depicts the potential use of several technologies such as Q.R. Code, GPS, cloud storage, and cellular networks that used as a technology base for designing fish tracking & tracing systems for the Province of Central Sulawesi. The idea is the distribution agent works as a sensor node by using a smartphone where interaction between nodes can do using a Q.R. Code so that the node can identify each other along with the object transacted, then use GPS for tracking distribution data. Data storage and synchronization between the distribution node to data-centre done by cloud and cellular-network technologies. Our hypotheses With this simplified design of the use of technology, the system design will meet the implementation requirement of the fish tracing & tracking system in Central Sulawesi Province.

3. Proposed Method

The proposed solution is to build a fish tracing & tracking system based on the Android mobile application. The proposed solution is utilizing QR-Code technology to identify transaction objects and transaction actors, GPS to identify transaction and fishing locations, cloud database for storing and processing information and cellular network technology for data synchronization needs. The proposed solution only requires a smartphone connected to the internet, built-in camera, and GPS as an operational device for all distribution nodes. All transaction processes from the fisherman, distribution agent, and the retailer will be recorded by the developed application for further data processing to get the tracing and tracking information of a fish commodity.

3.1. Research Method

The research is base on a case study conducted by Rombe [3]. Furthermore, after the problem was successfully mapping according to the apparent characteristics, a literature study was conducted on several technical papers to get an idea of what solutions might be applied to overcome the problem. The technical paper will be reviewed in the form of a simple review to compile a conceptual solution and technology used. The solution abstraction is further described in the form of UML diagrams to illustrate system interactions and tracing algorithms. Furthermore, the system abstraction realizes in the way of an android-based mobile application. Then the operation of the system will be tested to find out how accurately the system able to track fish distribution lines from the sea to the consumers. The research process depicted, as shown in Figure 1. The testing mechanism carried out by using a Black Box method—this method used to evaluate the software to identify a malfunction and a bug [15]. Black box testing considers the value of inputs on a study [16] and ignores the internal mechanisms of a system [17]. Our case definition is the ability to record catching and transaction data from every agent and the ability to track the distribution line and the origin of the fish.

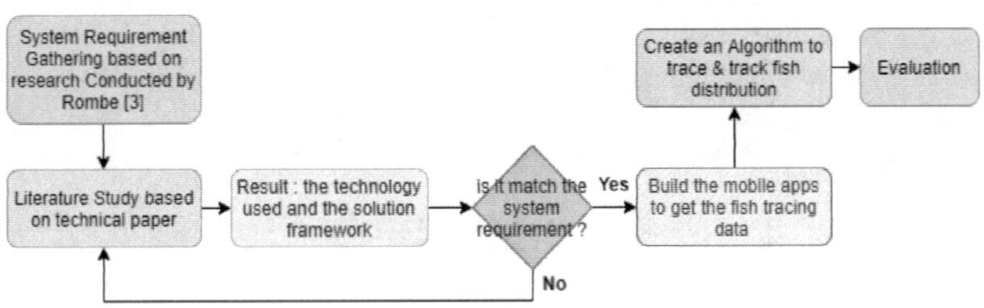

Fig. 1. Research Method.

3.2. Distribution Model

Distribution models were obtained based on research conducted by Rombe [3]. The distribution model depicted, as shown in Figure 2. The distribution model involves five actors (distribution agents), namely fishermen, collectors, wholesalers, retailers, and exporters. The model proposed that the fish distribution channel in Central Sulawesi has three distribution models. Through Wholesalers and then to retailers or from collectors directly to retailers. The proposed system built by adopting the three distribution models.

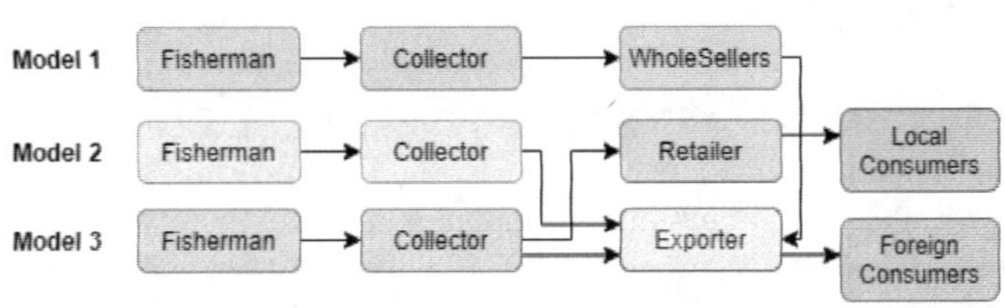

Fig. 2. Fish Distribution Model in Central Sulawesi.

3.3. System Abstraction

The proposed system has several actors who interact with this system, such as fishermen, exporters, and distributors. Distributors generally consist of collectors, wholesalers, and retailers.

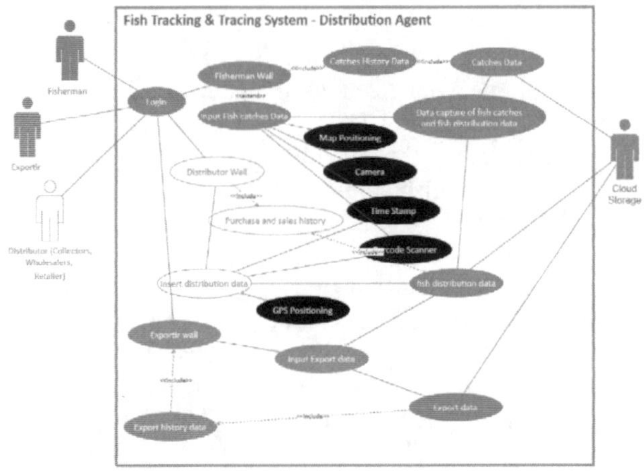

Fig. 3. UML Use-Case Distributor Agent System.

The proposed system can accommodate multi-level distributors such as in specific scenarios to ensure the availability of wholesaler stock must transact with fellow wholesalers or retailers with fellow retailers. Given that there are no rules relating to explicit distribution schemes from the local government so that the distribution model can be very dynamic. The proposed system consists of two different apps; it is an app for distributors agents and apps for consumers. The apps for consumers is built-in with fish tracing and tracking features. System interaction for distributor agents depicts, as shown in Figure 3 and system interactions for consumers, are represented, as shown in Figure 4.

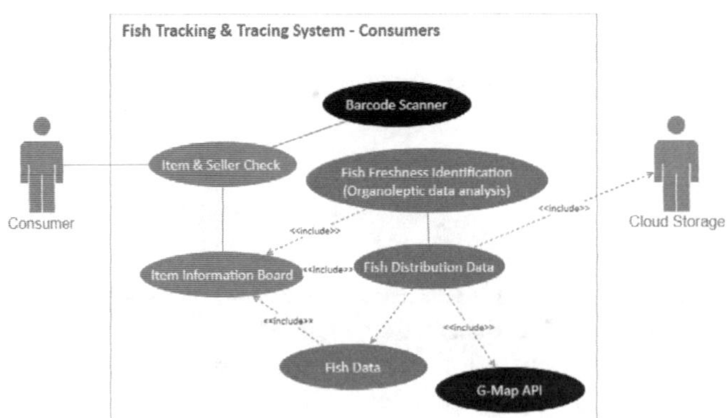

Fig. 4. UML Use-Case Consumers System.

Figure 3 shows the main features contained in the system for distributor agents. Each actor in the distributor agent has a wall to display the transaction history. The proposed system also allows fishers to input data even when they are not connected to cellular networks while in the ocean. The proposed system utilizes an integrated camera on a smartphone to read the Q.R. Code. Each entity, such as a fish commodity and distributor agent, has its unique Q.R. Code that is used to identify each other during the transaction process. Several different levels of users use the distributor agent system. Therefore access to the user pages must be through a user authentication mechanism first. The consumer system does not need an authentication mechanism, so the consumer only needs to activate the application. The mobile camera will be active so that it can use to scan the Q.R. Code for fish that is found on the market. Furthermore, the system will run the tracing & tracking algorithm based on the fish

IOP Conf. Series: Materials Science and Engineering 1212 (2022) 012045 doi:10.1088/1757-899X/1212/1/012045

transaction history and display the distribution channel mapping starting from the catch point to the point of sale on an electronic map on the mobile device, as shown in Figure 4.

3.4. Distribution Tracking Algorithm

Each fisher generally packs fish in a certain quantity, and then the fish package is then given a label in the form of a Q.R. Code. Furthermore, the data of the fish in the box, along with the Q.R. data, is recorded to the system. The consumer's system triggers the tracking & tracing process. The assumption is that when a fish commodity is on the market, it means that the product has gone through a distribution process. To enable the tracking of distribution channels, every time a product moves from one distributor agent to another, the data recorded by the system is the coordinate transaction data, transaction timestamp, transaction data agents (sellers & buyers), and commodities that are transacted. Based on this data set, the tracking algorithm will search the data tables in a database to get a distribution line from upstream to downstream. The system uses the timestamp to sort the transaction history, then draw the transaction point and direction on the map. Tracking the origin of fish catches is done by allowing fishers to record the coordinates and time of the catch when still in the ocean—the data stored in a local database on the mobile device. When fishers arrive on the port, the fishers then pack fish into the fish box and then input the fish data, Q.R. Code, along with the time and coordinates of the fish caught in the system. Then the system will synchronize data to the cloud. The catch time will be a benchmark for determining how long the fish have been caught. Tracing & tracking algorithm flow described in Figure 5.

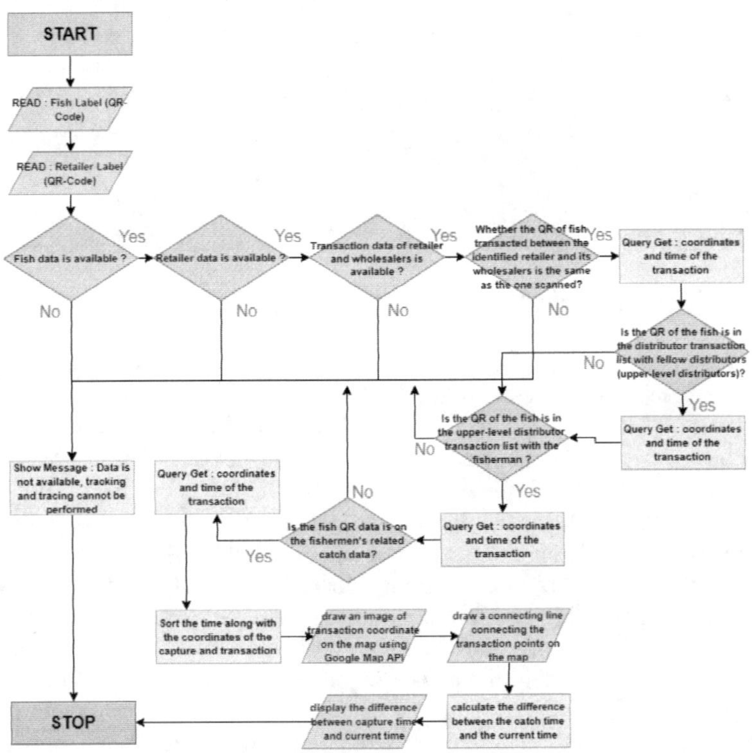

Fig. 5. Fish Tracking & Tracing Algorithm.

4. Result and Discussion

This section presents the result and the implication of the proposed method. Before the distribution lines traced, the proposed system requires distribution chain data as a reference. In our test scenario, fishers catch fish on the coastline of Donggala Regency. Every time a fisherman makes a catch, a fisher can record the coordinates of the location and time of the catch by utilizing the Global Positioning System (GPS) device integrated on a smartphone, as shown in figure 6. The recorded data is stored temporarily on the smartphone as a reference for the location and time of fish catch. After the fishermen arrive at the port, the fish they catch then grouped into fish boxes, and then each fish box is labelled with a QR-Code. The fish box then recorded in the system by adding a reference in the form of fisherman data along with coordinates and catch time. Each box of fish has the possibility of being purchased by several collectors in smaller portions. However, in our test scenario, the whole fish box is acquired by a collector. When the transaction process carried out, the buyer must scan the fisherman's QR-Code first, then input the transaction data, including the fish QR-Code box, as shown in figure 7.

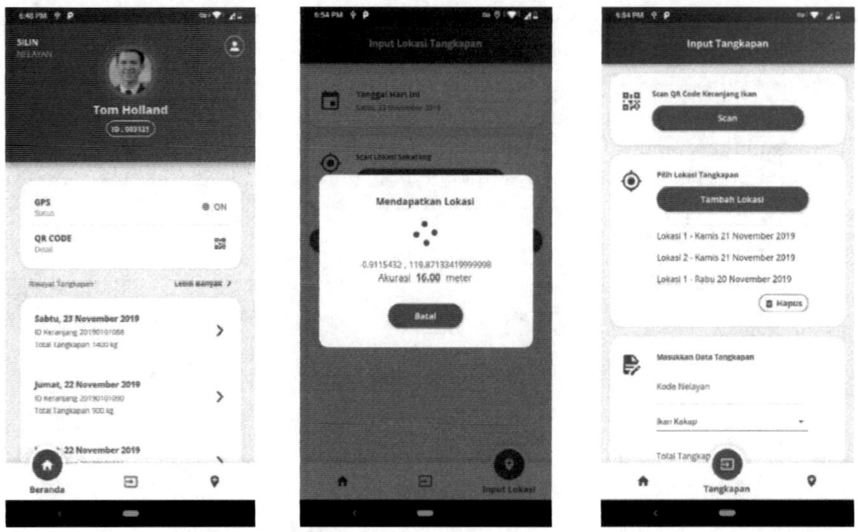

Fig. 6. Fishermen Agent Application.

Then the collectors carry out further distribution to wholesalers until they reach the final distribution point at the retailer. Each hop of the distribution transaction, the buyer, must record data related to the identification of the fish purchased and the identification of the seller. The system will record the coordinates of the transaction using the buyer's handheld device. This process will automatically update the sales data on the seller's side. Consumers who want to check the origin of fish can scan the QR-Code of the fish and the seller (retailer) on the consumer application, as shown in figure 8. This process will trigger the distribution line tracking algorithm along with the location and time of the fish catch, then displayed on an electronic map in the consumer application. Based on our test results, the proposed method can do catching and transaction data from every agent and the ability to track and trace the distribution line and the origin of the fish. Furthermore, we found two things that could cause the system to fail to operate, namely the device's ability to capture signals from the GPS satellite. It can cause the system to fail to record data to cloud storage if the device cannot lock the coordinate between 5 to 15 meters. Besides that, the timer setup on the handheld device must also be appropriate because otherwise, this could cause the tracking algorithm to fail to find the fish distribution path from upstream to downstream. We also need to carry out further testing of the proposed tracking algorithms in more complex distribution scenarios. For example, when the fish in a fish box identified by one Q.R. Code has branched distribution channels, goods are transacted between collectors and wholesalers. Up to several times before reaching the consumer, this must be done to ensure the algorithm's behaviour remains consistent in more complex distribution scenarios.

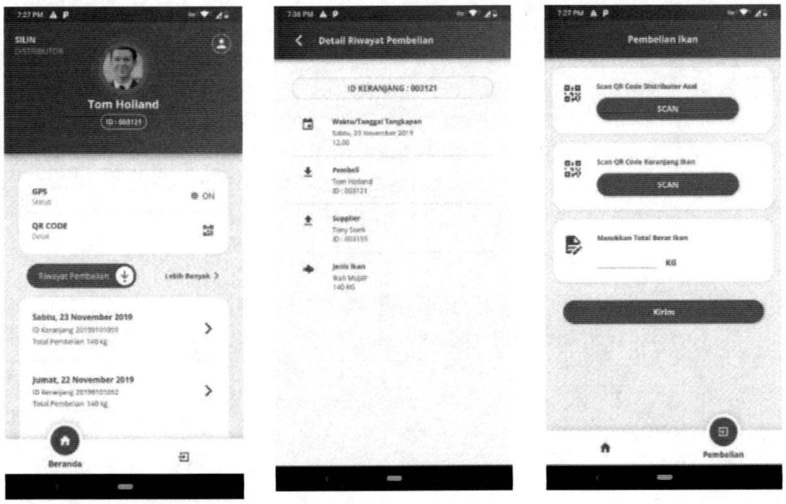

Fig. 7. Distributor Agent Application.

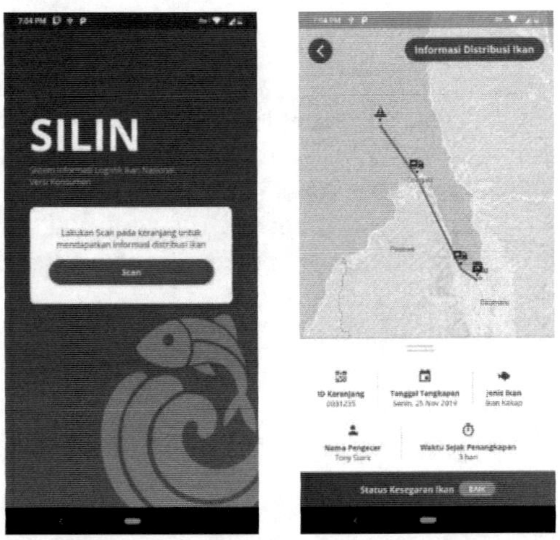

Fig. 8. Consumer Application.

For the regional fisheries ministry purpose at the later work, the fish catch data will be grouped based on the name of the port, the distribution of catch points anchored at the relevant port, the accumulation of fish commodities that distributed at each distribution endpoint. Central Sulawesi Province has two main ports, namely Donggala and Toli-toli Regencies port. Each district represented with a different dot colour. Furthermore, the points depicted in an electronic map on a web-based application, complete with statistical data on fish commodity distribution. The system for these purposes developed further.

5. Conclusion

The proposed system has been able to identify and track the distribution and origin of fish catches. Even though it has previously mentioned that this proposal does not require a significant overhaul of the existing distribution business model, it still needs adjustment. Therefore, the role of the government, in this case, is still necessary to manage the actors and distribution scenarios involving the proposed system. The proposed system has several technical issues related to the use of GPS, which in certain conditions and on specific mobile devices, have difficulty in locking coordinates. This issue will undoubtedly be quite difficult for the transaction process using the system because transaction data will not be saved to the database if the transaction coordinates cannot be locked. Another thing that is also still a problem is if the time setting on the cellphone does not match the actual time, it can undoubtedly disrupt the tracking mechanism of the fish distribution path. The proposed solution framework could be implemented and nearly fit with current implementation criteria. The framework later can be a base framework for developing a more advanced information system for SLIN in Central Sulawesi Province.

References

[1] R. Nattassha, Y. Handayati, G. Yudoko, T. M. Simatupang, A. Adhiutama, and N. B. Mulyono, "Performance measurement system for the cold fish supply chain: The case of National Fish Logistics System in Indonesia," Int. J. Agric. Resour. Gov. Ecol., vol. 15, no. 1, pp. 57–76, 2019.

[2] R. Mubaraq, E. Rombe, S. Hadi and R. Ardiansyah, "Strategic Information System, Supply Chain Performance and Operational Performance in the Fishing Industry : Conceptual Model," Korea, 2019. J. Clerk Maxwell, A Treatise on Electricity and Magnetism, 3rd ed., vol. 2. Oxford: Clarendon, 1892, pp.68–73.

[3] E. Rombe, R. Mubaraq, S. Hadi, R. Adriansyah, and G. Vesakha, "Barriers and Drivers for Applying Fish Traceability System in Emerging Market," Int. J. Eng. Technol., vol. 7, no. 4.15, p. 262, 2018.

[4] Shamsuzzoha, A., Helo, P.T., 2011. Real-time tracking and tracing system: Potentials for the logistics network, in: Proceedings of the 2011

[5] international conference on industrial engineering and operations management, pp. 22–24.

[6] M. A. M'hand, A. Boulmakoul, H. Badir, and A. Lbath, "A scalable real-time tracking and monitoring architecture for logistics and transport in RoRo terminals," Procedia Comput. Sci., vol. 151, no. 2018, pp. 218–225, 2019.K. Elissa, "Title of paper if known," unpublished.

[7] P. B. Purwandoko, K. B. Seminar, Sutrisno, and Sugiyanta, "Development of a smart traceability system for the rice agroindustry supply chain in Indonesia," Inf., vol. 10, no. 10, 2019.

[8] Yanuarti, A.R.; Afsari, M.D. Rice Commodity Profile; Ministry of Trade: Jakarta, Indonesia, 2016; pp. 1–37.

[9] Kumar, V.; Hallqvist, C.; Ekwall, D. Developing a framework for traceability implementation in the textile supply chain. Systems 2017, 5, 33.

[10] A. Shamsuzzoha and P. Helo, "Real-time tracking and tracing system: Potentials for the logistics network," Proc. 2011 Int. Conf. Ind. Eng. Oper. Manag., pp. 242–250, 2011.

[11] H. O. Egharevba, O. Fatokun, M. Aboh, O. O. Kunle, S. Nwaka, and K. S. Gamaniel, "Piloting a smartphone-based application for tracking and supply chain management of medicines in Africa," PLoS One, vol. 14, no. 7, pp. 1–14, 2019.

[12] M. A. M'hand, A. Boulmakoul, H. Badir, and A. Lbath, "A scalable real-time tracking and monitoring architecture for logistics and transport in RoRo terminals," Procedia Comput. Sci., vol. 151, no. 2018, pp. 218–225, 2019.

[13] P. Visconti, R. de Fazio, R. Velázquez, C. Del-Valle-soto, and N. I. Giannoccaro, "Development of sensors-based agri-food traceability system remotely managed by a software platform for optimized farm management," Sensors (Switzerland), vol. 20, no. 13, 2020.

[14] B. Yuan et al., "Fast QR code detection based on BING and AdaBoost-SVM," IEEE Int. Conf. High Perform. Switch. Routing, HPSR, vol. 2019-May, pp. 1–6, 2019.

[15] P. M. Jacob and M. Prasanna, "A Comparative analysis on Black box testing strategies," Proc. - 2016 Int. Conf. Inf. Sci. ICIS 2016, pp. 1–6, 2017.

[16] A. P. Estrada-Vargas, E. López-Mellado, and J. J. Lesage, "A Black-Box Identification Method for Automated Discrete-Event Systems," IEEE Trans. Autom. Sci. Eng., vol. 14, no. 3, pp. 1321–1336, 2017.

[17] M. A. Khan and M. Sadiq, "Analysis of black box software testing techniques: A case study," Proc. 2011 Int. Conf. Work. Curr. Trends Inf. Technol. CTIT'11, no. i, pp. 1–5, 2011.

Artificial intelligence in mobile communication: A Survey

Yingchi Mao [a], Andri Pranolo [a,b,*], Leonel Hernandez [c*], Aji Prasetya Wibawa[d], Zalik Nuryana [e]
[a] College of Computer and Information, Hohai University, Nanjing, China
[b] Informatics Department, Universitas Ahmad Dahlan, Indonesia
[c] Engineering Faculty, ITSA Institución Universitaria, Carrera 45# 48-31, Barranquilla, Colombia
[d] Universitas Negeri Malang, Indonesia
[e] Universitas Ahmad Dahlan, Indonesia
Email: maoyingchi@gmail.com, andri.pranolo@tif.uad.ac.id*, lhernandezc@itsa.edu.co

Abstract. In this paper, we elaborate on artificial intelligence (AI) techniques used to improve the performance of mobile communication. This article describes brief AI approaches in mobile communication, several classics AI techniques, and the current AI approaches in wireless communication. The techniques contain fuzzy logic, neural networks, reinforcement learning, and AI techniques implemented on mobile communication. Some keys or terms challenges between AI and future mobile communication, not only 5G generation issues but also how the sixth generation (6G) of mobile networks will be driven to give stable networks and service types on huge mobile devices and data.

Keywords—Mobile communication, 5G generation, 6G Generation, MIMO, Edge Computing

1. Introduction

Artificial Intelligence (AI) is one of the techniques to adopt the human brain or thinking [1], others act of animals [2], [3], biological systems, and species [4]. Especially, AI in the mobile communication system has an essential role, promising way to optimize its performance. In the general issue, AI techniques significantly contribute to the dynamic adaptation of mobile communication in an environment. Currently, the complicated network infrastructure needs to migrate from the traditional operation and management methods to the intelligence approach to reduce inefficient and expansiveness [5].

The next generation of wireless networks is more complex and needs more resources because of the need to improve a service requirement with various devices, complex networks, and various applications [6]. Furthermore, the network programmers must adapt their system to give the best and available resources to increase the service quality. Furthermore, In 2021, the industry forecasts that the IP traffic consumption will reach 3.3 zettabytes (1015 MB), and the smartphone traffic would outpacing PC traffic in the same year [7]. AI presents to build an adaptive system, providing a system and the environment better in performance. This big data era causes the increasingly massive datasets available from a mobile or wireless systems. In other words, applied AI on mobile communication will ensure communication systems are more efficient, better in performance, and increase key performance indicators (KPIs) [6], [8].

The increasing network infrastructure and hardware uptakes of mobile devices and their application will increase the need for mobile stations or exploding mobile traffic volumes and huge data to be processed. Zhang et al. [7] said that one of the important solutions is implementing advanced artificial intelligence techniques, such as machine learning and deep learning in mobile communication, to manage base stations' huge data and performance to provide full mobile stations.

Some researchers have elaborated AI approaches in wireless or mobile communications. Kibria et al. [6] discuss the relationship between data analytics and AI for wireless networks to be intelligent. Furthermore,

Zhang et al. [7] draw a review deep learning approach in mobile and wireless networking research, and Cayamcela and Lim [9] explore the AI approaches for 5G mobile and wireless communications technology. However, these surveys failed to elaborate on various AI approaches in mobile communication occasionally.

This paper aims to introduce Artificial Intelligence in a basic term to advance viewed and applied AI on the infrastructure, mobile management, security management, and software spectrum. We organize this paper into several sections. In section II, we discuss the challenges of artificial intelligence (AI) in mobile communication. In section III, we describe AI approaches and mobile communication applications, while the problem in mobile communication will be discussed in section IV. In the last section, we overview the AI applications in mobile communication.

2. Challenges of Artificial Intelligence in Mobile Communication

Artificial intelligence will have an essential role in managing big data as advanced data analytics and organizing various communication devices in the future of mobile or wireless networks. In short, Xhang et al. [6] stated that the mobile intelligence system should be self-adaptive, self-aware, proactive, and prescriptive. On the other hand, The infrastructure of mobile communication needs to be adaptive to diversified services and efficient and reliable, such as improving the performance of mobile broadband, minimize the peak-to-average power ratio (PAPR) [10], improving orthogonal frequency division multiple access (OFDM) [5], improved link quality [11].

In the 5G mobile communication system, the challenges will be appeared because of the huge nodes and mobile devices with fast data interchange and communication [5]. It was explained by Kibria et al. [6], who have overviewed data analytics, Machine learning, and AI in network communication. It supposed that the network management was and would be more complex within the new mobile communication generation network. Banupriya et al. [12] forecast that the traffic will rise to 1000 times in 5G mobile communication according to mobile end-user data speed and growth. Hence, traditional network management and techniques are not supported in the next generation of mobile communication [12]. Next, the traffic control issues are still open to topic research because the previous researchers still focused on the core network issue on the network layer side, such as applied AI on the routing problem [13]. However, few AI techniques are used for traffic control according to user experienced recommendations on the application and the semantic layers. Another challenge is how to apply AI to the next-generation mobile communication system. For example, in 5G, there are three types of challenges as hoe to provide broadband multimedia based on user experiences (enhanced mobile broadband(eMBB)), tactile internet application (ultra-reliable low latency service (URLLC)), and how to connect the amount of mobile device with real-time interconnection (Massive machine-type communications (mMTC)) [12].

The other AI challenges in the mobile communication system are decision making, network management, and resource optimization [8] with faced to mobile intelligent communication with has characterized proactive system, self-aware, self-adaptive, predictive, efficient, and cost-effective operation and optimization [6], [8]. Another essential in mobile communication is how AI can be applied to various wireless communication scenarios such as power control, radio resource management, mobile management, and interference management [12]. The evolution of mobile communication networks, from the 1st generation to the 6th generation, has been reported [10].

3. AI Approach and Application in Mobile Communication

There are some classic artificial intelligence approaches, such as fuzzy logic and neural network. Then, the neural network would be extended to be better performance techniques such as machine learning and deep learning approaches. The basic approach is Fuzzy logic, which is processed any values and resulting in true and false. Another term in AI is reinforcement learning, a technique to design a computer or machine for learning by itself instead of being precisely programmed [12]. One of the techniques is Neural networks (NN). This technique can be made by a machine or computer able to self-learning to solve a problem. NN process adopts the brain human system and behavior. In the current issues, also deep learning (DL) has been popular as an improved machine learning (ML). ML and DL are interesting approaches for advanced network traffic and management of future mobile communication. The two types of AI learning have been used in mobile communication: supervised learning and unsupervised learning.

3.1. Decision making in Mobile Communication

An AI decision making in mobile communication has been reported by several researchers such as Banupriya et al. [12], estimate QoE (Quality of Experience) for different types of services by using neural network approach, and also NN for classifying the KPIs, which is bridge a KPI's with QoE in mobile internet services.

AI assistant content retrieval on QoE. Fu et al. propose QoE of a personalized content retrieval service that is divided into two sections: 1) accurate to catch the user interest and experiences, and 2) The process to access convenience and suitable recommendation [13]. Data analytics, ML, and AI techniques can be used in analog, digital, and hybrid beamforming to generate the optimal beam patterns, dynamically selecting the most suitable beam, and performing the beam-steering operation [6].

To strengthen the evidence on research trends regarding "decision making" in mobile communication, data from scopus.com was used with a total of 188 articles. The keywords of the articles were analyzed based on using the VOSViewer application. The type of analysis used is co-occurrence with the author-to-words analysis unit. The minimum number of occurrences of a keyword is 2 of the 625 keywords, 82 meet the threshold. From Figure 1, it can be seen the analysis of the relationship between variable keywords, the circle shows the number of keywords used, the color shows the renewal of the keywords, and the line shows the relationship of the keywords.

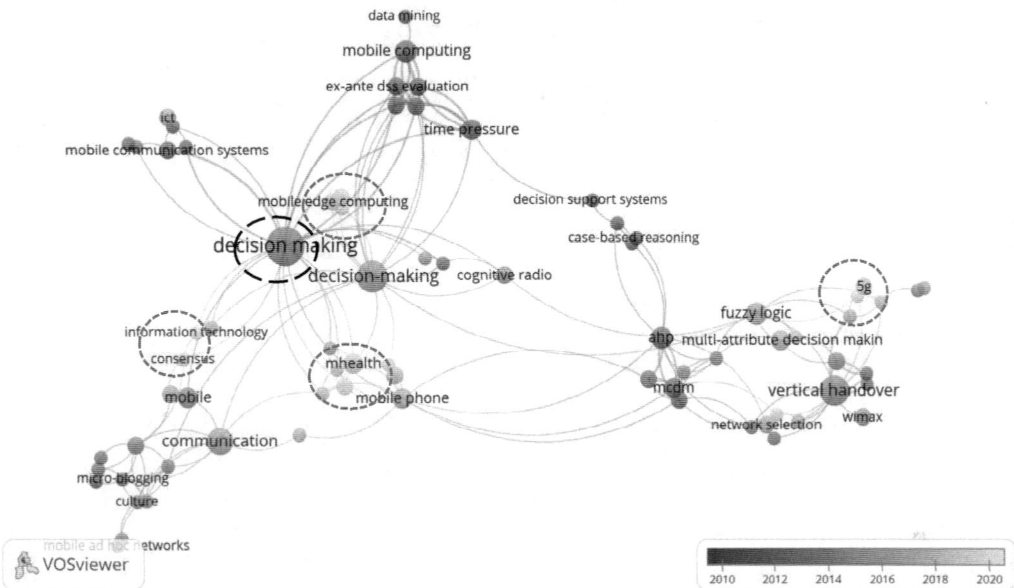

Figure 1. Overlay visualization keywords on Decision making in Mobile Communication

The "decision making" keyword becomes a motor theme that has a relationship with other keywords. Future research needs to pay attention to the aspect of the relationship between the keyword "Decision making" in Mobile Communication and the keywords circled in red with the keyword dot color in yellow. The brighter the dot color on keywords indicates the renewal of the keyword trend based on the article year..

3.2. Resource optimization in Mobile Communication

In resource optimization, Genetic algorithms have been used to optimize the construction of multicast trees mobile ad hoc networks. Additional objectives such as bounded end-to-end delay and energy efficiency are added to this optimization [9], [14]. Another research is neural networks and set-theoretic method used PAPR reduction problem which is used for online learning. A key ingredient in enhancing the efficiency of OFDM channel estimation can also be solved by Artificial intelligence techniques [14].

Considers a cognitive radio system with cooperative spectrum sensing where multiple secondary users cooperate in obtaining robust spectrum sensing results. To enable efficient and reliable opportunistic spectrum access. AI has also been proposed to tackle the inter-cell interference problem, which may have a detrimental impact on the performance of wireless users in mobile networks [14]. Existing AI algorithms integrate graphics processing units (GPU) and central processing units (CPU) to improve performance in edge computing [13]. Applied AI for mobile communication is about complex statistical methods and should take a look at devices, infrastructure, end-user, technology, and other resources [13].

To see the research trends on "resource optimization" in mobile communication, Biblioshiny application (R-Package) was used to analyze 130 articles from scopus.com. Figure 2 shows that the most used keywords in this trend are resource allocation (n=18), mobile edge computing (n=14), optimization (n=9), network optimization (n=6), quality of service (n= 6), wireless networks (n=5), Mobile communication (n=4), and QoS (n=4).

Figure 2. WordCloud analysis Resource optimization in Mobile Communication

3.3. Network management in Mobile Communication

An example of an application on network management in wireless communication is routing, a topic in communications. Some studies have already implemented AI for this topic. For instance, is a neural network is applied to realize self-configuration and self-optimization for both the radio resource and routing [12]. In another research, Machine learning methods have been used for tackling different types of routing problems in the past. It contained shortest path routing, adaptive routing, and multicasting routing [14]. Other networks management in Wireless communication is to monitor various network activities and detect anomalies, i.e., events that deviate from the current network behavior. AI also has been used for predicting traffic in a communication network. AI technologies can minimize traditional interventions in network traffic management and enable reliability, more adaptive systems, and better network performance.

To strengthen the research trends evidence on network management in mobile communication, the keywords from the 69 articles collected from scopus.com had been analyzed using the Biblioshiny application (R-Package). The unit of analysis is a document with the type word dynamics (Figure 3).

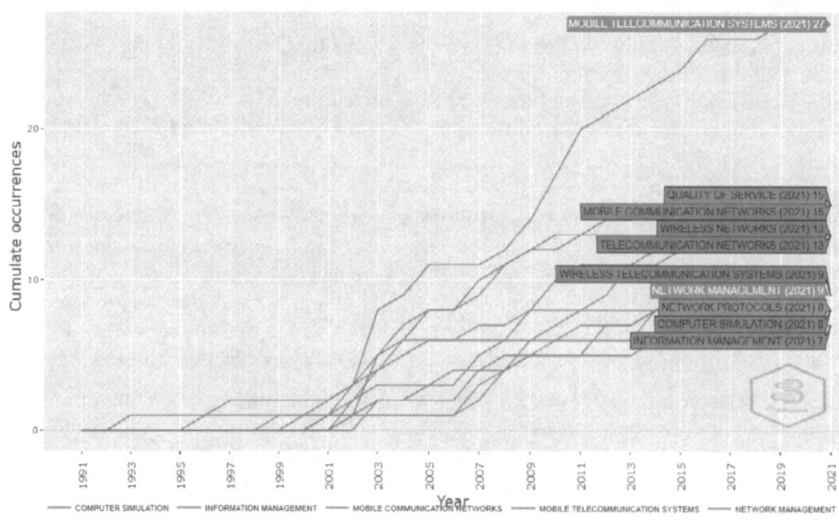

Figure 3. Word dynamics analysis on network management in mobile communication

From figure 3, it can be explained that since 1991 the keyword "mobile telecommunication systems" has been a word that often appears in the articles. After that, then "quality of service" is appeared in 2000 and has grown to this day. To sum up, "mobile telecommunication systems" theme is being the most popular keyword accompanied by other trends such as "mobile communication networks", "quality of service", "telecommunication networks", "wireless networks", "network management", "wireless telecommunication systems", "computer simulation", "network protocols", and "information management".

3.4. Other AI Application

As a sum, applied AI on Mobile Communication can be seen in Table I.

Table 1. Applied AI on Mobile communication

Mobile Communications Technology	Applied AI
Autonomous vehicles and medical assistance devices	Automation with inherent artificial intelligence (AI) [15] Current and major algorithms in the specific field of artificial intelligence for autonomous vehicles. Such systems are particularly suited for high-level decision making since they must, by definition, be able to perceive and react to their environment in order to reach given objectives [16]
Internet of Intelligent Things	Artificial intelligence techniques employed to create such intelligence, and network solutions to exploit the benefits brought by this capability [17]
Mobile Cloud Computing MCC	Resource intensive applications such as augmented reality, artificial intelligence, artificial vision, object tracking, image processing, and natural language processing are becoming popular to manage MCC[18]
5G Networks	AI and its subcategories like machine learning and deep learning have been evolving as a discipline, to the point that nowadays, this mechanism allows fifth-generation (5G) wireless networks to be predictive and proactive, which is essential in making the 5G vision conceivable [9]
Wireless Sensors Networks	In order to assist the smart radio channels, artificial intelligence tools are used. The tool called machine learning is considered to be the important tool in solving the above issue [19] Machine learning techniques for localization in WSNs using Received Signal Strength Indicator [20] The new approach in a wireless sensor network for selecting the cluster-head by making use of the artificial neural network in order to increase the network's lifetime [21]
Mobile heterogeneous networks (HetNets)	Machine learning, bio-inspired algorithms, fuzzy neural network, and so on, because AI techniques can naturally handle the problems of large-scale complex systems. [22] These promising architectures adopt artificial intelligence (AI) principles that incorporate learning, reasoning, and decision-making mechanism as natural choices for designing a tightly integrated network. [23]
Revealing mobile malware hidden communications	To spot malware covertly exchanging data using two detection methods based on artificial intelligence tools, such as neural networks and decision trees [24]
Mobile and Wireless Networking	Deep learning [7]
Vehicular mobile networks	Artificial intelligence (AI) based edge caching [25]

Next-generation wireless networks	Machine Learning, Artificial Neural Networks (ANNs) [26] Big data analytics for user-activity analysis and user-anomaly detection [27]
Mobile Multimedia	Deep Learning (DL) has become a crucial technology for multimedia computing [28]
Wi-Fi Based Indoor Location System	Artificial Neural Network ANN [29]
Cognitive Radio Networks	Development of cognitive routing protocols envisioned as routing protocols that fully and seamlessly incorporate artificial intelligence (AI)-based techniques into their design [30]
Data science and artificial intelligence (AI) for communications	The innovation in AI, machine learning (ML), and network data analytics provides a huge opportunity to revolutionize the world's communications systems and user experience [31]
Cellular Network System	AI concept and reviews its applications in cellular network design, operations, and optimization [32]

4. Conclusion

Artificial intelligence (AI) has an essential role in increasing the mobile communication system's performance: proactive system, self-aware, self-adaptive, predictive, efficient, and cost-effective operation and optimization. This article describes several classic AI techniques and the current AI approaches in wireless communication. The techniques contained fuzzy logic, neural networks, reinforcement learning, and some AI techniques implemented on mobile communication. Some keys or terms challenges between AI and future mobile communication are how to manage such as big data, data analytics, higher frequency transmission, device-to-device communication, the reliable architecture, ultra-dense network, Massive MIMO, 3D Beamforming, V2X, Drone Base Station, Multi-Access Connectivity, mm-wave, Cloud-RAN, Edge Network, and Micro Base Station. The challenges are 5G generation issues and how the sixth generation (6G) of mobile networks will be driven to give stable networks and service types on huge mobile devices and data.

References

[1] D. Hassabis, D. Kumaran, C. Summerfield, and M. Botvinick, "Neuroscience-Inspired Artificial Intelligence," 2017, doi: 10.1016/j.neuron.2017.06.011.

[2] A. M. Zador, "A critique of pure learning and what artificial neural networks can learn from animal brains," *Nat. Commun.*, 2019, doi: 10.1038/s41467-019-11786-6.

[3] M. Enquist and S. Ghirlanda, *Neural networks and animal behavior*, 2013, doi: 10.1515/9781400850785.

[4] B. Schölkopf, "Learning to see and act," *Nature*, 2015, doi: 10.1038/518486a.

[5] F. Long, N. Li, and Y. Wang, "Autonomic mobile networks: The use of artificial intelligence in wireless communications," in *2017 2nd International Conference on Advanced Robotics and Mechatronics, ICARM 2017*, 2018, doi: 10.1109/ICARM.2017.8273227.

[6] M. G. Kibria, K. Nguyen, G. P. Villardi, O. Zhao, K. Ishizu, and F. Kojima, "Big Data Analytics, Machine Learning, and Artificial Intelligence in Next-Generation Wireless Networks," *IEEE Access*, 2018, doi: 10.1109/ACCESS.2018.2837692.

[7] C. Zhang, P. Patras, and H. Haddadi, "Deep Learning in Mobile and Wireless Networking: A Survey," *IEEE Commun. Surv. Tutorials*, 2019, doi: 10.1109/comst.2019.2904897.

[8] X. Ge, J. Thompson, Y. Li, X. S. Liu, W. Zhang, and T. Chen, "Applications of Artificial Intelligence in Wireless Communications," 2019, doi: 10.1109/MCOM.2019.8663984.

[9] M. E. Morocho Cayamcela and W. Lim, "Artificial Intelligence in 5G Technology: A Survey," in *9th International Conference on Information and Communication Technology Convergence: ICT Convergence Powered by Smart Intelligence, ICTC 2018*, 2018, doi: 10.1109/ICTC.2018.8539642.

[10] S. J. Nawaz, S. K. Sharma, S. Wyne, M. N. Patwary, and M. Asaduzzaman, "Quantum Machine Learning for 6G Communication Networks: State-of-the-Art and Vision for the Future," *IEEE Access*, vol. 7, pp. 46317–46350, 2019, doi: 10.1109/ACCESS.2019.2909490, available at :https://ieeexplore.ieee.org/document/8681450/.

[11] T. S. Rappaport *et al.*, "Millimeter Wave Mobile Communications for 5G Cellular: It Will Work!," *IEEE Access*, vol. 1, pp. 335–349, 2013, doi: 10.1109/ACCESS.2013.2260813, available at :http://ieeexplore.ieee.org/document/6515173/.

[12] D. Banupriya, Pandi, S. Preetha, and R. R. Prathisha, "A Study on Use of Artificial Intelligence in Wireless Communications," *Asian J. Appl. Sci. Technol.*, vol. 2, no. 1, pp. 354–360, 2018.

[13] Y. Fu, S. Wang, C. X. Wang, X. Hong, and S. McLaughlin, "Artificial Intelligence to Manage Network Traffic of 5G Wireless Networks," *IEEE Netw.*, 2018, doi: 10.1109/MNET.2018.1800115.

[14] W. Samek, S. Stanczak, and T. Wiegand, "The Convergence of Machine Learning and Communications," *ITU J. ICT Discov.*, no. 1 (Special issue), pp. 1–10, 2017.

[15] M. Hengstler, E. Enkel, and S. Duelli, "Applied artificial intelligence and trust-The case of autonomous vehicles and medical assistance devices," *Technol. Forecast. Soc. Change*, 2016, doi: 10.1016/j.techfore.2015.12.014.

[16] L. Claussmann, M. Revilloud, S. Glaser, and D. Gruyer, "A study on ai-based approaches for high-level decision making in highway autonomous driving," in *2017 IEEE International Conference on Systems, Man, and Cybernetics, SMC 2017*, 2017, doi: 10.1109/SMC.2017.8123203.

[17] A. Arsénio, H. Serra, R. Francisco, F. Nabais, J. Andrade, and E. Serrano, "Internet of intelligent things: Bringing artificial intelligence into things and communication networks," *Stud. Comput. Intell.*, 2014, doi: 10.1007/978-3-642-35016-0_1.

[18] S. Deshmukh and R. Shah, "Computation offloading frameworks in mobile cloud computing : A survey," in *2016 IEEE International Conference on Current Trends in Advanced Computing, ICCTAC 2016*, 2016, doi: 10.1109/ICCTAC.2016.7567332.

[19] B. Baby and A. Professor, "A Brief Survey on Applying the Machine Learning Theories in Wireless Networks," 2018.

[20] H. Ahmadi and R. Bouallegue, "Exploiting machine learning strategies and RSSI for localization in wireless sensor networks: A survey," in *2017 13th International Wireless Communications and Mobile Computing Conference, IWCMC 2017*, 2017, doi: 10.1109/IWCMC.2017.7986447.

[21] S. Sharma, D. Sethi, and P. P. Bhattacharya, "Artificial Neural Network based Cluster Head Selection in Wireless Sensor Network," 2015.

[22] X. Wang, X. Li, and V. C. M. Leung, "Artificial intelligence-based techniques for emerging heterogeneous network: State of the arts, opportunities, and challenges," *IEEE Access*, vol. 3, pp. 1379–1391, 2015, doi: 10.1109/ACCESS.2015.2467174.

[23] T. E. Bogale, X. Wang, and L. B. Le, "Machine Intelligence Techniques for Next-Generation Context-Aware Wireless Networks," Jan. 2018.

[24] L. Caviglione, M. Gaggero, J. F. Lalande, W. Mazurczyk, and M. Urbański, "Seeing the unseen: Revealing mobile malware hidden communications via energy consumption and artificial intelligence," *IEEE Trans. Inf. Forensics Secur.*, vol. 11, no. 4, pp. 799–810, Apr. 2016, doi: 10.1109/TIFS.2015.2510825.

[25] K.-M. Liao, G.-Y. Chen, and Y.-J. Chen, "Artificial Intelligence based Edge Caching in Vehicular Mobile Networks: Architecture, Opportunities, and Research Issues," in *2019 20th Asia-Pacific Network Operations and Management Symposium (APNOMS)*, 2019, pp. 1–6, doi: 10.23919/APNOMS.2019.8893099.

[26] M. Chen, U. Challita, W. Saad, C. Yin, and M. Debbah, "Artificial Neural Networks-Based Machine Learning for Wireless Networks: A Tutorial," Oct. 2017.

[27] M. S. Parwez, D. B. Rawat, and M. Garuba, "Big data analytics for user-activity analysis and user-anomaly detection in mobile wireless network," *IEEE Trans. Ind. Informatics*, vol. 13, no. 4, pp. 2058–2065, Aug. 2017, doi: 10.1109/TII.2017.2650206.

[28] K. Ota, M. S. Dao, V. Mezaris, and F. G. B. De Natale, "Deep learning for mobile multimedia: A survey," Jun-2017, vol. 13, no. 3s, doi: 10.1145/3092831.

[29] I. Kirbas and A. Dukkanci, "Development of A Wi-Fi Based Indoor Location System Using Artificial Intelligence Techniques | Data Science and Applications," *Data Sci. Appl.*, vol. 2, no. 1, pp. 8–12, 2019.

[30] J. Qadir, "Artificial intelligence based cognitive routing for cognitive radio networks," *Artif. Intell. Rev.*, vol. 45, no. 1, pp. 25–96, Jan. 2016, doi: 10.1007/s10462-015-9438-6.

[31] I. Atov, K. C. Chen, and S. Yu, "Data Science and Artificial Intelligence for Communications," May-2019, vol. 57, no. 5, p. 56, doi: 10.1109/MCOM.2019.8713799.

[32] N. Eli-Chukwu, N. Clara Eli-Chukwu, J. M. Aloh, and C. Ogwugwuam Ezeagwu, "A Systematic Review of Artificial Intelligence Applications in Cellular Networks A Systematic Review of Artificial Intelligence Applications in Cellular Networks View project A Systematic Review of Artificial Intelligence Applications in Cellular Networks," 2019.

IOP Conf. Series: Materials Science and Engineering 1212 (2022) 012047 doi:10.1088/1757-899X/1212/1/012047

IoT-based temperature and humidity monitoring system for smart garden

YAnshori, D W Nugraha, and D Santi

Department of Informatics Engineering, Tadulako University, Palu, Indonesia

Email: iyus.jr@gmail.com, deny.wiria.nugraha@gmail.com, nyherbert@gmail.com

Abstract. The main objective of this paper is to design an IoT (Internet of Things) to monitor temperature and humidity for smart gardens. Temperature sensors and humidity sensors measure environmental conditions and are processed by a microcontroller. The actuator used is a spray pump that is used to spray water into the air to lower the temperature. Data from the sensors and status from the actuators are sent to the server and can be monitored via a smartphone. The data collected can be analyzed for various purposes. The result obtained is the effect of spraying on temperature reduction.

Keywords—Internet of Things, smart garden, spray pump.

1. Introduction

Internet of Things-IoT is part of Industry 4.0, which transforms conventional lifestyles into smart, dynamic, and modern lifestyles. In agriculture, several terms are used related to IoT, such as smart garden [1], smart farming [2], smart agriculture [3], and other similar terms. Smart Garden is a term that utilizes automation technology [4] in garden management using mechanical and electronic devices by using information technology and the internet [5]. Smart Garden is a solution for monitoring environmental conditions (temperature, humidity, soil moisture, light intensity, etc.) and maintaining ideal climate parameters that can be controlled to increase garden yield. The main agricultural product of Central Sulawesi province is shallots (Allium cepa L. Var. Aggregatum Group). The ideal environment for growing shallots is temperature at 27.7-30.0oC and humidity at 61.22-68.90%. The potential productivity of shallots is 9.7 tonnes/ha, but the productivity of farmers is lower. This is due to the application of cultivation techniques that are not according to standards and changes in ideal environmental parameters[7]. Several studies related to the application of IoT in agriculture are the implementation of IoT in agriculture for environmental monitoring with temperature sensors, soil pH, soil moisture, humidity, motion sensors and water volume sensors [2]. Other studies include monitoring and data transmission using the API service from Blynk which is used to monitor temperature, humidity, and control of watering plants [8], monitoring systems and controlling nutrients, temperature, and water levels in website-based hydroponic farming using NodeMCU [9] and IoT hydroponics management systems using Rapsberry Pi to monitor pH, water level, air temperature and air humidity [10]. Fuzzy systems are used for wireless sensor networks (WSN) for controller heating, cooling, irrigation, lighting, and shading in a greenhouse [11]. This study proposed a prototype of the Smart Garden system for shallots based on the Internet of Things (IoT) with the fuzzy logic controller method which functions to monitor temperature and humidity and perform mist spraying under certain conditions. The goal is to maintain the ideal temperature and humidity for the shallots. In contrast to other studies, in this study fuzzy logic functions as a spray pump control method for spraying mist water into the air to reduce the temperature.

2. Proposed Method

The monitoring of temperature and humidity uses sensors that will be read and processed by a microcontroller. The fuzzy method embedded in the microcontroller will activate the spray pump under certain conditions. Data from sensors and actuators are sent continuously over the internet to the server. The server receives a data reading request from the client so that monitoring can be seen and carried out anytime and anywhere. To maintain the ideal temperature and humidity conditions for shallots, a spray pump is used for spraying. The fuzzy method is used to regulate spraying. The smart garden system design is shown in Figure 1.

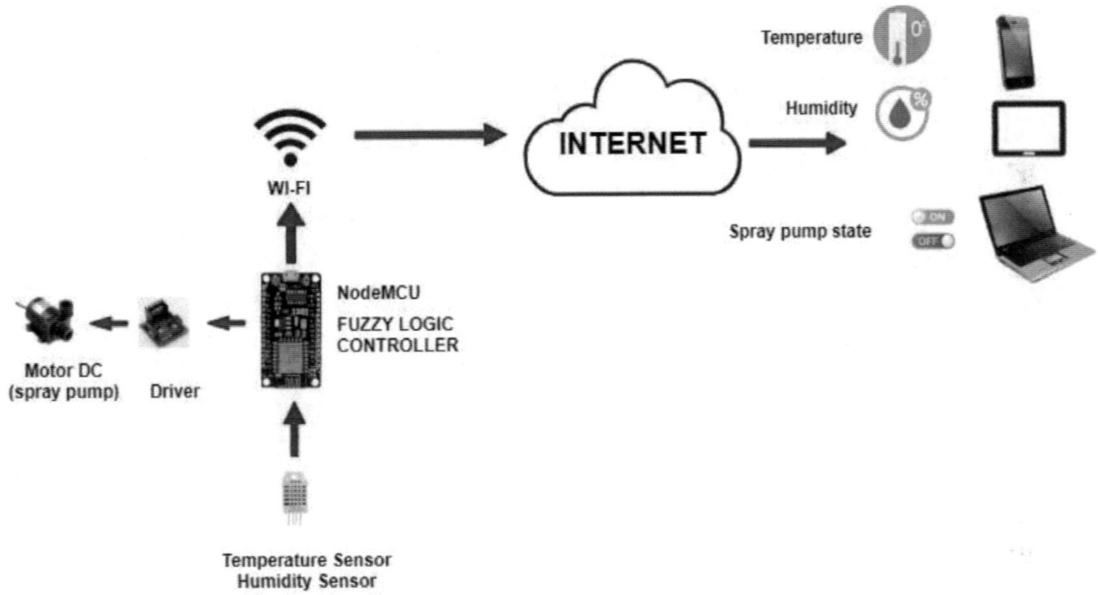

Figure 1. Smart Garden System

The application design for the proposed smart garden is shown in Figure 2.

Step 1: initialize the NodeMCU, temperature and humidity sensor (DHT22).
Step 2: connect NodeMCU to the internet via WIFI.
Step 3: NodeMCU reads the data sent by the sensor.
Step 4: NodeMCU processes the sensor data using fuzzy logic and controls the spraying when the conditions meet. Next, NodeMCU sends sensor reading data and readings of the state of spraying pump to the server.
Step 5: Smartphones, tablets, and computers read data from the server.

The hardware used consists of sensors, microcontrollers, and actuators. The sensors are temperature and humidity sensor (DHT22), microcontroller (NodeMCU ESP8266), and the actuator is spray pump. The control function is performed by fuzzy logic. Fuzzy logic is a computational method used for a complex nonlinear problem to model or an ambiguous problem [13]. Fuzzy logic follows the human way of thinking in natural linguistics of an object such as very low, low, medium, high, and very high. Fuzzy set is a set with each element that has degrees of membership which has an interval of 0 to 1.The architecture of fuzzy logic shown in Figure 3.

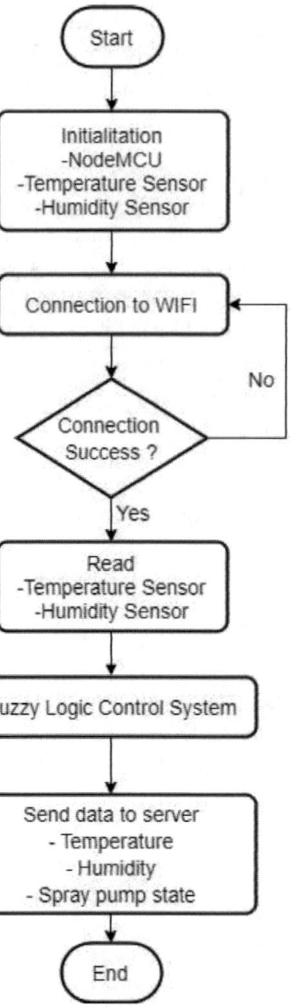

Figure 2. Flowchart proposes method

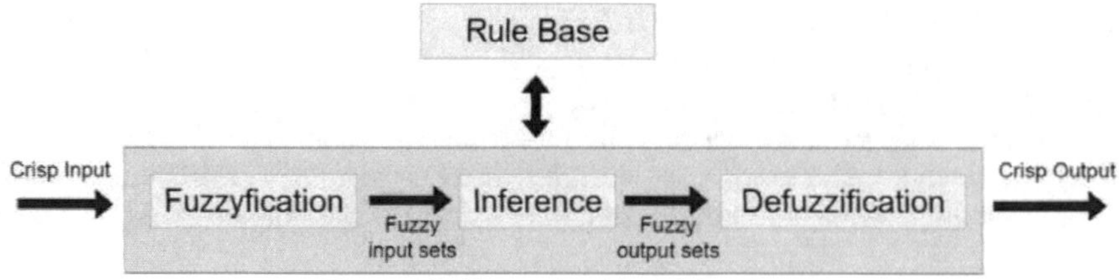

Figure 3. Fuzzy logic architecture

Fuzzy logic functions as a method of controlling the climate parameters (temperature and humidity) of a smart garden which has two inputs and one output. Input data consists of temperature (℃) and relative humidity (%), the output is a spray pump for spraying water in the air. Figure 4 shows a representation of the temperature variable and figure 5 shows a representation of the variable humidity. In the picture VL: Very Low, L: Low, M: Medium, H: High, and VH: Very High.

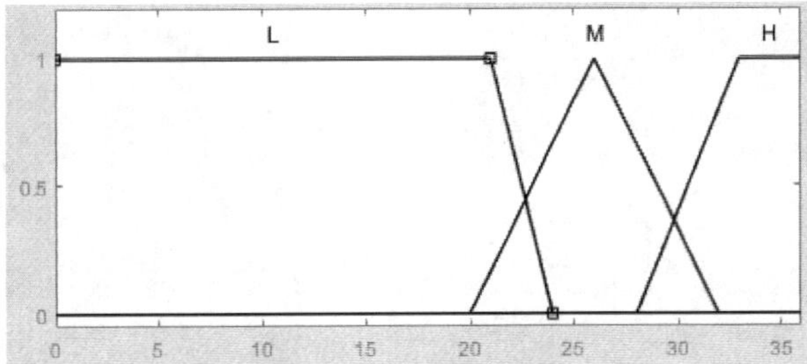

Figure 4. State variable temperature

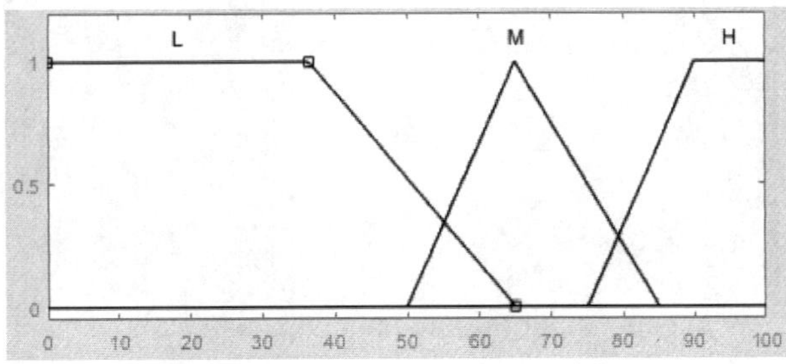

Figure 5. State variable humidity

After defining the membership function, the next is to set up a rule base. The relation and effects of temperature and humidity on the spray pump are shown in Table 1.

Table 1. Fuzzy logic rule tables

T \ H	L	M	H
L	S	Off	Off
M	M	Off	Off
H	L	L	M

where : Off, Short, Medium, Long

The rules set to regulate the motor state based on the input temperature and humidity. Based on the rule base, if the temperature is low and humidity is low, the spray pump will spray in a short time. This is aimed at increasing humidity. If the temperature is low and the humidity medium, the pump spray will be off due to the humidity in the medium. Likewise for conditions of low temperature and humidity high. For the next rule, if the temperature is medium and humidity is low, the motor will spray with medium time, this is to increase the humidity. For the rule base temperature, medium and high humidity, spray pump off. For the rule base temperature high and humidity low and medium, the spray pump will spray for a long duration because it aims to lower the temperature and increase the humidity. If the temperature is high and humidity high, the spray pump will spray with a medium duration because it aims to reduce the temperature.

3. Results

The application is designed using the Blynk server. There are three visual monitoring in the application, namely a gauge for temperature and humidity and visual information about the state of the motor. High for spray pump on and low for off. Hardware and applications made can be seen in Figure 6. and Figure 7.

Figure 6. Hardware

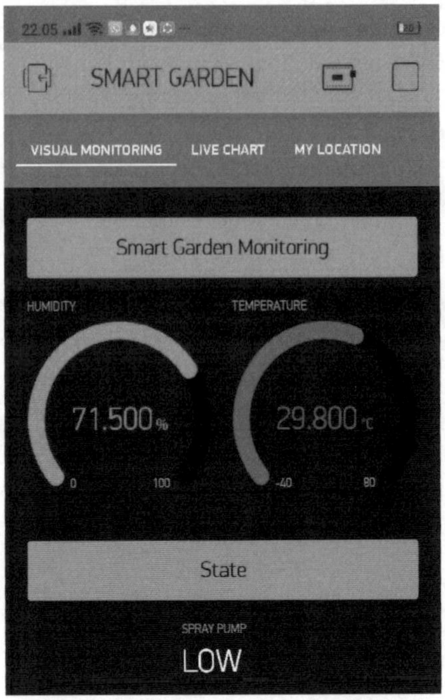

Figure 7. Smart garden monitoring

To check the performance of a smart garden with fuzzy is done by the measure of the temperature and humidity and getting the temperature and humidity values as in table 2.

Table 2. Temperature and humidity, before dan after spraying

Time	**Temp (C)**	**Humidity (%)**	**Spray Pump state (On/ Off)**	**Temp. after spraying (C)**	**Humidity after spraying (%)**
6	21	63	Off	21	63
7	27	57	On	27	60
8	32	45	On	27	55
9	32	42	On	28	57
10	34	35	On	28	62
11	34	34	On	28	62
12	35	32	On	28	63
13	34	32	On	28	60
14	33	34	On	27	63
15	31	34	On	27	64
16	29	37	On	27	66
17	28	41	On	26	68
18	28	43	On	26	69
AVERAGE	30.61	40.69		26.76	62.46

At a temperature of 21 °C and a humidity of 63%, the spray pump state turned off and this is by the predetermined fuzzy logic rule base. This condition is still considered ideal so that no spraying is carried out. When the temperature rises to 27 °C and the humidity drops to 57%, the spray pump turns on and starts spraying. Likewise, for temperatures above 27 °C, the spray pump remains on. After spraying, there is a decrease in temperature and an increase in humidity. From table 2, it is found that the temperature decrease between the temperature before spraying and the temperature after spraying is an average of 3.8 oC as shown in Figure 11 and the change in the increase in humidity is an average of 21.7 as shown in Figure 12. The average temperature after spraying is 26.76 oC and the average humidity after spraying is 62.4%. This shows that the presence of spraying is a decrease in the value of temperature and an increase in the value of humidity.

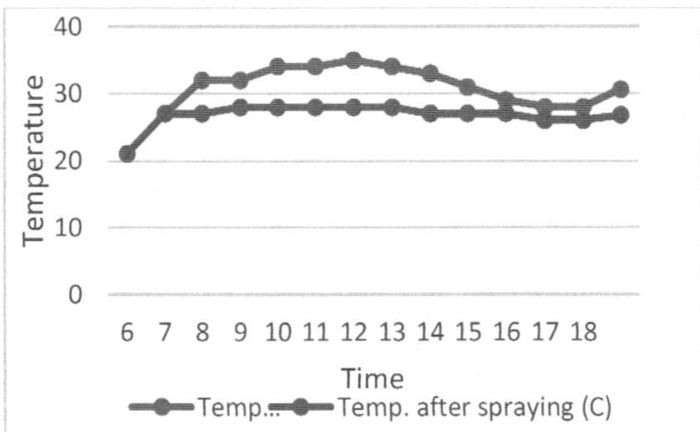

Figure 8. Temperature before and after spraying

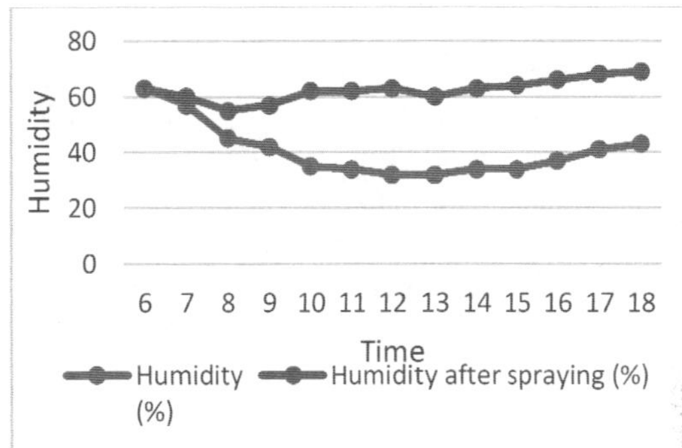

Figure 9. Humidity before and after spraying

The results prove that the smart garden with the fuzzy logic system can maintain the parameters of the ideal temperature and humidity for the growth of shallot

4. Conclusion

The prototype of smart garden with fuzzy logic system is made using NodeMCU as a controller and temperature sensor (DHT22). This smart garden prototype can monitor temperature and moisture parameters online and in real-time. This system also monitors the on / off state of the spray pump. The average temperature after spraying was carried out at 26.76 oC and the average humidity after spraying was 62.4%. Smart garden with fuzzy logic system can maintain ideal temperature and humidity parameters for growing.

References

[1] T. Okayama, "Future Gardening System : Smart Garden," *J. Dev. Sustain. Agric.*, 2014, doi: 10.11178/jdsa.9.47.

[2] R. Dagar, S. Som, and S. K. Khatri, "Smart Farming – IoT in Agriculture," in *2018 International Conference on Inventive Research in Computing Applications (ICIRCA)*, 2018, pp. 1052–1056, doi: 10.1109/ICIRCA.2018.8597264.

[3] N. Gondchawar and R. S. Kawitkar, "IoT based smart agriculture," *Int. J. Adv. Res. Comput. Commun. Eng.*, 2016, doi: 10.17148/IJARCCE.2016.56188.

[4] A. Al-Omary, H. M. AlSabbagh, and H. Al-Rizzo, "Cloud based IoT for smart garden watering system using Arduino Uno," in *IET Conference Publications*, 2018, vol. 2018, no. CP747, doi: 10.1049/cp.2018.1401.

[5] Sandhya.B.R, Pallavi.M, and Chandrashekar.M, *IoT Based Smart Home Garden Watering System Using Raspberry Pi 3*. 2017.

[6] M. Anshar, Tohari, B. H. Sunarminto, and E. Sulistyaningsih, "Pengaruh lengas tanah terhadap pertumbuhan dan hasil tiga varietas lokal bawang merah pada ketinggian tempat berbeda," *J. Agrol.*, vol. 18, no. 1, pp. 8–14, 2011.

[7] M. A. Pasigai, A. R. Thaha, B. Nasir, S. A. Lasmini, and Bahrudin, Teknologi *budidaya bawang merah varietas lembah palu. UNTAD Press. 190.*, 1st ed. Palu: Untad Press, 2016.

[8] W. Adi Prayitno, A. Muttaqin, and D. Syauqy, "Sistem Monitoring Suhu, Kelembapan, dan Pengendalian Penyiraman Tanaman Hidroponik menggunakan Blynk Andorid," *Circ. Res.*, 2017, doi: 10.1161/CIRCRESAHA.112.270033.

[9] Y. H. Putra, D. Triyanto, and Suhardi, "Sistem Pemantauan dan Pengendalian Nutrisi, Suhu, dan Tinggi Air Pada Pertanian Hidroponik," *J. Coding, Sist. Komput. Untan*, 2018.

[10] C. J. G. Aliac and E. Maravillas, "IOT Hydroponics Management System," in *2018 IEEE 10th International Conference on Humanoid, Nanotechnology, Information Technology,Communication and Control, Environment and Management (HNICEM)*, 2018, pp. 1–5, doi: 10.1109/HNICEM.2018.8666372.

[11] Ö. Alpay and E. Erdem, "The control of greenhouses based on fuzzy logic using wireless sensor networks," *Int. J. Comput. Intell. Syst.*, vol. 12, no. 1, pp. 190–203, 2018, doi: 10.2991/ijcis.2018.125905641.

[12] J. Fraden, *Handbook of modern sensors: Physics, designs, and applications*, 5th ed. Springer International Publishing, 2016.

[13] T. Tuncer, "Intelligent centroid localization based on fuzzy logic and genetic algorithm," *Int. J. Comput. Intell. Syst.*, vol. 10, no. 1, pp. 1056–1065, 2017, doi: 10.2991/ijcis.2017.10.1.70.

Evaluation of overall equipment effectivity in the cement manufacturing industry: a case study

F Mustakim, H M Asih*

Industrial Engineering Department, Faculty of Industrial Technology, Universitas Ahmad Dahlan. Jalan Ring Road Selatan, Tamanan, Banguntapan, Bantul, Yogyakarta 55166

*Corresponding author, Email: hayati.asih@ie.uad.ac.id

Abstract. *Overall equipment effectiveness is one of the important tools to measure the productivity in the company. This research is based on a case study of cement manufacturing industry in Indonesia, especially in kiln machine. The research aims to investigate the six big losses. In addition, the calculation of availability components, performance efficiency, and quality rate are carried out. The fishbone diagram is also employed to identify the root cause of the problem deeply. The results show the biggest losses in the kiln machine are the reduced speed losses. Therefore, the availability component and the performance efficiency should be improved to increase the productivity and the efficiency in the shop floor. Some suggestions are presented such as make a better surveillance system, improve the planning of preventive maintenance machine, and replace the more sophisticated maintenance tools to facilitate maintaining the machine.*

Keyworlds: Kiln machine; cement industry; overall equipment effectiveness; six big losses.

1. Introduction

In this 21st century, the development of the industry is so significant so that it does not rule out the possibility of growth in Indonesia. Hence, the competition is very tight. Therefore, companies want to increase their productivity because productivity is essential for companies to obtain maximum success [1]. One of the examples of increasing productivity is to evaluate the performance of production facilities at companies [2]. Overall equipment effectiveness (OEE) is a quantitative metric that has been increasingly used in manufactorysystems for controlling and monitoring the productivity of production equipment, and also as an indicator anddriver of process and performance improvements. The OEE is a key performance measure in mass-production environments that consists of three important components which are availability, productivity and quality [3].

XYZ company is engaged in the cement manufacturing industry located on Cilacap, Indonesia. When it comes to production at XYZ company, it processes a homogenization of raw materials in the raw mill. After the raw materials are mixed, the raw materials pass through the combustion process in the kiln machine, then the combustion results from the kiln machine are cooled on the cooler machine. After the cold clinker cement is crushed in the finish mill machine, then the crushed cement dust is packed in the packaging machine, after which the finished cement is stored in a storage warehouse. The production process is carried out 24 hours non-stop, so that apart from the problem of the effectiveness of the machine or equipment. Therefore, without a willingness to maintain the right method at its kiln, it will produce less optimal production so that the targeted quantity will be less, and it will not be achieved as it is desired.

Overall equipment effectiveness (OEE) plays a crucial role in increasing productivity. effectiveness is the measure of doing it properly to the extent to which the requirements are fulfilled [4]. OEE is a method used to implement total productive maintenance (TPM) to keep equipment in ideal condition by removing six big losses [5]. The OEE methodology is a proven approach to improve the overall performance of the equipment (rest time, lunch, scheduled maintenance, or periods in which nothing is produced [6].

Overall equipment effectiveness measures the gap between actual performance and the potential performance in the manufacturing. OEE includes three measuring metrics, such as availability, performance, and quality. It helps to measure the efficiency of the factory by categorizing the significant losses affecting the manufacturing process [7]. The primary purpose of overall equipment effectiveness (OEE) is to enhance and maintain equipment efficiency [8]. Most of the research involving OEE size is not only related to maintenance, but also for areas such as performance measurement and maintained productivity. The measure of OEE is carried out with attention to three critical things, the availability level, level of performance, and level of quality [9].

Table 1. Research Gap

Researcher	Objective	Problem	Methodology	Conclusion
Swomnya and Chetan [6]	To evaluate the Losses occurred	OEE values are below world class standards	OEE and Six Big Losses	Implementation of OEE in manufacturing problems cannot be a short-term goal. However, it is a joint improvement program where higher OEE levels near world-class scores can be achieved by reducing six significant losses and implementing TPM tools
Prabowo et al [10]	Evaluate the implementation of 8 Pillar TPM and measure the effect on manufacturing performance in the form of Overall equipment effectiveness (OEE) and Waste	Only 6 out of the 8 pillars affect manufacturing performance	TPM and OEE	74.3% of manufacturing performance variables can be explained/influenced by 8 Pillar TPM variables and the remaining 26.7% by other variables.
Sinaigar [11]	To determine and analyze each of the OEE values	Performance	TPM	The average OEE value of the kiln machine has met world-class results
Dawood and Abdullah [12]	To analyz the overall operating performance of the equipment	Performance and Quality	OEE	OEE values on all cement production machinery are below world-class standards
Proposed Research	To evaluate the OEE on kiln machine	Availability and Performance	OEE and Six Big Losses	OEE values on both machines are below world-class standards because they are caused by three losses, namely breakdown, idling stoppages, reduced speed.

The difference between this research and previous research is analyzing the OEE value of the machine and exploring the losses that affect the engine's low effectiveness value.

2. Method

This methodology describes the stages that will be performed in the study. This method identifies the issue until the last conclusion is taken through the flow diagram in the display in Figure 1.

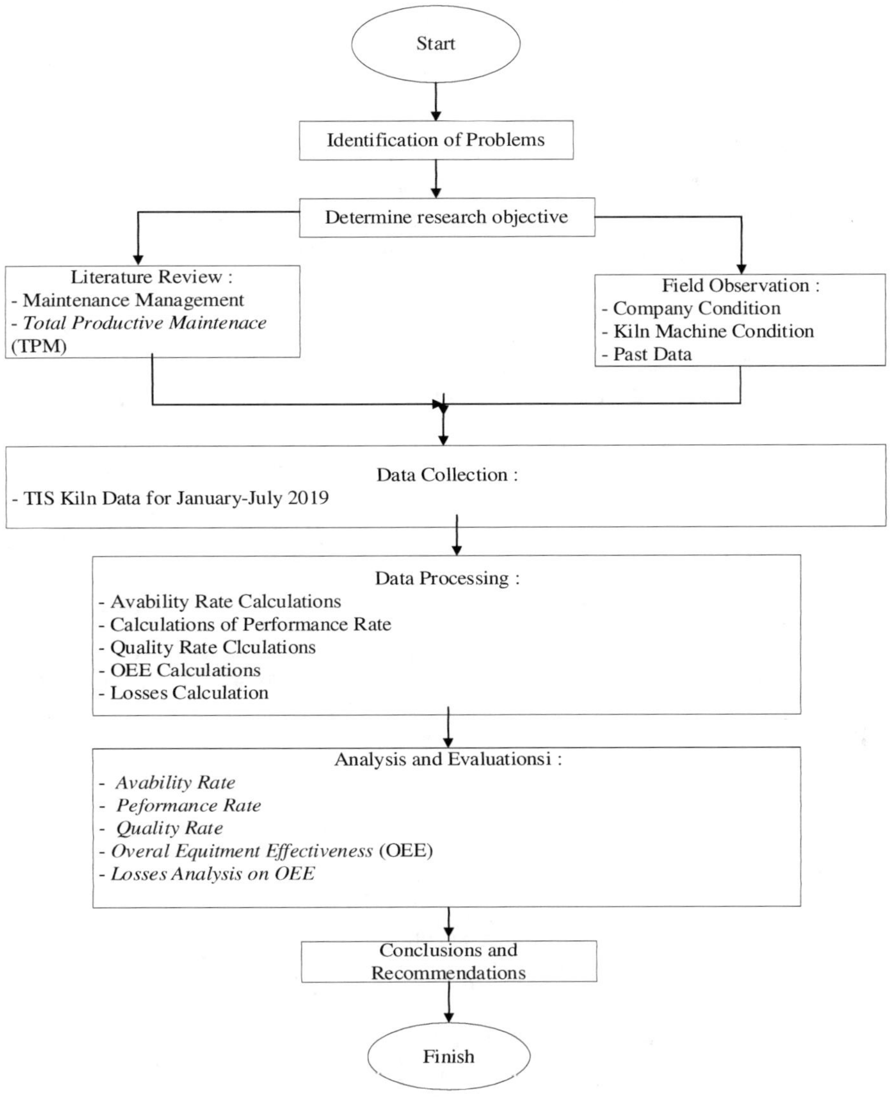

Figure 1. Flow Chart of research methods

These stages of research are the first Problem identification, identify the problem that occurs in the kiln machine, which is located at XYZ company can influence the effectiveness of kiln machine. Then research objectives, the purpose of this research is to know the value of overall equipment effectivity kiln machine and six big losses that influence the OEE kiln machine value. Then literature review, it studies the journals and related topics about the maintenance spells and total productive maintenance. Then field observations, it is conducted by observating at XYZ company to know field conditions, kiln machine conditions, and past data from the kiln machine. Then data collection, collecting previous data of kiln machines, which necessary to support the resolution to resolve the existing problems. Collecting data is done by observing the data from TIS kiln machine. Then data processing, data processing is done by calculating the value of availability, performance, quality, OEE, and losses in the kiln machine. Then analysis and evaluation, the result of its data processing depicts how the value of availability rate, performance rate, quality rate, OEE, and evaluate whether the OEE value obtained from the kiln machine already meets the World class standard. The final step of this study is conclusions and suggestions, the final step of this study is to conclude from the research results which have been done on the kiln machine located in XYZ company. It also discussed suggestions for XYZ company to increase the OEE value and reduce the Six Big Losses on the kiln machine.

3. Results and Discussion

3.1 OEE calculations

The availability of June is low because in the same month in XYZ company has an annual Schedule maintenance engine, and the engine is experiencing a breakdown. The calculation of the availability value of kiln machine can be seen in Table 2.

Table 2. The result of availability kiln machine

Month	Total *Availability Time* (h)	*Planed Downtime* (h)	*Loading Time* (h)	*Downtime* (h)	*Operating Time* (h)	*Availability* (%)
January	744	0	744	62.3	681.7	91.63
February	672	0	672	168.7	503.3	74.90
March	744	0	744	189.7	554.3	74.50
April	720	0	720	320.5	399.5	55.49
May	744	0	744	31.2	712.8	95.81
June	720	414.25	305.75	193.6	112.15	36.68
July	744	0	744	67.7	676.3	90.90
Average	726.86	59.18	667.68	147.67	520.01	74.27

The average kiln machine availability is 74.27%, which means that the kiln machine availability is still below the world standard of 90%. It is because the machines are often damaged, and the supervision is less effective at XYZ company. The result of the performance value in kiln machine can be seen in Table 3.

Table 3. The result of kiln machine performance

Month	*Operating* (h)	Total production (ton)	*Cycle time* (h)	*Peformance* %
January	644.71	205421	0.0018	54.24
February	467.09	149331	0.0018	53.41
March	528.96	173202	0.0018	56.25
April	373.01	124102	0.0018	55.92
May	690.76	226336	0.0018	57.16
June	301.52	38643	0.0018	62.02
July	626.42	214346	0.0018	57.05
Average	518.92	162318.86	0.0018	56.58

The average performance of kiln machine is 56.58%, which means the performance of kiln machine is still very far below the world standard, namely 95%. It is due to machines that are often damaged and less effective machine maintenance at XYZ company. The result of the quality value in kiln machine can be seen in Table 4.

Table 4. The result of kiln machine quality

Month	total production (ton)	The defect products (ton)	*Quality* %
January	205421	0	100
February	149331	0	100
March	173202	0	100
April	124102	0	100
May	226336	0	100
June	38643	0	100
July	214346	0	100
Average	162318,86	0	100

Kiln machine quality gets an average of 100%, which means the quality of kiln machine is above the world standard, namely 99.9%. This is due to the absence of defects when the kiln machine was in production. After calculating the value of availability, performance, and quality then calculate the value OEE kiln machine. Calculation of the value of overall equipment effectivity for kiln machines can be seen in table 5.

Table 5. Calculation result of OEE kiln machine

Month	Availability	Performance	Quality	OEE
January	91.63	54.24	100	49.7
Febuary	74.9	53.41	100	40
March	74.5	56.25	100	41.9
April	55.49	55.92	100	31.02
Mei	95.81	57.16	100	54.76
June	36.68	62.02	100	22.75
July	90.9	57.05	100	51.86
Average	74.27	56.58	100	41.71

The average OEE value for kiln machines is 41.71%, which means that the OEE value for kiln machines is far below World standards. It is because there are three losses in the kiln machine, which significantly affect the effectiveness of the machine.

3.2 Losses Calculation

In the kiln machine, three losses cause the OEE value of the small kiln machine: breakdown losses, idling and stoppages minor, and reduced losses. The result of idling & mirror stoppages losses can be seen in the Table 6.

$$Idling\ and\ Minor\ Stoppages = \frac{Non\ Productive}{Loading\ Time}\ X\ 100\% \tag{1}$$

Table 6. The result of idling & miror stoppages losses kiln machine

Month	*Stop line*	*Loading time* (h)	*Idling & Mirror Stoppages* (%)

January	0	744	0
february	0	672	0
March	180.8	744	24.3
April	0	720	0
Mei	4.93	744	0.66
Juni	0	305.75	0
Juli	0	744	0
Average	26.53	667.68	3.57

According to the table above, it can be noted that the average idling and minor stoppages of 3.57% of these losses have a relatively small value but also affect the amount of OEE. It is due to ineffective machine maintenance and supervision, so the performance value of the kiln machine is low. The result of the Reduced spees can be seen in Table 7.

$$Equipment\ Failure\ Losess = \frac{Total\ Breakdowns}{Loading\ Time}\ X\ 100\% \qquad (2)$$

Table 7. The result of reduced losses kiln machine

Month	*Operating* (h)	Total production (ton)	*Cycle time* (h)	*Loading time* (h)	*Reduced speed* (%)
January	681.7	205421	0.0018	744	41.93
February	503.3	149331	0.0018	672	34.90
March	554.3	173202	0.0018	744	32.60
April	399.5	124102	0.0018	720	24.46
May	712.8	226336	0.0018	744	41.05
June	112.15	38643	0.0018	305.75	13.93
July	676.3	214346	0.0018	744	39.04
Average	520.01	162318.8	0.0018	667.68	32.56

Based on the table, the average reduced losses are 32.56%, the losses are the most significant value so that the amount of OEE kiln machine under World Class. These losses are caused by a less effective or less stringent surveillance system so that the kiln machine's performance value is very low. The result of breakdown losess can be seen in table 8.

$$Recuded\ Speed = \frac{OT - (\ Cycletime\ x\ Processed\ Amount)}{Loading\ Time}\ X\ 100\%$$

Table 8. The result of breakdown losses kiln machine

Month	*Breakdown* (h)	*Loading time* (h)	Breakdwon losses (%)
January	62.3	744	8.37
February	168.7	672	25.10

March	8.9	744	1.20
April	320.5	720	44.51
May	26.27	744	3.53
June	193.6	305.75	63.32
July	67.7	744	9.10
Average	121.14	667.68	22.16

Based on the above table can be known that the average breakdown losses of 22.16%, these losses occurred due to the preventive maintenance plan at XYZ company is still less effective, so the value availability kiln machine under standard world class. It is due to ineffective machine maintenance and supervision, so the performance value of the kiln machine is low.

According to the table above, it obtained the average value of the availability kiln machine amounting to 74.27% of the value, it has not fulfilled the world-class standards of 90%, in June, it has a very small performance, it is because of the machine usually experiencing a break down in this month. For the value of Performance kiln machine of 56.58%, The value has not fulfilled world-class standards, and for the quality value of the kiln machine amounting to 100% the value already meets world-class standards. Thus the value of availability and performance of the kiln machine is still less good because still under world-class standards, thus causing the value of OEE under world-class standards of > 85%. For quality, value has been good because it exceeds the world-class standards.

From research in the field, the factors that affect the value of availability are below the standard due to the breakdown loss can be seen in Table 6. It is because the ineffective preventive maintenance which makes the value availability below the World class standards. The factors that affect the value performance below the standard of world class caused by idling and minor stop losess can be seen in Table 7. The reduced speed losess can be seen in Table 8. The performance value of kiln machine is below the standard of World class because the poor supervision and poor maintenance machine.

Figure 2 is a graph comparison of OEE values obtained by the kiln machine and the World class standard, from the above image can be concluded that the amount of the OEE kiln machine does not meet the standard World class.

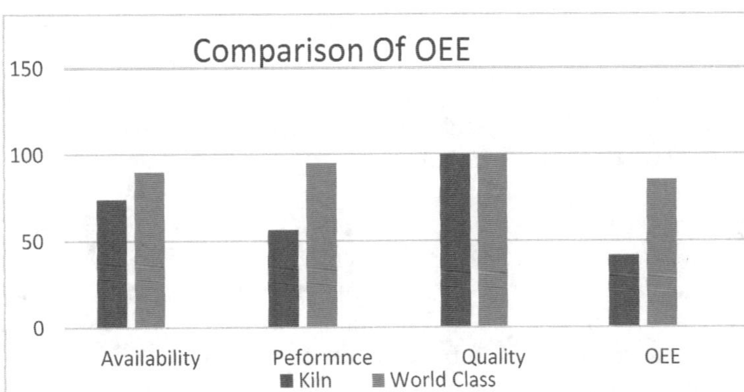

Figure 2. The comparison between real and the World class standard

3.3 Fishbone diagram

The fishbone diagram is employed to identify the problem of low OEE value, especially in kiln machine. It includes some factors, such as materials, machine, environment, and method.

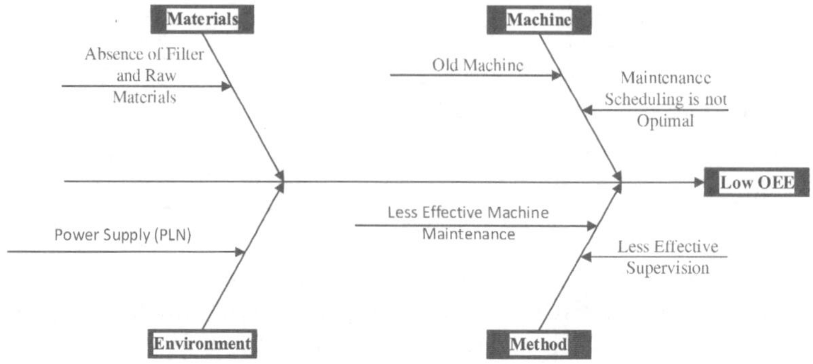

Figure 3. Fishbone diagram of kiln machine

The cause of the kiln's low OEE value is machine, one of the reasons for low OEE value is the machine is already old. It effects on the performance of that machine. In addition, the maintenance scheduling is very rare. It makes the condition become worse. Then method, another reason is the method employed in the production system. The controlling and monitoring that is conducted by the supervisor in the shop floor is poor. Next, there is no standard procedures in maintaining the machine. Then material, the raw materials such as sand and limestone that are from warehouse has no filtering procedures before entering into kiln machine. It makes some irons break the machine. In addition, the limestone are still in big size and there is no standardized size of limestone. It makes poor availability for that machine. Then environment, in the shop floor the electricity is easily down. It makes the machine break rapidly, and the machine's lifespan becomes short.

To overcome those problems, there are some suggestions in order to increase the OEE value while meeting the World Class standards. The supervisors should have effective and good surveillance system so that it helps on monitoring the performance of kiln machine. Then, the preventive maintenance planning should be improved in order to reduce the frequency of the kiln machine breakdown. Another suggestion is the sophisticated maintenance tools should be added in order to the machine maintenance becomes more effective and efficient.

4. Conclusion

This research is based on a case study in cement manufacturing industry in Cilacap, Indonesia. The research aims to investigate the six big losses. In addition, the calculation of availability components, performance efficiency, and quality rate are carried out. The fish bone diagram is also employed to identify the root cause of the problem deeply. The results show the value of availability 74.27%, the value of performance is 56.58%, and the value of quality is 100%, which means the component availability and performance are not meet the OEE world class standard, i.e. more than 85%. There are three losses that can cause low OEE values, such as reduced speed losses and idling and minor stops which affect the performance value of the kiln machine, and breakdown losses which affect the value availability of kiln machines. The fishbone diagram is employed to identify the root cause of the problem. Some suggestions are presented such as make a better surveillance system, improve the planning of preventive maintenance machine, and replace the more sophisticated maintenance tools to facilitate maintaining the machine.

Referefence

[1] D.H. Triwardani, A. Rahman, Farela, and M. Tantrika, "Analisis Overall Equipment Effectiveness (OEE) dalam Meminimalisasi Six Big Loses Pada Mesin Produksi Analysis Dual Filters DD07 (Studi Kasus: PT. Filtrona Indonesia, Surabaya, Jawa Timur) Analysis of Overall Equipment Effectiveness to Reduce Six Big Losses On Pro,"**Vol. 07**, PP.379-391

[2] H. Pratama, *"Peningkatan Produktivitas dan Kinerja Lingkungan Menggunakan Metode Green Productivity,"* J.Tek.Ind., **Vol.16**, No.2, P.63, 2017, DOI : 10.22219/JTIUMM.Vol 16.No.2.63-73

[3] P. H. Tsarouhas,"*Evaluation of Overall Equipment Effectiveness In The Beverage Industry: A Case Study ,"* Int.J.Prod Res., **Vol 51**, No.2, PP. 515-523, 2013, DOI : 10.1080/00207543.2011.653014

[4] M. Tech,"*Measurement of Overall Equipment Effectiveness (OEE) of a Manufacturing Industry: An Effective Lean Tool,*"Int.J.Recent Trends Eng. Res., **Vol 3**, No. 5, PP. 268-275, 2017, DOI: 10.23883/IJRTER.2017.3222.WCT10

[5] Mansur, R. Rayendra, and M. Mastur,"*Performance Acceleration On Production Machine Using The Overall Equipment Effectiveness (OEE) Approach*," Iop Conf. Ser.Mater.Sci.Eng., **Vol 105,** No.1, 2016, DOI : 10.1088/1757-899X/105/1/012019.

[6] Sowmya K and Chetan N," A Review On Effectiveness Of Resources Using Overall Equipment Effectiveness by Reducing Six Big Losses,"Int.J.Sci.Res.Sci.Eng. Technol., **Vol 2**, No.1, PP. 556-562, 2016

[7] S. Nallusamy, V.Kumar, V.Yadav, U.K. Prasad, and S.K. Suman," *Implementation Total Produvtive Maintenance to Enhance the Overall Equipment Effectiveness In Medium Scale Industries*,"Int.J.Mech. Prod. Eng, Res, Dev., **Vol.8,** No.1, PP. 1027-1038, 2018, DOI : 10.24247/*IJMPERDFEB*2018123

[8] R. Hedman, M.Subramaniyan, and P. Almstrom," *Analysis of Critical Factors For Automatic Measurement of OEE,"* Procedia CIRP, **Vol.57**, PP. 128-133, 2016, DOI : 10.1016/J.Procir.2016.11.023

[9] M.S. Hervian and C. Soekardi," Improving Productivity Based on Evaluation Score of Overall Equipment Effectiveness (OEE) Using DMAIC Approach on Blistering Machine," Int.J.Sci.Res., **Vol.5**, No.7, PP.736-739,2016, DOI: 10.21275/v5i7.ART2016204

[10] H.A. Prabowo, Y.B. Suprapto, and F.Farida," The Evaluation of Eight Pillars Total Productive Maintenance (TPM) Implementation and Their Impact on Overall Equipment Effectiveness (OEE) and Waste," Sinergi, **Vol.22**, No.1, P.13, 2018, DOI: 10.22441/Sinergi.2018.1.003

[11] R.I. Sinaigar,"Evaluasi Efektivitas Mesin Rotary Kiln Dengan Penerapan Total Produktif Maintenance (TPM) di PT Semen Holcim Indonesia," No. 16, PP.1-15, 2017

[12] L.M. Dawood and Z.H.Abdullah," *Study Impact of Overall Equipment Effectiveness Onto Cement Industry,"* J.Univ. Babylon Eng.Sci., **Vol. 26**, No. 3, PP. 187-198, 2018

IOP Conf. Series: Materials Science and Engineering 1212 (2022) 012049 doi:10.1088/1757-899X/1212/1/012049

Designing simulation model for minimizing coal train unloading time: A case study of Kereta Api Logistik Company.

H Wijaya[1], A Y Ridwan[1], and E B Setyawan[1]

School of Industrial Engineering, Telkom University, Indonesia

Email : wijayahervin@gmail.com

Abstract. The increase in coal production every year has influence the transport volume of coal trains of Kereta Api Logistik Company is getting higher. This increase causes the current number of train unloading equipment to be unable to keep up with this increase and has an impact on the poor performance of train unloading, which is indicated by the unachieved of train's waiting time target. The coal train unloading system is a very complex system and many uncertainties occur, so the appropriate method to use is discrete event simulation. The simulation model is designed using the Simulation Arena software. The results of the simulation method are 4 alternative scenarios will be selected by the Bonferonni test. Scenario 4 has the highest reduction in train unloading operating time, which is 30.7%. The results of this study recommend the addition of a tool with a combination of 1 unit of Gantry Crane integrated with coal traveling hopper and 5 units of Dump Truck and for management, this recommendation can reduce high overtime costs every month and increase coal transport capacity so that coal transport profits will increase.

1. Introduction

The high increasing of demand and the rapid evolution of technology is a big challenge for all companies in any sectors to improve their supply chain performance in order to meet the customer demands. Several studies have been conducted by researchers to provide the alternative solutions and suggestions at various planning levels and various streams. There are 3 main types of flow in the supply chain, they are information flow, financial flow and material flow. Research has been done on the management of information flows such as the development of a sustainable procurement monitoring system to support risk management of the food supply chain [1] and the development of reverse logistics to monitoring system in the leather tanning industry [2]. In financial flow, there are several studies to maximize the profit, such as the development of dynamic pricing in the hotel business sector [3] and online travel agency [4]. There are also several studies that have been done to improve the material flow performance such as the development of the e-Kanban application to reduce zero picks for pharmaceutical warehouses [5] and the development of warehouse picking models with genetic algorithm methods to minimize delays [6]. All research is carried out in order to improve supply chain performance to be better, as in this study which discusses the logistics on railways to improve and maintain material flow performance.

In the supply chain, railways system has an important role in the distribution of goods because it has a short travel time with a very large capacity. but in its implementation, the train is often delayed.

the cause of the delay is due to the unloading operation time which is too long, this is similar with the case that are raised in this study at the company PT. Kereta Api Logistik, especially in the coal transportation sector. This poor performance was influenced by the number of equipment that was unable to handle the increasing of demand for coal from PT. XYZ. The company's coal unloading system uses a combination of loading and unloading equipment with two systems. The train unloading system is influenced by time and distance. It is difficult to design such a system and the right instrument to be used is simulation model [7]. Designing simulation models in equipment systems has been widely researched. for example, a combination of transportation systems on dry bulk trains [7], simulations of using material handling equipment by considering utility equipment [8], simulation to analyze the method of throughput capacity terminal [10], development of simulation models to help decision making to reduce the ship waiting time and coal loading time where the simulation produces optimal utility tools [11]. It is clear that the simulation model can help the problems raised in research related to unloading and transportation equipment systems so this study will use a simulation model to design the system and solve the problem. Simulation modeling software will be used in this study is Arena Simulation.

2. Method

The coal train unloading operating system consists of several events such as the arrival of the train, the activity of loading and transporting by equipment so that the appropriate method is used is discrete event simulation (DES) to describe the behavior and system components. Simulation software used in this research is Arena Simulation. There are several procedures that must be followed to build a valid simulation model [12].

2.1. Formulate the problem.

Problem formulation is carried out to identify problems that occur, performance measures used for evaluation, and the objectives to be achieved in this study. Table 1 is a problem formulation which is translated into six problem elements.

Table 1. Six Problem Elements

Problem Element	Explanation on System
Decision Maker	Kereta Api Logistik Company
Decision Maker's Purpose	To minimize coal train unloading time
Decision Criteria	Reduce coal train unloading operating time by adding unloading and transport equipment
Perfomance Measures	Unloading operation time, unloading equipment utility
Controlled Input	the number of transport and unloading equipment
Context	Coal train unloading operation system

2.2. Collect data and define model.

The input data are needed in designing a train unloading operating system simulation model in logistics railway company is the number of equipments are used, the processing time for the unloading equipment, the travel time of transport equipment to the stockpile and back to unloading area, the train transport capacity with container units, shunting time, and train's interarrival times. The average of containers carrying capacity is around 120 containers and it must be shunted because the rail capacity in the unloading area can only load half of the train. The input data will be analyzed by the half-width test, uniformity test, independent test, and normality test. After that, the distribution of input data will be determined by input analyzer. The other data that are needed is flow chart of the coal train unloading operating system.

2.3. Translate model in Arena model simulation and verify.
The train unloading system model will be translated into a computer simulation model using Arena Simulation software. Arena simulation is easy to use because the modules in the software are easy to understand because the model is drawn like a flow chart. Verification of the translated model is done by checking the syntax errors that occur with the "Check Model" feature in the software.

2.4. Simulation model validation.
The validation of the simulation model requires several replication samples with the batch mean method because the unloading train operating system is a nonterminating system where the final state at a certain time becomes the initial state the next time. The number of initial replication samples in this study was 12 replications. To ensure that the number of replications can represent the real system, an analysis of the confidence interval or half width of the initial replication number is carried out whether it is appropriate or not.

2.5. Experimental and analysis results.
This stage will be carried out to develop alternative model solutions in the form of a scenario of allocation of additional tools based on experimental results. Then the existing scenario will be analyzed by statistically comparing the scenario with the real system. The parameters that will be analyzed are the time of the train unloading operation and the tool utility.

2.6. Selecting the best alternative
Selecting the best scenario will use a statistical system comparison based on the Bonferonni test. it compares each other of all developed scenarios.

3. Result and Discussion

3.1. Existing Condition
The existing conditions indicates that the average of coal train unloading operation time with 4.01 hours/train did not reach the target time of 3.3 hours/train shown by Figure 2. This condition causes the delay of empty trains to the coal loading terminal and causes workers to take overtime to complete the unloading process so that overtime costs increase and are detrimental to the company. The composition of equipment owned is 2 units of reach stacker, 2 units of gantry cranes, 11 units of dump trucks, and 1 unit of conveyor installed along the unloading area tracks. Overview of the coal train unloading operating system is shown in Figure 1.

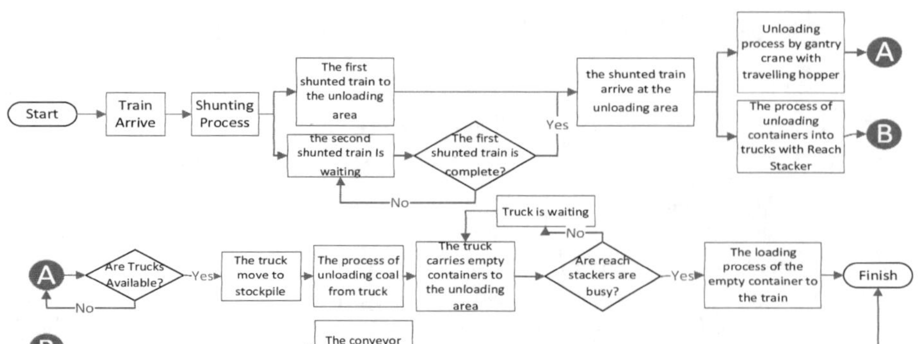

Figure 1. Flow Chart of Kereta Api Logistik Company's Coal Train Unloading System

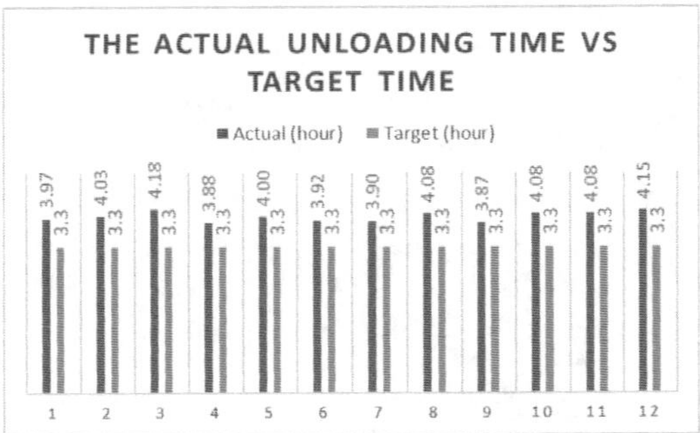

Figure 2. Graphic of actual unloading time vs target time

3.2. Model simulation

The model is made into 5 activity areas shown by A, B, C, D, and E in Figure 3 and figure 4. Part A is a model that describes the train arrival and the Shunting process. Parts B and C are the process of unloading the train by the gantry crane and reach stacker. Part D is the area where the coal dump truck is poured into the stockpile. Part E is an area of reintegration of trains that have been shunted and statistical recording of the time of unloading trains and utility equipment is recorded.

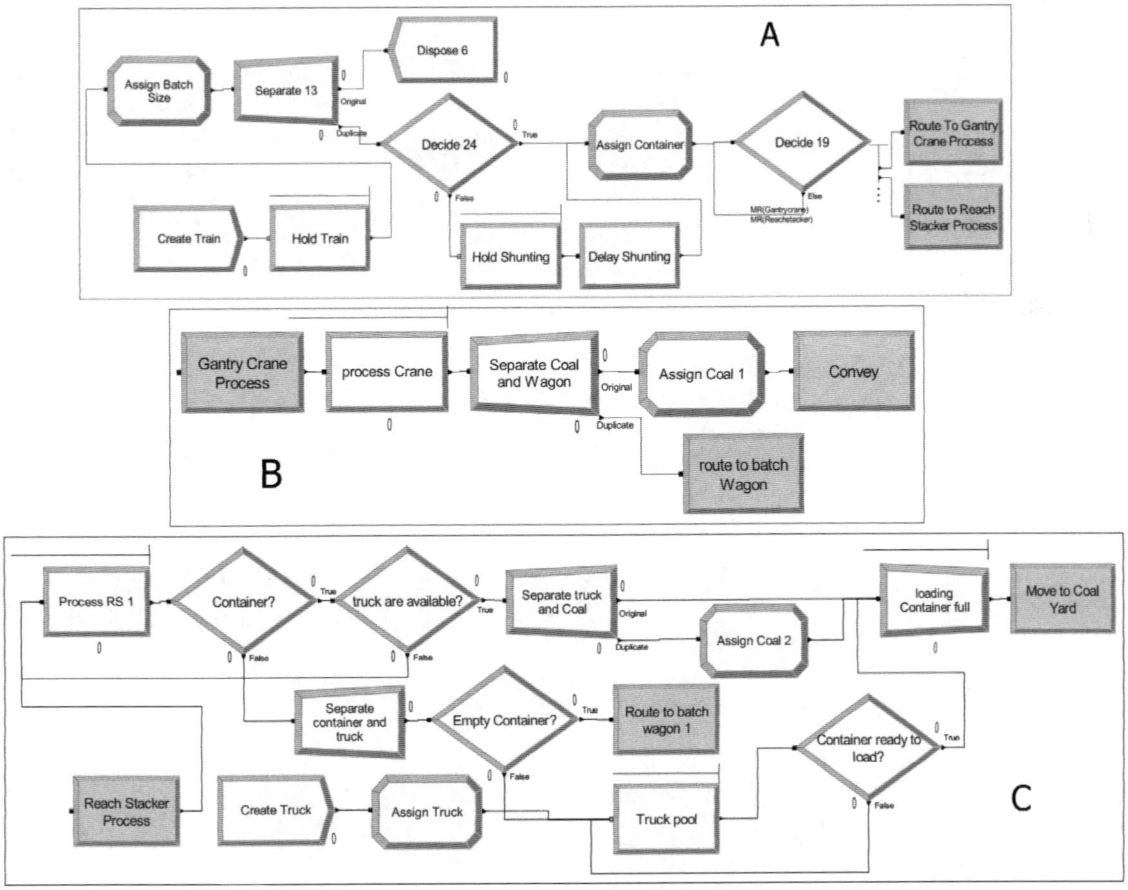

Figure 3. Translated Simulation Model with Arena Simulation (Continue)

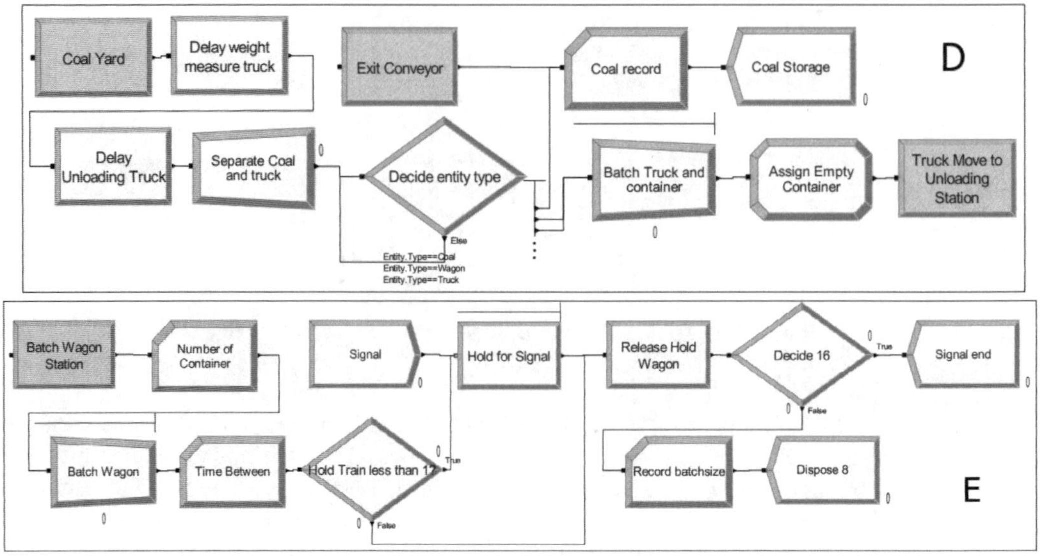

Figure 4. Translated Simulation Model with Arena Simulation

3.3. Model validation

Before performing the validation, it needs samples of replication from the simulation run for 2880 hours. The samples were collected using the batch mean method which begins by identifying the warm up period. The warm up period occurs up to the 200th hour. So, the simulation length is 3080 hours. The data are obtained under the steady state conditions will be grouped into 12 batches with each batch length is 240 hours. Table 2 is the result of calculations using the batch mean method.

Table 2. Summary Batch Mean Results

Batch	**Unloading Time (hour)**	
	Simulation	**Real System**
1	3.93	3.98
2	4.09	4.03
3	4.06	4.08
4	4.05	4.07
5	4.03	4.00
6	4.02	4.02
7	4.02	3.98
8	4.02	4.08
9	4.01	3.98
10	4.01	4.06
11	4.02	4.06
12	4.02	4.03
Mean	**4.02**	
Standard Deviation	**0.037**	
Variance	**0.001359**	
Alpha (0.05)	**2.201**	
C. Interval	**(3.9988,4.04572)**	

Batch sampling is the same as sampling replication. These 12 batches or replications are the initial replication sample size. To ensure the sample of replication can produce the expected confidence interval or half width, the following formula will be used.

$$n = \left(\frac{(z_{0.05/2}) \times s}{half\ width}\right)^2 \quad (1)$$

The value of n generated from the calculation is 9,523 ≈ 10. the initial replication size is 12 replications greater than the n the result of the calculation of the number of replications formula to produce the expected half width. So, 12 replications are enough to represent the simulation model. After that, the paired t test was carried out with the initial hypothesis that there was no significant difference between the simulation model and the real system. The initial hypothesis can be accepted if there is a zero value in the confidence interval of the paired-t calculation with help from the IBM SPSS software. based on the results, the confidence interval is (-0.03279, 0.01779). These results show the zero value in the interval which indicates the initial hypothesis is accepted. So, the simulation model is statistically valid.

3.4. Experimental design

Experiments were conducted by making several scenarios for the addition of loading and transport equipment. In existing condition, the number of Reach Stacker and Gantry Crane is 2 units of each equipment. The number of Dump Trucks in the existing condition is 11 units. For the Belt Conveyor is no additions are made because it is installed along the side of the unloading area track. The following is the experimental scenario design in this study shown by Table 3.

Table 3. Alternative Secenario Design

Scenario	Controlled Input		
	Reach Stacker	Gantry Crane	Dump Truck
Existing	2	2	11
1	2	3	11
2	3	2	16
3	4	2	16
4	2	3	16

3.5. Analysis result

Figure 5 is a graph that showing the comparison of train loading times between all scenarios. From this graph it can be seen that all alternatives scenario has reached the target time and even passed the target. The highest decreasing in train unloading operation time is scenario 4, by adding 1 gantry crane and 5 dump trucks. The utilization of unloading equipment describes the level of busy of equipment that are used in train unloading operations. The equipment utility for each scenario is shown by the graph in Figure 6. Based on this graph, the average utility of reach stacker of the entire scenario is higher than the utility of gantry crane. the reach stacker's high utility is due to double hauling. The double hauling is a dual activity performed by the reach stacker such as unloading containers from the train and putting empty containers back on the train. When the entire train is unloaded, the reach stacker is not finish because they must load the empty containers onto the remaining trains. Scenario 1 and scenario 4 are showing an increase in the utility of the gantry crane. In scenario 1, the addition of 1 gantry crane unit will decrease the reach stacker equipment utility, which means the reach stacker's level of busy is decreased. This condition occurs because the loading speed of the gantry crane is getting higher. It is different from scenario 4, when adding 1 unit of gantry crane and 5 units of dump truck, the reach stacker tool utility tends to be stable from scenario 3 and there is an increase in the utility of gantry crane which increases up to 76%. The best scenario based on unloading tool utility is scenario 4 with the reach stacker utility value of 76% and gantry crane of 76%.

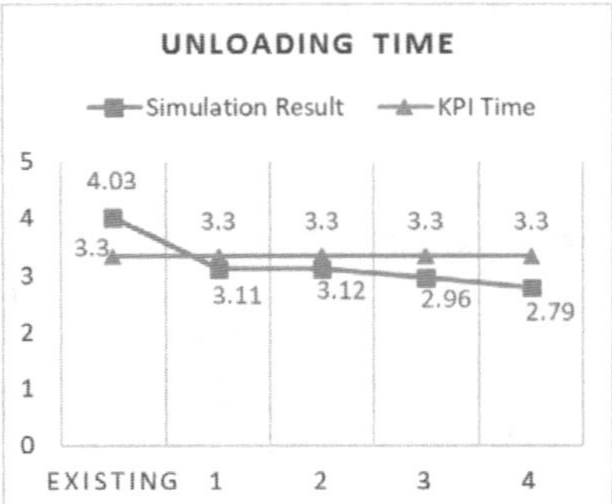

Figure 5. Unloading Time for Each Scenario

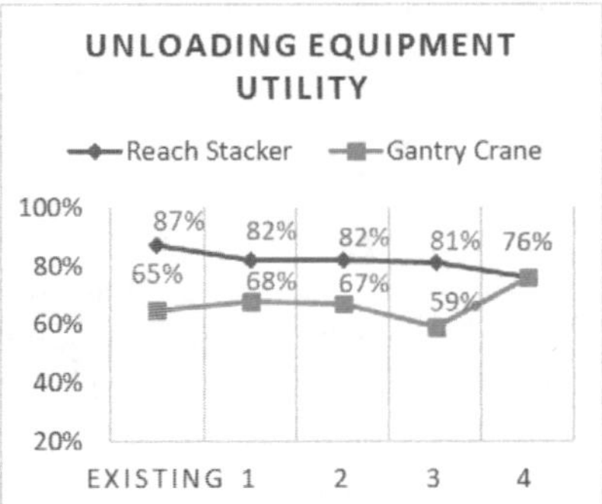

Figure 6. Unloading Equipment Utility for Each Scenario

3.6. Alternative selection

The scenario selection will be using the Bonferonni test to compare statistically between scenarios. The hypothesis used is as follows.

- H0: $\mu_1 = \mu_2 \dots = \mu_k = \mu$, for k is an alternative scenario.
- H1: $\mu_i = \mu_j$, for at least one pair i ≠ j.

The overall confidence level is 95% with a value of α equal to 0.05 and n is 12. based on the Benferonni test, there were 10 pairs so that for each pair it was 0.005 with a distribution value of $t_{11,0.0025}$ = 3.496.

Based on Table 4, almost all comparisons of both Existing and other scenarios so that H0 with the hypothesis that there is no significant difference between Existing and other scenarios can be rejected. Comparisons that do not show a significant difference are the paired comparison of scenario 1 and scenario 2. The highest confidence interval means that the most significant changes in train unloading time. It can be concluded that Scenario 4 is the best alternative scenario among all the other scenarios so that scenario 4 is chosen to be recommended with a scenario of adding of 1 gantry crane and 5 dump truck.

Table 4. Summary of Bonferonni Test

Pair	Average	Half Width	Interval
(*Existing* - Scen.1)	0.917	0.016	(0.901, 0.934)
(*Existing* - Scen.2)	0.920	0.032	(0.887, 0.952)
(*Existing* – Scen.3)	1.082	0.025	(1.058, 1.107)
(*Existing* – Scen.4)	1.258	0.041	(1.216, 1.299)
(Scen.1 – Scen.2)	0.002	0.026	(-0.024, 0.028)
(Scen.1 – Scen.3)	0.165	0.020	(0.145, 0.185)
(Scen.1 – Scen.4)	0.340	0.035	(0.305, 0.375)

(Scen.2 – Scen.3)	0.163	0.009	(0.153, 0.172)
(Scen.2 – Scen.4)	0.338	0.012	(0.326, 0.350)
(Scen.3 – Scen.4)	0.175	0.017	(0.158, 0.192)

4. Conclusion

The most effective model for unloading coal trains at Kertapati Station is by adding 1 gantry crane and 5 dump trucks. The addition of these tools has an effect on the performance of unloading trains in extreme conditions, namely the loading and unloading time on the existing conditions averagely 4.03 hours per year, which is 2.79 hours with a reduction rate of 30.7%, the unloading tool utility in the form of a reach stacker in existing conditions with 87% decreased to 76%, and the unloading tool utility in the form of gantry crane in existing conditions with a percentage of 65% increased to 76%. For management, this recommendation can reduce highly overtime costs every month and increase coal transport capacity so that coal transport profits will increases.

References

[1] D. A. Alfazah, A. Yanuar Ridwan, F. Yulianti, and P. G. Artha Kusuma, "Designing Procurement Process Monitoring Dashboard for Supporting Food Security Supply Chain Risk Management System in Indonesian Bureau of Logistics," *13th International Conference on Telecommunication Systems, Services, and Applications (TSSA)*, 2019.

[2] R. Y. Kuswandi, A. Y. Ridwan, and E. Hadi, "Development of Monitoring Reverse Logistic System for Leather Tanning Industry using Scor Model," *12th International Conference on Telecommunication Systems, Services, and Applications (TSSA)*, 2018.

[3] M. Fadly, A. Y. Ridwan, and D. Akbar, "Hotel room price determination based on dynamic pricing model using nonlinear programming method to maximize revenue,"*2nd International Conference on Applied Information Technology and Innovation: Exploring the Future Technology of Applied Information Technology and Innovation*, 2019.

[4] S. Shadiqurrachman, A. Y. Ridwan, and P. G. Artha Kusuma, "Online Travel Agency Channel Pricing Policy based on Dynamic Pricing Model to Maximize Sales Profit Using Nonlinear Integer Programming Approach," 2019.

[5] R. Razafuad, A. Y. Ridwan, and B. Santosa, "Development of e-Kanban application using stock-needs rule prioritizing policy to reduce 0-pick for pharmaceutical warehousing," *6th International Conference on Information and Communication Technology*, 2018.

[6] E. B. Setyawan, D. D. Damayanti, L. Andrawina, and B. Santosa, "Warehouse Picking Model for Single Picker Routing Problem in Multi Dimensional Warehouse Layout Using Genetic Algorithm Approach to Minimize Delay," *International Conference on Soft Computing and Data Mining*, 2018.

[7] M. Straka, J. Šaderová, P. Bindzár, T. Małkus, and M. Lis, "Computer simulation as a means of efficiency of transport processes of raw materials in relation to a cargo rail terminal: A case study," *Acta Montanistica Slovaca*, vol. 24, no. 4, pp. 307–317, 2019.

[8] Miriam Andrejiova, Anna Grincova, and D. Marasova, "Measurement and simulation of impact wear damage to industrial conveyor belts," *Wear*, 2016.

[9] B. Zhu, Q. Zhou, Y. Tian, and K. Chen, "Analysis method of terminal throughput capacity for coal export terminals," *Journal of Waterway, Port, Coastal and Ocean Engineering*, vol. 144, no. 1, pp. 325–332, 2018, doi: 10.1061/(ASCE)WW.1943-5460.0000422.

[10] S. Pratap, Y. Daultani, M. K. Tiwari, and B. Mahanty, "Rule based optimization for a bulk handling port operations," *Journal of Intelligent Manufacturing*, pp. 287–311, 2015.

[11] Kelton et al, *Simulation with Arena Sixth Edition*. 2015.

A review of oil palm fruit fiber reinforced composites

B Bakri*, Naharuddin, Mustafa, A Medi, L Padang

Mechanical Engineering Department, Tadulako University-Palu, Indonesia

*Corresponding author: bakri@untad.ac.id; bakriuntad@gmail.com

Abstract. Oil palm fibers have been developed as reinforcement in the composite. These fibers can be produced from fruit, trunk, and frond of oil palm. In this review, the oil palm fruit fiber for reinforcing composite was focused. Oil palm fruit fibers consist of empty fruit bunch (EFB) and mesocarp fruit (MF) fibers. The chemical composition and characteristics of oil palm fruit fiber are described. Furthermore, the mechanical properties of the composite are reported to be related to the surface treatment of EFB and MF fibers. Applications of such fiber composite are included in this review. From some researches, the surface treatment methods for MF and EFB fibers as reinforcement composite was conducted with using alkali, silane, acryilic acid, acetic anhydride, hydogen peroxide, microwave, and superheated steam. The effect of these surface treatments on oil palm EFB and MF fibers displayed the improvement of the mechanical properties (tensile, flexural and impact strengths) of the composite due to enhance the interface adhesion between fiber and matrix after treatment of fibers.

1. Introduction

Oil palm (*Elaeis guineensis*) is a tropical plant coming from West Africa. It has spread to tropical countries, including Indonesia and Malaysia . Oil palm arrived in Indonesia by the Dutch in 1848, and some seeds were planted in the Bogor Botanical Gardens [2]. Since 1991, Indonesia has planted oil palm commercially [3]. According to the data from the United States Department of Agriculture (USDA) in 2019, Indonesia has the largest area of oil palm plantations, about 11.75 million ha, followed by Malaysia, Nigeria, Thailand, Colombia, Cote D'Ivoire, Ghana, Guinea, Ecuador, Guatemala and produced palm oil about 42.5 million metric tons. The palm oil sector has contributed to the increase in foreign exchange of these countries. From this data, dry by-products of oil palm can be predicted by approximately 170 million tons based on the estimation that production of one metricton palm oil can generate about 4 tons dry by-products [4-5]. These dry by-products of oil palm are from trunk, frond, and fruit. These can lead to a heavy problem on the environment [6-9] because of the utilization of those has not been seriously considered. It is still focused on the main product of oil palm (crude palm oil (CPO) and palm kernel oil (PKO)) [3]. This problem can be solved by utilizing these by-products for various applications, such as for reinforcement of composite [10].

Natural fibers like ramie, jute, kenaf, coir, and sisal fibers have been used in composite materials as reinforcement. Natural fiber composites are made of fiber component as reinforcement and matrix component as a binder where the fibers used are natural fibers. These composites have been developed as an alternative to change the synthetic fiber composites because natural fibers are low density, abundance, low cost, biodegradability and eco-friendly [11-13]. One of natural fibers which become a concern as filler or reinforcement in the composite is oil palm fiber due to abundant in the area of the oil palm industry. Oil palm fiber can be extracted from trunk, frond, and fruit. The use of palm oil fibers as reinforcement of composite requires surface treatment like the treatment carried out by other natural fibers to improve fiber and matrix compatibility in the composite. This requirement is due to the hydrophilic property of natural fibers and the hydrophobic property of non-polar polymers [11-13]. The physical and chemical treatments are often used to improve that compatibility. Improving fiber-matrix adhesion in composites consequently enhances the mechanical properties of composites [11,14].

In this paper, the review focuses on oil palm fruit fibers consisting of empty fruit bunch (EFB) and mesocarp fruit (MF) fibers. The influence of fiber surface treatment on the mechanical properties of composites are discussed. Chemical composition and mechanical properties of those fibers are also presented.

2. Chemical Composition and Mechanical Properties of Oil Palm Fruit Fiber

Oil palm EFB fiber is taken from oil palm fruit bunch which extracted by retting method [9]. Retting methods include chemical, heat, mechanical decortication, dew, water and enzymatic methods [9,15]. Those methods have advantages and disadvantages in utilization. According to Tahir et al. [15], water retting can be used to obtain the best quality of fibers; meanwhile, it needs a long time process and also can lead to water pollution. Then, chemical retting can produce cleaner fibers, but it reduced tensile properties of those. Besides, enzymatic retting is a retting process which does not need a long duration for yielding high quality of fibers, but it is the relatively high cost and reduces fiber strength [15-16]. The retting process which is safe to the environment and in a short time produces short fiber in large quantities is mechanical retting; however, it is more expensive for processing and also produces the low quality of fiber [9,16]. Meanwhile, the mesocarp fruit (MF) fiber is the fiber gained from the residual milling of palm fruit oil during crude palm oil production [13].

Like other natural fibers, the main chemical composition of EFB and MF fibers is cellulose, hemicellulose and lignin. Chemical composition of EFB and MF fibers have been studied by some researchers (Table 1). Oil palm fruit fiber (EFB and MF) are lignocellulosic fibers which are binded by lignin with cellulose and hemicellulose as reinforcement [17].

Table 1. Chemical Composition of oil palm fruit fibers

Oil Palm Fiber	Chemical composition (%)			Ref.
	Cellulose	Hemicellulose	Lignin	
EFB	42.7-65	17.1-33.5	13.2-25.31	[6,8-9,18-19]
MF	32.2-60	21.7-31.62	11-23.89	[6,20-21]

Regarding the mechanical properties of EFB fiber, Yousif and El-Tayeb [22-23] have investigated tensile strength, tensile modulus and elongation of untreated and alkali-treated EFB fibers. They obtained results that the tensile strength and tensile modulus before and after alkali treatment of fibers enhanced from 70 MPa to 80 MPa and 0.59 GPa to 1.7 respectively, but elongation reduced from 16.2 % to 14.2%. Then, Gunawan et al. [24] have measured the density, tensile modulus and tensile strength of EFB fiber. They found that the diameter of EFB fiber is various between 0.4 mm and 0.72 mm and influences the tensile strength of fiber where the smaller diameter of fiber, the higher tensile strength of that. The tensile strength of fiber with 0.44 mm diameter is 253 MPa, and the tensile modulus is about 16 GPa.

Furthermore, the tensile modulus and density of mesocarp fruit (MF) fiber based on nut variety are about 1.8 – 3.6 GPa and 1.268 g/cm^3 [26], respectively. This fiber has diameter of 0.22 mm for fresh fiber and 0.19 mm for dried fiber [25]. Hanipah et al. [27] have investigated the mechanical properties of MF fiber and found that the higher tensile stress, elongation and tensile modulus of fiber with 0.1 to 0.25 mm diameter are 373 MPa, 0.21 for 1.0 m/s test speeds, and 4.84 GPa for 0.1 m/s test speed, respectively.

The mechanical properties of EFB and MF fiber are competitive to other natural fibers like kenaf, sugar palm, coir, and sisal for reinforcement of polymer composite (Table 2).

Table 2. Comparison of mechanical properties of EFB and MF fibers with other natural fibers

Fibers	Tensile strength (MPa)	Tensile modulus (GPa)	Diameter (mm)	Ref.
EFB	253	16	0.44	[24]
MF	373	3.87	0.1-0.25	[27]
Kenaf	282.6	7.13	0.083	[28]
Coir	218.3	6.1	-	[29]
Sisal	312.5	13.19	0.268	[30]
Sugar Palm	133.3	3.85	0.22	[31]

3. Oil palm fruit fiber composite and its applications

3.1. Oil palm empty fruit bunch (EFB) fiber composite

Oil palm fruit (EFB and MF) fiber composites have been studied by researchers which are related to surface treatment of fiber, including chemical and thermal treatments (Table 3). Chemical treatment of EFB fiber with alkali, silane (γ-Methacryloxypropyltri-methoxysilane (MTS) and vinyltriethoxysilane (VTS)) and acrylic acid has been performed by Arif et al. [11] to improve the mechanical behaviour of high density polyethylene (HDPE) composite. They obtained that tensile strength and modulus of composite increase after 10% NaOH, 3%MTS, 3%VTS and 5% acrylic acid treatments of fibers compared to untreated fibers. The highest tensile strength of the composite is after 3% MTS treatment of fiber. Meanwhile, the highest tensile modulus of the composite is 5% acrylic acid treatment of fiber. Alkali and silane treatments of fibers also enhanced flexural strength and flexural modulus of the composite. Then, alkali, silane and acrylic acid treatments can improve the water absorption of the composite, which displayed the water absorption is reduced after these treatments. Dhandapani et al. [12] has also studied the effect of alkali, silane and acetic anhydride treatments of EFB fibers on mechanical, thermal and morphological properties of EFB fiber/polyester composites. Tensile strength, tensile modulus, flexural strength and flexural modulus of the composite enhanced after alkali, silane and acetic anhydride treatments due to excellent adhesion between the fiber and matrix [12]. The highest tensile and flexural strengths of the composite are 28.11 MPa for silane treatment and 46.15 MPa for acetic anhydride treatment, respectively. Then, EFB fiber-polyester composite after alkali (6% NaOH) treatment of fiber showed better wear property than untreated fiber [23]. Anggawan and Mohamad [32] investigated the various concentration (2%, 5% and 10%) of alkali (NaOH) treatment of EFB fiber for reinforcing composite. The 2% NaOH treatment of EFB fiber was the best concentration of NaOH for improving tensile and thermal properties of the composite.

Moreover, hydrogen peroxide treatment of EFB fiber and addition of coupling agent was obtained by enhancing the tensile and flexural strength of composite due to the improvement of interfacial adhesion of fiber-matrix [33]. The similar hydrogen peroxide treatment of EFB fiber was employed by Rayung et al. [34] for reinforcing poly(lactic acid) composite. The improvement of mechanical properties of EFB/PLA composite was obtained after this treatment. In addition, surface treatment of EFB fiber with acetic, propionic and succinic anhydride was found that it can increase the physical and mechanical properties of EFB fiber-polyester composite [10].

Table 3. The mechanical properties of EFB and MF fibers composites at different surface treatment of fibers

Property	Surface treatment	Composite	Value	Ref.
Tensile strength	Untreated NaOH(10%)	EFB/HDPE	~21 ~21	[11]

(MPa)	Silane (3%) Acryilic acid (5%)		~25 ~23	
	Untreated NaOH Silane Acetic anhydride	EFB/Polyester	23.78 25.61 28.11 25.49	[12]
	Untreated NaOH (2%)	EFB/Polyester	11.24 21.88	[32]
	Untreated Hydogen peroxide	EFB/ Polypropylene	~22 ~40	[33]
	Untreated Microwave-alkali	EFB/GF/ Polypropylene	28.3 39.5	[14]
	Untreated Superheated steam-alkali	MF/PBS	13.86 23.44	[13]
	Untreated Alkaline-peroxide	MF/PBS	13.86 21.39	[20]
Flexural Strength (MPa)	Untreated NaOH(10%) Silane (3%) Acryilic acid (5%)	EFB/HDPE	~34 ~36 ~37 ~33	[11]
	Untreated NaOH Silane Acetic anhydride	EFB/Polyester	33.91 40.97 46.10 46.15	[12]
	Untreated Hydogen peroxide	EFB/ Polypropylene	~45 ~60	[33]
	Untreated Microwave + alkali	EFB/GF/ Polypropylene	23.5 33.3	[14]
	Untreated Alkaline-peroxide	MF/PBS	27.25 31.04	[20]
Impact strength (J/m)	Untreated NaOH Silane Acetic anhydride	EFB/Polyester	19.49 31.22 34.71 30.81	[12]
	Untreated Hydogen peroxide	EFB/ Polypropylene	~42 ~49	[33]
	Untreated Alkaline-peroxide	MF/PBS	65.75 82.19	[20]

Beside chemical treatment of EFB fiber, heat treatment of fiber was also conducted to improve physical and mechanical properties of the composite. Heat treatment of EFB fiber at 180^{0}C influenced the resistance of EFB fiber-high density polyethylene composite on the soil burial and indoor environment condition. Mechanical and thermal properties of untreated fiber composite are moderately lower than treated fiber composite after soil burial conditioning effect [35]. Then, microwave treatment has also used to modify the surface of EFB fiber for reinforcement of hybrid composite [14]. This treatment was combined by 12.5% NaOH solution where fiber was immersed in this solution in the microwave oven at 70, 80 and 90°C for 60 minutes and at 90°C for 60, 90 and 120 minutes. Its result showed that the EFB fiber treatment at 90°C for 90 minutes has higher mechanical properties (tensile, flexural and impact strengths) of EFB fiber/glass fiber (GF) hybrid composite than others. Related to thermal properties, EFB fiber and treated EFB fiber increase the melting point of composites.

3.2. Oil palm mesocarp fruit (MF) fiber composite

Oil palm mesocarp fruit (MF) fiber has been used as reinforcement in composite with linear low-density polyethylene as a matrix [36]. The improvement of tensile modulus and hardness of composite took place when weight fraction of fibers increase in the composite, but tensile strength, impact strength and strain of composite reduced. Then et al. [13] studied surface modification of oil palm MF fiber with superheated steam treatment followed by various NaOH composition treatment for reinforcement of poly(butylene succinate) (PBS) composite. Their results displayed that the mechanical properties of composite enhanced after treatment of MF fiber compared to untreated fiber. The highest tensile strength and strain of composite were 23.44 MPa and 3.40 % respectively after fiber treatment with the combination of superheated steam and 2%NaOH for 3 h soaking time, whereas the highest elastic modulus of composite was 700.60 MPa after fiber treatment with combination of superheted steam and 2%NaOH for 1 h soaking time. Other work on MF fiber with alkaline peroxide treatment found that its treatment can improve the mechanical properties like tensile, flexural and impact strengths of poly(butylene succinate) composite [20]. The improvement of tensile, flexural and impact strengths achieved about 54%, 14%, and 25% respectively compared to untreated MF fiber of composite (Table 3). Then, the surface of oil palm MF fiber was modified by methacrylate silane for improving fiber-matrix interface bonding in hybrid composite [37]. Its hybrid composite is composed to clay, MF fiber, polylactic acid (PLA) and polycaprolactone (PCL). Treated MF fiber hybrid composite has higher mechanical properties than untreated MF fiber hybrid composite.

3.3. Application of oil palm fruit fiber composite

The natural fiber composites have been applied in automotive, construction and furniture, and packaging. Henry Ford firstly introduces the use of natural fiber composite for producing the automotive component in the 1940s followed by the East German Trabant, Daimler-Benz and Mercedes [38]. Regarding oil palm fiber, this fiber has been utilized as reinforcement of polymer composite for automotive application by Proton Car (car producer from Malaysia) [17]. Then, EFB fiber for furniture application has been described by Suhaily et al. [39]. Such fiber has been fabricated in medium density fiberboard (MDF) which is usually used for furniture. MDF from this fibers is very competitive with plywood, particleboard and hardboard [39]. In addition, this fiber has also been applied in particle boards, plywood, and hybrid composite [17]. The utilization of EFB fiber composite has a great potential in the structural building like EFB fiber as reinforcement of concrete roof slates [40]. Momoh and Osofero [18] have studied EFB fiber as reinforcement of cement composite and concluded that its composite has the potential application for building materials like bricks, roofing tiles, building cladding and facades. Also, oil palm fiber composite was used for windows, door frames, panel and so on [41].

4. Conclusions

Oil palm fruit fibers consist of EFB and MF fibers. The EFB fiber can be extracted from oil palm fruit bunch by retting method, and MF fiber can be acquired from the crude palm oil extraction of palm fruit oil. These fibers have mechanical properties that are relatively similar to other natural fibers used as composite reinforcement. The use of these fibers as reinforcement in composites requires surface treatment to increase the interface bonding between fiber and matrix, consequently improving the mechanical properties of the composite. Researchers have found that surface treatments (chemical and physical treatments) of EFB and MF fiber can improve mechanical properties, including tensile, flexural, and impact strengths). Based on this, these fibers have the potential to be developed as a reinforcement for composites in various applications like other natural fiber composites.

References

[1] Wahid M B, Abdullah S N A, and Henson I E 2004 4th Int. Crop Sci. Congress (Brisbane, Australia)
[2] Baudoin A, Bosc P M, Bessou C, and Levang P 2017 *Bogor, Indonesia: CIFOR.* **Working Paper 219**
[3] Geng A 2014 Upgrading of oil palm biomass to value-added products *Biomass and Bioenergy: Applications* (Springer International Publishing) chapter 10 pp 187–209
[4] Yasim-Anuar T A T, Ariffin H, Norrahim M N F, and Hassan M A 2017 *BioResources* **12** 715-34.
[5] Sreekala M S, Kumaran M G, and Thomas S 1997 *J. Appl. Polym. Sci.* **66** 821–35
[6] Law K N, Wan Daud W R, and Ghazali A 2007 *BioResources* **2** 351-62
[7] Abdul Khalil H P S, Nurul Fazita M R, Jawaid M, Bhat A H, and Abdullah C K 2010 *J. Compos. Mater.* **45** 219–36
[8] Shinoj S, Visvanathan R, Panigrahi S, and Kochubabu M 2011 *Indust. Crops Produc.* **33** 7–22
[9] Budidarsono S, Susanti A, and Zoomers A 2013 Oil palm plantations in Indonesia: The implications for migration, settlement/resettlement and local economic development *Biofuels* (IntechOpen) chapter 6 pp 173-93
[10] Abdul Khalil H P S, Ismail H, Ahmad M N, Ariffin A, and Hassan K 2001 *Polym. Inter.* **50** 395–402
[11] Arif M F, Megat-Yusoff P S M, and Ahmad F 2009 *J. Reinfor. Plast. Compos.* **29** 2105–18
[12] Dhandapani S, Nayak S K, and Mohanty S 2015 *J. Elastom. Plast.* **48** 456–79
[13] Then Y Y, Ibrahim N A, Zainuddin N, Ariffin H, and Wan Yunus W M Z 2014 *Sci. Resea. J.* **11** 1–12
[14] Islam M., Gupta A, Rivai M, and Beg M 2015 *J. Thermop. Compos. Mater* **30** 986–1002
[15] Tahir P M, Basher A B, Azry S O A S, and Ahmed Z 2011 *BioResources* **6** 5260–81
[16] Lee C H, Khalina A, Lee S H, and Liu M 2020 *Adv. Mater. Sci. Eng.* **2020** 1–27
[17] Abdul Khalil H P S, Jawaid M, Hassan A, Paridah M T, and Zaidon A 2012 Oil palm biomass fibres and recent advancement in oil palm biomass fibres based hybrid biocomposites *Composites and Their Applications* (IntechOpen) chapter 8 pp 187-220
[18] Momoh E O and Osofero A I 2020 *Front. Struct. Civil Eng.* **14** 94–108
[19] Abdul Khalil H P S, Siti Alwani M, Ridzuan R, Kamarudin H, and Khairul A 2008 *Polym.-Plast. Tech. Eng.* **47** 273–80
[20] Then Y Y, Ibrahim N A, Zainuddin N, Chieng B W, Ariffin H, and Wan Yunus W M Z 2015 *BioResources* **10** 1730–46
[21] Bonnet-Masimbert P A, Gauvin F, Brouwers H J H, and Amziane S 2020 *Resul. Eng.* **7** 100150
[22] Yousif B F and El-Tayeb N S M 2007 *Surf. Rev. Lett.* **14** 1095–102
[23] Yousif B F and El-Tayeb N S M 2008 *Proc. Inst. Mech. Eng. Part J: Journal of Engineering Tribology* **222** 637–46
[24] Gunawan F E, Homma H, Brodjonegoro S S, Hudin A B B, and Zainuddin A B 2009 *J. Sol. Mech. Mater. Eng.* **3** 943–51
[25] Talib A T, Chang Jie C, Mohammed M A P, Samsu Baharuddin A, Mokhtar M N, and Wakisaka M 2019 *Biosys. Eng.* **186** 307–22
[26] Mejouyo P W H, Harzallah O, Sikame Tagne N R, Ndapeu D, Tchemou G, Drean J Y, Njeugna E 2019 *J. Natur. Fiber.*, 1–17
[27] Hanipah S H, Yu Xiang L, Mohammed M A P, and Samsu Baharuddin A 2017 *J. Natul. Fiber.* **14** 153–65
[28] Asim M, Jawaid M, Abdan K, and Ishak M R 2016 *J. Bion. Eng.* **13** 426–35
[29] Bakri B and Eichhorn S J 2010 *Cellulose* **17** 1–11.
[30] Fiore V, Scalici T, Nicoletti F, Vitale G, Prestipino M, and Valenza A. *Compo.B Eng.* **85** 150–60
[31] Atiqah A, Jawaid M, Ishak M R, and Sapuan S M 2018 *J. Natur. Fiber.* **15** 251–61
[32] Anggawan A D and Mohamad N 2020 *AIP Conf. Proc.* **2233** 020002
[33] Suradi S S, Yunus R M, and Beg M D H 2010 *J. Compos. Mater.* **45** 1853–61

[34] Rayung M, Ibrahim N A, Zainuddin N, Saad W Z, Razak N I, and Chieng B W 2014 *Int. J. Molec. Sci.* **15**. 14728-42
[35] Nordin N A, Rahman N M M A, and Hassan A 2020 *High Perfor. Polym.* **32** 158–67
[36] Olusunmade O F, Adetan D A, and Ogunnigbo C O 2016 *J. Compos.* **2016** 3137243
[37] Eng C C, Ibrahim N A, Zainuddin N, Ariffin H, and Wan Yunus W M Z 2016 *BioResources* **11** 861-72
[38] Peças P, Carvalho H, Salman H, and Leite M 2018 *J. Compos. Sci.* **2** 1-20
[39] Suhaily S, Jawaid M, Abdul Khalil H P S, Mohamed A R, and Ibrahim F 2012 *BioResources* **7** 4400-23
[40] Kaliwon J, Sh Ahmad S, and Abdul Aziz A 2010 *Int. Conf. on Science and Social Research* (CSSR 2010), 528–31
[41] Mohammed L, Ansari M N M, Pua G, Jawaid M, and Islam M S 2015 *Int. J. Polym. Sci.* **2015** 243947

Acknowledgments
The authors would like to thank to Engineering Faculty Tadulako University for funding support to oil palm fruit fiber composite research project.

IOP Conf. Series: Materials Science and Engineering 1212 (2022) 012051 doi:10.1088/1757-899X/1212/1/012051

Application of jacket cooling water system recovery for fuel heating on diesel engine to reduce emissions

Barokah[1], Semin[1*], B Cahyono[1], Bambang Sampurno[1], M Tappy[2], J Huwae[2], J Ratela[2], and Y Wibisono[3]

[1]Department of Marine Engineering, Institut Teknologi Sepuluh Nopember, Surabaya, Indonesia

[2]Department of Marine Engineering, Politeknik Kelautan dan Perikanan Bitung, Bitung, Indonesia

[3]Engine Department, Politeknik Pelayaran Surabaya, Surabaya, Indonesia

E-mail: semin@its.ac.id

Abstract. The existence of a diesel engine is very necessary for life today. In development, the diesel engine is experienced many developments in all systems. In line with the development of the diesel engine can't be separated from the impact of fuel use. This research was carried out to make a fuel heating system by utilizing jacket cooling water system recovery as a fuel heater and the method used in this research is an experimental method including making installation of a fuel heating system, testing on the diesel engine generator with specifications 4 strokes, 4 cylinders, 1500 rpm, 18 kVA using biodiesel (B20) fuel. The data were collected using a smoke opacity meter. The experiment was carried out by heating treatment of fuel starting from 30°C - 50°C at the interval of 5°C. Data analysis by describing experimental data. The result is showed that heating fuel can reduce the emission of the diesel engine.

1. Introduction

The diesel engine is an internal combustion engine which is a machine whose work process is through a compression process [1]. In the process of working, the internal combustion engine converts the chemical energy of the fuel into mechanical energy. The chemical energy of the fuel is converted into heat energy through combustion [2]. Diesel engines are machines that are widely used in the industrial and transportation world today. The problems with the use of diesel engines to date include the problem of exhaust gas emissions.

Exhaust gas emissions are the residual results of the combustion of fuel in the internal combustion engine that is removed through the engine exhaust system. The prevention of exhaust emissions from ships is an indispensable requirement to reduce the level of environmental pollution in air pollution. Related to this problem, air pollution prevention measures must be taken to minimize the dangers due to air pollution or pollution. At this time, many researchers in all corners of the world are intensifying research related to air pollution issues originating from engine exhaust gases operating both on land and on ships.

IMO regulations on reducing hazardous emissions and ship waste are contained in (Marine Pollution) MARPOL Annex VI, which consists of regulations to reduce ship exhaust emissions, first adopted in 1997, entered into force on 19th May 2005 and revised in October 2008 The main focus is on SOx and NOx emission but also concern ozone-depleting substance, volatile organic compound, ship incineration, and others [3]. The main objective is to reduce the sulfur content of ship fuels from 4.5% to 0.5% by 2020 globally and from 1.5% to 0.1% in 2015 for Sulfur Emission Control Areas (SECAs) [4].

Based on research throughout 2014 in the Ningbo-Zhousan Port area of China, air pollution from ship exhaust emission reached 80%[5], besides that pollution is related to ship is a significant problem worldwide. So efforts to reduce emission and develop innovative solutions are very important [4]. The phenomenon of environmental pollution, especially air pollution, is getting higher and higher. Proactive steps to reduce air pollution researchers carry out various studies related to strategies, methods, techniques for preventing or reducing air pollution originating from ships. In his research related to the measurement of exhaust gas emission [6], he formulated a method of measuring exhaust gas emission from a diesel engine using the Tunable Diode Laser Absorption Spectroscopy (TDLAS) technique. This technique is quite simple, with a high level of precision and high time resolution. Exhaust gas emission can be reduced by increasing the exhaust gas after-treatment inlet flow temperature above 250°C by adjusting the initial and final closing of the intake valve after the treatment temperature of the exhaust gas from the engine[7].

Besides, NOx emissions are also limited based on engine speed on board [4]. NOx limitation can also be done using the Transient Torque Control method [8]. And now the Hydrogen Energy Share (HES) method by dual fuel are hydrogen and diesel has been found which is useful in reducing harmful emission at low and medium load [9]. Currently, marine diesel engine operations use a lot of fuel oil as fuel which is a source of exhaust gas emission. These exhaust emissions must comply with the provisions of the International Maritime Organization (IMO) and the World Bank. IMO set the limit for nitrogen oxide emission (NOx) = 45.n-2 in g / kWh where n is the engine speed in rpm. Meanwhile, the emission limit for NOx gas emission is following the provision of the Word Bank NOx = 940 ppm. With 15% oxygen in the exhaust gas. And the particulate PM emission limit = 50 mg / m3 [10].

Greenhouse gas emission from the transportation sector and how to reduce them are a major challenge for policymakers [11]. When the ship maneuver in the port area, that is when there is an increase in the concentration of NOx and SOx emission [12]. CO2 emission from maritime transport represents approximately 3% of total annual anthropogenic greenhouse gas (GHG) emission. These emissions are assumed to increase by 150-250% in 2050. GHG emissions contribute to international ambitions to limit the effects of climate change [13]. that the shipping sector can always contribute to efficient global emission reduction and thus can always achieve global cost-saving, but also that the size of the contribution and the measure of cost savings are highly dependent on the assumed MACC case [14].

The International Maritime Organization (IMO) has recently proposed several operational and technical measures to increase shipping efficiency and reduce of Green House Gas (GHG) emission. The reduction potential estimated for these measures has been further used by many organizations to project future GHG emission reduction and plot for Marginal Abatement Cost Curves (MACC) [15]. Motor vehicle pollution and environmental degradation are on the rise with the increase of vehicles and to stop these strict regulations have imposed motor vehicle emission. Also, it is getting thinner [9]. The remainder of the combustion product from a diesel engine is in the form of several substances which are known, among others, CO is also called poisonous carbon monoxide, CO2 is also called carbon dioxide which is a greenhouse gas, NOx nitrogen oxide compound, HC is a charcoal hydrate compound as a result of imperfect combustion processes and loose particles.

The research that has been done is similar to that of [16], where preheating biodiesel fuel, namely CPO, Jatropha, and WCO. Research produces data on changes in emissions. Heating the fuel can change the composition of NOx and CO to a smaller one in the mixture of about 5%. Heating fuel shows the presence and reduces emission, carbon monoxide, and hydrocarbon [17][18]. Research on fuel with preheated fuel treatment using magnetic fuel has resulted in a change in exhaust emission levels. From the comparison between preheated and untreated fuel, it shows lower levels of CO, HC, and CO2 [19]. The same result is shown by research from [20] where Jatropha fuel oil treated with preheater produces a lower emission level than Jatropha fuel oil that is not preheated. The test parameters are CO2, Smoke Opacity, CO, and HC. To heat fuel, generally using a heater or heat exchanger.

Related to the above research, it is necessary to research the emissions produced by biodiesel which is currently being widely used by another method. However, fuel combustion in naturally aspirated diesel engines produces 100% energy, but what is actually used to power the engine is 35%. While the rest of the energy is distributed at 8% radiation, 5% lubrication, 15% cooling and 37% exhaust gas [23]. Based on these references, the researchers conducted experiments on a heat recovery jacket cooler to heat the fuel to reduce emissions.

2. Method

To support the research, several tools and materials are needed including Diesel engine, Generator Unit, and Gas Analyzer, Fuel. Heater unit consists of Shell and Tube Heat Exchanger, Piping Equipment Material, Computer, Camera, Thermometer, Digital Flow Rate, Manometer, Thermostat, Solenoid, Measuring Cup [21] and Fuel (Biodiesel B20) produced by Pertamina. Standard of Biodiesel (B20) according to table 1.

Table 1. Properties of Biodiesel B20

No	Characteristic	Unit	Min	Max
1	Cetane Number	-	48	-
	Cetane Index	-	45	-
2	Density @15°C	Kg/m³	815	860
3	Viscosity @40°C	mm³/sec	2.0	4.5
4	Sulphur Content	% m/m	-	0.25
5	Distillate 90% evaporation	°C	-	370
6	Flash Point	°C	52	-
7	Pour Point	°C	-	18
8	Carbon Residue	% m/m	-	0.1
9	Water Content	mg/kg	-	500
10	Biological Growth	-	-	-
11	FAME Content	% v/v	-	-
12	Methanol Content	% v/v	-	-
13	Ash Content	% v/v	-	0.01
14	Sediment Content	% m/m	-	0.01
15	Strong Acid Number	mgKOH/gr	-	0
16	Total Acid Number	mgKOH/gr	-	0.6
17	Particulate	mg/l	-	-
18	Visual Appearance	-	Bright and clear	
19	Color	No.ASTM	-	30
20	Lubricity	micron	-	460

IOP Conf. Series: Materials Science and Engineering 1212 (2022) 012051 doi:10.1088/1757-899X/1212/1/012051

Experiment activity using four-stroke and four-cylinder diesel generator engines. The diesel engine uses a direct combustion system where the settings are shown in figure 3. Generator diesel engine specifications are listed in table 2.

Table 2. Data Generator Set

No	Specification	Parameter
1	Merk	Krisbow
2	Type	KW26-1052
3	Model	KF 820
4	Series	A12091918
5	KVA	18
6	HZ	50
7	RPM	1500
8	Volt	230/400
9	Phase	3
10	Ampere	24.5
11	Combustion System	Direct Injection

The research procedure is a sequence of activities carried out so that the research was done regularly and systematically so that the research objectives can be achieved. The procedure is carried out from the beginning to the end of the research process including:

a. Formulate the problem as related to current issues related to a global problem, namely the diminishing reserves of energy, especially petroleum.
b. Tracing literature studies on several similar studies were conducted to enrich the literature review that underlies research on fuel heating techniques and reducing emission.
c. Various kinds of equipment are used in advance to be prepared and confirmed to function or operate properly.
d. Before taking data, the equipment is calibrated first to avoid error or deviation that is too large than it should be.
e. Retrieval of data by conducting direct trial experiments. By installing a shell and tube heat exchanger on the fuel system and cooling system. Adjusting the heating of the fuel at a temperature of 30ºC, 35ºC, 40ºC, 45ºC, 50ºC by controlling the jacket cooling flow that comes out of the engine and then enters the heat exchanger. In this system, a thermostat and solenoid are installed, as shown in figure 1. The engine operates at 1500 rpm.
f. Measure Emission by using opacity smoke meter from exhaust gas manifold, as figure 2.

The data obtained in the field were analyzed using the descriptive data analysis method. The data described are in the form of a percentage, tabulation, and graphical form.

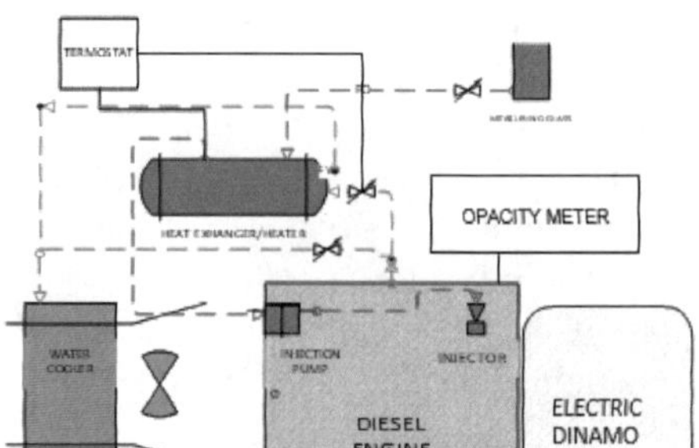

Figure 1. Schematic of Experiment

Figure 2. Opacity Meter

Figure 3. Engine Setting

3. Discussion

Exhaust gas emission is emitted as a result of the combustion of fuels such as natural gas, gasoline, petrol, biodiesel blends, diesel fuel, fuel oil, or coal. According to the type of engine, it is discharged into the atmosphere through an exhaust pipe. The emission testing is conducted to collect data on the exhaust gas opacity from the diesel engine. The test set at temperature variation ranging from 30°C to 50°C with an interval of 5°C. The gas smoke opacity is calculated in%. In the test carried out, it is set up at 1500 RPM engine speed at no load engine condition. Exhaust smoke opacity data are presented in Figure 4.

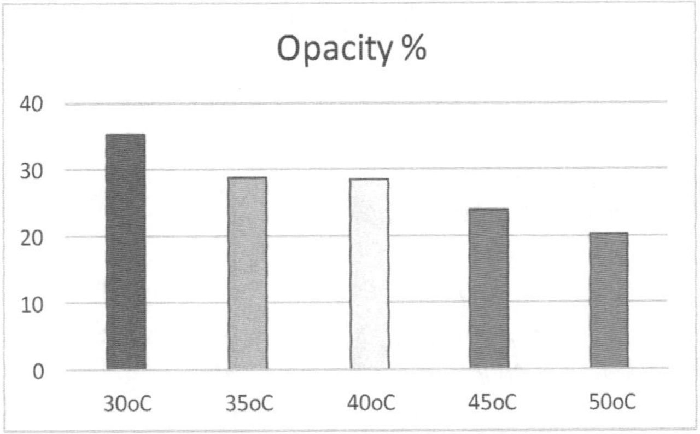

Figure 4. Graph of Smoke Opacity

The graph in Figure 4 shows that the opacity at 30°C is the highest at 35.3%, then at 35°C, it decreases to 28.8%. At a temperature of 40°C, there was a slight decrease, is 28.6%, then at a temperature of 45°C it dropped a bit sharply, is 23.8, while at 50°C it was the lowest opacity, is 20.3% From the description, it can show that there is a decrease in exhaust emission opacity in line with the increase in fuel temperature. So it can be concluded that the increase in fuel temperature affects the decrease in opacity in exhaust emission.

According to the results of research conducted by several researchers related to emission [22] that the implementation of preheating can reduce emissions. Likewise [19] stated that fuel with preheated fuel treatment has resulted in a change in the level of exhaust emission to be lower when compared to without heating. The same result is shown by research from [20]where heated fuel oil produces a lower emission level than fuel oil that is not heated.

4. Conclusion

The fuel heating system using a shell and tube heat exchanger with the jacket cooler recovery method installed on the diesel generator engine at the Docking & Ship Yard at Polytechnic of Marine and Fisheries of Bitung can be done well. In stationary conditions, it can increase the heat of the fuel from a current temperature of 30°C to 50°C.

Fuel heating at 50°C can reduce exhaust gas opacity by 15% compared to the normal temperature, which is 30°C with an opacity level of 35.3%,

References

[1] Bakar R A, Semin, and Ismail A R 2007 The Internal Combustion Engines Diversification Technology and Fuel Research for The Future: a Review. *AEESEAP Reg. Conf. Eng. Educ.*, pp. 57–62, 2007.

[2] Heywood J B 1998 *Internal Combustion Engine Fundamental*. McGraw-Hill, Inc.

[3] MARPOL 2006 *Marpol Consolidated Edition 2006*, vol., no. Regulation 8. IMO, 2006, pp. 353–362.

[4] Bilgili L and Celebi U B 2018 Developing a New Green Ship Approach for Flue Gas Emission Estimation of Bulk Carriers. *Meas. J. Int. Meas. Confed.*, vol. **120**, no. November 2017, pp. 121–127.

[5] Huang l, Wen Y, Geng X, Zhou C, and Xiao C 2018 Integrating Multisource Maritime Information to Estimate Ship Exhaust Emissions Under Wind, Wave and Current Conditions, *Transp. Res. Part D Transp. Environ.*, vol. **59**, no. January, pp. 148–159.

[6] Wang Z, Liu X, Mu Y, Yang X, Yang L, and Jiang Z 2018 Optik The Exhaust Emission Online Detection onThe Diesel Engine, *Opt. - Int. J. Light Electron Opt.*, vol. **164**, pp. 126–131.

[7] Ustun H and Azmi O 2017 Effects of Application of Variable Valve Timing on The Exhaust Gas Temperature Improvement in a Low Loaded Diesel Engine, *Appl. Therm. Eng.*, vol. **122**, pp. 758–767.

[8] Grondin O, Thibault L, Qu'erel C 2012 Transient Torque Control of a Diesel Hybrid Powertrain for NOx Limitation," pp. 286–295.

[9] Sharma P and Dhar A 2018 Science Direct Effect of Hydrogen Supplementation on Engine Performance and Emissions, *Int. J. Hydrogen Energy*, vol. **43**, no. 15, pp. 7570–7580.

[10] Hagenow G, Reders K, Heinze H E, Steiger W, Zigan D, and Mooser D 2010 Fuels, in *Handbook of Diesel Engines*, K. Mollenhauer and H. Tschoeke, Eds. Berlin Heidelberg: Springer, 2010, pp. 87–125.

[11] Styhre L, Winnes H, Black J, Lee J, and Le-Griffin H 2017 Greenhouse Gas Emissions from Ships in Ports – Case Studies in Four Continents, *Transp. Res. Part D Transp. Environ.*, vol. **54**, pp. 212–224.

[12] Pan W, Yao C, Han G, Wei H, and Wang Q 2015 The impact of intake air temperature on performance and exhaust emissions of a diesel methanol dual fuel engine, *Fuel*, vol. **162**, pp. 101–110.

[13] Bouman E A, Lindstad E, Rialland A I, and Strømman A H 2017 State of The-Art Technologies, Measures, and Potential for Reducing GHG Emissions from Shipping – A review . *Transp. Res. Part D Transp. Environ.*, vol. **52**, pp. 408–421.

[14] Heitmann N and Peterson S 2014 The Potential Contribution of The Shipping Sector to an Efficient Reduction of Global Carbon Dioxide Emissions, *Environ. Sci. Policy*, vol. **42**,

pp. 56–66.

[15] Yuan J, Ng S H, and Sou W S 2016 Uncertainty Quantification of CO2 Emission Reduction for Maritime Shipping," *Energy Policy*, vol. **88**, pp. 113–130.

[16] Tajuddin A S A, Khalid A, Hadi S A, and Nursal R S 2016 Effects of Preheat Biodiesel Fuel Derive From Crude Palm Oil, Jatropha Oil and Waste Cooking Oil on Influences of Preheat Biodiesel Fuel on Performance and Emissions of Diesel Engine, in *International Conference on Automotive Innovation Green Energy Vehicle 2016*, pp. 1–8.

[17] Martin M L J, Geo V E, and Nagalingam B 2017 Effect of Fuel Inlet Temperature on Cottonseed Oil–Diesel Mixture Composition and Performance in a DI Diesel Engine, *J. Energy Inst.*, vol. **90,** no. 4, pp. 563–573.

[18] Barokah, . Semin, Cahyono B, Sampurno B, and Rompas P T D 2018 The Effect of Fuel Temperature on Characteristics, Combustion Process, Emissions and Performance of Diesel Engine: A Review," *Proc. 7th Eng. Int. Conf. Educ. Concept Appl. Green Technol.*, pp. 54–60,

[19] Salih A M and Al-rawaf M A 2015 The Effect of Increasing of Diesel Fuel Temperature Upon the Engine Performance By Using Two Magnetic Fields University of Technology - Baghdad-Iraq Al-Mansour University College - Baghdad-Iraq," no. 35.

[20] Agarwal D and Agarwal A K 2007 Performance and Emissions Eharacteristics of Jatropha Oil (Preheated and Blends) in a Direct Injection Compression Ignition Engine," *Appl. Therm. Eng.*, vol. **27**, no. 13, pp. 2314–2323.

[21] Barokah, Semin, Cahyono B, and Sampurno B 2020 Effect of heating B20 biodiesel fuel to increase performance using heat recovery of cooling jacket on diesel engine, *Int. Rev. Mech. Eng.*, vol. **14**, no. 3, pp. 192–197.

[22] Nanthagopal K, Ashok B, Varatharajan V, Anand V, and Kumar D 2017 Study on The Effect of Exhaust Gas - Based Fuel Preheating Device on Ethanol – Diesel Blends Operation in a Compression Ignition Engine, *Clean Technol. Environ. Policy*, vol. **19**, no. 10, pp. 2379–2392.

[23] Doug W 2004 Pounder's Marine Diesel Engine and Gas Turbine. 8th edn. Edited by W. Doug. Burlington. MA: Elsevier Butterworth-Heinemann.

Acknowledgment
We would like to thank the Department of Marine Engineering ITS of Surabaya, Center for Marine and Fisheries Education of Ministry Marine and Fisheries, and Polytechnic of Marine and Fisheries of Bitung, who has facilitated this research, therefore research activities were carried out well.

Failure analysis of excavator hydraulic pump

H Hidayat[1,*], D Aviva[1], A Muis[1], A Halik[1], S Sudarsono[2], S Pranoto[3] and D Cahyadi[4]

[1] Department of Mechanical Engineering, Politeknik Negeri Samarinda, Kalimantan Timur, 75131, Indonesia
[2] Department of Mechanical Engineering, Universitas Halu Oleo, Kendari, 93132, Indonesia
[3] Department of Electrical Engineering, Politeknik Negeri Ujung Pandang, Makassar, 90245, Indonesia
[4] Department of Design, Politeknik Negeri Samarinda, Kalimantan Timur, 75131, Indonesia

*Corresponding author: hidayat@polnes.ac.id

Abstract. Hydraulic pump failures may be related to hardware or problem in oil. In this study, the excavator hydraulic pump failures were investigated by using visual observed and measuring the part component. The disassembled process of the pump was performed considering the manual part book of the excavator hydraulic pump. The abrasive wear on the pump slipper and swash plate was observed by comparing the guidelines for the reusable part. The value of more than 1.07 mm clearance within piston and cylinder bore was measured then the results over the allowable limit considering the manual part book. Properly analyzing of component failure can provide valuable information about what caused the failure and thus can be to avoiding future unscheduled downtime.

1. Introduction

All pumps can be classified into two, namely positive and non-positive displacement. A hydraulic pump is a class of positive displacement machine in fluid applications to provide hydraulic flow to fluid powered devices such as heavy equipment, heavy vehicle, tractor, etc. For heavy equipment in construction or mining such as excavator, hydraulic power provides to move booms, arm or bucket. It is often easy the main equipment in local good mining breakdown due to hydraulic system failure. A hydraulic pump failure can be caused by several factors for example, poor system design, using low quality fluid and poor contamination control.

Several researchers on the failure analysis and diagnostic of hydraulic pumps have been reported. A fault diagnosis method of an intelligent hydraulic pumps considering a non-linear unknown input observer had been proposed [1]. The method was great significance in improving the reliability of the intelligent hydraulic pump system. The other method is a diagnosis ball bearing of hydraulic pump applied using a vibration signal for the maintenance [2]. An engineering driven performance degradation analysis method considering the nature of mechanical wear of hydraulic pumps had been proposed [3]. Another study concerned with early fault diagnosis of hydraulic pump that are the heart of hydraulic tube tester [4]. The result indicate that the proposed approach detecting the signal of hydraulic pump. A study on the causes, reasons and solution due to effects of the exaggerated elastic deformation and the fracture, such as the slipper and retainer that leads to the loss of performance of the axial piston pump, had been carried out [5]. A work based on the study of the thermoelastic deformation mechanism of the slipper retainer which is revealed by its deformation and fracture, had been done [6]. A method on the basis of the elastohydrodynamic lubrication model to analyze the wear behavior of the swash plate had been presented [7]. The new intelligent fault diagnosis scheme based on many signal processing techniques had been proposed [8]. A study presents the mathematical model of a hydraulic axial piston pump developed in order to replicate the dynamic behavior of the swash plate for PHM application [9].

As far as the previously published from the other researcher, there is no study on failure hydraulic pumps on main equipment such as an excavator. The purpose of this study is to analyze the failure of the hydraulic pump on heavy mining equipment, in this case, is the excavator. The hydraulic pump was dissembled to investigate the condition of the part inside the hydraulic pump, such as the piston, retainer and cylinder block. All component was observed by visual inspection or measuring the dimension considering the manual part book.

2. Method

2.1. Specification

The hydraulic pump used in this research is taken from Excavator Hitachi 1200-6. This excavator is a revolutionary hydraulic excavator in which takes advantage of the latest developments technology by Hitachi. The Excavator Hitachi 1200-6 was designed and built to apply in major works such as mines and civil engineering construction projects. The specifications of the engine are shown below.

Merk	:	Hitcahi Zaxis 1200-6
Serial Number	:	10Z10536
Model	:	CumminsQSK23 C
Cylinder	:	6 Cylinder, inline
Fuel System	:	Direct Injection Type
Firing Order	:	1-5-3-6-2-4
Speed	:	1800 rpm
Oil Engine	:	18.5 gal

Firstly, failure analysis of the hydraulic pump was performing by disassembling and assembling procedures to find out the component of the pump is still usable or has to be replaced. Secondly, the component was measured, considering the specification range measurement of manual parts.

2.2 Observation (Visual and Measurement)

The disassembly hydraulic pump had been done considering the procedure from the manual part book. The hydraulic pump component was disassembled by using specific tools.

Figure 1 shows the disassemble hydraulic pump component. The component of the hydraulic pump was released by using some special tool kit. Then, the swash plate which is used as controller oil flow to the system, was removed from the top surface of the hydraulic pump. Then, the plunger was removed from the cylinder block located inside the hydraulic pump.

Figure 1. Schematic of Experiment

The cylinder block in which component of the hydraulic pump used as a house of the driven shaft is shown in figure 2.

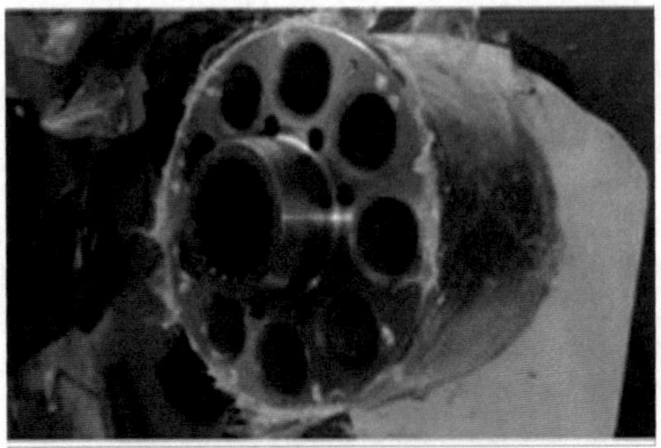

Figure 2. Cylinder Block

The cylinder block is also used to compress the oil by plunger and barrel so that the oil from the hydraulic pump has high pressure. Finally, removed all spring which located under the housing of plunger. All components inside the hydraulic pump were investigated by measurement or visual inspection. Measurement of component is the process of observing and recording the observations collected as part of this research. The component was measured to compare with the manual part book in order to define that the component is still reusable or not. As for the visual inspection, the condition of the component was observed by viewing or touching, considering the manual parts book.

3. Result and Discussion

Figure 3 shows the abrasive on the piston pump. Abrasive wear usuallny occurs when a hard rough surface slides across the other surface. In this case, abrasive wear in this area due to frictional force between slipper and swash plate. The swashplate axial piston is a common hydraulic pump used in heavy equipment unit. High pressure operation condition and also interference between valve plate and cylinder block causes abrasive wear on slipper of piston pump.

Figure 3. Abrasive wear on piston pump slipper

Figure 4 shows the cracking on slipper retainer plate. The crack of the slipper retainer is generally occure by phenomenom of the slipper getting stuck in the cylinder block. The motion of slipper retainer in axial direction is related to the parts of piston and cyliner block.

Figure 4. Cracking on slipper retainer plate

Figure 5 shows the abrasive wear on the swash plate. Swash plate is one of the components in the hydraulic pump referred to as the rotator group. The displacement of an axial piston pump is controlled by the swash plate angle. The most common case of failure accompanied with swash plates is abrasive wear. The slipper of piston slide along the surface of swash plate. Even though the slipper of the piston is made of the same steel as swash plate, there is no carbide detachment and wear comes from contaminant of the fluid.

Figure 5. Abrasive wear on a swash plate surface

Figure 6 shows measuring the clearance of the piston pump. The clearance between the piston pump and the cylinder bore was measured. The allowable clearance based on the manual part book is 0.94 mm. Then, the measuring result between piston and cylinder bore was found more than 1.07 mm. The measuring result of clearance within the cylinder bore is shown in table 1. Increasing nominal clearances between the rate and stroking piston-cylinder pairs exhibited acceptable pump performance. In this case, further increase in clearance within the cylinder bore and piston pump has indicated a major degradation of the performance of the pump.

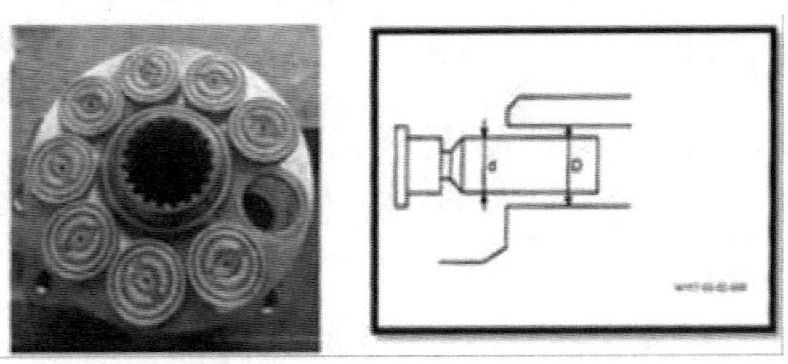

Figure 6. Measuring the clearance of piston pump

IOP Conf. Series: Materials Science and Engineering 1212 (2022) 012052 doi:10.1088/1757-899X/1212/1/012052

Table 1. Measuring result of clearance within cylinder bore

No.	Piston	Measuring result
1	Piston 1	1.09
2	Piston 2	1.07
3	Piston 3	1.07
4	Piston 4	1.08
5	Piston 5	1.08
6	Piston 6	1.07
7	Piston 7	1.07
8	Piston 8	1.09
9	Piston 9	1.09

4. Conclusions

The failure analysis of the excavator hydraulic pump had been investigated by using visual inspection and measuring each part of the pump. The visual inspection of the pump had been performed considering the manual part book from the dealer. The most common of failure in hydraulic pump results from oil contamination, blocked or restricted pump inlet, low level of oil and pump case over-pressurization.

References

[1] M. A. Zhonghai, W. Shaoping, S. H. I. Jian, L. I. Tongyang, and W. Xingjian, "Fault diagnosis of an intelligent hydraulic pump based on a nonlinear unknown input observer," *Chinese J. Aeronaut.*, vol. 31, no. 2, pp. 385–394, 2018.

[2] T. Kim, Y. Jeon, and M. G. Lee, "A Study on Failure Diagnosis System for a Hydraulic Pump in Injection Molding Machinery Using Vibration Analysis," *J. Korean Soc. Manuf. Technol. Eng.*, vol. 22, no. 3, pp. 343–348, 2013.

[3] Z. Ma, S. Wang, H. Liao, and C. Zhang, "Engineering-driven performance degradation analysis of hydraulic piston pump based on the inverse Gaussian process," *Qual. Reliab. Eng. Int.*, vol. 35, no. 7, pp. 2278–2296, 2019.

[4] Z. Zhao, M. Jia, F. Wang, and S. Wang, "Intermittent chaos and sliding window symbol sequence statistics-based early fault diagnosis for hydraulic pump on hydraulic tube tester," *Mech. Syst. Signal Process.*, vol. 23, no. 5, pp. 1573–1585, 2009.

[5] G. Haidak, D. Wang, and E. L. E. Awong, "Modelling of deformation and failure of slipper-retainer assembly in axial piston machine," *Eng. Fail. Anal.*, vol. 111, p. 104490, 2020.

[6] G. Haidak, D. Wang, and E. Shiju, "Research on the thermo-elastic deformation and fracture mechanism of the slipper retainer in the axial piston pumps and motors," *Eng. Fail. Anal.*, vol. 100, pp. 259–272, 2019.

[7] J. Ma, J. Chen, J. Li, Q. Li, and C. Ren, "Wear analysis of swash plate/slipper pair of axis piston hydraulic pump," *Tribol. Int.*, vol. 90, pp. 467–472, 2015.

[8] Y. Lan *et al.*, "Fault diagnosis on slipper abrasion of axial piston pump based on extreme learning machine," *Measurement*, vol. 124, pp. 378–385, 2018.

[9] A. Bedotti, M. Pastori, F. Scolari, and P. Casoli, "Dynamic modelling of the swash plate of a hydraulic axial piston pump for condition monitoring applications," *Energy Procedia*, vol. 148, pp. 266–273, 2018.

AUTHOR INDEX